Papers contributed to the international symposium on foundations of plasticity

Monographs and textbooks on mechanics of solids and fluids

editor-in-chief G. Æ. Oravas

Mechanics of plastic solids

editor J. Schroeder

Papers contributed to the international symposium on

Foundations of Plasticity

Warsaw, August 30 – September 2, 1972

Edited by

A. SAWCZUK

Polish Academy of Sciences, Warsaw

NOORDHOFF INTERNATIONAL PUBLISHING
LEYDEN

A division of A. W. Sijthoff International Publishing Company B.V., Leyden,
The Netherlands

ISBN 90 01 77570 5

Library of Congress Catalog Card Number 72-76786

Printed in the Netherland by
Koninklijke drukkerij Van de Garde – Zaltbommel

List of contents

List of contents

List of contents

Preface

This volume contains papers selected for presentation at the International Symposium on Foundations of Plasticity in Warsaw, August 30–September 2, 1972.

The symposium was organized by the Polish Academy of Sciences and its aim was to bring together engineers, physicists and mathematicians in order to discuss the actual state and prospects of the theory and applications of plasticity.

The symposium invited lectures and the proceedings of the symposium discussions will be published in a separate volume.

The advise and assistance of the International Scientific Committee in shaping the symposium program is gratefully acknowledged.

Antoni Sawczuk

SYMPOSIUM ORGANIZED BY THE POLISH ACADEMY OF SCIENCES

Warsaw, August 30 – September 2, 1972

INTERNATIONAL SCIENTIFIC COMMITTEE

J. M. Alexander, London
D. C. Drucker, Urbana
L. Finzi, Milan
A. E. Green, Oxford
P. G. Hodge, Jr., Minneapolis
H. G. Hopkins, Manchester
A. A. Ilyushin, Moscow
W. T. Koiter, Delft
J. Mandel, Paris
P. M. Naghdi, Berkeley
W. Nowacki, Warsaw
F. K. G. Odqvist, Djursholm
E. T. Onat, New Haven
W. Olszak, Udine
W. Prager, Providence
Yu. N. Rabotnov, Moscow
M. Reiner, Haifa
L. I. Sedov, Moscow
V. V. Sokolovsky, Moscow
T. Y. Thomas, Los Angeles
C. Truesdell, Baltimore
H. Ziegler, Zürich

ORGANIZING COMMITTEE

M. Janas, S. Kajfasz, R. Kowalczyk, M. Kwieciński, P. Perzyna, A. Sawczuk (chairman), S. Zahorski, M. Życzkowski.

Dedication

Symposium organized in connection with Professor Wacław Olszak's 70th anniversary.

Fundamental relations of plasticity derived from hypo-elasticity

Tatsuo Tokuoka

Kyoto University, Kyoto, Japan

A plasticity theory is defined within the theory of hypo-elasticity. The constitutive equation of hypo-elasticity is regarded as a linear transformation relation on a six-dimensional inner product space of symmetric tensor. The yield surface and plastic stretching are defined, respectively, by the singularity and non-zero null space of the linear operator of the constitutive equation. The domain of the linear operator is decomposed into the invariant direct sum. Thus follows a natural classification of plastic yield into two types: normal and shear. $\mathscr{T}$-space is introduced, which has a six-dimensional orthonormal basis composed of the proper vectors of stress. Then the yield criteria and their associated flow rules are given in explicit form.

Nomenclature

$\boldsymbol{T}$	stress tensor
$\boldsymbol{D}$	stretching tensor
$\boldsymbol{W}$	spin tensor
$\mathring{\boldsymbol{T}}$	$\dot{\boldsymbol{T}}-\boldsymbol{WT}+\boldsymbol{TW}$, co-rotational stress rate tensor
$\mathbf{H}(\boldsymbol{T})$	hypo-elasticity, linear operator
${}_{\mathrm{P}}\boldsymbol{D}$	plastic stretching tensor
$\{\boldsymbol{v}_k\}$	orthonormal basis of proper vectors of $\boldsymbol{T}$
$\boldsymbol{A}$, $\boldsymbol{B}$	any symmetric second-order tensors
$\mathscr{V}$	six-dimensional inner product space, which has inner product $\mathrm{tr}(\boldsymbol{AB})$
$\mathscr{V}_{\mathrm{N}}$, $\mathscr{V}_{\mathrm{S}}$	subspaces of $\mathscr{V}$
$\mathbf{H}_{\mathrm{N}}(\boldsymbol{T})$, $\mathbf{H}_{\mathrm{S}}(\boldsymbol{T})$	linear operators on $\mathscr{V}_{\mathrm{N}}$, $\mathscr{V}_{\mathrm{S}}$, respectively
$\mathscr{N}$	null space of $\mathbf{H}(\boldsymbol{T})$
$\mathscr{N}_{\mathrm{N}}$, $\mathscr{N}_{\mathrm{S}}$	null spaces of $\mathbf{H}_{\mathrm{N}}(\boldsymbol{T})$, $\mathbf{H}_{\mathrm{S}}(\boldsymbol{T})$, respectively
$\boldsymbol{D}_{\mathrm{N}}$, $\boldsymbol{D}_{\mathrm{S}}$	stretchings on $\mathscr{V}_{\mathrm{N}}$, $\mathscr{V}_{\mathrm{S}}$, respectively
$\{\boldsymbol{e}_\alpha\}$	six-dimensional orthonormal basis framed by $\{\boldsymbol{v}_k\}$

$\mathscr{T}$	six-dimensional inner product space defined by $\{\boldsymbol{e}_\alpha\}$
$\mathscr{T}_3$	subspace of $\mathscr{T}$, principal stress space
$\boldsymbol{t}, \boldsymbol{d}, \mathring{\boldsymbol{t}}, {}_{\mathrm{P}}\boldsymbol{d}$, $\boldsymbol{d}_{\mathrm{N}}, \boldsymbol{d}_{\mathrm{S}}, \boldsymbol{a}, \boldsymbol{b}$	six-dimensional vectors on $\mathscr{T}$ corresponding to $\boldsymbol{T}, \boldsymbol{D}, \mathring{\boldsymbol{T}}$, ${}_{\mathrm{P}}\boldsymbol{D}, \boldsymbol{D}_{\mathrm{N}}, \boldsymbol{D}_{\mathrm{S}}, \boldsymbol{A}, \boldsymbol{B}$, respectively
$\boldsymbol{H}(\boldsymbol{t}), \boldsymbol{H}_{\mathrm{N}}(\boldsymbol{t}), \boldsymbol{H}_{\mathrm{S}}(\boldsymbol{t})$	six-dimensional second-order tensors on $\mathscr{T}$ corresponding to $\mathsf{H}(\boldsymbol{T}), \mathsf{H}_{\mathrm{N}}(\boldsymbol{T}), \mathsf{H}_{\mathrm{S}}(\boldsymbol{T})$, respectively
ℷ's	scalar functions of principal invariants of stress
t_k	proper numbers of $\boldsymbol{T}$
λ, μ, a's, b's, c's,	material constants
p	$-\frac{1}{3}\mathrm{tr}\, \boldsymbol{T}$, pressure
x, y, z	rectangular Cartesian coordinates on $\mathscr{T}_3$
$\boldsymbol{T}^*, \boldsymbol{t}^*$	stress deviator tensor, and vector, respectively

Introduction

Until now numerous *yield criteria* have been advocated. Among them most famous and important ones are those of Tresca and v. Mises [1, 2]. The so-called *flow rule* was at first proposed by St. Venant [3] and Lévy [4]. The yield criteria can be represented as geometrical surfaces in the principal stress space which was introduced by Haigh [5] and Westergaard [6].

In contrast to empirical formulae, a striking theoretical reduction of yield criteria from an assumed flow rule was made by Thomas [7]. The constitutive equations in implicit form were studied by Morgan [8] and Fox [9].

Truesdell [10] advocated the remarkable rate-type hypo-elastic constitutive equation. In this paper an attempt is made to deal with plasticity within the scope of hypo-elasticity [11].

Definitions of yield criterion and flow rule

The constitutive equation of hypo-elasticity is defined by

$$\mathring{\boldsymbol{T}} = \mathsf{H}(\boldsymbol{T})[\boldsymbol{D}] \quad \text{or} \quad \mathring{T}_{km} = \mathsf{H}_{kmpq} D_{pq}, \tag{1}$$

where $\mathring{\boldsymbol{T}} \equiv \dot{\boldsymbol{T}} - \boldsymbol{W}\boldsymbol{T} + \boldsymbol{T}\boldsymbol{W}$ is the co-rotational stress rate and $\boldsymbol{D}$ and $\boldsymbol{W}$ are respectively the stretching and spin tensors.

Let us now regard $\mathsf{H}(\boldsymbol{T})$ as a linear operator on six-dimensional inner product space $\mathscr{V}$ and (1) as a transformation relation from $\boldsymbol{D}$ into $\mathring{\boldsymbol{T}}$, where

$\mathscr{V}$ has the inner product $(\boldsymbol{A}, \boldsymbol{B}) = \operatorname{tr}(\boldsymbol{AB})$ for any symmetric tensors $\boldsymbol{A}$ and $\boldsymbol{B}$.

Let a relation of $\boldsymbol{T}$ for which $\mathbf{H}(\boldsymbol{T})$ is singular be called the *yield criterion* and the null space $\mathscr{N}$ of $\mathbf{H}(\boldsymbol{T})$ be called the *plastic stretching* ${}_{\mathrm{P}}\boldsymbol{D}$.

Attention will be focused on the yield criterion and the plastic stretching and it is not the subject of this work to consider the transition process from an *unyield region,* which is defined by the set of $\boldsymbol{T}$ for which $\mathbf{H}(\boldsymbol{T})$ is non singular, to a *yield region,* which is defined by the set of $\boldsymbol{T}$ satisfying a yield criterion. Thomas [12] attempted special types of the constitutive equation (1) intended to hold for such transition. However we must mention the analyses of Truesdell [13] and Green [14], who showed that such transitions in simple shear and extension may only be done in the limit.

Classification of yield

The domain of linear operator $\mathbf{H}(\boldsymbol{T})$ is proved to be decomposable into the invariant direct sum [11]

$$\mathscr{V} = \mathscr{V}_{\mathrm{N}}(\boldsymbol{T}) \oplus \mathscr{V}_{\mathrm{S}}(\boldsymbol{T}), \tag{2}$$

where $\mathscr{V}_{\mathrm{N}}(\boldsymbol{T})$ is the set of the elements having the same proper vectors as $\boldsymbol{T}$, and $\mathscr{V}_{\mathrm{S}}(\boldsymbol{T})$ is the set of the elements being orthogonal to $\boldsymbol{v}_k \otimes \boldsymbol{v}_k$ for every proper vector $\boldsymbol{v}_k$ of $\boldsymbol{T}$.

With this invariant direct sum decomposition of $\mathscr{V}$, we have

$$\mathbf{H}(\boldsymbol{T})[\boldsymbol{D}] = \mathbf{H}_{\mathrm{N}}(\boldsymbol{T})[\boldsymbol{D}_{\mathrm{N}}] + \mathbf{H}_{\mathrm{S}}(\boldsymbol{T})[\boldsymbol{D}_{\mathrm{S}}], \tag{3}$$

$$\mathscr{N}(\boldsymbol{T}) = \mathscr{N}_{\mathrm{N}}(\boldsymbol{T}) \oplus \mathscr{N}_{\mathrm{S}}(\boldsymbol{T}), \tag{4}$$

where $\boldsymbol{D} = \boldsymbol{D}_{\mathrm{N}} + \boldsymbol{D}_{\mathrm{S}}$, $\boldsymbol{D}_{\mathrm{N}} \in \mathscr{V}_{\mathrm{N}}(\boldsymbol{T})$, $\boldsymbol{D}_{\mathrm{S}} \in \mathscr{V}_{\mathrm{S}}(\boldsymbol{T})$, $\mathbf{H}_{\mathrm{N}}(\boldsymbol{T})$ and $\mathbf{H}_{\mathrm{S}}(\boldsymbol{T})$ are, respectively, linear operators on $\mathscr{V}_{\mathrm{N}}(\boldsymbol{T})$ and $\mathscr{V}_{\mathrm{S}}(\boldsymbol{T})$ and $\mathscr{N}_{\mathrm{N}}(\boldsymbol{T})$ and $\mathscr{N}_{\mathrm{S}}(\boldsymbol{T})$ are, respectively, null spaces of $\mathbf{H}_{\mathrm{N}}(\boldsymbol{T})$ and $\mathbf{H}_{\mathrm{S}}(\boldsymbol{T})$.

The yield criteria defined by $\mathbf{H}_{\mathrm{N}}(\boldsymbol{T})$ and $\mathbf{H}_{\mathrm{S}}(\boldsymbol{T})$ are called the *normal* and *shear yield criteria*, respectively. When $\boldsymbol{T}$ satisfies one of those criteria, the material is said to be in *normal* or *shear yield.* From (4) we see that *the stress may do work in normal yield while it does not work in shear yield*, where the work is defined by $\operatorname{tr}(\boldsymbol{T}_{\mathrm{P}}\boldsymbol{D})$.

Six-dimensional representation

Consider the six-dimensional inner product space $\mathscr{T}$ which has the orthonormal basis $\{e_\alpha\}$

$$e_1 = v_1 \otimes v_1, \qquad e_4 = \frac{1}{\sqrt{2}}(v_2 \otimes v_3 + v_3 \otimes v_2) \quad \text{etc}, \tag{5}$$

where $\{v_k\}$ is the orthonormal basis of the proper vectors of a given $\boldsymbol{T}$. The inner product on $\mathscr{T}$ is defined in the usual way, and then we have

$$(\boldsymbol{A}, \boldsymbol{B}) = (\boldsymbol{a}, \boldsymbol{b}), \tag{6}$$

where $\boldsymbol{a}$ and $\boldsymbol{b}$ are the six-dimensional vectors on $\mathscr{T}$ corresponding to $\boldsymbol{A}$ and $\boldsymbol{B}$, respectively. So $\mathscr{V}$ and $\mathscr{T}$ are isometric.

In this notation the constitutive equation (1) is transformed into

$$\mathring{\boldsymbol{t}} = \boldsymbol{H}(\boldsymbol{t})\boldsymbol{d}, \tag{7}$$

where $\boldsymbol{H}$ is the six-dimensional second-order tensor on $\mathscr{T}$ and $\boldsymbol{t}$, $\boldsymbol{d}$ and $\mathring{\boldsymbol{t}}$ are six-dimensional vectors corresponding to $\boldsymbol{T}$, $\boldsymbol{D}$ and $\mathring{\boldsymbol{T}}$, respectively.

From (7) we have the explicit expression of the yield criterion

$$\det \boldsymbol{H}(\boldsymbol{t}) = 0 \tag{8}$$

and the plastic stretching is given by the nontrivial solution of

$$\boldsymbol{H}(\boldsymbol{t})_{\mathrm{P}}\boldsymbol{d} = 0. \tag{9}$$

The plastic stretching does not, in general, necessarily correspond to a motion of the material particle. When ${}_{\mathrm{P}}\boldsymbol{D}$, which is the second-order tensor corresponding to ${}_{\mathrm{P}}\boldsymbol{d}$, satisfies the condition of compatibility

$${}_{\mathrm{P}}D_{km,pq} + {}_{\mathrm{P}}D_{pq,km} - {}_{\mathrm{P}}D_{kp,mq} - {}_{\mathrm{P}}D_{qm,pk} = 0, \tag{10}$$

at a point concerned, the plastic stretching may be a symmetric part of certain velocity gradient and the material particle may flow at the point.

According to the principle of frame-indifference $\mathsf{H}(\boldsymbol{T})[\boldsymbol{D}]$ is a symmetric tensor-valued isotropic function and then by a special case of representation theorem of Wang [15] for isotropic function we have

$$\begin{aligned}\mathsf{H}(\boldsymbol{T})[\boldsymbol{D}] = {} & [\beth_1 \operatorname{tr} \boldsymbol{D} + \beth_2 \operatorname{tr}(\boldsymbol{T}\boldsymbol{D}) + \beth_3 \operatorname{tr}(\boldsymbol{T}^2\boldsymbol{D})]\mathbf{1} + \\ & + [\beth_4 \operatorname{tr} \boldsymbol{D} + \beth_5 \operatorname{tr}(\boldsymbol{T}\boldsymbol{D}) + \beth_6 \operatorname{tr}(\boldsymbol{T}^2\boldsymbol{D})]\boldsymbol{T} + \\ & + [\beth_7 \operatorname{tr} \boldsymbol{D} + \beth_8 \operatorname{tr}(\boldsymbol{T}\boldsymbol{D}) + \beth_9 \operatorname{tr}(\boldsymbol{T}^2\boldsymbol{D})]\boldsymbol{T}^2 + \\ & + \beth_{10}\boldsymbol{D} + \beth_{11}(\boldsymbol{D}\boldsymbol{T} + \boldsymbol{T}\boldsymbol{D}) + \beth_{12}(\boldsymbol{D}\boldsymbol{T}^2 + \boldsymbol{T}^2\boldsymbol{D}),\end{aligned} \tag{11}$$

where $\beth_1, \ldots, \beth_{12}$ are functions of the three principal invariants of stress. In $\mathscr{T}$-representation we have

$$\mathsf{H}_{kmpq} = A_{kp}\delta_{km}\delta_{pq} + \tfrac{1}{2}B_{km}(\delta_{kp}\delta_{mq} + \delta_{kq}\delta_{mp}), \tag{12}$$

where

$$A_{kp} \equiv \beth_1 + \beth_4 t_k + \beth_2 t_p + \beth_7 t_k^2 + \beth_5 t_k t_p + \\ + \beth_3 t_p^2 + \beth_6 t_k t_p^2 + \beth_8 t_k^2 t_p + \beth_9 t_k^2 t_p^2, \tag{13}$$

$$B_{km} \equiv \beth_{10} + \beth_{11}(t_k + t_m) + \beth_{12}(t_k^2 + t_m^2), \tag{14}$$

where t_k ($k = 1, 2, 3$) are the proper numbers of stress and indices k, m and p are not summed.

Then we have the 6×6 matrix representation

$$[\boldsymbol{H}] = \left\| \begin{matrix} [\boldsymbol{H}_\mathrm{N}] & [\boldsymbol{0}] \\ [\boldsymbol{0}] & [\boldsymbol{H}_\mathrm{S}] \end{matrix} \right\|, \tag{15}$$

where

$$[\boldsymbol{H}_\mathrm{N}] \equiv \left\| \begin{matrix} A_{11}+B_{11} & A_{12} & A_{13} \\ A_{21} & A_{22}+B_{22} & A_{23} \\ A_{31} & A_{32} & A_{33}+B_{33} \end{matrix} \right\|, \tag{16}$$

$$[\boldsymbol{H}_\mathrm{S}] \equiv \left\| \begin{matrix} B_{23} & & 0 \\ & B_{31} & \\ 0 & & B_{12} \end{matrix} \right\|, \tag{17}$$

$[\boldsymbol{0}]$ is the 3×3 null matrix and $\boldsymbol{H}_\mathrm{N}$ and $\boldsymbol{H}_\mathrm{S}$ are the images on $\mathscr{T}$ of $\boldsymbol{H}_\mathrm{N}$ and $\boldsymbol{H}_\mathrm{S}$, respectively.

Therefore the yield criterion (8) is separated into two types

$$\det \boldsymbol{H}_\mathrm{N}(\boldsymbol{t}) = 0 \quad \text{and} \quad \det \boldsymbol{H}_\mathrm{S}(\boldsymbol{t}) = 0 \tag{18}$$

and we have two types of plastic stretchings which are the nontrivial solutions of

$$\boldsymbol{H}_\mathrm{N}(\boldsymbol{t})\,{}_\mathrm{P}\boldsymbol{d}_\mathrm{N} = 0 \quad \text{and} \quad \boldsymbol{H}_\mathrm{S}(\boldsymbol{t})\,{}_\mathrm{P}\boldsymbol{d}_\mathrm{S} = 0. \tag{19}$$

Yield surface

The stress vector $\boldsymbol{t}$ on $\mathscr{T}$ lies in the three-dimensional space $\mathscr{T}_3$, which is identical with the Haigh-Westergaard principal stress space. Then the yield condition (8) constitutes a two-dimensional surface on $\mathscr{T}_3$.

Any permutation of t_1, t_2 and t_3 in (13) and (14) causes same interchanges of rows and columns in (16) and (17). Thus we can say that *the yield surface has threefold symmetry around pressure axis*, which passes through the origin of $\mathscr{T}_3$ and makes equal angle with the coordinate axes.

When $\mathbf{H}(\boldsymbol{T})$ is independent of pressure $p \equiv -\frac{1}{3}\operatorname{tr}\boldsymbol{T}$, the material is said to be pressure-insensitive, and *the yield surface of the pressure-insensitive material is a cylinder generated by lines parallel to the pressure axis.* When $\mathbf{H}(\boldsymbol{T})$ is an even function with respect to the stress deviator, we say that the material is free from the Bauschinger effects, and then *the yield surface of a material of free of Bauschinger effects has sixfold symmetry around the pressure axis.*

Let us introduce the rectangular coordinates x, y and z on $\mathscr{T}_3$ such that

$$\begin{aligned} t_1 &= \sqrt{\frac{2}{3}}\,y + \frac{1}{\sqrt{3}}\,z, \\ t_2 &= -\frac{1}{\sqrt{2}}\,x - \frac{1}{\sqrt{6}}\,y + \frac{1}{\sqrt{3}}\,z, \\ t_3 &= \frac{1}{\sqrt{2}}\,x - \frac{1}{\sqrt{6}}\,y + \frac{1}{\sqrt{3}}\,z, \\ p &= -\frac{1}{\sqrt{3}}\,z. \end{aligned} \tag{20}$$

Then the yield criteria (18) can be expressed in terms of x, y and z.

Normal yield

Consider a special case:

$$\begin{aligned} A_{kp} &= \lambda + a_1 p + a_2 p^2 + a_3 t_k^* t_p^*, \\ B_{km} &= 2\mu + b_1 p + b_2 p^2, \end{aligned} \tag{21}$$

where λ, μ, a's and b's are material constants and $t_k^* \equiv t_k + p$ is the stress deviator vector. The proper numbers of $\boldsymbol{H}_N$ are

$$\begin{aligned} H_N^{(1)} &= 2\mu + b_1 p + b_2 p^2 + a_3 \sum_{k=1}^{3} t_k^{*2}, \\ H_N^{(2)} &= 3\lambda + 2\mu + (3a_1 + 2b_1)p + (3a_2 + 2b_2)p^2, \\ H_N^{(3)} &= 2\mu + b_1 p + b_2 p^2. \end{aligned} \tag{22}$$

The plastic stretching corresponding to $H_N^{(1)} = 0$ is $\psi \boldsymbol{t}^*$, where ψ has any value. Then we arrive at the *St. Venant-Lévy flow rule*: ${}_{\mathrm{P}}\boldsymbol{D} = \psi \boldsymbol{T}^*$.

The yield surface $H_N^{(1)} = 0$ is

$$a_3(x^2+y^2)+\tfrac{1}{3}b_2z^2 - \frac{1}{\sqrt{3}}\,b_1z+2\mu = 0. \tag{23}$$

When the material is pressure-insensitive, we must put $a_1 = a_2 = b_1 = b_2 = 0$ and then (23) is reduced to

$$a_3(x^2+y^2)+2\mu = 0 \tag{24}$$

or

$$\sum_{k=1}^{3} t_k^{*2} = -\frac{2\mu}{a_3}, \tag{25}$$

which denotes the *v. Mises yield criterion.*

Shear yield

Restricting $\mathbf{H}_S(\boldsymbol{T})$ to the grade two in $\boldsymbol{T}$, we have the yield surface $B_{23}=0$:

$$c_1x^2+c_2y^2+c_3z^2+2c_4yz+2c_5y+2c_6z+2\mu = 0, \tag{26}$$

where μ and c's are material constants.

When the material is pressure-insensitive, we have

$$c_1x^2+c_2y^2+2c_5y+2\mu = 0. \tag{27}$$

When the material is free of Bauschinger effects, we have

$$c_1x^2+c_2y^2+c_3z^2+2c_6z+2\mu = 0. \tag{28}$$

The plastic stretching associated with $B_{23} = 0$ is D_{23}. It is natural to assume that the direction of the plastic shear stretching coincides with that of the maximum or minimum shear stress. From this assumption we may put $t_2 \geqq t_1 \geqq t_3$ or $t_3 \geqq t_1 \geqq t_2$ on $B_{23} = 0$. From (20) y denotes the intermediate stress. When B_{23} is independent of y, the material is said to be insensitive to intermediate stress effects. In that case we have

$$c_1x^2+c_3z^2+2c_6z+2\mu = 0. \tag{29}$$

Moreover, when the material is pressure-insensitive and free of intermediate stress effects, at the same time, we have

$$c_1 x^2 + 2\mu = 0 \tag{30}$$

or

$$|t_1 - t_3| = 2\sqrt{\left(-\frac{\mu}{c_1}\right)}, \tag{31}$$

and which represents the *Tresca yield criterion.*

References

[1] Tresca, H., Mémoire sur l'écoulement des corps solides. *Mém. prés. par div. savants 18* (1870) 733–799.

[2] Mises, R. v., Mechanik der festen Körper im plastisch deformablen Zustand. *Göttingen Nachr. math. phys. K 1913* (1913) 582–592.

[3] Venant, B. de St., Mémoire sur l'établissement des équations différentielles des mouvements intérieurs opérés dans les corps solides ductiles au delà des limites où l'élasticité pourrait les ramener à leur premier état. *C. R. Acad. Sci. Paris 70* (1870) 473–480.

[4] Lévy, M., Mémoire sur les équations générales des mouvements intérieurs des corps solides ductiles au delà des limites où l'élasticité pourrait les ramener à leur premier état. *C. R. Acad. Sci. Paris 70* (1870) 1323–1325.

[5] Haigh, B. P., The strain energy function and the elastic limit. *Engineering 109* (1920) 158–160.

[6] Westergaard, H. M., On the resistance of ductile materials to combined stresses in two or three directions perpendicular to one another. *J. Franklin Inst. 189* (1920) 627–640.

[7] Thomas, T. Y., Interdependence of the yield condition and the stress-strain relations for plastic flow. *Proc. Acad. Sci. U.S.A. 40* (1954) 593–597.

[8] Morgan, A. J. A., Some properties of media defined by constitutive equations in implicit form. *Int. J. Engng Sci. 4* (1966) 155–178.

[9] Fox, N., On the derivation of the constitutive equations of ideal plasticity. *Acta Mech. 7* (1969) 248–251.

[10] Truesdell, C., Hypo-elasticity. *J. Rational Mech. Anal., 4* (1955) 88–133; 1019–1020.

[11] Tokuoka, T., Yield conditions and flow rules derived from hypo-elasticity. *Arch. Rational Mech. Anal. 42* (1971) 239–252.

[12] Thomas, T. Y., Combined elastic and Prandtl-Reuss stress-strain relations. *Proc. Nat. Acad. Sci. U.S.A. 41* (1955) 720–726; Combined elastic and von Mises stress-strain relations. Ibid, 908–910.

[13] Truesdell, C., Hypo-elastic shear. *J. Appl. Phys. 27* (1956) 441–447.

[14] Green, A. E., Hypo-elasticity and plasticity II. *J. Rational Mech. Anal. 5* (1956) 725–734.

[15] Wang, C. — C., A new representation theorem for isotropic functions, Part I and II. *Arch. Rational Mech. Anal. 36* (1970) 166–223.

Some solutions for a class of plastic-elastic solids

A. J. M. Spencer and J. E. Ferrier

University of Nottingham, England

We consider a class of plastic-elastic constitutive equations. In formulating these it is assumed that the strain-rate can be decomposed into a plastic strain-rate and an elastic strain-rate. The plastic strain-rate is derived from a von Mises plastic potential and the elastic strain-rate is linear in one of a class of properly defined stress-rate measures which are linear in the stress and velocity gradients. The theory is applied to simple shear and simple torsion of incompressible materials. It is found that for these problems varying the choice of stress-rate measure produces only second-order effects in the stress components; however the sign as well as the magnitude of these effects depends on a parameter which distinguishes stress-rate measures. The limiting values of the stress components for large shear or torsion are compared with the corresponding solutions for a plastic-rigid body, and the status of the plastic-rigid theory discussed in the light of these results.

1. Introduction

Although the mathematical theory of plasticity has been in use for many years, its correct formulation is still a matter of discussion. The classical approach is to decompose the strain or strain-rate into plastic and elastic parts, and postulate constitutive equations for both parts separately, the elastic strain-rate being assumed linear in a measure of the stress-rate. More recently, alternative formulations have been given by Pipkin and Rivlin [1], Green and Naghdi [2] and Thomas [3]. Lee [4] has proposed a theory in which the strain decomposition is multiplicative rather than additive.

From the point of view of solving boundary value problems the classical formulation in terms of strain-rates, or equivalently strain-increments, seems to lead to the simplest theory to apply. However the older work using this approach has the defect that it uses a measure of stress-rate which is not properly defined, except for small strains. This defect is easily remedied, as for example by Hill [5, 6], by introducing a stress-rate measure with the correct invariance properties, but there remains a problem in that there is no unique way of defining such a stress-rate measure.

The first objective of this paper is to investigate, in two simple cases, the effect of varying the stress-rate measure within a fairly broad family of such measures which have the required properties. The problems considered are those of simple shear and pure torsion. In order to obtain explicit solutions to these problems we make a number of simplifying assumptions, for example that the material is perfectly plastic, satisfies von Mises yield condition, and that the ratio of the shear yield stress k to elastic shear modulus μ is a small parameter. Within the limitations imposed by these approximations we find that the choice of stress-rate measure has no first-order (in k/μ) effect but does have a second-order effect so that, for example, either a positive or negative plastic-elastic 'Poynting effect' may be obtained depending on the value of a coefficient in the constitutive equation.

A secondary objective is to investigate, again in these simple examples, the relationship of plastic-elastic theory to the much simpler plastic-rigid theory. If plastic-rigid theory is to have value plastic-rigid solutions must, in some sense, be asymptotic solutions of the plastic-elastic equations. Such behaviour is often assumed, but seems never to have been proven. We show here that in the examples considered, for large strains the plastic-elastic and plastic-rigid solutions differ only by terms of second order in k/μ, and that the solutions coincide in the limit $k/\mu \to 0$.

The theory employed is purely mechanical, and we take no account of thermodynamic restrictions. For the most part we consider incompressible materials, but the theory is formulated for compressible materials also.

2. Notation

We refer quantities to a fixed system of rectangular cartesian coordinates. The coordinates of a typical particle are denoted by X_R $(R = 1, 2, 3)$ in a reference configuration at a reference time $t = 0$, and by x_i $(i = 1, 2, 3)$ at time t. Thus the deformation is described by the dependence

$$x_i = x_i(X_R, t) \tag{2.1}$$

of x_i on X_R and t. Here t is not necessarily the physical time, but may be any parameter which determines the sequence of events. Components of velocity are denoted by $v_j(x_i, t)$. The components of strain-rate d_{ij} are

$$d_{ij} = \tfrac{1}{2}\left(\frac{\partial v_i}{\partial x_j} + \frac{\partial v_j}{\partial x_i}\right) \tag{2.2}$$

and the components of the spin ω_{ij} are

$$\omega_{ij} = \tfrac{1}{2}\left(\frac{\partial v_i}{\partial x_j} - \frac{\partial v_j}{\partial x_i}\right). \tag{2.3}$$

The components of the Cauchy stress tensor are denoted t_{ij}. The deviatoric stress is denoted by s_{ij} and the mean hydrostatic pressure by p. Thus

$$t_{ij} = s_{ij} - p\delta_{ij}, \tag{2.4}$$

where

$$s_{ii} = 0, \quad p = -\tfrac{1}{3}t_{ii}. \tag{2.5}$$

When convenient, we will use matrix notation, and denote

$$\begin{aligned} &\boldsymbol{X} = \|X_R\|, \quad \boldsymbol{x} = \|x_i\|, \quad \boldsymbol{D} = \|d_{ij}\|, \quad \omega = \|\omega_{ij}\|, \\ &\boldsymbol{T} = \|t_{ij}\|, \quad \boldsymbol{S} = \|s_{ij}\|. \end{aligned} \tag{2.6}$$

3. Yield condition

For simplicity, it is assumed that the material is homogeneous, isotropic and satisfies von Mises yield condition

$$s_{ij}s_{ij} \leqq 2k^2 \tag{3.1}$$

where k is the shear yield stress of the material. In general k is a scalar parameter which depends on the deformation history. In the applications considered in sect. 8 and sect. 10 we assume, again for simplicity, that the material is non-hardening, so that k is a constant. Since the deviatoric stress rather than the total stress is used in (3.1), there is included an assumption that yielding is not affected by a superposed hydrostatic pressure.

4. Strain-rate decomposition

We now assume that the strain-rate can be expressed as the sum of two parts

$$d_{ij} = d_{ij}^{(e)} + d_{ij}^{(p)}, \tag{4.1}$$

or, equivalently, in matrix notation

$$\boldsymbol{D} = \boldsymbol{D}_e + \boldsymbol{D}_p. \tag{4.2}$$

For convenience we term $d_{ij}^{(e)}$ and $d_{ij}^{(p)}$ components of elastic and plastic strain-rate respectively, but note that $d_{ij}^{(e)}$ and $d_{ij}^{(p)}$ individually are not derived from velocities and need not satisfy strain-rate compatibility conditions. It is assumed that $d_{ij}^{(e)}$ and $d_{ij}^{(p)}$ transform as components of symmetric second-order tensors under orthogonal transformations of the coordinates x_i, and are independent of orthogonal transformations of the coordinates X_R.

Lee [4] has proposed an alternative decomposition of the deformation into elastic and plastic deformations. If the elastic deformation is small in an appropriate sense then, as Lee [4] indicates, this alternative method of decomposition is equivalent to (4.1).

5. Plastic strain-rate. Constitutive equation

The constitutive assumption for the plastic strain-rate is that it obeys the flow rule associated with the yield condition (3.1), namely

$$d_{ij}^{(p)} = \lambda s_{ij}. \tag{5.1}$$

For a hardening material λ depends on the deformation history. For a non-hardening material λ is an arbitrary non-negative scalar multiplier, which can take non-zero values only when the equality sign holds in (3.1). Thus

$$\begin{aligned} \lambda = 0 \quad &\text{when} \quad s_{ij}s_{ij} < 2k^2, \\ \lambda \geqq 0 \quad &\text{when} \quad s_{ij}s_{ij} = 2k^2. \end{aligned} \tag{5.2}$$

It is implicit in (5.1) that

$$d_{ii}^{(p)} = 0 \tag{5.3}$$

so that no volume changes are associated with the plastic strain-rate.

6. Elastic strain rate. Constitutive equation

As constitutive equation for the elastic strain-rate it will be assumed that the elastic strain-rate is a linear function of a suitable measure of the stress-rate. The stress-rate measure must transform as a symmetric second-order tensor under orthogonal transformations of the coordinates x_i and be independent of superposed rigid body motions of the body. It is well-known (e.g. [7–9]) there are an infinite number of stress-rate measures which satisfy these requirements, and although plausible arguments can be advanced (e.g. [10–13]) for the adoption of one rather than another, there appears to be no compelling reason why a particular stress-rate should be selected. In any event, terms which represent the difference between two possible stress-rates are terms which may appear in constitutive equations. We therefore begin by constructing a general class of stress-rate measures.

For convenience and definiteness we base these on the 'Jaumann' derivative of t_{ij}. This we denote $\mathscr{D}_J t_{ij}/\mathscr{D}t$, and it is defined by

$$\mathscr{D}_J t_{ij}/\mathscr{D}t = Dt_{ij}/Dt + t_{ik}\omega_{kj} + t_{jk}\omega_{ki}, \tag{6.1}$$

where Dt_{ij}/Dt denotes the convected derivative

$$Dt_{ij}/Dt = \partial t_{ij}/\partial t + v_k \partial t_{ij}/\partial x_k. \tag{6.2}$$

In matrix notation we write (6.1) in the form

$$\mathscr{D}_J \boldsymbol{T} = D\boldsymbol{T} + \boldsymbol{T}\boldsymbol{\omega} - \boldsymbol{\omega}\boldsymbol{T}. \tag{6.3}$$

An acceptable measure of stress-rate is obtained by adding to $\mathscr{D}_J t_{ij}/\mathscr{D}t$ any symmetric second-order tensor function of d_{ij} and t_{ij} with the dimensions of (stress) × (time)$^{-1}$. We consider only tensor functions which are linear in d_{ij} and t_{ij}. The most general expression of this kind is, in matrix notation

$$\alpha(\boldsymbol{DT}+\boldsymbol{TD}) + \beta\boldsymbol{T}\,\mathrm{tr}\,\boldsymbol{D} + \gamma\boldsymbol{D}\,\mathrm{tr}\,\boldsymbol{T} + \boldsymbol{I}(\delta'\,\mathrm{tr}\,\boldsymbol{DT} + \varepsilon'\,\mathrm{tr}\,\boldsymbol{D}\,\mathrm{tr}\,\boldsymbol{T}) \tag{6.4}$$

where α, β, γ, δ', ε' are numerical coefficients, tr $\boldsymbol{A}$ denotes the trace of the matrix $\boldsymbol{A}$, and $\boldsymbol{I}$ is the unit matrix. We therefore form the stress-rate $\mathscr{D}\boldsymbol{T}$, with components $\mathscr{D}t_{ij}/\mathscr{D}t$, by the relation

$$\mathscr{D}\boldsymbol{T} = \mathscr{D}_J\boldsymbol{T} + \alpha(\boldsymbol{DT}+\boldsymbol{TD}) + \beta\boldsymbol{T}\,\mathrm{tr}\,\boldsymbol{D} + \gamma\boldsymbol{D}\,\mathrm{tr}\,\boldsymbol{T} + \boldsymbol{I}(\delta'\,\mathrm{tr}\,\boldsymbol{DT} + \varepsilon'\,\mathrm{tr}\,\boldsymbol{D}\,\mathrm{tr}\,\boldsymbol{T}). \tag{6.5}$$

For $\gamma = \delta' = \varepsilon' = 0$, various special cases of (6.5) appear in the literature. Among them are $\alpha = 0$, $\beta = 0$ (Jaumann [14]); $\alpha = 1$, $\beta = 0$ (Cotter and Rivlin [15]); $\alpha = -1$, $\beta = 0$ (Oldroyd [16], Green [17]); and $\alpha = -1$, $\beta = 1$ (Truesdell [18]). Hill [19 and subsequent papers] first introduced the concept of a family of measures (Hill's family corresponds to our case $\beta = 1$, α variable) and has extensively investigated its implications for elasticity and plasticity theory.

By taking the trace of (6.5) we obtain

$$\mathrm{tr}\,\mathscr{D}\boldsymbol{T} = D(\mathrm{tr}\,\boldsymbol{T}) + (2\alpha+3\delta')\,\mathrm{tr}\,\boldsymbol{DT} + (\beta+\gamma+3\varepsilon')\,\mathrm{tr}\,\boldsymbol{D}\,\mathrm{tr}\,\boldsymbol{T}, \tag{6.6}$$

where we have used the relation, which follows readily from (6.3)

$$\operatorname{tr} \mathscr{D}_J \boldsymbol{T} = D(\operatorname{tr} \boldsymbol{T}). \tag{6.7}$$

We now postulate the constitutive equation

$$2\mu \boldsymbol{D}_e = \mathscr{D}\boldsymbol{T} - \eta \boldsymbol{I} \operatorname{tr} \mathscr{D}\boldsymbol{T}. \tag{6.8}$$

where μ and η are constants. For *infinitesimal purely elastic deformations*, and for any choice of the stress-rate measure, (6.8) integrates to give the usual stress-strain relation of linear isotropic elasticity if μ is interpreted as the elastic shear modulus and $\eta/(1-\eta)$ is interpreted as Poisson's ratio ν. On combining (6.5), (6.6) and (6.8) we obtain

$$2\mu \boldsymbol{D}_e = \mathscr{D}_J \boldsymbol{T} - \eta D(\operatorname{tr} \boldsymbol{T}) + \alpha(\boldsymbol{DT} + \boldsymbol{TD}) + \beta \boldsymbol{T} \operatorname{tr} \boldsymbol{D} + \gamma \boldsymbol{D} \operatorname{tr} \boldsymbol{T} + \\ + \boldsymbol{I}(\delta \operatorname{tr} \boldsymbol{DT} + \varepsilon \operatorname{tr} \boldsymbol{D} \operatorname{tr} \boldsymbol{T}). \tag{6.9}$$

where

$$\delta = \delta' - \eta(2\alpha + 3\delta'), \quad \varepsilon = \varepsilon' - \eta(\beta + \gamma + 3\varepsilon'). \tag{6.10}$$

As a further simplification we observe that since

$$\boldsymbol{D} = \boldsymbol{D}_e + \boldsymbol{D}_p$$

the term $\gamma \boldsymbol{D} \operatorname{tr} \boldsymbol{T}$ may be neglected if $\mu \gg \operatorname{tr} \boldsymbol{T}$. In this case (6.9) reduces to

$$2\mu \boldsymbol{D}_e = \mathscr{D}_J \boldsymbol{T} - \eta \boldsymbol{I} D(\operatorname{tr} \boldsymbol{T}) + \alpha(\boldsymbol{DT} + \boldsymbol{TD}) + \beta \boldsymbol{T} \operatorname{tr} \boldsymbol{D} + \\ + \boldsymbol{I}(\delta \operatorname{tr} \boldsymbol{DT} + \varepsilon \operatorname{tr} \boldsymbol{D} \operatorname{tr} \boldsymbol{T}) \tag{6.11}$$

where η, α, β, δ and ε are dimensionless constants.

It is clearly possible to generalise (6.11), for example by permitting the coefficients to be functions of the stress components t_{ij}. However we envisage applications to materials and situations in which $t_{ij} \ll \mu$, and then (6.11) may be regarded as a permissible linearisation of a more general relation. This point of view is also consistent with the neglect of the term involving γ.

An alternative way of regarding (6.11) is as a constitutive equation for $\boldsymbol{D}_e$ in which $\boldsymbol{D}_e$ is postulated to be a linear function of a *particular* stress-rate (for example the Jaumann stress-rate) and to depend linearly on $\boldsymbol{T}$ and $\boldsymbol{D}$.

Since the constitutive equation is given in terms of rates, initial conditions are required to complete its specification. For simplicity we assume that the reference configuration is a stress-free state, so that $t_{ij} = 0$ when $t = 0$.

All the constitutive equations of this and the previous section satisfy the usual invariance requirements for isotropic materials.

7. Incompressible material

When the material is incompressible we have tr $\boldsymbol{D} = 0$, and it follows from (5.3) that tr $\boldsymbol{D}_e = 0$. The hydrostatic pressure p becomes an arbitrary scalar function. It is therefore convenient to first express (6.11) in terms of $\boldsymbol{S}$ and p. Setting tr $\boldsymbol{D} = 0$, this gives

$$2\mu\boldsymbol{D}_e = \mathscr{D}_J\boldsymbol{S} - \boldsymbol{I}Dp/Dt + 3\eta\boldsymbol{I}Dp/Dt + \alpha(\boldsymbol{DS}+\boldsymbol{SD}) - 3\alpha p\boldsymbol{D} + \delta\boldsymbol{I}\,\mathrm{tr}\,\boldsymbol{DS}. \tag{7.1}$$

By taking the trace of (7.1), we have

$$0 = 3(-1+3\eta)Dp/Dt + (2\alpha+3\delta)\,\mathrm{tr}\,\boldsymbol{DS}. \tag{7.2}$$

Hence, multiplying (7.2) by $\boldsymbol{I}/3$ and subtracting from (7.1)

$$2\mu\boldsymbol{D}_e = \mathscr{D}_J\boldsymbol{S} + \alpha(\boldsymbol{DS}+\boldsymbol{SD}) - 3\alpha p\boldsymbol{D} - \tfrac{2}{3}\alpha\boldsymbol{I}\,\mathrm{tr}\,\boldsymbol{DS}. \tag{7.3}$$

The term $3\alpha p\boldsymbol{D}$ is discarded for the same reason as that used to reject the term in γ in section 6, so we obtain

$$2\mu\boldsymbol{D}_e = \mathscr{D}_J\boldsymbol{S} + \alpha(\boldsymbol{DS}+\boldsymbol{SD}) - \tfrac{2}{3}\alpha\boldsymbol{I}\,\mathrm{tr}\,\boldsymbol{DS}. \tag{7.4}$$

This is the constitutive equation for $\boldsymbol{D}_e$ for incompressible material. The right-hand side involves only the single coefficient α.

8. Simple shear

As a first example we consider simple shearing deformations defined by

$$x_1 = X_1, \quad x_2 = X_2 + stX_1, \quad x_3 = X_3, \tag{8.1}$$

where s is constant. The corresponding velocity components are

$$v_1 = 0, \quad v_2 = sx_1, \quad v_3 = 0 \tag{8.2}$$

and the only non-zero components of strain-rate and rotation are

$$d_{12} = \tfrac{1}{2}s, \quad \omega_{12} = -\omega_{21} = -\tfrac{1}{2}s. \tag{8.3}$$

Since the deformation is homogeneous, we seek solutions in which the stress components are functions of t only. The equilibrium equations are then satisfied. Derivatives with respect to t are denoted by superposed dots.

For brevity only deformations of incompressible material satisfying the constitutive equation (7.4) are considered, but the solution may also be found in a similar manner for compressible materials governed by the constitutive equation (6.11).

Elastic solution

For sufficiently small values of t the behaviour is elastic, the plastic strain-rate is zero, and the elastic strain-rate coincides with the total strain-rate. The constitutive equations (7.4) become, for this deformation

$$\begin{aligned}
2\mu d_{11} &= \dot{s}_{11}+s(1+\tfrac{1}{3}\alpha)s_{12} = 0,\\
2\mu d_{22} &= \dot{s}_{22}+s(-1+\tfrac{1}{3}\alpha)s_{12} = 0,\\
2\mu d_{33} &= \dot{s}_{33}-s(\tfrac{2}{3}\alpha)s_{12} = 0,\\
2\mu d_{12} &= \dot{s}_{12}+\tfrac{1}{2}s\{(-1+\alpha)s_{11}+(1+\alpha)s_{22}\} = \mu s,\\
2\mu d_{13} &= \dot{s}_{13}+\tfrac{1}{2}s(1+\alpha)s_{23} = 0,\\
2\mu d_{23} &= \dot{s}_{23}+\tfrac{1}{2}s(-1+\alpha)s_{13} = 0.
\end{aligned} \tag{8.4}$$

The solution of these, subject to the initial conditions $s_{ij} = 0$ when $t = 0$, is

$$\begin{aligned}
s_{11} &= \frac{\mu}{m^2}(1+\tfrac{1}{3}\alpha)(\cos mst-1), \quad s_{22} = -\frac{\mu}{m^2}(1-\tfrac{1}{3}\alpha)(\cos mst-1),\\
s_{33} &= -\tfrac{2}{3}\frac{\mu\alpha}{m^2}(\cos mst-1), \quad s_{12} = \frac{\mu}{m}\sin mst,\\
s_{13} &= 0, \quad s_{23} = 0,
\end{aligned} \tag{8.5}$$

where

$$m^2 = 1-\tfrac{1}{3}\alpha^2. \tag{8.6}$$

The apparent periodic nature of the solution is not real because, as we show below, the elastic solution applies only for small values of st. To second order in st, (8.5) are

$$\begin{aligned}
s_{11} &= -\tfrac{1}{2}\mu(1+\tfrac{1}{3}\alpha)s^2t^2, \quad s_{22} = \tfrac{1}{2}\mu(1-\tfrac{1}{3}\alpha)s^2t^2,\\
s_{33} &= \tfrac{1}{3}\mu\alpha s^2t^2, \quad s_{12} = \mu st, \quad s_{13} = 0, \quad s_{23} = 0.
\end{aligned} \tag{8.7}$$

Note that the constitutive equation predicts non-zero normal stress components s_{11}, s_{22}, s_{33}, but that these are of second order in st. We observe also that the normal stress difference $s_{11}-s_{22}$ is, to this order, independent of α.

The hydrostatic pressure p may be any function of t only. The total stress t_{ij} is of course given by (2.4).

Yielding

When $s_{13} = 0$, $s_{23} = 0$, the yield condition (3.1) becomes

$$s_{11}^2 + s_{22}^2 + s_{33}^2 + 2s_{12}^2 \leqq 2k^2. \tag{8.8}$$

The elastic solution applies for $t \leqq t_0$, where t_0 is the value of t which satisfies the *equality* (8.8) when the stress components are given by (8.5). By substituting (8.5) into (8.8) and solving for $\cos mst$, we find

$$\cos mst_0 = \frac{3}{2\alpha^2}\left\{(1+\tfrac{1}{3}\alpha^2) - m^2\left(1+\tfrac{2}{3}\alpha^2\frac{k^2}{\mu^2}\right)^{\frac{1}{2}}\right\}. \tag{8.9}$$

We are interested in materials for which $k/\mu \ll 1$. We therefore expand (8.9) in powers of k/μ and neglect $(k/\mu)^4$ and higher powers. This gives

$$\cos mst_0 \approx 1 - \tfrac{1}{2}m^2\frac{k^2}{\mu^2},$$

so that

$$\sin mst_0 \approx mst_0 \approx mk/\mu. \tag{8.10}$$

Thus the elastic solution applies only for values of $st \leqq k/\mu$, which justifies the statement made above that st is small in the elastic range. The values of the non-zero stress components at the yield value $t = t_0$ are, to order $(k/\mu)^2$

$$\begin{aligned} s_{11}/\mu &= -\tfrac{1}{2}(1+\tfrac{1}{3}\alpha)\left(\frac{k}{\mu}\right)^2, & s_{22}/\mu &= \tfrac{1}{2}(1-\tfrac{1}{3}\alpha)\left(\frac{k}{\mu}\right)^2, \\ s_{33}/\mu &= \tfrac{1}{3}\alpha\left(\frac{k}{\mu}\right)^2, & s_{12}/\mu &= k/\mu. \end{aligned} \tag{8.11}$$

Plastic-elastic solution

For $t > t_0$ we have

$$\boldsymbol{D} = \boldsymbol{D}_e + \boldsymbol{D}_p$$

where $\boldsymbol{D}_e$ and $\boldsymbol{D}_p$ are given by (7.4) and (5.1). The constitutive equations therefore become

$$
\begin{aligned}
2\mu d_{11} &= \dot{s}_{11}+s(1+\tfrac{1}{3}\alpha)s_{12}+\lambda s_{11} = 0,\\
2\mu d_{22} &= \dot{s}_{22}+s(-1+\tfrac{1}{3}\alpha)s_{12}+\lambda s_{22} = 0,\\
2\mu d_{33} &= \dot{s}_{33}+s(-\tfrac{2}{3}\alpha)s_{12}+\lambda s_{33} = 0,\\
2\mu d_{12} &= \dot{s}_{12}+\tfrac{1}{2}s\{(-1+\alpha)s_{11}+(1+\alpha)s_{22}\}+\lambda s_{12} = \mu s,\\
2\mu d_{13} &= \dot{s}_{13}+\tfrac{1}{2}s(1+\alpha)s_{23}+\lambda s_{13} = 0,\\
2\mu d_{23} &= \dot{s}_{23}+\tfrac{1}{2}s(-1+\alpha)s_{13}+\lambda s_{23} = 0.
\end{aligned}
\tag{8.12}
$$

The last two equations are satisfied if $s_{13} = 0$, $s_{23} = 0$, and the yield condition is then (8.8) with the equality sign. Since

$$
s_{11}+s_{22}+s_{33} = 0,
\tag{8.13}
$$

this yield condition may be written

$$
s_{12}^2+s_{11}^2+s_{22}^2+s_{11}s_{22} = k^2.
\tag{8.14}
$$

We make the substitutions

$$
\begin{aligned}
s_{11} &= -k\sin\phi\left[\frac{1}{\sqrt{3}}\sin\psi + \cos\psi\right],\\
s_{22} &= -k\sin\phi\left[\frac{1}{\sqrt{3}}\sin\psi - \cos\psi\right],\\
s_{33} &= \frac{2k}{\sqrt{3}}\sin\phi\sin\psi, \quad s_{12} = k\cos\phi,
\end{aligned}
\tag{8.15}
$$

where ϕ and ψ are functions of t for $t > t_0$. These expressions identically satisfy (8.13) and (8.14). The values ϕ_0 and ψ_0 of ϕ and ψ at $t = t_0$ are chosen so that the stress components are continuous at $t = t_0$. From (8.11) and (8.15) these values are found to be

$$
\sin\phi_0 = \tfrac{1}{2}\frac{k}{\mu}(1+\tfrac{1}{3}\alpha^2)^{\frac{1}{2}}, \quad \tan\psi_0 = \frac{\alpha}{\sqrt{3}}.
\tag{8.16}
$$

By eliminating λ from (8.12_1) and (8.12_2) and making the substitutions (8.15) we obtain

$$
\dot{\psi}+s\cot\phi\left(\sin\psi-\frac{\alpha}{\sqrt{3}}\cos\psi\right) = 0.
\tag{8.17}
$$

This is satisfied, irrespective of the value of ϕ, by

$$\tan\psi = \tan\psi_0 = \alpha/\sqrt{3} \tag{8.18}$$

and this solution also satisfies the initial condition at $t = t_0$.

Next eliminate λ from (8.12_3) and (8.12_4), make the substitutions (8.15) and set $\psi = \psi_0$. This gives, after reduction

$$\dot{\phi}+s\left[\sin\phi\left(\frac{\mu}{k}+\frac{\sin\phi}{\cos\psi_0}\right)-\frac{\cos^2\phi}{\cos\psi_0}\right]=0. \tag{8.19}$$

Since $\mu/k \gg 1$, it is consistent with our previous approximations to neglect the term $\sin\phi/\cos\psi_0$ compared to μ/k, so we approximate (8.19) by

$$\dot{\phi}+s\left(\frac{\mu}{k}\sin\phi-\frac{\cos^2\phi}{\cos\psi_0}\right)=0. \tag{8.20}$$

This equation is separable and can be solved explicitly. It turns out however, that its solution subject to the initial condition (8.16) shows that ϕ is of order k/μ for all $t > t_0$. It is therefore sufficient to set $\sin\phi = \phi$, $\cos\phi = 1$, so that (8.20) becomes

$$\dot{\phi}+\frac{\mu}{k}s\phi = s\sec\psi_0. \tag{8.21}$$

The solution of this subject to (8.16) is

$$\begin{aligned}\phi &= \frac{k}{\mu}\sec\psi_0+\left(\phi_0-\frac{k}{\mu}\sec\psi_0\right)\exp\left\{-\frac{\mu}{k}s(t-t_0)\right\}\\ &= \frac{k}{\mu}(1+\tfrac{1}{3}\alpha^2)^{\frac{1}{2}}[1-\tfrac{1}{2}\exp\{1-\mu st/k\}].\end{aligned} \tag{8.22}$$

The stress components are obtained by substituting (8.18) and (8.22) into (8.15). Note that the normal stress components are of second order in k/μ and that, to first order in k/μ, s_{12} is independent of the choice of α.

Asymptotic behaviour as $t \to \infty$

Because the argument of the exponential function in (8.22) contains the large factor μ/k this term tends rapidly to zero as $t \to \infty$. Hence

$$\phi \to \phi_1 = \frac{k}{\mu}\sec\psi_0 = \frac{k}{\mu}(1+\tfrac{1}{3}\alpha^2)^{\frac{1}{2}}. \tag{8.23}$$

The corresponding limiting values of the non-zero stress components are, to second order

$$s_{11} = -\frac{k^2}{\mu}(1+\tfrac{1}{3}\alpha), \quad s_{22} = \frac{k^2}{\mu}(1-\tfrac{1}{3}\alpha), \quad s_{33} = \tfrac{2}{3}\frac{k^2\alpha}{\mu}, \quad s_{12} = k. \tag{8.24}$$

Thus normal stress effects occur in the plastic-elastic solution, but are always of second order. Note however, that not only the magnitude but also the sign of the normal stress components is dependent on the choice of α.

The *plastic-rigid* solution of this problem is

$$s_{12} = k, \quad s_{11} = s_{22} = s_{33} = s_{13} = s_{23} = 0.$$

Thus the asymptotic plastic-elastic solution differs from the plastic-rigid solution only by terms of second order in k/μ. Since the plastic-rigid material may be regarded as the limit of a plastic-elastic material as $\mu/k \to \infty$ there is no inconsistency in these results, for this simple problem.

Unloading by shear in the negative x_1 direction may be treated in a similar manner, but we omit details.

9. Cylindrical polar coordinates

For the next example we require the constitutive equations expressed in terms of cylindrical polar coordinates (R, Θ, Z) in the reference configuration and (r, θ, z) in the deformed configuration. Thus

$$X_1 = R\cos\Theta, \quad X_2 = R\sin\Theta, \quad X_3 = Z, \tag{9.1}$$

$$x_1 = r\cos\theta, \quad x_2 = r\sin\theta, \quad x_3 = z. \tag{9.2}$$

Components of velocity in the r, θ, z directions are denoted v_r, v_θ, v_z. The orthogonal matrix of transformation from x_1, x_2, x_3 coordinates to (r, θ, z) coordinates is denoted $\boldsymbol{R}$, so that

$$\boldsymbol{R} = \begin{bmatrix} \cos\theta & \sin\theta & 0 \\ -\sin\theta & \cos\theta & 0 \\ 0 & 0 & 1 \end{bmatrix}. \tag{9.3}$$

The matrices of the components of deviatoric stress, strain-rate, elastic strain-rate and plastic strain-rate and spin referred to (r, θ, z) coordinates are denoted $\boldsymbol{S}^*$, $\boldsymbol{D}^*$, $\boldsymbol{D}_e^*$, $\boldsymbol{D}_p^*$, $\boldsymbol{\omega}^*$ respectively, so that

$$\boldsymbol{S}^* = \boldsymbol{RSR}^T, \quad \boldsymbol{D}^* = \boldsymbol{RDR}^T, \quad \boldsymbol{D}_e^* = \boldsymbol{RD}_e\boldsymbol{R}^T,$$
$$\boldsymbol{D}_p^* = \boldsymbol{RD}_p\boldsymbol{R}^T, \quad \boldsymbol{\omega}^* = \boldsymbol{R\omega R}^T. \tag{9.4}$$

Also

$$\boldsymbol{D}^* = \tfrac{1}{2}(\boldsymbol{F}^*+\boldsymbol{F}^{*T}), \quad \boldsymbol{\omega}^* = \tfrac{1}{2}(\boldsymbol{F}^*-\boldsymbol{F}^{*T}), \tag{9.5}$$

where F^* is the matrix of velocity gradients referred to (r, θ, z)

$$\boldsymbol{F}^* = \boldsymbol{RFR}^T = \begin{bmatrix} \dfrac{\partial v_r}{\partial r} & \dfrac{1}{r}\dfrac{\partial v_r}{\partial \theta} - \dfrac{v_\theta}{r} & \dfrac{\partial v_r}{\partial z} \\ \dfrac{\partial v_\theta}{\partial r} & \dfrac{1}{r}\dfrac{\partial v_\theta}{\partial \theta} + \dfrac{v_r}{r} & \dfrac{\partial v_\theta}{\partial z} \\ \dfrac{\partial v_z}{\partial r} & \dfrac{1}{r}\dfrac{\partial v_z}{\partial \theta} & \dfrac{\partial v_z}{\partial z} \end{bmatrix}. \tag{9.6}$$

For the plastic strain-rate we have immediately from (5.1) and (9.4)

$$\boldsymbol{D}_p^* = \lambda \boldsymbol{S}^*. \tag{9.7}$$

For brevity, for the elastic strain-rate we consider only the constitutive equation (7.4) for incompressible material. The equation (6.11) for compressible material can be treated similarly. From (7.4) and (9.4)

$$2\mu \boldsymbol{D}_e^* = \boldsymbol{R}(\mathscr{D}_J\boldsymbol{S})\boldsymbol{R}^T + \alpha(\boldsymbol{D}^*\boldsymbol{S}^*+\boldsymbol{S}^*\boldsymbol{D}^*) - \tfrac{2}{3}\alpha \boldsymbol{I}\,\mathrm{tr}\,\boldsymbol{D}^*\boldsymbol{S}^*. \tag{9.8}$$

From (6.3), (9.3) and (9.4) we obtain after a little manipulation

$$\boldsymbol{R}(\mathscr{D}_J\boldsymbol{S})\boldsymbol{R}^T = D\boldsymbol{S}^* + \boldsymbol{S}^*(\boldsymbol{\omega}^*+\boldsymbol{\Omega}) - (\boldsymbol{\omega}^*+\boldsymbol{\Omega})\boldsymbol{S}^* \tag{9.9}$$

where in cylindrical polar coordinates D denotes the operator

$$D = \frac{\partial}{\partial t} + v_r\frac{\partial}{\partial r} + \frac{v_\theta}{r}\frac{\partial}{\partial \theta} + v_z\frac{\partial}{\partial z} \tag{9.10}$$

and $\boldsymbol{\Omega}$ is the skew symmetric matrix

$$\boldsymbol{\Omega} = \begin{bmatrix} 0 & \dfrac{v_\theta}{r} & 0 \\ -\dfrac{v_\theta}{r} & 0 & 0 \\ 0 & 0 & 0 \end{bmatrix}. \tag{9.11}$$

10. Simple torsion

Now suppose that an incompressible circular cylinder with radius a of plastic-elastic material is subjected to the simple torsion

$$\begin{aligned} r &= R, \quad \theta = \Theta+\tau RZt, \quad z = Z, \\ v_r &= 0, \quad v_\theta = \tau RZ, \qquad v_z = 0. \end{aligned} \tag{10.1}$$

The elastic solution for sufficiently small t is obtained by substituting (10.1) into the equations of sect. 9 and integrating. The solution is very similar to the elastic solution of sect. 8, and is

$$\begin{aligned} s_{rr} &= -\tfrac{2}{3}\frac{\mu\alpha}{m^2}(\cos m\tau rt-1) \approx \tfrac{1}{3}\mu\alpha\tau^2r^2t^2, \\ s_{\theta\theta} &= -\frac{\mu}{m^2}(1-\tfrac{1}{3}\alpha)(\cos m\tau rt-1) \approx \tfrac{1}{2}\mu(1-\tfrac{1}{3}\alpha)\tau^2r^2t^2, \\ s_{zz} &= \frac{\mu}{m^2}(1+\tfrac{1}{3}\alpha)(\cos m\tau rt-1) \approx -\tfrac{1}{2}\mu(1+\tfrac{1}{3}\alpha)\tau^2r^2t^2, \\ s_{\theta z} &= \frac{\mu}{m}\sin m\tau rt \approx \mu\tau rt, \quad s_{r\theta} = 0, \quad s_{rz} = 0, \end{aligned} \tag{10.2}$$

where m is given by (8.6). The pressure p is determined by the equilibrium equations. If $t_{rr} = 0$ when $r = a$, they give

$$p = s_{rr}+\int_a^r (s_{rr}-s_{\theta\theta})r^{-1}dr. \tag{10.3}$$

The resultant couple M and normal force N acting on a surface z = constant are

$$M = 2\pi\int_0^a r^2 s_{\theta z}dr, \quad N = 2\pi\int_0^a rt_{zz}dr. \tag{10.4}$$

By inserting (10.2) in the yield condition (3.1) it is found that yielding first occurs at $r = a$ when t has the value t_0 given by

$$\tau at_0 = k/\mu+0(k^3/\mu^3). \tag{10.5}$$

For $t > t_0$ we seek solutions in which the cylinder contains an inner elastic region $0 \leqq r < c(t)$ surrounded by an outer plastic region $c(t) < r < a$. It is assumed that the solution (10.2) continues to hold in the elastic region.

The interface $c(t)$ is determined by the condition that material on the elastic side of the interface is about to yield, which gives

$$\tau ct = k/\mu + 0(k^3/\mu^3). \tag{10.6}$$

The solution in the plastic region also closely resembles the simple shear solution of sect. 8. The yield condition is satisfied by

$$s_{rr} = \frac{2k}{\sqrt{3}} \sin\phi \cos\psi, \quad \left.\begin{matrix} s_{\theta\theta} \\ s_{zz} \end{matrix}\right\} = -k \sin\phi \left\{ \frac{1}{\sqrt{3}} \sin \mp \cos\psi \right\}, \tag{10.7}$$

$$s_{\theta z} = k \cos\phi.$$

With approximations analogous to those made in sect. 8, the appropriate solution in the plastic region $c < r < a$ is

$$\tan\psi = \alpha/\sqrt{3}, \quad \phi = \frac{k}{\mu}(1+\tfrac{1}{3}\alpha^2)^{\frac{1}{2}} \left\{ 1 - \tfrac{1}{2}\exp\left(1 - \frac{\mu}{k}\tau rt\right) \right\}. \tag{10.8}$$

To second order in k/μ, (10.7) and (10.8) give for $c < r < a$

$$\begin{aligned} s_{rr}/\mu &= \tfrac{2}{3}\frac{k^2}{\mu^2}\alpha \left\{ 1 - \tfrac{1}{2}\exp\left(1 - \frac{\mu}{k}\tau rt\right) \right\}, \quad s_{\theta z}/\mu = k/\mu, \\ \left.\begin{matrix} s_{\theta\theta}/\mu \\ s_{zz}/\mu \end{matrix}\right\} &= \pm \frac{k^2}{\mu^2}(1 \mp \tfrac{1}{3}\alpha) \left\{ 1 - \tfrac{1}{2}\exp\left(1 - \frac{\mu}{k}\tau rt\right) \right\}. \end{aligned} \tag{10.9}$$

The hydrostatic pressure p is given as before by (10.3). Note that, as in sect. 8, the normal stress components are of second order in k/μ and that, to second order in k/μ, $s_{\theta z}$ is independent of the choice of α. As before, the resultant couple M and axial force N are given by (10.4).

Asymptotic behaviour as $t \to \infty$

From (10.6), $c \to 0$ as $t \to \infty$, but an elastic region remains in the centre of the cylinder for any finite value of t. In the limit $t \to \infty$, the stress components (10.9) become

$$s_{rr}/\mu = \tfrac{2}{3}\alpha(k/\mu)^2, \quad \left.\begin{matrix} s_{\theta\theta}/\mu \\ s_{zz}/\mu \end{matrix}\right\} = \pm(1 \mp \tfrac{1}{3}\alpha)(k/\mu)^2, \quad s_{\theta z}/\mu = k/\mu.$$

In this limit M and N take the values

$$M = \tfrac{2}{3}\pi k a^3, \quad N = -\frac{3\pi k^2}{4\mu}(1+\tfrac{1}{3}\alpha)a^2.$$

These results differ from those given by plastic-rigid theory only by terms of second order in k/μ, so again the plastic-rigid theory agrees with the present theory in the limit $k/\mu \to 0$.

The axial force required to maintain the deformation is positive or negative according to whether α is less or greater than -3. It is to be expected that a cylinder subjected to torsion without being constrained in the axial direction will extend or contract according to whether N is negative or positive. There therefore exists the possibility of either a positive or negative plastic-elastic 'Poynting effect', depending on whether α is greater or less than 3.

Unloading by torsion in the reverse sense may be treated in a similar manner.

Acknowledgement

One of the authors (J.E.F.) thanks the Science Research Council for the award of a Research Studentship.

References

[1] Pipkin, A. C. and R. S. Rivlin, *ZAMP 16* (1965) 313.
[2] Green, A. E. and P. M. Naghdi, *Arch. ratl. Mech. Anal. 18* (1965) 251.
[3] Thomas, T. Y., *Plastic flow and fracture of solids*. Academic Press, New York, (1961).
[4] Lee, E. H., *J. Appl. Mech. 36* (1969) 1.
[5] Hill, R., *J. Mech. Phys. Solids. 7* (1959) 209.
[6] Hill, R., *J. Mech. Phys. Solids. 10* (1962) 1.
[7] Green, A. E. and J. E. Adkins, *Large elastic deformations*. 1st Edn. Clarendon Press, Oxford, (1960).
[8] Prager, W., *Introduction to mechanics of continua*. Ginn & Co. Boston, (1961).
[9] Sedov, L. I., *Foundations of the non-linear mechanics of continua*. English translation, Pergamon, London, (1966).
[10] Prager, W., *Quart. Appl. Math. 18* (1961) 403.
[11] Masur, E. F., *Quart. Appl. Math., 19* (1961) 160.
[12] Zhong-Heng, Guo, *Arch. Mech. Stos. 15* (1963) 131.
[13] Naghdi, P. M. and W. L. Wainwright. *Quart. Appl. Math. 19*, (1961) 106.
[14] Jaumann, G., *Sitzungsberichte Akad. Wiss. Wien (IIa) 120* (1911) 385.
[15] Cotter, B. A. and R. S. Rivlin. *Quart. Appl. Math. 13* (1955) 177.
[16] Oldroyd, J. G., *Proc. Roy. Soc. A, 200* (1950) 523.
[17] Green, A. E., *Proc. Roy. Soc. A, 234* (1956) 46.
[18] Truesdell, C., *J. Ratl. Mech. Anal. 4* (1955) 83.
[19] Hill, R., *Recent progress in applied mechanics* (*The Folke Odquist Volume*), 241. Wiley, New York, (1967).

Solved and unsolved problems in elastic-plastic torsion of solid bars

Tsuan Wu Ting

University of Illinois at Urbana-Champain
Urbana, Illinois, U.S.A.

1. Two equivalent variational formulations

Consider a simply connected cylindrical bar twisted by terminal couples. We shall denote by G the cross sections of the bar. Since the stress function depends on the geometric nature of G in a rather sensitive way, we state our restrictions on G explicitly as follows[1]) [21]: (i) G is a Jordan domain satisfying the well-known cone condition, (ii) ∂G, boundary of G, possesses continuous non-zero radius of curvature except at a finite number of corners and (iii) between any two neighboring corners the curvature of ∂G achieves only a finite number of maxima and minima.

We shall suppose that the mechanical behavior of the bar is governed by any one of the classical theories of elasticity and plasticity under either the von Mises or the Tresca yield condition [17, 18, 27]. Under these hypotheses and by using the semi-inverse method, the problem has been formulated mathematically and can be stated as follows, [14, 17]: Find the Prandtl stress function $\psi(x, y)$ in $C_0^0(G+\partial G) \cap C^1(G)$ such that $|\text{grad}\,\psi|$ *is less than or equal to a positive yield constant k in G and that wherever* $|\text{grad}\,\psi| < k$, ψ *must be twice differentiable and satisfies the Poisson equation,* $\Delta\psi = -2\mu\theta$, *there*. Here as well as what follows, θ stands for the angle of twist per unit length and μ is the shear modulus.

Except in some special cases [1], the problem just stated has not been successfully attacked directly. As methods of attack, we now state two e-quivalent variational formulations, [4, 9, 22]. To this end, let $H_0^1(G)$ be the Sobolev space consisting of functions which together with their distribution

[1]) Numbers in brackets designate References at end of paper.

derivatives are in $L^2(G)$, square summable in G. Let $\Psi(x, y)$ be the distance function,

$$\Psi(q) = k\rho(q, \partial G) \text{ for all points } q \text{ in } G+\partial G,$$

where k is the yield constant and $\rho(q, \partial G)$ stands for the distance from the point q to ∂G. Then $\Psi(x, y)$ is in $H_0^1(G)$ and it is the stress function which solves the completely plastic torsion problem, [21]. Let F be the closed convex subset of $H_0^1(G)$ such that

$$F = \{u | u \leqq \Psi \quad \text{a.e. in} \quad G, u \in H_0^1(G)\}.$$

Our first variational formulation is the following, [22]: *Find the function $\psi(x, y)$ in $H_0^1(G)$ such as to minimize the functional,*

$$J[u] \equiv \iint_G [|\text{grad}\, u|^2 - 4\mu\theta u] \mathrm{d}x\mathrm{d}y,$$

among all functions u in F. This formulation includes the purely elastic torsion as well as completely plastic torsion as special cases, [22].

Let K be the closed convex subset of $H_0^1(G)$ such that

$$K \equiv \{u | u \in H_0^1(G), |\text{grad}\, u| \leqq k \quad \text{a.e. in} \quad G\}.$$

Our second formulation is this, [2, 3, 9]: *Find the minimizing extremal ψ for the functional $J[u]$ in the subset K of $H_0^1(G)$.* Although K is a proper subset of F, it has been proved that *the two variational formulations are equivalent in the sense that a solution of one problem is a solution of the other and conversely*, [4]. For convenience, we shall use the first formulation in the following discussions.

2. Existence and uniqueness of the extremal

For the existence of the extremal, we first note that a minimizing sequence is actually a Cauchy sequence in $H_0^1(G)$ because of the parallelogram law. Secondly, the completeness of $H_0^1(G)$ and the closedness of F ensures that the limit function ψ is a minimizing extremal. The uniqueness of the extremal ψ follows from the convexity of the functional $J[u]$ in u, [22–25]. However, much work remains to be done in order to show that $\psi(x, y)$ solves the elastic-plastic torsion problem in the strict sense.

3. Hölder continuity of extremal and a natural partition of cross sections

It is really amazing that so much information can be derived from the minimizing property of the extremal ψ. As a first consequence, we have the fine result that *ψ can be identified by a function Hölder continuous in $G+\partial G$*, [24, 25]. For by choosing suitable variations one can transform the variational condition into a Dirichlet growth condition for the extremal ψ in G. Once the continuity of ψ is established, we can define the elastic region E and the plastic portions P of $G+\partial G$ as follows:

$$E \equiv \{q | q \in G,\ \psi(q) < \Psi(q)\},\ P \equiv \{q | q \in \bar{G},\ \psi(q) = \Psi(q)\}.$$

Then, E and P constitute a partition of $G+\partial G$. As we shall see later on E is, indeed, the elastic region and P are the plastic portions of $G+\partial G$ in the ordinary sense, [14, 17].

4. Existence of an elastic core

Consider the points γ in G, for which there are points s_1 and s_2 on ∂G such that

$$\rho(\gamma, s_1) = \rho(\gamma, s_2) = \rho(\gamma, \partial G).$$

The closure of these points γ in G forms the *ridge Γ* of the Jordan domain G, [21]. It can be proved in an elementary way that Γ consists of curves which are smooth except at a finite number of points on Γ, [21]. Also, Γ is precisely the set of points where grad Ψ suffers jump discontinuities. By an elastic core, we mean an open neighborhood of Γ, in which ψ is strictly less than the majorant function Ψ.

It turns out that for each point γ in $\Gamma \cap G$ there is an $\varepsilon > 0$ such that in the disk $D(\gamma, \varepsilon)$ with center γ and radius ε the solution ψ^* of the Dirichlet problem,

$$\Delta\psi^* = -2\mu\theta \quad \text{in} \quad D(\gamma, \varepsilon),\ \psi^* = \psi \quad \text{on} \quad \partial D(\gamma, \varepsilon),$$

is strictly less than Ψ in $D(\gamma, \varepsilon)$, [25]. From this fact we can conclude that $\psi < \Psi$ there. This means that $D(\gamma, \varepsilon) \subset E$ and hence the existence of an elastic core is assured. Moreover, *an elastic core always exists no matter how large the angle θ of twist per unit length may be*. Accordingly, *a completely plastic torsion can never be realized in practice.*

5. 'Elasticness' of E

Since the elastic region always has positive area and since ψ is strictly less than Ψ there, small variations of the extremal can be chosen arbitrarily in E. Based on this fact, one can transform the variational condition into certain integral representation for the extremal ψ, which then shows that $\psi \in C^2(E)$ and it satisfies the Poisson equation, $\Delta\psi = -2\mu\theta$, there, [22]. Thus, *the material in E is governed by the classical linear theory of elasticity.*

6. 'Unyieldness' of E

Actually, ψ is analytic in E. Consequently, one can perform computations to check that $|\text{grad}\,\psi|^2$ is subharmonic in E. Moreover, $\psi = \Psi$ in P and hence $|\text{grad}\,\psi| = k$ in $G-E$. Thus, we can apply the maximum principle for subharmonic functions to conclude that $|\text{grad}\,\psi| < k$ in E. This shows that *the stress in E stays below the yield point.* Accordingly, the region E is, indeed, elastic in the ordinary sense, [22].

7. Hölder continuity of stress components in G

We have seen that the Prandtle stress function ψ is Hölder continuous in $G+\partial G$. Moreover, it is analytic in E. Also, if the interior of the plastic portions P is non-empty, then ψ is differentiable there. In fact, grad $\psi =$ grad Ψ in $G-(E+\partial E)$. Thus, to establish the continuity of grad ψ in the entire domain G, it suffices to show that it is continuous across the elastic-plastic boundary,

$$l \equiv \partial E \cap G \equiv \partial P \cap G.$$

However, so far we have no information about the free boundary l. We are even not sure whether l has two dimensional measure zero. To overcome this difficulty, we take a neighborhood $N(\Gamma)$ of the ridge Γ with $\partial N(\Gamma)$ being sufficiently smooth so that Kellogg's regularity theorem for harmonic functions on the boundary of a region can be applied. Now, under our restrictions on ∂G, the majorant function Ψ possesses second derivatives in $L_p(G-N(\Gamma))$ for all $p > 0$. Accordingly, for all even integers $p > 2$, the non-linear Dirichlet problem,

$$\varepsilon\Delta\psi_\varepsilon = (\psi_\varepsilon - \psi)^{1/(p-1)} + \varepsilon\Delta\Psi \quad \text{in} \quad G-N(\Gamma),$$

$$\psi_\varepsilon = \psi \quad \text{on} \quad \partial(G-N(\Gamma)),$$

always has a solution ψ_ε in $H_p^2(G-N(\Gamma))$ for all $\varepsilon > 0$. As a consequence of Sobolev's lemma and Morrey's theorem on Dirichlet growth condition, ψ_ε has Hölder continuous first derivatives with exponent $1-2/p$ and with same Hölder constant. Accordingly, ψ_ε together with its first derivatives converges, as $\varepsilon \to 0$, uniformly to a limit function with Hölder continuous first derivatives. Actually, ψ_ε converges to the extremal ψ in $G-N(\Gamma)$. Hence, ψ possesses Hölder continuous first derivatives in G. It should be said that the details of this proof are far from simple, we refer interested readers to [25] for more complete account.

8. The elastic-plastic boundary

(i) *An intersection property of P*. By definition, $\psi = \Psi$ in the plastic portions P of G. Consider an inward normal to ∂G issuing from a point s on ∂G ending at a point γ on the ridge Γ. *This inward normal intersects P in a single segment sq*. For if this were not the case, then we could derive a contradiction to the fact that $|\text{grad}\,\psi| \leqq k$ in G. However, it is possible that the segment sq may degenerate into a single point s on ∂G. Nevertheless, we have proved that *the plastic portions P of G always adhere to the boundary ∂G and hence that yielding always starts from the boundary.*

(ii) *A natural parametric representation of the free boundary*. Denote by s the arc length of ∂G and by sq the intersection segment of P with an inward normal to ∂G. Then the collection of all such points q constitutes the elastic-plastic boundary l. Now, we formally represent l in the form:

$$x = f(s)+R(s)n_x(s, \omega), \quad y = g(s)+R(s)n_y(s, \omega),$$

where $(f(s), g(s))$, (x, y) are respectively the coordinates of points s on ∂G and the points q on l, and where n_x and n_y are the x- and y-components of the unit inward normal of ∂G at the point s and $R(s)$ is the length of the intersection segment sq. In general, n_x and n_y are functions of the arc length s of ∂G. If s is a reentrant corner of ∂G, then $n_x(s, \omega)$, $n_y(s, \omega)$ are functions of ω which varies from zero to the exterior angle of ∂G at s. It turns out that the function $R(s)$ is a continuous function of s, [24]. Hence, the equations we have just written down are parametric representations of l. Moreover, *the continuity of R(s) ensures that the free boundary l does* not contain any segment normal to ∂G.

(iii) *Rectifiability of free boundary*. It is desirable to know whether the elastic-plastic boundary l is rectifiable. For in order to check the dynamic compatibility conditions across l, [17, 19], we have to be sure that the surfaces separating the elastic and plastic materials have finite area. Although in the cases for which ∂G consists of line segments, it is known that l consists of analytic arcs, [22, 23], *yet this is an open question for the general case.* In fact, we could raise the question *how the smoothness of ∂G implies the smoothness of the free boundary*?

(iv) *Simple connectivity of elastic region.* We have seen that the ridge Γ is properly contained in the elastic region E. For every point q in E, let s be a point in ∂G such that $\rho(q, s) = \rho(q, \partial G)$. Let γ be the intersection of the extension of the segment sq with Γ. According to the intersection property of P, the segment γq is contained in E. Since Γ is known to be connected [21], it follows that E *is simply connected.* The fact that $\Gamma \subset E$ *describes qualitatively the relative position of E as well as its extent in G.*

9. Existence of a strict solution to elastic-plastic torsion problem

We have seen in § 3 and § 7 that both the extremal and its partial derivatives are Hölder continuous in G. Also, we have seen in § 3, 5, 6, and 8 that the extremal ψ partitions $G+\partial G$ into a simply connected elastic region E and plastic portions P and that both E and P have the usual meaning in the sense that

$$\Delta\psi = -2\mu\theta \quad \text{in} \quad E, \qquad |\text{grad}\,\psi| < k \quad \text{in} \quad E, \qquad |\text{grad}\,\psi| = k \quad \text{in} \quad P.$$

Hence, ψ *solves elastic-plastic torsion problem in the strict sense.* We proceed to derive further informations from the extremal property of ψ.

10. How the plastic portions grow as load increases

When the angle of twist per unit length is sufficiently small, the linear theory of elasticity applies. That is the torsion is purely elastic. As the load increase to certain critical value yielding begins at some points on the boundary, ∂G. The plastic portions P will grow as the load further increases. What has just been said is not just based on physical reasoning, but a consequence of the minimizing property of ψ. In fact, let $\psi(x, y; \mu\theta_1, k)$ and $\psi(x, y; \mu\theta_2, k)$ be stress functions corresponding to the angles of twist θ_1 and θ_2, respectively. Let E_1, P_1 and E_2, P_2 be respectively, the elastic-plastic partitions of G

corresponding to θ_1 and θ_2. Suppose that $\theta_2 > \theta_1$. Then, as proved in [22],

$$P_1 \subset P_2, \quad \psi(x, y; \mu\theta_2, k) \geqq \psi(x, y; \mu\theta_1, k) \quad \text{in} \quad G+\partial G.$$

The inclusion relation, $P_1 \subset P_2$, means that *the plastic portions grow as the angle of twist per unit length increases*; while the inequality implies that *the stress function is a nondecreasing function of* θ. Moreover, we have [22] the sharp inequality, $\psi(x, y; \mu\theta_2, k) > \psi(x, y; \mu\theta_1, k)$ in E_1, and hence

$$\iint_G \psi(x, y; \mu\theta_2, k)\mathrm{d}x\mathrm{d}y > \iint_G \psi(x, y; \mu\theta_1, k)\mathrm{d}x\mathrm{d}y,$$

which says that *the applied torque is a strictly increasing function of the angle of twist per unit length and conversely.*

As was mentioned before, it is desirable to know whether the elastic-plastic boundary is smooth. Suppose that the free boundaries l_1 and l_2 corresponding, respectively, to θ_1 and θ_2 are smooth. Then we can prove that for every point q on $l_2 \cap G$ there is a disk $D(q, \varepsilon)$ with center at q and radius $\varepsilon > 0$ such that $D(q, \varepsilon)$ is properly contained in E_1, [22]. This means that *l_1 and l_2 have no point of contact in G*, and it gives us a picture of how the plastic portions P grow as the angle of twist per unit length increases.

11. Inequalities of Harnack type

We continue to look into the effects of the yielding strength k and the angle of twist θ upon the stress function $\psi(x, y; \mu\theta, k)$. Needless to say that the effect of μ is precisely the same as that of θ and hence its effect needs not be mentioned.

Although the linear theory of elasticity has been adopted in this problem, we can not expect that ψ depends on θ and k linearly because of the non-linear nature of the plasticity theory. On the other hand, we would expect, in view of the yield conditions, that $\psi(x, y; \mu\theta, k)$ with (x, y) being fixed is bounded from above by a linear function in θ and k. In fact, what has been said is also a consequence of the extremal property of ψ.

First, we consider the dependence of ψ upon the yielding strength k. Let $k_2 > k_1 > 0$ be two yielding constants. Let E_1, P_1 and E_2, P_2 be the elastic-plastic partitions of the cross section G corresponding to the yielding constants k_1 and k_2, respectively. Suppose further that the area of P_2 is greater than zero. Then,

$$P_1 \supset P_2 \quad \text{and hence} \quad E_2 \supset E_1.$$

Moreover, we have the Harnack inequalities,

$$\frac{k_1}{k_2}\,\psi(x,y;\mu\theta,k_2) \leqq \psi(x,y;\mu\theta,k_1) \leqq \psi(x,y;\mu\theta,k_2),$$

where the first strict inequality sign holds throughout E_2 and the second strict inequality sign holds through out P_2. We refer to [25] for the details of a proof of the above statements. Since the torsion problem corresponding to the yielding constant k_2 is assumed not to be purely elastic, the area of P_2 is positive. Also the existence of an elastic core is always assured and hence the area of E_2 is positive. Thus, we have the following Harnack inequalities,

$$\frac{k_1}{k_2}\iint_G \psi(x,y;\mu\theta,k_2)\mathrm{d}x\mathrm{d}y \leqq \iint_G \psi(x,y;\mu\theta,k_1)\mathrm{d}x\mathrm{d}y$$
$$\leqq \iint_G \psi(x,y;\mu\theta,k_2)\mathrm{d}x\mathrm{d}y,$$

which says that *the applied torque for fixed angle of twist θ is a strictly increasing function of the yielding strength. However it is bounded from above by a linear function in k.*

Naturally, the question arises whether a similar statement can be made for the stress energy, $\iint_G |\mathrm{grad}\,\psi|^2\,\mathrm{d}x\mathrm{d}y$. *This is an open question if G is an arbitrary simply connected domain under the restrictions stated at the beginning of this article.* Nevertheless, when G is convex, it is known [25] that there hold the Harnack inequalities,

$$\left[\frac{k_1}{k_2}\right]^2\iint_G |\nabla\psi(x,y;\mu\theta,k_2)|^2\,\mathrm{d}x\mathrm{d}y < \iint_G |\nabla\psi(x,y;\mu\theta,k_1)|^2\,\mathrm{d}x\mathrm{d}y$$
$$\leqq \iint_G |\nabla\psi(x,y;\mu\theta,k_2)|^2\,\mathrm{d}x\mathrm{d}y.$$

For the effect of the angle of twist θ, an analogus analysis shows that *if $\theta_2 > \theta_1 > 0$, then*

$$\psi(x,y;\mu\theta_1,k) \leqq \psi(x,y;\mu\theta_2,k) \leqq \frac{\theta_2}{\theta_1}\,\psi(x,y;\mu\theta_1,k),$$

where the first equality sign only holds in P_1 and the second strict inequality sign holds throughout P_1. Also, if G is convex, then

$$\iint_G |\nabla\psi(x,y;\mu\theta_2,k)|^2\,\mathrm{d}x\mathrm{d}y > \iint_G |\psi\nabla(x,y;\mu\theta_1,k)|^2\,\mathrm{d}x\mathrm{d}y$$

$$\geqq \left[\frac{\theta_1}{\theta_2}\right]^2 \iint_G |\nabla\psi(x,y;\mu\theta_2,k)|^2\,\mathrm{d}x\mathrm{d}y.$$

12. Dependence upon the domain

In this problem, the material is supposed to be described completely by the two constants μ and k. Accordingly, for fixed angle of twist θ, the stress function is determined by the geometry of the domain G. It is natural to ask the question *how the stress function varies with the domain G*. Since this is a geometric problem in the large, it is certainly not simple. For convex domains, certain outer and inner approximation theorems have been established. The same proof given in [24] can also be generalized so that the same results hold for simply connected domains.

One of the most significant questions along this line was posed by St. Venant and he conjectured that *among all solid bars with the same cross-sectional area the circular shaft gives the maximum torsional rigidity*. This profound isoperimetric problem was latter solved and generalized by Pólya and al, [15]. Not too long ago, a similar problem was solved for completely plastic torsion, [12]. It is probable that St. Venant-Pólya theorem can be generalized for elastic-plastic torsion. However, in despite of my repeated attempt, I am unable to prove or disprove it up to the present time.

Another question arised when I tried to establish certain properties for the free boundary. It is my feeling that *if G is convex then the stress surface for elastic-plastic torsion*, i.e.,

$$z = \psi(x,y), \quad (x,y) \in G + \partial G,$$

is concave. Again, I can not prove or disprove it. In fact, we can even *ask the same question for purely elastic torsion*. To my knowledge, there is no existing answer to it.

Acknowledgement

The work was partially supported by U.S. N.S.F. Grant, GP-19590 and GP-33174X.

References

[1] Annin, B. D., Existence and uniqueness of the solution of the elastic-plastic torsion problem for a cylindrical bar of oral cross section. *J. Appl. Math. Mech. 29* (1965) 1038–1047.

[2] Amosov, G. Ja., On the torsion of prismatic bars in the case of elastic-plastic deformation. *Vestn. Mosk. Univ. ser. 1. Math. Mech. 21* (1966) 98–106.

[3] Brezis, H. R. and G. Stampacchia, Sur la régularité de la solution d'inéquations elliptiques. *Bull. Soc. Math. France, 96* (1969) 791–794.

[4] Brezis, H. R. and M. Sibony, Equivalence de deux inéquations variationnelles et applications, *Arch. Rat. Mech. Anal. 41* (1971) 254–265.

[5] Cherepanov, G. P., On the solution of certain problems with an unknown boundary in the theory of elasticity and plasticity. *J. Appl. Math. Mech.* (PMM), *28* (1964) 162–167.

[6] Gajewski, H., Ein konstruktiver Existenz und Einzigkeitsnachweise der Lösung des elastisch-plastischen Torsionproblems für prismatic Stäbe. *Math. Nachr., 35* (1967) 153–169.

[7] Garabedian, P. R., H. Lewy and M. Schiffer, Axially symmetric cavitational flow. *Ann. Math. 56* (1952) 560–602.

[8] Geiringer, H., The ideal plastic body. *Handbuch der Physik*, Vol. *6*, pp. 322–433, Berlin, Springer, 1958.

[9] Hodge, P. G. Jr., Elastic-plastic torsion as a problem in non-linear programing. *Internat. J. Solid Structures, 3* (1967) 989–999.

[10] Lanchon, H., Solution du problème de torsion élasto-plastique d'une barre cylindrique de section quelconque, *Compt. Rend. 269* (1969) 791–794.

[11] Lanchon, H. and G. Duvaut, Sur la torsion du problème de torsion élasto-plastique d'une barre cylindrique de section quelconque. *Compt. Rend. 264* (1967) 520–523.

[12] Leavitt, J. and P. Ungar, Circle supports the largest sandpile. *Comm. Pur. Appl. Math. 15* (1962) 35–37.

[13] Lewy, H. and G. Stampacchia, On the regularity of the solution of a variational inequality. *Comm. Pur. Appl. Math. 22* (1969) 153–188.

[14] Mises, R. von, Three remarks on the theory of the ideal plastic body. *Reissner Anniversary Vol.* 415–419, Ann. Arbor, Mich., Edwards, 1949.

[15] Pólya, G. and G. Szegö, Isoperimetric inequalities in mathematical physics. *Ann. Math. Studies*, No. 27, Princeton, N.J., Princeton Univ. Press, (1951).

[16] Ponter, A. R. S., On plastic torsion. *Internat. J. Mech. Sci. 8* (1966) 227–235.

[17] Prager, W. and P. G. Hodge Jr., *Theory of perfectly plastic solids.* John Wiley and Sons, New York, (1951).

[18] Thomas, T. Y., *Plastic flow and fracture in solids.* Academic Press, New York, (1961), Chap. IV.

[19] Thomas, T. Y., The general theory of compatibility conditions. *Internat. J. Engrg. Sci. 4* (1966) 207–233.

[20] Thomas, T. Y., *Concepts from tensor analysis and differential geometry.* Academic Press, New York, (1965).

[21] Ting, T. W., The ridge of a Jordan domain and completely plastic torsion. *J. Math. Mech. 15* (1966) 15–48.
[22] Ting, T. W., Elastic-plastic torsion of a square bar. *Trans. Am. Math. Soc. 123* (1966) 369–401.
[23] Ting, T. W., Elastic-plastic torsion problem II, III. *Arch. Rat. Mech. Anal. 25* (1967) 342–366, *34* (1969) 22–244.
[24] Ting, T. W., Elastic-plastic torsion of convex cylindrical bars. *J. Math. Mech. 19* (1969) 531–551.
[25] Ting, T. W., Elastic-plastic torsion of simply connected cylindrical bars. *Indiana Univ. Math. J. 20* (1971) 1047–1076.
[26] Ting, T. W., Topics in the mathematical theory of plasticity. *Handbuch der Physik*, Bd. *VIa/3*, Springer-Verlag, Heidelberg, (1972).
[27] Truesdell, C. and W. Noll, The non-linear field theory of mechanics. *Enclopedia of physics*, Vol. *III/3*, Springer, Berlin-Heidelberg-New York, 1965.
[28] St. Venant, A. J. C. de, Mémoire sur la torsion de prismes. *Mém. Divers Savants Acad. Sci.* Paris, *14* (1855) 233–560.

Comportement élasto-plastique des milieux granulaires

P. Stutz

Université de Grenoble, France

1. Introduction

L'étude rhéologique des milieux pulvérulents met en évidence un comportement complexe: ainsi la loi de ce matériau est-elle le plus souvent idéalisée par une loi élastique linéaire, ou par une loi de plasticité parfaite.

Nous avons tenté de décrire le comportement de ces milieux par une loi rhéologique unique permettant le passage continu du stade des déformations 'élastiques' au stade des grandes déformations 'plastiques'. En outre, la loi proposée décrit ce phénomène si caractéristique des milieux pulvérulents: la variation de volume provoquée par le cisaillement.

Soulignons tout d'abord quelques propriétés caractéristiques des milieux pulvérulents:

a) Le comportement a un caractère essentiellement non visqueux. Le temps n'intervient que pour repérer deux états successifs de la déformation; la réponse du matériau est indépendante de la vitesse d'application des sollicitations auxquelles on le soumet.

b) L'expérience montre qu'au cours d'un premier chargement il n'est pas possible de définir un domaine de déformation parfaitement réversible. La partie irréversible de la déformation augmente avec la grandeur de celle-ci.

c) Les variations de volume sont dues non seulement à la variation de la pression isotrope, mais encore, et ce phénomène est le plus caractéristique, aux déformations de cisaillement.

d) Dans la phase des grandes déformations où l'écoulement est possible à volume constant et à état de contrainte constant (état critique), la densité du matériau (densité critique) est fonction de la pression moyenne S_1 [1)]

$$\rho_c = \rho_c(S_1).$$

[1)] Symboles sont expliqués en Appendice.

Elle est indépendante de la densité initiale de l'échantillon. La relation $\rho_c = \rho_c(S_1)$ doit être considérée comme caractéristique du matériau.

e) L'allure des courbes de charge (correspondant par exemple à un essai triaxial drainé) de deux échantillons, différant uniquement par la valeur de leur densité initiale, dépendra essentiellement de la différence existant entre la densité initiale de l'échantillon et la densité critique.

Le cisaillement sera accompagné d'une augmentation ou d'une diminution de volume suivant que sa densité sera supérieure ou inférieure à la densité critique. En conséquence, le matériau atteindra l'état critique en subissant soit un amollissement, qui se traduit par un pic dans les courbes de charge à vitesse imposée, soit un durcissement.

f) Quand l'état critique est atteint, l'état de contrainte vérifie simultanément les relations:

$$\bar{S}_2 = f\left(\frac{\bar{S}_3}{\bar{S}_2}\right) \cdot S_1, \qquad \rho = \rho_c(S_1)$$

qui représentent la surface de plasticité infinie, que l'on appellera également surface critique. Comme il est d'usage en mécanique des sols, l'état critique pourra être caractérisé par la valeur de l'angle de frottement mesuré par exemple par un essai triaxial de compression ou de traction.

De ces constatations, on peut penser que la loi de comportement des milieux pulvérulents devra satisfaire les conditions suivantes:

(1) Pour des déformations infinitésimales, la loi doit être tangente à une loi élastique.

(2) Pour de déformations, où l'état critique peut être atteint, la loi doit être équivalente à une loi de plasticité parfaite, il semble que la loi de déformation plastique la plus plausible soit: $E_1 = 0$, $\bar{\boldsymbol{d}} = \lambda \bar{\boldsymbol{\sigma}}$.

(3) Entre la phase des déformations infinitésimales et celle des grandes déformations, le comportement dépend essentiellement de l'écart existant entre la densité du matériau et sa densité critique.

Pour prendre en compte les deux premières exigences, nous devons envisager une relation entre les tenseurs vitesse de déformation, contrainte et la dérivée de la contrainte par rapport au temps, [1], [2]. L'influence de la densité sera introduite pour prendre en compte la troisième exigence.

2. Loi de comportement

Le principe d'objectivité sera respecté en utilisant la dérivée par rapport au temps du tenseur $\boldsymbol{\sigma}$ sous la forme de Jaumann $\hat{\boldsymbol{\sigma}}$.

La forme la plus générale de la loi de comportement est:

$$\hat{\boldsymbol{\sigma}} = f(\boldsymbol{d}, \boldsymbol{\sigma}, \rho).$$

Le principe d'objectivité montre que $f(\boldsymbol{d}, \boldsymbol{\sigma}, \rho)$ doit être une fonction tensorielle isotrope de ses arguments.

Une telle relation ne pourra traduire le comportement décrit que si la condition d'homogénéité d'ordre -1 par rapport au temps lui est imposée.

2.1 *Ecriture de la loi*

De la forme générale [3], nous ne retiendrons que la forme quasi linéaire par rapport à $\boldsymbol{d}$ et $\boldsymbol{\sigma}$, [4]–[5].

$$\hat{\boldsymbol{\sigma}} = a_0\boldsymbol{\delta} + a_1\boldsymbol{d} + a_2\boldsymbol{\sigma} \tag{1}$$

où les fonctions a_i dépendent des 10 invariants construits avec les tenseurs $\boldsymbol{\sigma}$ et $\boldsymbol{d}$, et de la densité ρ.

Les fonctions a_0 et a_2 devant être homogènes de degré 1 par rapport à $\boldsymbol{d}$ et a_1, de degré zéro, on peut satisfaire cette condition, en prenant:

$$\begin{aligned} a_0 &= \alpha_0 E_1 + \alpha_3 \overline{M} + \alpha_5 \overline{N}, \\ a_1 &= \alpha_1, \\ a_2 &= \alpha_2 E_1 + \alpha_4 \overline{M} + \alpha_6 \overline{N} \end{aligned} \tag{2}$$

où les α_i ne sont plus que des fonctions des invariants de $\boldsymbol{\sigma}$ et de ρ.

Avec le choix qui est fait pour les fonctions a_i, la loi ne fera plus intervenir que six invariants: les trois invariants de l'état de contrainte S_i, $i = 1, 2, 3$, et les trois invariants E_1, $\overline{M}$ et $\overline{N}$ qui dépendent de la vitesse de déformation.

La décomposition en partie isotrope et déviatoire conduit à:

$$\begin{aligned} \hat{\sigma}_{ii} &= (\alpha_1 + \alpha_2 S_1 + 3\alpha_0)E_1 + (3\alpha_3 + \alpha_4 S_1)\overline{M} + (3\alpha_5 + \alpha_6 S_1)\overline{N}, \\ \hat{\bar{\sigma}}_{ij} &= \alpha_1 \bar{d}_{ij} + (\alpha_2 E_1 + \alpha_4 \overline{M} + \alpha_6 \overline{N})\bar{\sigma}_{ij}. \end{aligned} \tag{3}$$

La loi de comportement ne peut se décomposer en parties isotrope et déviatoire indépendentes: il y a couplage entre les deux parties.

2.2 *Interprétation*
Si l'état de contrainte initial est isotrope, on constate que la loi proposée peut être interprétée comme une loi tangente à la loi élastique linéaire qui aurait pour coefficients de Lamé: $\alpha_1 = 2G$, $3\lambda = 3\alpha_0 + \alpha_2 S_1$.

On peut constater que les variations de volumes sont dues d'une part, à la variation de la pression isotrope:

$$E_{1_i} = \frac{\hat{\sigma}_{ii}}{3\alpha_0 + \alpha_1 + \alpha_2 S_1} \tag{4}$$

d'autre part, à la partie déviatoire de l'état de contrainte

$$E_{1_d} = \frac{(3\alpha_3 + \alpha_4 S_1)\overline{M} + (3\alpha_5 + \alpha_6 S_1)\overline{N}}{3\alpha_0 + \alpha_1 + \alpha_2 S_1}. \tag{5}$$

Sous certaines conditions que nous allons examiner, la loi peut décrire un comportement plastique.

2.2.1 *Forme invariante.* Des relations (3) on peut obtenir les relations scalaires reliant les dérivées des invariants de $\boldsymbol{\sigma}$, aux six invariants de $\boldsymbol{\sigma}$ et $\boldsymbol{d}$

$$\begin{aligned} \hat{S}_1 &= (3\alpha_0 + \alpha_1 + \alpha_2 S_1)E_1 + (3\alpha_3 + \alpha_4 S_1)\overline{M} + (3\alpha_5 + \alpha_6 S_1)\overline{N}, \\ \frac{\hat{\bar{S}}_2^2}{2} &= \alpha_2 \bar{S}_2^2 E_1 + (\alpha_1 + \alpha_4 \bar{S}_2^2)\overline{M} + \alpha_6 \bar{S}_2^2 \overline{N}, \\ \frac{\hat{\bar{S}}_3^3}{3} &= \alpha_2 \bar{S}_3^3 E_1 + \alpha_4 \bar{S}_3^3 \overline{M} + (\alpha_1 + \alpha_6 \bar{S}_3^3)\overline{N}. \end{aligned} \tag{6}$$

Les formes des fonctions α_i, comme fonctions des invariants $\bar{S}_i$ et de la densité ρ, étant supposées connues, le système (6) est un système de trois équations linéaires pour les inconnues E_1, $\overline{M}$ et $\overline{N}$.

Si le déterminant de (6) n'est pas nul,

$$\frac{\Delta}{\alpha_1} = (3\alpha_0 + \alpha_1)\cdot(\alpha_1 + \alpha_4 \bar{S}_2^2 + \alpha_6 \bar{S}_3^3) + \alpha_2(\alpha_1 S_1 - 3\alpha_5 \bar{S}_3^3 - 3\alpha_3 \bar{S}_2^2) \tag{7}$$

les variables E_1, $\overline{M}$, $\overline{N}$ seront parfaitement déterminées en fonction de l'état de contrainte $\bar{S}_i$ et de sa variation instantanée $\hat{S}_i$. La densité ρ du matériau est obtenue en chaque point du trajet de charge de l'espace des S_i, par intégration de l'équation de continuité $-(1/\rho)(\mathrm{d}\rho/\mathrm{d}t) = E_1$. Dans le cas où $\Delta \neq 0$, un écoulement du matériau ne peut donc avoir lieu à l'état de contrainte constant.

Examinons alors le cas où le déterminant Δ s'annule, la relation $\Delta = 0$ peut s'interpréter comme représentant l'équation d'une surface dans l'espace à quatre dimensions des $\bar{S}_i$ et de ρ. Le système (6) ne pourra admettre de solution que si le déterminant caractéristique Δ_c s'annule, soit:

$$\Delta_c = \hat{S}_1 \cdot (\alpha_1 \alpha_2 \bar{S}_3^3) + \frac{\hat{\bar{S}}_2^2}{2} \cdot \bar{S}_3^3 \cdot ([3\alpha_0 + \alpha_1]\alpha_4 - 3\alpha_2\alpha_3) - \frac{\hat{\bar{S}}_3^3}{3} \Delta' = 0 \tag{8}$$

où Δ' est le mineur formé avec les coefficients de E_1 et $\bar{M}$ des équations (6_1) et (6_2)

$$\Delta' = (3\alpha_0 + \alpha_1 + \alpha_2 S_1)\alpha_1 + \bar{S}_2^2[\alpha_4(3\alpha_0 + \alpha_1) - 3\alpha_2\alpha_3]. \tag{9}$$

Si on admet que $\Delta' \neq 0$, le système (6) admet la solution

$$E_1 = \frac{\hat{S}_1(\alpha_1 + \alpha_4 S_2^2) - (\hat{\bar{S}}_2^2/2)(3\alpha_3 + \alpha_4 S_1) - \bar{N}[3\alpha_5(\alpha_1 + \alpha_4 \bar{S}_2^2) + \alpha_6(\alpha_2 S_1 - 3\alpha_3 \bar{S}_2^2)]}{\Delta'}, \tag{10}$$

$$\bar{M} = \frac{(3\alpha_0 + \alpha_1 + \alpha_2 S_1)(\hat{\bar{S}}_2^2/2) - \alpha_2 \bar{S}_2^2 \hat{S}_1 - \bar{N}(\bar{S}_2^2)[(3\alpha_0 + \alpha_1)\alpha_6 - 3\alpha_2\alpha_5]}{\Delta'}$$

où $\bar{N}$ est arbitraire. Ces relations représentent les valeurs de E_1 et $\bar{M}$ quand $\Delta = 0$, le chemin suivi étant situé sur cette hypersurface.

L'interprétation de ces résultats conduit alors à certaines restrictions pour les fonctions α_i.

2.2.2 *Ecoulement à l'état de contrainte constant.* Des relations (10), on déduit qu'un écoulement à l'état de contrainte constant serait possible à la condition que $\Delta = 0$ et $\Delta' \neq 0$. Or la relation $\Delta = 0$ détermine la densité ρ comme fonction de l'état de contrainte: on en déduit qu'un écoulement à état de contrainte constant s'effectuera à volume constant. Le coefficient de $\bar{N}$ dans (10_1) doit donc être nul. Le système (6) devient dans ce cas un système de 3 équations homogènes à 2 inconnues, $\bar{M}$ et $\bar{N}$, qui n'admettra de solution non nulle que si:

$$\begin{aligned} &\alpha_1 + \alpha_4 \bar{S}_2^2 + \alpha_6 \bar{S}_3^3 = 0, \\ &3\alpha_3\alpha_6 \bar{S}_2^2 - 3\alpha_5(\alpha_1 + \alpha_4 \bar{S}_2^2) - \alpha_1\alpha_6 S_1 = 0 \end{aligned} \tag{11}$$

ou, en raportant (11_1) dans (11_2)

$$\alpha_1 S_1 - 3\alpha_3 \bar{S}_2^2 - 3\alpha_5 \bar{S}_3^3 = 0. \tag{12}$$

La condition $\Delta = 0$ doit donc être interprétée comme représentant l'intersection des hypersurfaces (11_1) et (12). Si cette intersection satisfait à certaines conditions de régularité, elle représente une surface dans l'espace à trois dimensions:

$$G(S_1, \bar{S}_2, \bar{S}_3) = 0. \tag{13}$$

Puisque en chaque point de cette surface, un écoulement à état de contrainte constant et à volume constant est possible, cette surface peut être interprétée comme la surface limite de plasticité du matériau. En chacun de ses points, la densité du matériau sera donnée en fonction de l'état de contrainte par la résolution de (11) ou (12).

La relation (10_2), compte tenu des relations (11_1) et (12), se réduit à:

$$\frac{\bar{M}}{\bar{N}} = \frac{\bar{S}_2^2}{\bar{S}_3^3}. \tag{14}$$

Les équations de comportement (3) se réduisent alors aux relations de plasticité:

$$E_1 = 0 \quad \bar{d}_{ij} = \lambda \bar{\sigma}_{ij} \quad \text{avec} \quad \lambda = \frac{\bar{M}}{\bar{S}_2^2} = \frac{\bar{N}}{\bar{S}_3^3}. \tag{15}$$

2.2.3 *Trajet sur la surface limite.* Avec les restrictions (11) qui doivent être imposées aux fonctions α_i, les relations (10) se réduisent à:

$$E_1 = \frac{\hat{S}_1(\alpha_1 + \alpha_4 \bar{S}_2^2) - (\hat{\bar{S}}_2^2/2)(3\alpha_3 + \alpha_4 S_1)}{\Delta'},$$

$$\bar{M} = \frac{(3\alpha_0 + \alpha_1 + \alpha_2 S_1)(\hat{\bar{S}}_2^2/2) - \alpha_2 \bar{S}_2^2 \hat{S}_1}{\Delta'} + \bar{N}\,\frac{\bar{S}_2^2}{\bar{S}_3^3} \tag{16}$$

avec:

$$\Delta' = (3\alpha_0 + \alpha_1)\cdot(\alpha_1 + \alpha_4 \bar{S}_2^2) + \alpha_2 [\alpha_1 S_1 - 3\alpha_3 \bar{S}_2^2].$$

Les fonctions α_i peuvent être reliées à l'équation de la surface critique $G(S_i) = 0$. En effet, quand l'état de contrainte est assujetti à rester sur la surface critique, on peut écrire:

$$\frac{\partial G}{\partial S_1}\hat{S}_1 + \frac{\partial G}{\partial \bar{S}_2^2}\hat{\bar{S}}_2^2 + \frac{\partial G}{\partial \bar{S}_3^3}\hat{\bar{S}}_3^3 = 0. \tag{17}$$

En comparant à la relation (8), on obtient:

$$\frac{\partial G/\partial S_1}{\alpha_1\alpha_2\bar{S}_3^3} = \frac{2(\partial G/\partial \bar{S}_2^2)}{[(3\alpha_0+\alpha_1)\alpha_4-3\alpha_2\alpha_3]\bar{S}_3^3} = -\frac{3\partial G/\partial \bar{S}_3^3}{\Delta'}. \tag{18}$$

Ce qui conduit en particulier à:

$$(3\alpha_0+\alpha_1+\alpha_2 S_1) = -\alpha_2 \frac{2\bar{S}_2^2(\partial G/\bar{S}_2^2)+3\bar{S}_3^3(\partial G/\partial \bar{S}_3^3)}{\partial G/\partial S_1}. \tag{19}$$

On peut encore préciser ces fonctions:

a) Dans le cas où la fonction G est une fonction homogène par rapport à S_1, $\bar{S}_2$ et $\bar{S}_3$

$$\frac{\partial G}{\partial S_1} S_1 + \frac{2\partial G}{\partial \bar{S}_2^2}\bar{S}_2^2 + \frac{3\partial G}{\partial \bar{S}_3^3}\bar{S}_3^3 = 0 \tag{20}$$

(19) conduit à $3\alpha_0+\alpha_1+\alpha_2 S_1 = \alpha_2 S_1$ soit $\alpha_1+3\alpha_0 = 0$, quand l'état critique est atteint. Les relations (18) donnent alors:

$$\alpha_3 = \frac{-\frac{2}{3}\alpha_1(\partial G/\partial \bar{S}_2^2)}{\partial G/\partial S_1}, \qquad \alpha_5 = \frac{-\alpha_1(\partial G/\partial \bar{S}_3^3)}{\partial G/\partial S_1}. \tag{21}$$

b) Si la densité critique ρ_c ne dépend que de la pression moyenne, d'après (16)

$$\alpha_4 = \frac{-3\alpha_3}{S_1} \quad \text{et d'après (5)} \quad \alpha_6 = \frac{-3\alpha_5}{S_1}. \tag{22}$$

Dans ce cas:

$$E_1 = \frac{\hat{S}_1}{\alpha_2}, \tag{23}$$

$$\alpha_4 = \frac{+2\alpha_1}{S_1}\,\frac{\partial G/\partial \bar{S}_2^2}{\partial G/\partial S_1}, \quad \alpha_6 = \frac{+3\alpha_1}{S_1}\cdot\frac{\partial G/\partial \bar{S}_3^3}{\partial G/\partial S_1}. \tag{24}$$

2.3 *Choix des fonctions*

Les relations que nous venons d'obtenir permettent de déterminer la forme de certaines fonctions α_i quand le matériau a atteint l'état critique. En dehors de cet état, des hypothèses supplémentaires doivent être faites.

Il faut tout d'abord exclure l'hypothèse très simple qui consisterait à retenir pour fonctions les valeurs prises à l'état critique. En effet, compte tenu des relations (16), les variations de volume ne seraient dues qu'à la variation de S_1, même en dehors de l'état critique, ce qui ne correspond pas au comportement que nous voulons décrire. Nous voyons d'après les relations (5) que le signe de la variation de volume sera donné par le signe des quantités

$$(3\alpha_3+\alpha_4 S_1) \quad \text{et} \quad (3\alpha_5+\alpha_6 S_1). \tag{25}$$

Les contraintes de compression étant comptées positives, E_1 positif ou négatif correspond à une augmentation de densité (durcissement) ou à une diminution de densité (amollissement). Les quantités (21) doivent donc être >0 ou <0 suivant que la densité du matériau est $>$ ou $<$ à la densité critique. Si on remarque également que seules les fonctions α_4 et α_6 interviennent dans la partie déviatoire de la loi et que dans les parties isotropes, $\overline{M}$ et $\overline{N}$ ont pour multiplicateurs les quantités (25), nous sommes amenés à faire les hypothèses suivantes:

1) Les fonctions α_3 —, α_6 ont, en dehors de l'état critique, même expression formelle que dans l'état critique.

2) Les fonctions α_3 et α_5 doivent être reliées à la densité critique et les fonctions α_4 et α_6 à la densité actuelle du matériau.

3) Remarquons que l'état de contrainte actuel apparaît dans les expressions (3) soit directement par les composantes de σ_{ij} soit dans $(\alpha_4\overline{M}+\alpha_6\overline{N})$ et $(\alpha_3+\alpha_4 S_1)\overline{M}+(\alpha_5+\alpha_6 S_1)\overline{N}$, par les invariants $\overline{M}$ et $\overline{N}$, ou par les invariants S_i, arguments des fonctions α_i. Puisque nous avons choisi dans le développement de $f(\boldsymbol{d}, \boldsymbol{\sigma})$ une forme linéaire par rapport à $\boldsymbol{\sigma}$, on imposera aux expressions précédentes d'être de degré un par rapport à l'état de contrainte actuel. Ceci nous conduit à demander aux fonctions α_3, α_4 et α_5, α_6 d'être respectivement des fonctions homogènes de degré 0, -1, et -2 par rapport à l'état de contrainte actuel.

3. Application a un milieu pulvérulent

3.1 *Surface limite*

Pour caractériser la résistance mécanique des sols, il est classique d'utiliser le paramètre ϕ, angle de frottement interne, et d'admettre l'existence d'une courbe intrinsèque pour les sols. On obtient alors pour les essais triaxiaux de compression ou de traction les relations:

$$\bar{S}_2 = 2\sqrt{\tfrac{2}{3}}\,\frac{\sin\phi}{3 \pm \sin\phi}\,S_1. \tag{26}$$

(Les signes ∓ sont relatifs à l'essai de compression et à l'essai de traction.)

Abandonnant l'hypothèse de la courbe intrinsèque, nous avons supposé que la surface critique de plasticité était de la forme:

$$\bar{S}_2 = \frac{AS_1}{[1 - B(\bar{S}_3/\bar{S}_2)^3]^m} \tag{27}$$

où A et B sont fonction de la densité.

Les relations (26) peuvent être obtenues à partir de (27) en posant:

$$K = \frac{3 + \sin\phi}{3 - \sin\phi} \tag{28}$$

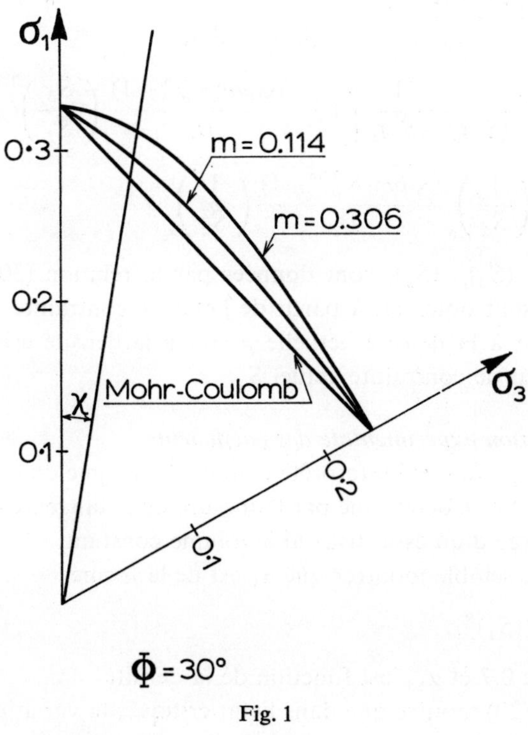

Fig. 1

et

$$A = (K^{1/m} - 1)\cdot(K^{1/m} + 1), \quad B = \sqrt{\tfrac{2}{3}}\cdot 2^m (K-1)(1+K^{1/m}). \tag{29}$$

La condition (27) s'écrit alors:

$$\bar{S}_2 = \frac{\sqrt{\frac{2}{3}}\, 2^m (K-1) S_1}{[K^{1/m} + 1 - \sqrt{6}(\bar{S}_3/\bar{S}_2)^3 (K^{1/m} - 1)]^m} \tag{30}$$

où K peut être considéré comme une fonction de la densité ρ.

La forme de la surface limite proposé est illustrée sur la Figure 1.

3.2 *Forme des fonctions*

Les hypothèses que nous avons formulées en (2, 3) et la forme particulière de (30) que nous retenons pour les milieux pulvérulents, conduisent à:

$$\begin{aligned}
\alpha_3 &= \frac{\alpha_1}{3}\left(\frac{S_1}{\bar{S}_2}\right)_a \frac{1}{(\bar{S}_2)_c}\left[1 + \frac{3\sqrt{6}(K_c^{1/m} - 1)}{D_c}\left(\frac{\bar{S}_3}{S_{2a}}\right)^3\right],\\
\alpha_5 &= \frac{-\alpha_1\sqrt{6}m(K_c^{1/m} - 1)}{D_c}\cdot\frac{1}{(S_2)_c}\cdot\frac{S_1}{(\bar{S}_2^2)_a},\\
\alpha_4 &= -\alpha_1 \frac{1}{(\bar{S}_2)_\rho}\,\frac{1}{(\bar{S}_2)_a}\left[1 + \frac{3\sqrt{6}m(K_\rho^{1/m} - 1)}{D_\rho}\left(\frac{\bar{S}_3}{\bar{S}_2}\right)_a^3\right],\\
\alpha_6 &= 3\alpha_1\left(\frac{1}{\bar{S}_2}\right)_\rho \frac{\sqrt{6}m(K_\rho^{1/m} - 1)}{D_\rho}\left(\frac{1}{\bar{S}_2^2}\right)_a.
\end{aligned} \tag{31}$$

Les valeurs de $(\bar{S}_2)_\rho$, $(\bar{S}_2)_c$ sont données par la relation (30) dans laquelle S_1 et $(\bar{S}_3/S_2)$ sont obtenues à partir de l'état de contrainte actuel, mais où K est relié soit à la densité actuelle ρ soit à la densité critique ρ_c correspondant à l'état de contrainte actuel S_1.

3.3 *Détermination expérimentale des coefficients*

Le coefficient α_1 ayant été interprété comme l'analogue du coefficient d'élasticité $2G$, il peut être déterminé par la mesure de la tangente à l'origine de la courbe de charge d'un essai triaxial à volume constant.

L'expérience semble montrer que α_1 est de la forme

$$\alpha_1 = \alpha_{1\,0}\cdot(S_1)^R, \tag{32}$$

R est voisin de 0.7 et $\alpha_{1\,0}$ est fonction de la densité.

La relation (20) montre que dans l'état critique, la variation de la densité critique obéit à la loi différentielle

$$\frac{d\rho}{\rho} = \alpha_2 \cdot dS_1 \tag{33}$$

ce qui permet de déterminer le coefficient α_2 dans l'état critique.

Le coefficient α_0 peut être déterminé par un essai de compression isotrope, pour lequel (4) conduit à:

$$dS_1 = (3\alpha_0 + \alpha_1 + \alpha_2 S_1) \frac{d\rho}{\rho}. \tag{34}$$

Au cours de cet essai et pour un domaine de pression S_1 n'entrainant pas de

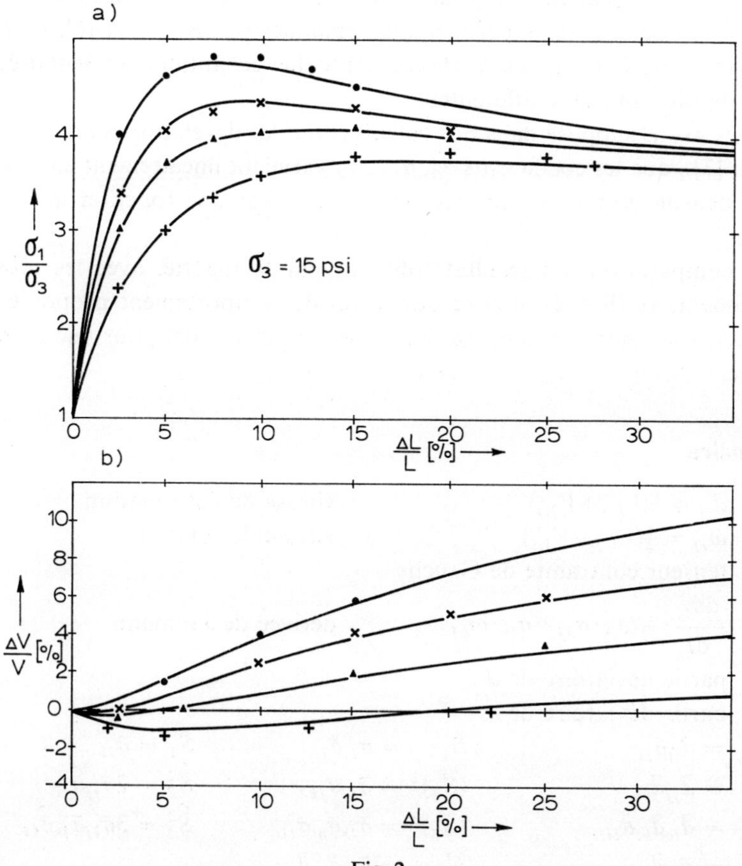

Fig. 2

fracture des grains, les variations de densité sont très faibles: α_0, α_2 peuvent alors être assimilés à des constantes.

La comparaison de la solution de (34), où on considère α_0 et α_2 comme des constantes et de la courbe expérimentale permet de déterminer α_0 et α_2. Une série d'essais effectués sur des échantillons d'un même matériau préparés à différentes densités permettent de déterminer les fonctions $\alpha_{0(\rho)}$ et $\alpha_{2(\rho)}$. La connaissance de l'état critique permet de déterminer les constantes K_c, m et $\rho_c(S_1)$. La fonction K sera obtenue par comparaison des courbes de charge et des courbes de variation de densité que l'on obtient par exemple dans les essais triaxiaux à pression latérale constante ou à pression moyenne constante.

Exemple. Nous avons essayé de retrouver par intégration des relations (3) les courbes de charges expérimentales obtenues par Boutwell [6], pour des essais triaxiaux à pression latérale constante. Les échantillons étaient préparés à des densités initiales différentes.

Nous avons supposé dans nos calculs (dont les développements sont donnés en [7]), que les coefficients α_0, $\alpha_{1\,0}$, α_2 variaient linéairement en fonction de la densité initiale, et que (K) était également une fonction linéaire de $(\rho_a - \rho_c)$.

La comparaison des résultats obtenus par la théorie, avec les résultats expérimentaux (Fig. 2) montre que la loi de comportement proposée peut approcher le comportement réel du matériau pulvérulent pour une première charge.

Appendice

$\boldsymbol{d}$	$d_{ij} = \frac{1}{2}(V_{i,j} + V_{j,i})$	vitesse de déformation
$\boldsymbol{\omega}$	$\omega_{ij} = \frac{1}{2}(V_{i,j} - V_{j,i})$	vitesse de rotation
$\boldsymbol{\sigma}$	tenseur contrainte de Cauchy	
$\hat{\boldsymbol{\sigma}}$	$\dfrac{d\sigma_{ij}}{dt} - \omega_{ih} \cdot \sigma_{hj} + \sigma_{ih} \cdot \omega_{hj}$	dérivée de Jaumann
$\bar{\boldsymbol{d}}$	partie déviatoire de $\boldsymbol{d}$	
$\bar{\boldsymbol{\sigma}}$	partie déviatoire de $\boldsymbol{\sigma}$	

$E_1 = d_{ij}\delta_{ji}$ $\quad$ $S_1 = \sigma_{ij}\delta_{ji}$ $\quad$ $\hat{S}_1 = \hat{\sigma}_{ii}$

$(\bar{E}_2)^2 = \bar{d}_{ij}\bar{d}_{ji}$ $\quad$ $(\bar{S}_2)^2 = \bar{\sigma}_{ij}\bar{\sigma}_{ji}$ $\quad$ $\hat{\bar{S}}_2^2 = 2\bar{\sigma}_{ij}\hat{\bar{\sigma}}_{ji}$

$(\bar{E}_3)^3 = \bar{d}_{ih}\bar{d}_{hl}\bar{d}_{li}$ $\quad$ $(\bar{S}_3)^3 = \bar{\sigma}_{ih}\bar{\sigma}_{hl}\bar{\sigma}_{li}$ $\quad$ $\hat{\bar{S}}_3^3 = 3\bar{\sigma}_{ij}\bar{\sigma}_{jk}\hat{\bar{\sigma}}_{ki}$

$\bar{M} = \bar{\sigma}_{ij}\bar{d}_{ji}$ $\quad$ $\bar{N} = \bar{\sigma}_{ij}\bar{\sigma}_{jl}\bar{d}_{li}$

a_0, a_1, a_2: fonctions scalaires des invariants de $\boldsymbol{\sigma}$ et $\boldsymbol{d}$, et de ρ
α_i fonctions scalaires des invariants de $\boldsymbol{\sigma}$ et ρ
ρ densité
ϕ angle de frottement interne

Références

[1] Green, A. E., Hypo-elasticity and plasticity. *Proc. Roy. Soc. London. A 254* (1956) 46–50.

[2] Thomas, T. Y., Combined elastic and Prandtl-Reuss stress-strain relations. *Proc. Nat. Acad. Sci. USA*, *41* (1955) 720–726.

[3] Rivlin, R. S. and J. L. Ericksen, Stress deformation relations for isotropic materials. *J. Rat. Mech. Anal. 4* (1955) 324–425.

[4] Sawczuk, A. and P. Stutz, Contribution to the finite deformation theory of elastic plastic behaviour of soils. *Bull. Acad. Pol. Sci., Sér. Sci. tech. 19* (1970) 287–291.

[5] Stutz, P., Comportement élasto-plastique des sols en grande déformation. *Archivum Mechaniki stosowanej 6*, *22* (1970).

[6] Boutwell, G., *On the yield behaviour of cohesionless materials*. Ph. D. Thesis, Duke University, (1968).

[7] Stutz, P., Thèse de Doctorat d'Etat, Grenoble (Mai 1972).

Sur le comportement élasto-plastique et viscoplastique des monocristaux et polycristaux métalliques de structure cubique à faces centrées

H. D. Bui

Ecole Polytechnique, Paris et Electricité de France, Centre des Renardières

A. Zaoui et J. Zarka

Ecole Polytechnique et Centre National de la Recherche Scientifique, Paris, France

L'étude s'appuie sur l'analyse des phénomènes plastiques au niveau du cristal et des interactions entre cristaux pour déterminer le comportement du polycristal.

Deux étapes sont considérées:
– dans la première étape, on construit un modèle physique représentant le monocristal. Cela permet de donner la forme explicite des relations de comportement en viscoplasticité et en élastoplasticité.
– dans la deuxième étape, on commente et on développe différents modèles qui donnent le passage du cristal au polycristal.

Sur cette base on déduit pour le polycristal le comportement en élastoplasticité et en viscoplasticité. En étudiant le cas particulier de la traction simple, on met en évidence certaines propriétés propres au polycristal.

Etude du comportement viscoplastique du monocristal cubique à faces centrées

Modèle physique représentant le monocristal aux contraintes et températures peu élevées.

On sait qu'il existe de nombreux défauts d'empilement dans la structure cristalline et que ce sont les déplacements de ces défauts qui provoquent les déformations plastiques. Aux températures peu élevées les défauts sont principalement les dislocations. Les déplacements des dislocations sont gênés par les actions des atomes (force de Peierls-Nabarro) et les actions des autres dislocations qui peuvent très rapidement être en nombre important grâce à l'activation de certaines d'entre elles (Sources de Frank-Read). Des observations au microscope électronique ont montré que les dislocations sont presque toutes décomposées en petits segments rectilignes disposés selon certaines directions privilégiées et bloqués devant ou sur d'autres dislocations. Aux températures de l'ordre de 20°C et aux contraintes peu élevées, les dislocations pour le système C.F.C. se trouvent pour la plupart dans les plans denses [111] et leur vecteur de Burgers est suivant l'une des directions denses [101]; le glissement simple dans leur plan est le plus probable.

A l'aide de ces quelques éléments fondamentaux nous pouvons construire le modèle physique simple suivant, qui est une généralisation du modèle unidimensionnel et isotrope de G. Saada [1].

Nous définissons un mécanisme de glissement K comme l'ensemble de tous les segments de dislocations qui se trouvent dans des plans parallèles (de normale unitaire n_i^K) et ont même vecteur de Burgers B_i^K. Le mécanisme K est caractérisé par le nombre $\mathcal{N}_K$ de ces petits segments par unité de volume et leur longueur moyenne l_K.

Avec certaines hypothèses statistiques sur la distribution des dislocations, nous avons montré (J. Zarka [2]) qu'une dislocation d'un mécanisme (K) est susceptible de glisser dans son plan si la force par unité de longueur qui lui est appliquée sur ce plan et qui est donnée par la formule de Peach-Koelher[1]):

$$F^K = B_i^K n_j^K \sigma_{ij} \equiv |\boldsymbol{B}| b_i^K n_j^K \sigma_{ij} \equiv |\boldsymbol{B}| \alpha_{ij}^K \sigma_{ij} \equiv |\boldsymbol{B}| \tau^K \tag{1}$$

est supérieure à:

$$\mathscr{F}^K = \mathscr{F}_{PN}^K + \mu |\boldsymbol{B}|^2 \sum_R D^{KR} (\mathcal{N}_R l_R)^{\frac{1}{2}}. \tag{2}$$

Dans ces formules, $|\boldsymbol{B}|$ est le module des vecteurs de Burgers constant pour tous les mécanismes, et b_i^K est le vecteur unitaire suivant B_i^K;

$$\alpha_{ij}^K = \tfrac{1}{2}(b_i^K n_j^K + b_j^K n_i^K) \tag{3}$$

[1]) Dans tout ce texte, nous adoptons la convention suivante:
... sommation, s'il s'agit d'indices minuscules, ex:

$$x^i y^i = x_i y_i = \sum_i x^i y^i$$

... sans sommation, s'il s'agit d'indices majuscules, ex:

$x^K y_K = x^K y_K$ (sans sommation).

α_{ij}^K est la matrice d'orientation pour le mécanisme K;
τ^K est la cission réduite appliquée sur le mécanisme K;
$\mathscr{F}_{PN}^K$ est la force de Peierls-Nabarro, généralement négligeable pour les mécanismes que nous considérons ici;
μ est le module de cisaillement élastique;
D^{KR} est la matrice des coefficients d'interaction que nous avons évalués [2].

Nous avons aussi évalué en fonction de la distribution des dislocations et des forces appliquées, le temps moyen nécessaire à une dislocation K pour franchir une autre dislocation:

$$t_{FD}^K \simeq \frac{C_0^K}{F^K - \mathscr{F}^K} \tag{4}$$

où C_0^K est une constante physique du métal (si la température est maintenue constante).

Pendant le temps dt, certaines dislocations K vont être sources; elles émettront $d\mathscr{N}_K$ boucles de dislocations par unité de volume; les boucles, de longueur moyenne $2\pi L_K$ balayeront une aire moyenne πL_K^2 par glissement simple avant d'être bloquées sur les dislocations qui percent leur plan; chaque boucle aura franchi pendant son expansion en moyenne x_K dislocations. Nous avons ainsi, pour un mécanisme K actif:

$$\dot{\gamma}_K dt = |\boldsymbol{B}| d\mathscr{N}_K \pi L_K^2 \tag{5}$$

(accroissement de glissement moyen suivant B_i^K entre les deux plans K distants de l'unité);

$$\dot{\overline{(\mathscr{N}_K l_K)}} dt = 2\pi L_K d\mathscr{N}_K \tag{6}$$

(variation de longueur totale des dislocations K);

$$x_K = \pi L_K^2 \rho_K \tag{7}$$

(nombre d'intersections d'une boucle K avec les dislocations qui percent le plan);
et

$$d\mathscr{N}_K \simeq \omega_K \mathscr{N}_K \frac{dt}{x_K t_{FD}^K} \tag{8}$$

(accroissement de la densité de dislocations),
ω_K représente la proportion des dislocations susceptibles d'être sources;
ρ_K densité de dislocations par unité de surface, est déterminée à partir de la distribution des dislocations $\mathscr{N}_K$, l_K;

x_K est fonction uniquement de ρ_K;
et ω_K est pris constant lorsque le chargement est progressif et sans choc.

Potentiel viscoplastique pour le monocristal

La notion de potentiel plastique reste inchangée en viscoplasticité.

Les accroissements de glissement $\dot{\gamma}_K \mathrm{d}t$ entraînent un accroissement du tenseur de déformation viscoplastique qui s'exprime par:

$$v_{ij}^p \mathrm{d}t = \sum_K \alpha_{ij}^K \dot{\gamma}_K \mathrm{d}t. \tag{9}$$

Comme à chaque instant t, $\dot{\gamma}_K$ n'est fonction que des contraintes appliquées actuelles et de la distribution des dislocations, en posant sur chaque mécanisme K:

$$f_K(\sigma_{ij}, \mathcal{N}_S, l_S) \equiv \int_{\mathscr{F}^K}^{F^K} \dot{\gamma}_K(F - \mathscr{F}^K, \mathcal{N}_S, l_S) Y(F^K - \mathscr{F}^K) \frac{\mathrm{d}F}{|\boldsymbol{B}|} \qquad (10)^1)$$

et pour le monocristal (J. R. Rice [3]):

$$f(\sigma_{ij}, \mathcal{N}_S, l_S) = \sum_K f_K(\sigma_{ij}, \mathcal{N}_S, l_S) \tag{11}$$

il est facile de montrer que (9) est équivalent à:

$$v_{ij}^p \equiv \left(\frac{\partial f}{\partial \sigma_{ij}}\right)_t = \sum_K \left(\frac{\partial f_K}{\partial \sigma_{ij}}\right)_t = \sum_K \alpha_{ij}^K \frac{\mathrm{d}f_K}{\mathrm{d}F} |\boldsymbol{B}| = \sum_K \alpha_{ij}^K \dot{\gamma}_K \tag{12}$$

f est le potentiel viscoplastique multiple du monocristal. La vitesse v_{ij}^p est normale à la surface équipotentielle $f = C$, tracée dans l'espace des contraintes. En reportant les différentes formules (2), (4), (7) et (8) dans (5) on trouve une équation du type:

$$\sum_K A_K(\mathcal{N}_S, l_S)\{F^K - \mathscr{F}^K\}^2 = C \tag{13}$$

où A_K est uniquement fonction de la distribution des dislocations.

C'est donc une quadrique convexe sans point singulier lorsque la constante C est différente de 0. Comme la partie sphérique du tenseur de contrainte n'intervient pas, cette quadrique est un cylindre. La figure 1 représente la surface élastique qui correspond à $C = 0$, une surface équipotentielle, ainsi que la composition des différents mécanismes actifs pour un point de chargement.

1) $Y(x)$ est la fonction échelon unité d'Heaviside.

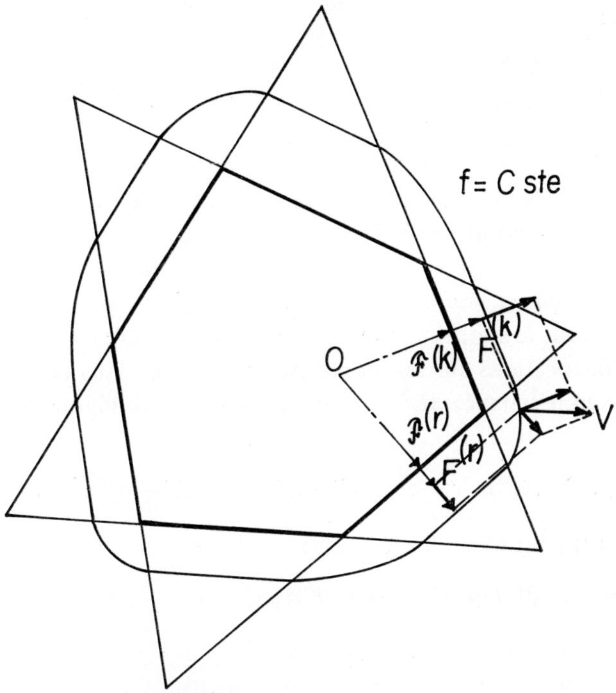

Fig. 1. Frontière élastique et potentiel viscoplastique du monocristal

Cas particulier où les vitesses de chargement sont négligeables. Lors d'un chargement continu à vitesse de déformation comprise entre 10^{-5} et 10^{-3}/s, nous avons pratiquement à tout instant, pour chaque mécanisme actif K: $F^K \simeq \mathscr{F}^K$ et:

$$\dot{F}^K = |\boldsymbol{B}|\alpha_{ij}^K\dot{\sigma}_{ij} \simeq \dot{\mathscr{F}}^K = \mu|\boldsymbol{B}|^2\tfrac{1}{2}\sum_R D^{KR}\dot{\overline{(\mathcal{N}_R l_R)}}(\mathcal{N}_R l_R)^{-\frac{1}{2}}. \tag{14}$$

A partir des formules (5) et (6) nous pouvons écrire (14) sous la forme:

$$|\boldsymbol{B}|\left\{\alpha_{ij}^K\dot{\sigma}_{ij} - \sum_R \frac{\mu D^{KR}}{L_R(\mathcal{N}_R l_R)^{\frac{1}{2}}}\,\dot{\gamma}_R\right\} \simeq 0 \tag{15}$$

ou encore:

$$\alpha_{ij}^K\dot{\sigma}_{ij} = \sum_R h^{KR}\dot{\gamma}_R \tag{16}$$

si l'on pose:

$$h^{KR} = \mu \frac{D^{KR}}{(\mathcal{N}_R l_R)^{\frac{1}{2}} L_R}. \tag{17}$$

La relation (16) avait été écrite par J. Mandel [4]; nous en donnons ici la forme explicite. Nous pouvons immédiatement constater que la matrice h^{KR} n'est pas en général symétrique.

Cas particulier où un seul mécanisme K est actif. L'indice K étant ici seul à intervenir, nous le supprimons pour toutes les variables indicées. La distribution des dislocations sur les autres systèmes reste, dans ce cas, inchangée et l'aire moyenne balayée πL^2 est donc constante. L'intégration de (5) et (6) conduit à:

$$\gamma = |\boldsymbol{B}|(\mathcal{N} - \mathcal{N}^0)\pi L^2 \quad \text{et} \quad \mathcal{N} l - \mathcal{N}^0 l^0 = (\mathcal{N} - \mathcal{N}^0) 2\pi L. \tag{18}$$

La formule (2) peut être réécrite ainsi:

$$\mathscr{F} = \mathscr{F}_0 + \mu |\boldsymbol{B}|^2 D\{(\mathcal{N} l)^{\frac{1}{2}} - (\mathcal{N}^0 l^0)^{\frac{1}{2}}\}. \tag{19}$$

Lorsque γ est encore petit, nous avons:

$$(\mathcal{N} l)^{\frac{1}{2}} - (\mathcal{N}^0 l^0)^{\frac{1}{2}} \simeq \frac{\gamma}{|\boldsymbol{B}| L (\mathcal{N}^0 l^0)^{\frac{1}{2}}}. \tag{20}$$

En combinant les relations (8), (4), (20) et (19) nous obtenons en portant dans (5):

$$\dot{\gamma} \simeq \frac{\omega \mathcal{N}^0 |\boldsymbol{B}|}{\rho C_0} \left\{ F - \mathscr{F}_0 - \frac{\mu D |\boldsymbol{B}| \gamma}{(\mathcal{N}^0 l^0)^{\frac{1}{2}} L} \right\} \tag{21}$$

soit encore:

$$\eta \dot{\gamma} = \tau - \tau_0 - h\gamma \tag{22}$$

où nous avons posé:

$$F = |\boldsymbol{B}|\tau, \qquad \mathscr{F}_0 = |\boldsymbol{B}|\tau_0,$$

$$\eta = \frac{\rho C_0}{\omega \mathcal{N}^0 |\boldsymbol{B}|^2}, \quad h = \frac{\mu D}{(\mathcal{N}^0 l^0)^{\frac{1}{2}} L}. \tag{23}$$

La relation (22) est analogue à celle de Bingham-Kepes; le potentiel viscoplastique qui lui est associé par la formule (10) vaut:

$$f(\sigma_{ij}, \gamma) = \frac{1}{2\eta}\{\tau-\tau_0-h\gamma\}^2 Y(\tau-\tau_0-h\gamma). \tag{24}$$

Relations entre grandeurs microscopiques et grandeurs macroscopiques dans le polycristal

Relations générales

Pour décrire l'état mécanique du polycristal on a recours à des grandeurs macroscopiques qui doivent être rattachées aux grandeurs microscopiques propres aux cristaux dont il est constitué. L'élément de volume V considéré dans le polycristal contient de nombreux cristaux d'orientations, de formes et d'états internes différents. Les grandeurs mécaniques étant définies de façon analogue dans le monocristal et le polycristal, le passage des unes aux autres ne peut se faire, en général, par une opération directe de moyenne.

Cependant, R. Hill [5] a montré que, lorsque les déformations sont infiniment petites et lorsque certaines hypothèses de macro-homogénéité sont réalisées, indépendamment des relations de comportement du cristal, dans l'élément de volume V, à partir de tout champ de contrainte $\sigma_{ij}(\boldsymbol{x})$ statiquement admissible et de tout champ de déformation $\varepsilon_{ij}(\boldsymbol{x})$ cinématiquement admissible, ces deux champs n'étant pas nécessairement associés, on peut définir la contrainte macroscopique Σ_{ij}, la déformation macroscopique E_{ij}, et le travail de déformation sous forme de moyenne directe:

$$\Sigma_{ij} = \frac{1}{V}\int_V \sigma_{ij}(\boldsymbol{x})\mathrm{d}v_x \equiv \langle\sigma_{ij}(\boldsymbol{x})\rangle, \tag{25}$$

$$E_{ij} = \langle\varepsilon_{ij}(\boldsymbol{x})\rangle, \tag{26}$$

$$\Sigma_{ij}E_{ij} = \langle\sigma_{ij}(\boldsymbol{x})\varepsilon_{ij}(\boldsymbol{x})\rangle. \tag{27}$$

Hill a montré que la formule (25) est exacte lors d'un essai à contrainte homogène et que la formule (26) l'est lors d'un essai de déformation homogène. On ne peut pas écrire des relations semblables pour les parties élastiques et plastiques de la déformation dont les définitions sont rappelées ici:

$$\begin{aligned} \varepsilon^e_{ij}(\boldsymbol{x}) &= s_{ijhk}(\boldsymbol{x})\sigma_{hk}(\boldsymbol{x}), & \varepsilon^p_{ij}(\boldsymbol{x}) &= \varepsilon_{ij}(\boldsymbol{x})-\varepsilon^e_{ij}(\boldsymbol{x}), \\ E^e_{ij} &= S_{ijhk}\Sigma_{hk}, & E^p_{ij} &= E_{ij}-E^e_{ij} \end{aligned} \tag{28}$$

où $s_{ijhk}(x)$ et S_{ijhk} sont les tenseurs, microscopiques et macroscopiques, des compliances élastiques. Les relations correctes entre ces grandeurs ont été établies par J. Mandel [6] puis par H. D. Bui [7], ainsi que par R. Hill [8] et J. R. Rice [3]. Pour cela, introduisons le tenseur 'concentration des contraintes élastiques' $A_{ijhk}(x)$ qui fait correspondre à la contrainte Σ_{hk} la contrainte locale $A_{ijhk}(x)\Sigma_{hk}$ en supposant le milieu purement élastique (R. Hill [9]). Partant des relations précédentes, on peut établir les relations suivantes:

$$E_{hk} = \langle \varepsilon_{ij}(x) A_{ijhk}(x) \rangle, \tag{29}$$

$$E^e_{hk} = \langle \varepsilon^e_{ij}(x) A_{ijhk}(x) \rangle \tag{30}$$

soit par soustraction, la relation fondamentale:

$$E^p_{hk} = \langle \varepsilon^p_{ij}(x) A_{ijhk}(x) \rangle. \tag{31}$$

La formule (29) est strictement équivalente à (26), mais les formules (30) et (31) ne peuvent se ramener en général à des formes aussi simples que (26), sauf si on suppose l'isotropie élastique, c'est à dire si $A_{ijhk}(x) \equiv I_{ijhk}$.

Parmi les principaux problèmes posés par l'étude du comportement du polycristal à partir de celui du monocristal, citons celui de la détermination de $A_{ijhk}(x)$ et celui de la détermination de $\sigma_{ij}(x)$ lorsque les déformations plastiques $\varepsilon^p_{ij}(x)$ sont connues. Ces problèmes présentent de grandes difficultés mais peuvent être résolus au prix de certaines simplifications (modèles de Sachs, Taylor, Kröner, etc.). Nous nous limiterons à l'étude du modèle particulièrement important dit 'autoéquilibré' de Kröner-Budiansky-Wu.

Commentaires sur la validité du modèle 'autoéquilibré'

Relation de Kröner-Budiansky-Wu. Ce modèle repose sur le calcul de J. D. Eshelby [10], d'une inclusion de forme ellipsoidale c, à élasticité isotrope, ayant subi une déformation homogène $\varepsilon^p_{ij}(c)$ et plongée dans une matrice infinie élastique et isotrope. On trouve que la contrainte $\sigma_{ij}(c)$ dans l'inclusion est homogène ainsi que la déformation $\varepsilon_{ij}(c)$ et que, si Σ_{ij} est le tenseur contrainte appliquée à l'infini, on a la relation suivante pour une inclusion sphérique:

$$\sigma_{ij}(c) = \Sigma_{ij} - 2\mu(1-\beta)\varepsilon^p_{ij}(c) \tag{32}$$

où $\beta = 2(4-5\nu)/15(1-\nu)$ est une constante ne dépendant que du coefficient de Poisson ν. A partir de ce calcul rigoureux, E. Kröner [11] puis T. T. Wu et B. Budiansky [12], ont construit un modèle polycristallin où la matrice est cette fois supposée subir une déformation plastique E^p_{ij} homogène, moyenne des déformations plastiques $\varepsilon^p_{ij}(c)$ propres à chaque cristal c considéré comme inclusion.

La formule (32) devient alors:

$$\sigma_{ij}(c) = \Sigma_{ij}+2\mu(1-\beta)(E^p_{ij}-\varepsilon^p_{ij}(c)). \tag{33}$$

Dans ce modèle, le champ de contrainte $\sigma_{ij}(c)$ n'est pas statiquement admissible, le champ de déformation $\varepsilon_{ij}(c)$ n'est pas cinématiquement admissible. Cependant, puisque $E^p_{ij} = \langle\varepsilon^p_{ij}(c)\rangle$ la propriété d'autoéquilibre (25) est vérifiée: la relation (27) n'est pas nécessairement vérifiée.

Il est possible d'étendre la validité de la relation de Kröner-Budiansky-Wu ou d'en modifier l'expression en levant trois des hypothèses de base du modèle: isotropie élastique, homogénéité de la déformation plastique dans la matrice, et forme sphérique des inclusions.

Prise en compte de l'anisotropie élastique. Le problème précédent peut être également résolu en s'affranchissant de l'hypothèse d'isotropie élastique dans l'inclusion. On montre que, en remplaçant les déformations plastiques, supposées homogènes dans chaque cristal, par des forces de masse fictives équivalentes et en utilisant le tenseur de Kelvin-Somigliana pour un milieu infini isotrope, on obtient le champ de contrainte $\sigma_{ij}(c)$ dans le cristal c sous la forme:

$$\sigma_{ij}(c) = A_{ijhk}(c)\{\Sigma_{hk}+2\mu(1-\beta)(E^p_{hk}-\varepsilon^p_{hk}(c))\} \tag{34}$$

qui diffère de (33) par la présence du tenseur de concentration des contraintes élastiques $A_{ijhk}(c)$.

Pour le modèle d'inclusion, le tenseur A est donné par les formules:

$$A_{ijhk}(c) = s^{-1}_{ijpq}(c)B_{pqlm}(c)S_{lmhk}, \tag{35}$$

$$B^{-1}_{pqlm}(c) = I_{pqlm}+\left\{\frac{4-5\nu}{15(1-\nu)}I_{pqhk}-\frac{\delta_{pq}\delta_{hk}}{30(1-\nu)}\right\}[s^{-1}_{hklm}(c)-S^{-1}_{hklm}] \tag{36}$$

où $B_{pqlm}(c)$ est le tenseur concentration des déformations en élasticité [9].

Pour certains métaux, caractérisés par une forte anisotropie élastique au niveau du cristal, l'adoption de (33) au lieu de (34) peut entraîner d'importantes erreurs. Par exemple, pour le cuivre pur, la contrainte résiduelle σ_{ij} rapportée à $2\mu(1-\beta)(E^p_{ij}-\varepsilon^p_{ij})$ vaut 1 pour toutes les composantes, avec l'hypothèse d'isotropie élastique (formule (33) avec $\Sigma_{ij} = 0$). Dans l'hypothèse d'anisotropie, selon (34), nous avons les valeurs 0.4 pour les composantes diagonales ($i = j$) et 2.0 pour les autres composantes ($i \neq j$). Ce rôle important de l'anisotropie élastique est également signalé par W. Hutchinson [13].

Prise en compte de la relaxation viscoplastique des contraintes internes dans l'agrégat. (*Elasticité isotrope*). La formule de Kröner repose sur l'hypothèse que l'accomodation entre la déformation plastique de l'inclusion ε_{ij}^p et la déformation plastique E_{ij}^p supposée homogène de la matrice hors de l'inclusion est purement élastique. Cela entraîne une surestimation des contraintes internes. Si l'on s'affranchit de cette hypothèse simplificatrice de base, on doit résoudre un problème intégral dont la solution permet de formuler une loi d'interaction a posteriori. L'expression d'une loi générale d'interaction en viscoplasticité soulève de grandes difficultés. Afin de voir cependant de quelle manière la non-homogénéité de la déformation de la matrice et la viscosité modifient l'évaluation des contraintes internes donnée par la formule de Kröner-Wu-Budiansky, il est utile d'étudier un cas simple qui peut être entièrement résolu: celui de la viscoélasticité linéaire anisotrope en traction simple.

Envisageons un agrégat de cristaux sphériques C.F.C. incompressibles, à élasticité isotrope, à viscoélasticité linéaire, et où l'écoulement visqueux ne peut se produire que suivant le mécanisme de glissement principal[1]) initial pour la traction simple: cet agrégat est soumis à la contrainte constante Σ suivant l'axe Ox_3.

Ici, c'est le caractère anisotrope de la viscoélasticité qui est source de contraintes internes. La résolution de ce problème de fluage permet d'avoir une idée de l'évolution des contraintes internes en viscoplasticité lorsque les seuils d'écoulement plastique sont négligeables par rapport aux contraintes appliquées.

Les principales étapes du calcul sont les suivantes:

a. décomposons les déformations totales $\varepsilon_{ij}(t)$ et $E_{ij}(t)$ en partie élastique ($\varepsilon_{ij}^e(t)$, $E_{ij}^e(t)$) et en partie visqueuse ($\varepsilon_{ij}^v(t)$ et $E_{ij}^v(t)$). L'inclusion étant sphérique, $\sigma_{ij}(t)$, $\varepsilon_{ij}(t)$, $\varepsilon_{ij}^e(t)$ et $\varepsilon_{ij}^v(t)$ y sont homogènes. Chaque grain est caractérisé par une matrice d'orientation α_{ij} liée à son mécanisme principal. $J(t)$ étant la fonction de fluage pour le glissement principal (supposée identique pour tous les grains), γ^v désignant le glissement unitaire et

[1]) Le mécanisme de glissement principal est celui pour lequel la cission réduite appliquée est maximale.

$\tau = \alpha_{ij}\sigma_{ij}(t)$ la cission réduite principale, nous avons:

$$\gamma^v(t) = \int_0^t J(t-\theta)\mathrm{d}\tau(\theta), \tag{37}$$

$$\varepsilon_{ij}^v(t) = \int_0^t \alpha_{ij}\alpha_{hk}J(t-\theta)\mathrm{d}\sigma_{hk}(\theta). \tag{38}$$

b. considérons le problème élastique parallèle fictif d'une matrice infinie élastique incompressible et isotrope de module de cisaillement μ_1, contenant une inclusion sphérique de comportement quelconque qui a subi la déformation totale $\varepsilon_{ij}(t)$. La matrice étant soumise au même état de charge que précédemment, on a d'après Z. Hashin [14], à tout instant t et si $E_{ij}^{\infty}(t)$ désigne la déformation à l'infini de la matrice pour la contrainte $\Sigma_{ij}(t)$:

$$\sigma_{ij}(t) = \Sigma_{ij}(t) + 3\mu_1[E_{ij}^{\infty}(t) - \varepsilon_{ij}(t)]. \tag{39}$$

c. nous ramenons alors le problème viscoélastique à un problème élastique en appliquant aux équations du problème la transformation de Laplace-Carson. (p étant une variable complexe, $\overset{+}{G}$ est l'image de la fonction $G(t)$ par cette transformation si

$$\overset{+}{G} = p\int_0^{\infty} \exp(-pt)G(t)\mathrm{d}t).$$

On a maintenant:

$$\overset{+}{E}{}_{ij}^v = \langle \overset{+}{\varepsilon}{}_{ij}^v \rangle \equiv \langle \alpha_{ij}\alpha_{kl}\overset{+}{s}_{kl} \rangle \overset{+}{J} \equiv \tfrac{1}{2}\overset{+}{S}_{ij}\overset{+}{F} \tag{40}$$

où $s_{ij}(t)$ et $S_{ij}(t)$ sont les déviateurs de $\sigma_{ij}(t)$ et $\Sigma_{ij}(t)$; $F(t)$ est la fonction de fluage de la matrice pour la déformation purement visqueuse.

Si μ désigne le module de cisaillement élastique, on a:

$$\left.\begin{aligned} \overset{+}{E}{}_{ij}^e &= \langle \overset{+}{\varepsilon}{}_{ij}^e \rangle = \left\langle \frac{1}{2\mu}\overset{+}{s}_{ij} \right\rangle = \frac{1}{2\mu}\overset{+}{S}_{ij} \\ \overset{+}{E}_{ij} &= \langle \overset{+}{\varepsilon}_{ij} \rangle = \frac{1+\mu\overset{+}{F}}{2\mu} S_{ij} \end{aligned}\right\}. \tag{41}$$

Transformons la formule (39) avec $\overset{+}{\mu}_1 = \mu/(1+\mu\overset{+}{F})$; nous obtenons simplement:

$$\overset{+}{\sigma}_{ij} = \overset{+}{\Sigma}_{ij} + \frac{3\mu}{1+\mu\overset{+}{F}}(\overset{+}{E}_{ij} - \overset{+}{\varepsilon}_{ij}). \tag{42}$$

Pour un agrégat isotrope de cristaux C.F.C. nous avons, en posant $m = \alpha_{33}$:

$$\overline{m\alpha_{ij}} \equiv \overline{m^2} A_{ij} \quad \text{où} \quad \overline{m^2} \simeq 0{,}206 \tag{43}$$

peut être calculé exactement, et:

$$(A_{ij}) = \begin{bmatrix} -\frac{1}{2} & 0 & 0 \\ 0 & -\frac{1}{2} & 0 \\ 0 & 0 & 1 \end{bmatrix}. \tag{44}$$

Nous obtenons alors (A. Zaoui [15]):

$$\overset{+}{\sigma}_{ij} = \overset{+}{\Sigma}_{ij} + \frac{6\mu\overset{+}{X}\overset{+}{J}\Sigma(\overline{m^2}A_{ij} - \overline{m\alpha_{ij}})}{5 + 6\mu\overline{m^2}\overset{+}{J}\overset{+}{X}} \tag{45}$$

$\overset{+}{X}$ étant solution de l'équation:

$$6\mu\overline{m^2}\overset{+}{J}\overset{+}{X}{}^2 + (5 - 3\mu(5\overline{m^2} - 1)\overset{+}{J})\overset{+}{X} - 5 = 0. \tag{46}$$

d. pour déterminer les contraintes $\sigma_{ij}(t)$ il faut appliquer la transformation inverse de Laplace-Carson à (45). Dans le cas d'une viscosité newtonienne pure où $\overset{+}{J} = (1/p\eta)$, ($\eta$ coefficient de viscosité) on montre [15] alors que la loi d'interaction a posteriori s'écrit:

$$\sigma_{ij} = \Sigma_{ij} + 2\mu(1 - \beta_{\text{Inc}})(E_{ij}^{v} - \varepsilon_{ij}^{v})\lambda^{ve}(t) \tag{47}$$

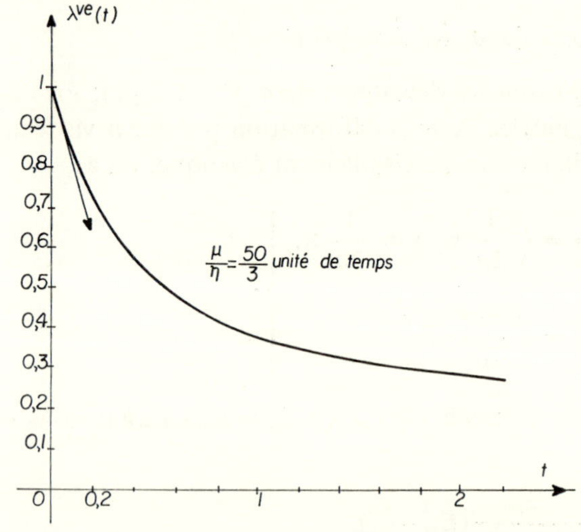

Fig. 2. Facteur d'atténuation pour la formule de Budiansky-Kröner-Wu.

où: $\beta_{\mathrm{Inc}} = \frac{2}{5}$ (pour $\nu = \frac{1}{2}$: milieu incompressible), et $\lambda^{ve}(t)$ peut être calculé numériquement, pour $\mu\eta^{-1}$ fixé, sans difficulté (voir figure 2). On peut voir que la non-homogénéité de la déformation et la viscosité modifient simplement la loi de Kröner-Budiansky par le facteur d'atténuation $\lambda^{ve}(t)$.

Dans le cas de la viscoplasticité, la conclusion doit être qualitativement la même: la relation de Kröner-Budiansky-Wu n'est adaptée que pour les très petites déformations. On peut à partir des considérations précédentes postuler une loi d'interaction viscoplastique a priori et empirique sous la forme:

$$\sigma_{ij} = \Sigma_{ij} + 2\mu(1-\beta)(E^p_{ij} - \varepsilon^p_{ij})\lambda^{vp}(\Sigma_{ij}, E^p_{ij}, P^*) \tag{48}$$

où P^* représente la proportion de grains actifs.

λ^{vp} est choisi de telle façon qu'il soit proche de l'unité pour les faibles valeurs de E^p_{ij} et P^*, et décroisse en restant positif quand ces paramètres augmentent. Une telle formule approchée ne peut représenter qu'un comportement moyen: en toute rigueur elle devrait être exprimée sous la forme d'une relation fonctionnelle.

Prise en compte d'un point de vue statistique de la forme non sphérique des cristaux. (*Elasticité isotrope*). L'application stricte de la relation (33) à chaque cristal de l'agrégat suppose que les cristaux ont tous une forme sphérique. En général cela n'est pas vérifié et les contraintes et déformations ne sont pas homogènes dans le cristal. On peut montrer que la relation de Kröner-Budiansky garde cependant sa validité, mais seulement d'un point de vue statistique.

Supposons $\Sigma_{ij} = 0$, la contrainte résiduelle $\sigma_{ij}(\boldsymbol{x})$ en tout point $\boldsymbol{x}$ est une fonctionnelle linéaire des déformations plastiques $\mathscr{L}_{ij}(\varepsilon^p_{kl}(\boldsymbol{x}))$, si l'on suppose que l'accomodation s'effectue élastiquement.

Considérons un ensemble c de N cristaux $c_1 \ldots, c_N$ de volume total v dispersés dans le volume V. Evaluons la contrainte moyenne dans cet ensemble:

$$\sigma_{ij}(c) = \frac{1}{V}\sum_{k=1}^{N}\int_{c_k}\sigma_{ij}(\boldsymbol{x})\mathrm{d}v_x = \frac{1}{V}\sum_{k=1}^{N}\int_{c_k}\mathscr{L}_{ij}[\varepsilon^p_{kl}(\boldsymbol{x}')]\mathrm{d}v_x. \tag{49}$$

Soit $\boldsymbol{g}_k$ le centre de gravité de c_k; posons $\boldsymbol{x}' = \boldsymbol{y}' + \boldsymbol{g}_k$, et définissons la fonction:

$$\varepsilon^*_{ij}(\boldsymbol{y}') = \frac{1}{N}\sum_{k=1}^{N}\varepsilon^p_{ij}(\boldsymbol{y}' + \boldsymbol{g}_k). \tag{50}$$

Utilisons les coordonnées sphériques (ρ, θ, φ) dans chaque cristal rapportées à l'origine $\boldsymbol{g}_k$ et définissons la fonction scalaire $\phi(\rho, \theta, \varphi)$ égale au nombre de cristaux de c contenant le point de coordonnées (ρ, θ, φ) rapporté au nombre total N de cristaux de c. Si ρ est supérieur à la taille maximale R des cristaux de c, ϕ est nul; si ρ est inférieur à la taille minimale r, ϕ est égal à 1. Il est. raisonnable de supposer que, lorsque N est grand, ϕ ne dépend que de ρ: cela signifie qu'il n'y a aucune forme privilégiée. Le volume v est alors égal

$$v = 4\pi N \int_{\rho<R} \phi(\rho)\rho^2 \mathrm{d}\rho. \tag{51}$$

De même, si l'on suppose la distribution plastique $\varepsilon^p_{ij}(\boldsymbol{x}')$ complètement aléatoire et les cristaux de c en très grand nombre et répartis au hasard dans tout le solide, on peut admettre que $\varepsilon^*_{ij}(\boldsymbol{y}')$ ne dépend en fait que de $\rho = |\boldsymbol{y}'|$. (Hypothèse de 'macro-isotropie').

Plus précisément désignons par $\varepsilon^p_{ij}(c)$ la déformation plastique moyenne dans l'ensemble c et par E^p_{ij} la déformation plastique moyenne dans l'ensemble de tous les cristaux de V. On peut montrer alors que, lorsque N est très grand, $\varepsilon^*_{ij}(y')$ tend vers l'expression:

$$\varepsilon^*_{ij}(\rho) = \phi(\rho)\varepsilon^p_{ij}(c) + [1-\phi(\rho)]E^p_{ij}. \tag{52}$$

La conclusion est la même lorsque l'expression (50) est étendue à un sous-ensemble de N' cristaux de c, pourvu que N' soit grand. A partir de ces remarques, il est possible d'écrire l'équation (49) sous la forme:

$$\sigma_{ij}(c) = \frac{N}{V}\int_{\rho<R} 4\pi\rho^2 \mathscr{L}_{ij}[\varepsilon^*_{kl}(\rho')]\mathrm{d}\rho. \tag{53}$$

En effet on peut montrer que la différence entre les formules (49) et (53) est nulle pour N grand à cause de la propriété suivante: la contrainte résiduelle moyenne dans une couronne sphérique $\rho_1 < \rho < \rho_2$ due à une distribution plastique $\varepsilon^*_{ij}(\rho)$ de révolution, et constante pour $\rho > \rho_1$ est nulle. Les propriétés de la fonctionnelle $\mathscr{L}_{ij}$ permettent de soustraire de $\varepsilon^*_{kl}(\rho')$ la distribution constante E^p_{kl} sans changer le résultat de (53) et d'exprimer $\sigma_{ij}(c)$ sous la forme (en tenant compte de (52)):

$$\begin{aligned}\sigma_{ij}(c) &= \frac{N}{V}\int_{\rho<R} 4\pi\rho^2 \mathscr{L}_{ij}[\varepsilon^*_{kl}(\rho') - E^p_{kl}]\mathrm{d}\rho = \\ &= \frac{N}{V}\int_{\rho<R} 4\pi\rho^2 \mathscr{L}_{ij}\{\phi(\rho')[\varepsilon^p_{kl}(c) - E^p_{kl}]\}\mathrm{d}\rho = \\ &= \frac{N}{V}\, 2\mu(1-\beta)(E^p_{ij} - \varepsilon^p_{ij}(c))\int_{\rho<R} 4\pi\rho^2\phi(\rho)\mathrm{d}\rho. \end{aligned} \tag{54}$$

Compte-tenu de la relation (51), cette expression s'écrit donc simplement:

$$\sigma_{ij}(c) = 2\mu(1-\beta)[E^p_{ij} - \varepsilon^p_{ij}(c)] \tag{55}$$

ce qui n'est autre que la formule de Kröner-Budiansky pour $\Sigma_{ij} = 0$. Mais cette fois $\sigma_{ij}(c)$ et $\varepsilon^p_{ij}(c)$ désignent les valeurs moyennes des quantités correspondantes sur un sous-ensemble c de grains non sphériques, et de déformation plastique non homogène.

Etude du polycristal cubique à faces centrées

J. R. Rice [3] a récemment démontré que le potentiel viscoplastique macroscopique Ω du polycristal est la valeur moyenne des potentiels viscoplastiques f des cristaux.

Nous avons en effet d'après les formules (31) et (12):

$$\dot{E}^p_{ij} = \langle \dot{\varepsilon}^p_{hk}(\boldsymbol{x}) A_{hkij}(\boldsymbol{x}) \rangle = \left\langle \left(\frac{\partial f(\boldsymbol{x})}{\partial \sigma_{hk}}\right)_t \left(\frac{\partial \sigma_{hk}(\boldsymbol{x})}{\partial \Sigma_{ij}}\right)_t \right\rangle = \tag{56}$$

$$= \left(\frac{\partial}{\partial \Sigma_{ij}} \langle f(\boldsymbol{x}) \rangle\right)_t = \left(\frac{\partial \Omega}{\partial \Sigma_{ij}}\right)_t$$

avec:

$$\Omega = \langle f(\boldsymbol{x}) \rangle. \tag{57}$$

Le calcul de Ω est, en général, en raison des problèmes évoqués plus haut très difficile. (cf. Brown [16]). Nous allons étudier deux cas particuliers:

a. les vitesses de chargement sont négligeables (élastoplasticité)
b. un seul mécanisme est actif dans tous les cristaux lors d'une traction simple (viscoplasticité).

Expression des relations de comportement d'un agrégat de cristaux C.F.C. en élastoplasticité.

Nous avons obtenu en (16) dans le cas où les vitesses de chargement sont négligeables, les relations de comportement du cristal sous la forme:

$$\alpha^K_{ij} \dot{\sigma}_{ij} = \sum_R h^{KR} \dot{\gamma}_R \qquad \text{(pour un mécanisme } K \text{ actif).} \tag{58}$$

Nous avons vu que la surface limite élastique, formée d'un ensemble de plans, était convexe. Lorsque les glissements γ_R sont petits, le domaine peut être défini par l'intersection des zones:

$$\alpha_{ij}^K \sigma_{ij} - h^K(\gamma_1 \ldots \gamma_{24}) \leqq 0 \qquad \text{(J. Mandel [4])} \tag{59}$$

h^{KR} (formule 46) est défini par $\partial h^K / \partial \gamma_R$.

Nous cherchons ici à écrire, pour l'agrégat, des relations macroscopiques analogues à (58) et (59).

Nous présentons pour cela (H. D. Bui [7]) un modèle élastoplastique fondé sur la loi de Kröner considérée d'un point de vue statistique (formule (55)); c'est à dire qu'on peut envisager des sous-ensembles de grains de forme non nécessairement sphérique et d'états internes différents.

Considérons un agrégat de cristaux C.F.C. formé de sous-ensembles $c(\delta)$, $\delta = 1, n$, chaque sous-ensemble $c(\delta)$ étant formé de grains répartis au hasard dans le polycristal, mais de même orientation, de volume total v supposé indépendant de δ ($v = V/n$), et possédant 24 mécanismes possibles de déformation plastique, caractérisés par les matrices α_{hk}^R, où R varie de δ_1 à δ_{24}. Afin de simplifier les raisonnements, mais sans restreindre en rien la généralité du modèle, nous supposons que, à un mécanisme (R), correspond une orientation (δ) et une seule. Tous les mécanismes sont ainsi classés en sous-ensembles disjoints (1, ..., 24), (25, ..., 48), ... Si R est S appartiennent à un même sous-ensemble ils seront dits compatibles et on notera $(R) = (S)$. Dans le cas contraire, ils seront dits incompatibles et on notera $(R) \neq (S)$.

Les seuils initiaux peuvent varier de grain à grain, y compris dans l'ensemble $c(\delta)$. Nous pouvons faire les remarques préliminaires suivantes.

a. Le mécanisme R n'est actif que dans le sous-ensemble C_R de volume v_R de la famille $c(\delta)$ possédant ce mécanisme. La concentration de cristaux actifs selon le mécanisme R est donc:

$$p_R = \frac{v_R}{V} \leqq \frac{1}{n}. \tag{60}$$

b. Les cristaux simultanément déformés suivant deux mécanismes R et S appartiennent à l'intersection C_{RS} des deux ensembles C_R et C_S. La concentration de ces cristaux, relativement à C_R est donc:

$$p_{RS} = \frac{v_{RS}}{v_R} \leqq 1. \tag{61}$$

Naturellement, si $(R) \neq (S)$, l'intersection est vide et nous avons $P_{RS} = 0$

et si $R = S$, nous avons exactement $P_{RS} = 1$. Remarquons que le tableau $M_{RS} = P_{RS}/P_S$ est symétrique en R et S et que les termes diagonaux sont toujours plus importants que les autres éléments.

Nous noterons par $\langle Q \rangle^R$ la valeur moyenne limitée au volume C_R d'une fonction quelconque $Q(\boldsymbol{x})$ définie dans tout le volume:

$$\langle Q \rangle^R = \frac{1}{v_R} \int_{C_R} Q(\boldsymbol{x}) \mathrm{d}v_x.$$

Nous supposerons que chaque ensemble non vide contient suffisamment de grains pour que la moyenne d'une fonction dans cet ensemble coincide avec la moyenne de la même fonction dans un sous-ensemble quelconque. Moyennant cette hypothèse nous pouvons écrire par exemple:

$$\langle \dot{\gamma}_S \rangle^R = p_{RS} \langle \dot{\gamma}_S \rangle^S \qquad \text{(pas de sommation sur } S\text{)}$$

La vitesse de déformation plastique, dans l'hypothèse d'isotropie élastique est donnée par:

$$\dot{E}^p_{ij} = \frac{1}{V} \int_V \dot{\varepsilon}^p_{ij} \mathrm{d}v = \sum_S \alpha^S_{ij} \dot{\Gamma}_S \tag{62}$$

en posant

$$\dot{\Gamma}_S = p_S \langle \dot{\gamma}_S \rangle^S.$$

Nous pouvons obtenir des relations analogues à (58) en faisant la moyenne relativement à C_R des deux membres de (55) après multiplication par α^R_{ij} et en utilisant les relations suivantes, qui sont faciles à établir:

$$\langle \alpha^R_{ij} \dot{\varepsilon}^p_{ij} \rangle^R = \alpha^R_{ij} \sum_S \alpha^S_{ij} M_{RS} \dot{\Gamma}_S$$

$$\langle \sum_S h^{RS} \dot{\gamma}_S \rangle^R = \sum_S h^{RS} M_{RS} \dot{\Gamma}_S \qquad (h^{RS} \text{ constant})$$

Pour les mécanismes actifs, nous obtenons alors la loi d'écoulement:

$$\alpha^R_{ij} \dot{\Sigma}_{ij} - \sum_S H^{RS} \dot{\Gamma}_S = 0 \tag{63}$$

avec:

$$H^{RS} = M_{RS} h^{RS} + 2\mu(1-\beta)(M_{RS}-1)\alpha^R_{ij}\alpha^S_{ij}. \tag{64}$$

Dans l'expression (64) il y a un grand nombre de termes nuls correspondant aux indices R et S incompatibles, les éléments diagonaux sont plus importants ($P^{RR} = 1$ et $\alpha_{ij}^{R}\alpha_{ij}^{S} \leqq \alpha_{ij}^{R}\alpha_{ij}^{R} = \frac{1}{2}$). Cette propriété de la matrice H donne une certaine justification à la théorie des mécanismes indépendants de W. T. Koïter [17].

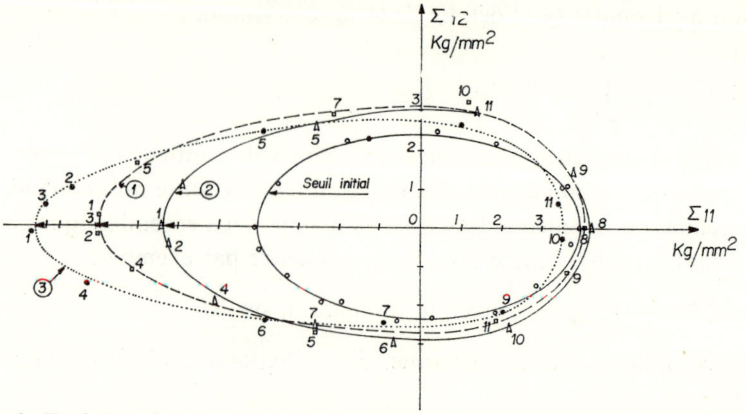

Fig. 3. Evolution de la frontière élastique avec l'écrouissage: étude expérimentale. Aluminium 99.5.

L'étude de l'évolution du domaine élastique, peut être faite sur la base des relations (64), mais il est nécessaire de se donner a priori des lois de variations p_S et p_{RS}, suggérées par les expériences. Les essais effectués par H. D. Bui [7] ont montré le caractère anisotrope de l'écrouissage avec des perturbations rappelant le point conique [18], l'effet Bauschinger [19] et l'effet d'expansion de la surface opposée au point de charge [7]. (Fig. 3) L'effet Bauschinger est interprété de façon satisfaisante par tous les modèles, y compris le modèle

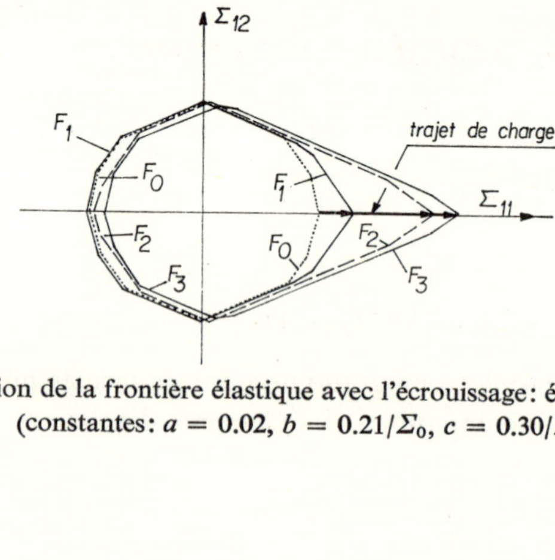

Fig. 4. Evolution de la frontière élastique avec l'écrouissage: étude théorique (constantes: $a = 0.02$, $b = 0.21/\Sigma_0$, $c = 0.30/\Sigma e$).

de Kröner-Budiansky-Wu que nous retrouvons ici, en prenant $P_S = 1$ et $P_{RS} = 1$, $(R) = (S)$. Nous retrouvons l'effet d'expansion en prenant pour P_S et P_{RS} des fonctions croissantes des paramètres d'écrouissage, respectant les valeurs limites du modèle de Kröner-Budiansky.

Ainsi, la figure 4 montre les surfaces limites élastiques successives obtenues lors d'un écrouissage en traction simple pour un modèle comprenant trois orientations différentes (une famille de cristaux est suivant Ox_1, x_2, x_3, les deux autres se déduisent par rotation de + ou $-\pi/3$ autour de l'axe Ox_3) avec $h^{RS} = 0$ et les lois adoptées pour P_S et P_{RS} jusqu'à leurs valeurs limites respectives, sont de la forme:

$$p_S = a + b(\Sigma - \Sigma_0), \qquad p_{RS} = \begin{cases} 0 & \text{si} \quad (R) \neq (S) \\ a + b(\Sigma - \Sigma_0) + c(\Sigma - \Sigma_0)^2 & \text{si} \quad (R) = (S) \text{ et } R \neq S \\ 1 & \text{si} \quad R = S \end{cases}$$

où Σ_0 est la limite élastique initiale et *a*, *b*, et *c* sont des constantes.

Etude de comportement viscoplastique d'un agrégat de cristaux C.F.C.: cas de la traction simple, de l'élasticité isotrope et du glissement unique

Définissons un modèle statistique polycristallin par la donnée de :
une loi de distribution statistique des orientations des cristaux; une loi d'interaction a priori entre les cristaux; une loi de comportement visco-plastique du monocristal.

Limitons-nous aux conditions et hypothèses suivantes:
– glissement unique dans le monocristal suivant le système principal initial; on a vu (22) que, pour les petites déformations, il est légitime d'adopter une loi d'écoulement de type Bingham-Képès:

$$\eta\dot{\gamma} = \tau - \tau_0 - h\gamma \quad \text{pour} \quad \tau > \tau_0 + h\gamma \tag{65}$$

– distribution isotrope des orientations des cristaux, indépendant de l'état de déformation;
– cas de la traction simple ($\Sigma_{33} = \Sigma$ seul différent de 0). Avec ces hypothèses, on peut déjà montrer que:

$$E^p_{ij} = \langle \varepsilon^p_{ij} \rangle = \langle \alpha_{ij}\gamma \rangle = E^p A_{ij} = \langle m\gamma \rangle A_{ij} \tag{66}$$

où m et A_{ij} sont définis dans la formule (43); on a des relations analogues pour $\dot{E}^p_{ij}$, $\dot{\varepsilon}^p_{ij}$, $\dot{E}^p$ et $\dot{\gamma}$;
– loi d'interaction de Kröner-Budiansky dans le cas de l'élasticité isotrope.

On a alors, suivant la relation (33) et compte tenu de (66):

$$\tau = \alpha_{ij}\sigma_{ij} = m[\Sigma + 3\mu(1-\beta)E^p] - \mu(1-\beta)\gamma \tag{67}$$

– caractéristiques τ_0, h, η identiques pour tous les cristaux.

A partir de là, il est nécessaire de calculer la densité de probabilité $P(m)$, représentant la fraction de cristaux dont le coefficient d'orientation m est compris entre m et $(m + \mathrm{d}m)$. Le calcul explicite (A. Zaoui [15]) fait intervenir des intégrales elliptiques. m est compris entre $2/(3\sqrt{6})$ et $\frac{1}{2}$ (figure 5).

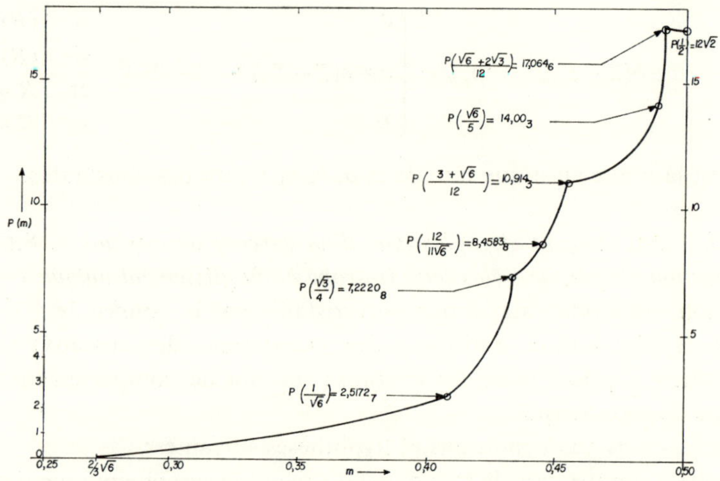

Fig. 5. Variation de la densité de probabilité $P(m)$.

Il est alors possible de déterminer le comportement du polycristal, par exemple à partir du calcul du potentiel viscoplastique. En portant (66) dans (24) et en utilisant (57), on trouve:

$$\Omega = \int_{m^*}^{\frac{1}{2}} \frac{\{m[\Sigma + 3\mu(1-\beta)E^p] - \tau_0 - [h + \mu(1-\beta)]\gamma\}^2}{2\eta} P(m)\mathrm{d}m \tag{68}$$

où m^* est la valeur de m qui sépare, à l'instant t, l'ensemble des grains actifs de l'ensemble des grains inactifs. On calcule m^* à partir de (67) en exprimant que:

$$\tau(m^*) = \tau_0 + h\gamma(m^*). \tag{69}$$

La vitesse de déformation viscoplastique macroscopique est alors donnée par:

$$\dot{E}^p = \left[\frac{\partial \Omega}{\partial \Sigma}\right]_t =$$

$$= \int_{m^*}^{\frac{1}{2}} \frac{m^2[\Sigma + 3\mu(1-\beta)E^p] - m\{\tau_0 + [h + \mu(1-\beta)]\gamma\}}{\eta} P(m)\mathrm{d}m \qquad (70)$$

Ces relations permettent de préciser certaines propriétés mécaniques viscoplastiques propres au polycristal. Un exemple significatif est celui de 'l'effet d'entraînement viscoplastique': il s'agit d'un phénomène de plastification progressive à contrainte appliquée constante, de cristaux initialement inactifs, sous l'effet de l'écoulement viscoplastique des cristaux actifs et des contraintes internes liées à cette différence de comportement.

Une telle propriété est en fait contenue dans les formules classiques d'interaction, telles que celles de Kröner, Taylor, etc. Par exemple pour un essai de fluage ($\dot{\Sigma}_{ij} = 0$) et si, pour une raison quelconque, il existe une certaine proportion de cristaux inactifs ou bloqués tandis que d'autres sont en écoulement viscoplastique, ces formules d'interaction indiquent que dans les cristaux inactifs ($\dot{\varepsilon}^p_{ij} = 0$), $\dot{\sigma}_{ij}$ a le même signe que $\dot{E}^p_{ij}$. Ainsi, pour $\dot{E}^p_{ij} > 0$, la contrainte augmente progressivement dans les cristaux inactifs et peut, si elle atteint un niveau suffisant, y déclencher le glissement plastique.

On peut, en reprenant le modèle précédent, dégager plus précisément cette propriété en étudiant la réponse à une mise en charge instantanée en traction simple ou à une décharge partielle en cours de fluage, et la confronter avec l'expérience; pour la simplicité des calculs, on adoptera la loi d'interaction de Kröner-Budiansky, mais on précisera les modifications introduites par l'adoption d'une loi du type de (48).

Mise en charge instantanée en traction simple ($\Sigma_{33} = \Sigma_1$ constante). Différents cas sont à envisager suivant la valeur de Σ_1

- Si $\Sigma_1/\tau_0 < 2$ = l'écoulement plastique ne s'amorce dans aucun des cristaux.
- Si $\Sigma_1/\tau_0 < (3\sqrt{6})/2$ = l'écoulement plastique se fait d'emblée dans tous les cristaux.

On doit prendre $m^* = 2/(3\sqrt{6})$ et (70) fournit la loi de fluage macroscopique:

$$\eta\dot{E}^p + [h + \mu(1-\beta)(1 - 3\overline{m^2})]E^p = \overline{m^2}\Sigma_1 - \overline{m}\tau_0 \qquad (71)$$

où $\overline{m^2}$ est donné par (43) et $\overline{m} \simeq 0{,}452$ peut être calculé exactement.

On obtient donc un fluage limité et le coefficient d'écrouissage statique macroscopique est de l'ordre de μ. Ces deux conclusions, contredites par l'expérience, sont dues aux insuffisances du modèle (glissement unique, loi de Kröner), mais peuvent être corrigées, même dans l'hypothèse du glissement unique, en utilisant une loi d'interaction telle que (48) – sous réserve que le fluage soit non accéléré.

– Si $2 < \Sigma_1/\tau_0 < (3\sqrt{6})/2$ = l'écoulement plastique se produit aussitôt dans les seuls cristaux d'orientation telle que $m > (\tau_0/\Sigma_1)$. Mais progressivement, sous l'effet de l'écoulement dans les cristaux actifs, certains cristaux, initialement inactifs, se déformeront plastiquement: la valeur frontière m^* décroît avec le temps. On a, d'après (67) et (69), et en exprimant que $\gamma(m^*) = 0$:

$$m^* = \frac{\tau_0}{\Sigma_1 + 3\mu(1-\beta)E^p}. \tag{72}$$

L'équation de fluage macroscopique s'obtient, d'après (70) et (72):

$$\eta \dot{E}^p + [h + \mu(1-\beta)]E^p = \tau_0[M^*|m^2\rangle - |m\rangle] \tag{73}$$

où:

$$M^* = \frac{1}{m^*}, \quad |m\rangle = \int_{m^*}^{\frac{1}{2}} mP(m)\mathrm{d}m, \quad |m^2\rangle = \int_{m^*}^{\frac{1}{2}} m^2 P(m)\mathrm{d}m. \tag{74}$$

La variation de M^* est donnée par (lorsque h est négligé devant μ):

$$\eta \dot{M}^* = \mu(1-\beta)\left[\frac{\Sigma_1}{\tau_0} - M^* + 3|m^2\rangle M^* - 3|m\rangle\right]. \tag{75}$$

Ainsi, l'entraînement sera total si $(\Sigma_1/\tau_0) > M_l^*$, avec:

$$M_l^* = \frac{3\sqrt{6}}{2} + 3\overline{m} - \frac{9\sqrt{6}\overline{m^2}}{2} \simeq 2{,}757. \tag{76}$$

Il ne sera que partiel pour des valeurs inférieures de Σ_1/τ_0.

i) Dans le premier cas, on peut définir et calculer un 'temps d'entraînement' T_e, au bout duquel l'écoulement est généralisé (en posant $M_0^* = \Sigma_1/\tau_0$):

$$T_e = \frac{\eta}{\mu(1-\beta)} \int_{M_0^*}^{(3\sqrt{6})/2} \frac{\mathrm{d}M^*}{M_0^* - M^* + 3|m^2\rangle M^* - 3|m\rangle}. \tag{77}$$

Au-delà de T_e, la déformation se poursuit suivant la loi (71).

ii) Dans le second cas, la déformation plastique limite s'écrit:

$$E^p(\infty) = \frac{\tau_0(M^*_\infty - M^*_0)}{3\mu(1-\beta)} \tag{78}$$

où M^*_∞ se calcule aisément d'après (75). Cela correspond à une portion convexe sur la courbe d'écrouissage statique, qui se raccorde à la partie linéaire pour $\Sigma_1 = M^*_l\tau_0$. Ces résultats sont à rapprocher des observations expérimentales de B. Jaoul [20] et de A. Zaoui [15].

Dans le cas d'une 'loi de Kröner modifiée' telle que (48), on obtient des résultats analogues; mais on a de plus la possibilité d'un fluage limité pour $\Sigma_1 < M^*_l\tau_0$, et illimité au delà, propriété souvent observée expérimentalement (A. Zaoui [15]).

Décharge partielle en cours de fluage en traction simple. On impose Σ_1 jusqu'à l'instant θ, puis la décharge partielle $\Delta\Sigma = \Sigma_1 - \Sigma_2$, où $\Sigma_2 < (3\sqrt{6}/2)\tau_0$. Les phénomènes sont alors très semblables à ceux dégagés dans le cas précédent: pour toute une gamme d'amplitudes $\Delta\Sigma$, le glissement se bloque dans une partie des cristaux jusque là actifs; mais il peut reprendre, partiellement

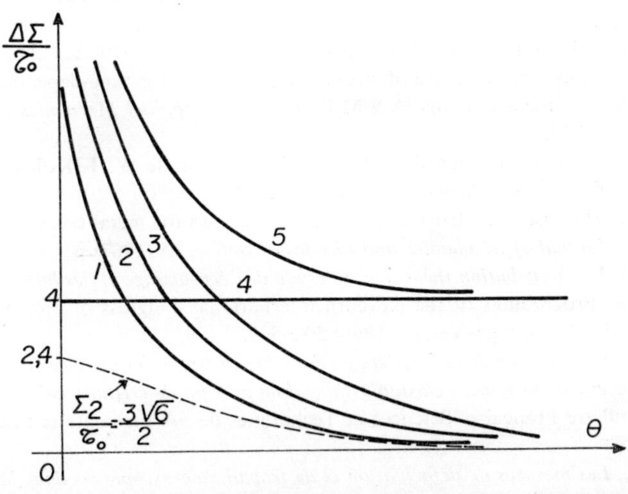

Fig. 6. Réponse à une décharge partielle en cours de fluage (Σ_2/τ_0 fixé). 1. non-hésitation (pas de blocage initial). 2. blocage initial partiel, puis reprise totale (hésitation). 3. blocage initial partiel, puis reprise partielle. 4. blocage initial total (pas de reprise). 5. renversement du glissement.

ou totalement sous l'effet de l'écoulement des grains restés actifs ('hésitation au fluage').

La valeur frontière m^* est cette fois donnée, d'après (67) et (69), par:

$$m^* = \frac{1}{M^*} = \frac{\tau_0 + [h+\mu(1-\beta)]\gamma^*(\theta)}{\Sigma_2 + 3\mu(1-\beta)E^p} \tag{79}$$

où $\gamma^*(\theta)$ pour $t = \theta$.

C'est cette fois $M^*(\theta)$ qui joue le rôle de paramètre principal, au lieu de M_0^*; les différents cas possibles s'obtiennent en comparant $M^*(\theta)$ aux valeurs 2, M_l^* et $(3\sqrt{6})/2$.

L'ensemble des éventualités est représenté sur la figure 6 dans le plan $(\Delta\Sigma/\tau_0, \theta)$ pour Σ_2 fixé.

Les conclusions globales sont analogues dans le cas d'une loi d'interaction telle que (48), mais permettent un meilleur accord avec l'expérience (A. Zaoui [15]).

Références

[1] Saada, V. G., *Etude théorique de l'écrouissage à partir d'un modèle de dislocations.* Publication de l'I.R.S.I.D. – Série A, no. 25–1. Saint-Germain-en-Laye, (1961).

[2] Zarka, J., *Sur la viscoplasticité des Métaux.* Mémorial de l'Artillerie Française. Sciences et Techniques de l'Armement, 2ème fascicule, (1970) 223–292.

[3] Rice, J. R., On the structure of stress-strain relations for time-dependent plastic deformations in metals. Trans. A.S.M.E. *Journal of Applied Mechanics 37*, september (1970) 728–737.

[4] Mandel, J., Généralisation de la théorie de plasticité de W. T. Koiter. *International Journal of Solids and Structures, 1* (1965) 273–296.

[5] Hill, R., The essential structure of constitutive laws for metal composites and polycrystals. *Journal of Mechanics and Physics of Solids, 15* (1967) p. 79.

[6] Mandel, J., *Contribution théorique à l'étude de l'écrouissage et des lois de l'écoulement plastique.* Proceedings of the eleventh international congress of applied mechanics. Munich 1964, Springer-Verlag (1966) 502–509.

[7] Bui, H. D., *Evolution de la frontière du domaine élastique des métaux avec l'écrouissage plastique et comportement élastoplastique d'un agrégat de cristaux cubiques.* Mémorial de l'Artillerie Française. Sciences et Techniques de l'Armement, ler fascicule, (1970) 141–165.

[8] Hill, R., *Les mesures de déformation et de travail macroscopiques dans la déformation d'un milieu micro-non-homogène.* (En Russe) – Mathématique et Mécanique Appliquée, Tome 35, (1971) 31–39.

[9] Hill, R., Elastic properties of reinforced solids: some theoretical principles. *Journal of Mechanics and Physics of Solids 11* (1963) 357.

[10] Eshelby, J. D., The determination of the elastic field of an ellipsoidal inclusion and related problem. *Proceedings of the Royal Society, Série A, 241* (1957) 376.

[11] Kröner, E., Zur plastischen Verformung des Viel-Kristalls. *Acta Metallurgica, 9* (1961) 155.

[12] Budiansky B. et Tai Te Wu, Theoretical prediction of plastic strains of polycrystals. Proceedings of the Fourth U.S. Nat. Congress of Applied Mechanics, (1962) 1175.

[13] Hutchinson, J. W., Elastic-plastic behavior of polycrystalline metals and composites. *Proceedings of the Royal Society*, A, *319* (1970) 247–271.

[14] Hashin, Z., The Inelastic inclusion problem. *International Journal of Engineering Science*, *7* (1969) 11.

[15] Zaoui, A., *Etude de l'influence propre de la désorientation des grains sur le comportement viscoplastique de métaux polycristallins* (*système C.F.C.*). Mémorial de l'Artillerie Française. Sciences et Techniques de l'Armement, (1972) (à paraître).

[16] Brown, G. M., A self-consistent polycrystalline model for creep under combined stress states. *Journal of Mechanics and Physics of Solids 18* (1970) 367–381.

[17] Koiter, W. T., *General theorems for elastic-plastic solids. Progress in Solid Mechanics, 1* (1960), North-Holland, Pub. Comp. Amsterdam.

[18] Bertsch, P. K. et W. Findley, An experimental study of subsequent yield surface. Proceedings of the Fourth U.S. *Nat. Congress of Applied Mechanics*, ASME, New York, (1962) 893.

[19] Naghdi, P. M., E. Essenburg et W. Koff, An experimental study of initial and subsequent yield surfaces in plasticity. Transactions of the A.S.M.E. *Journal of Applied Mechanics 80* (1958) 201–209.

[20] Jaoul, B., Etude de la forme des courbes de déformation plastique. *Journal of Mechanics and Physics of Solids 5* (1957) 85.

Microstress fields of slip bands and inhomogeneity of plastic deformation of metals

T. H. Lin

University of California, Los Angeles, U.S.A.

A slip band with uniform plastic strain in the most favorably oriented crystal in an aggregate is considered. The microstress field caused by this plastic strain tends to increase the width but keeps the original thickness of the band. The resolved shear stress field caused by equal slip in two closely located parallel slip bands shows that the relief of resolved shear stress due to this equal slip is more in the region between the two bands than in the bands themselves. Hence such bands may continue sliding while the region between them may remain unslid. A more general case is considered, in which the slip bands may not be closely located. Formation of slip bands in a plastically deformed crystal in an elastic matrix is calculated. This calculation shows concentration of plastic strain in few slip bands and seems to explain large inhomogeneity of plastic deformation in metals.

Introduction

Rosenhain [1] showed in 1905 that dark bands on the surface of a plastically deformed metal crystal appear to be steps formed through shearing of slabs of atoms over each other along well defined planes just like a deck of cards. The thickness of each slab measured by the spacing of the steps is of the order of a few microns for most metals. There seems to be no relative movement of the atomic planes within a slab. Heidenreich and Shockley [2] found that under a light microscope the slip bands resolved into clusters of finer lines. This was taken to indicate that the steps found before were terraces of smaller steps formed by slip over several parallel slip planes spaced some 100 atoms apart. These slip planes have been considered as the boundaries of elementary slip lamellae. Each lamella has been found to have slid over its neighbors a distance of about 2000 angstroms (10^{-4} microns). Using transparent crystals, Nye [3] has found coarse slip bands which run right through the crystal. It is clear that concentrated slip occurs throughout the crystal in slip bands spaced few microns apart. Slip bands have also been observed on the surface of polycrystals. These bands may also occur in the interior of the polycrystal which may lead to a large inhomogeneity of plastic deformation in metals.

One way to explain this inhomogeneity is to assume that some weak slip plane exists in a crystal and slip occurs in this plane under an applied stress. The fact that slip does not continue on this plane until fracture occurs but shifts to other planes, has been commonly explained by work hardening, i.e. the lattice around the active slip plane is supposed to be severely distorted, so that a higher stress is required to cause slip to continue on the original plane. Brown [4] objected to this assumption and has pointed out that slip occurs by visible amounts of some 1000 atoms or so on each plane and it does not seem reasonable that so much movement should be possible before the current slip plane becomes harder than some other planes which are undamaged. It is hoped that the present micromechanics analysis will show some cause of this slip in slip bands in metals.

Slip and resolved shear stresses

As pointed out by Dorn and Mote [5], translation glide i.e. slip, qualifies as the principal process among the different mechanisms of plastic deformation, in face-centered-cubic metals at low and intermediate temperatures. It is well known [6] that slip is caused by the movement of dislocations and the force causing the dislocation to slide is directly proportional to the resolved shear stress i.e., the shear stress along the slip direction on the slip plane. This agrees well with single crystal tests which show that slip depends only on the resolved shear stress and is independent of the normal stress on the slip plane [7].

Since all metals have imperfections there is always an initial stress field τ_{ij_I}. After the metal is stressed beyond the elastic limit, plastic strain occurs in the more favorably oriented crystals. If the metal is unloaded [8, 9], plastic strain remains and causes a residual stress field τ_{ij_R} and if the load is applied again an applied stress field τ_{ij_A} is generated. During this loading and unloading which may be imaginary, no additional plastic strain occurs. The resultant stress field in the metal is then the sum of these stress fields

$$\tau_{ij} = \tau_{ij_I} + \tau_{ij_R} + \tau_{ij_A}. \tag{1}$$

Assuming that active slip occurs only in one slip system and denoting the resolved shear stress in this slip system by $\tau_{\alpha\beta}$, we have

$$\tau_{\alpha\beta} = \tau_{\alpha\beta_I} + \tau_{\alpha\beta_R} + \tau_{\alpha\beta_A}. \qquad (2)$$

The resolved shear stress to initiate slip or to cause slip to continue is called the critical shear stress τ_c. Active sliding occurs when $\tau_{\alpha\beta} = \tau_c$ in some region in the metal.

Stress field caused by a uniform slip in a thin slice

Let us consider an aluminum polycrystal. Face-centered-cubic crystals like aluminum have twelve slip systems. To simplify the calculation, the crystal is assumed to have only one main sliding plane, like hexagonal crystals. The elastic anisotropy of an aluminum crystal is small [10] and is neglected in the present study. Consequently the aluminum polycrystal is considered to be an elastically homogeneous and isotropic body. Under uniform loading, this polycrystal is exposed to a uniformly applied stress field and hence to a uniform, resolved shear stress field $\tau_{\alpha\beta_A}$ in a particular crystal. $\tau_{\alpha\beta_I}$ is not uniform but does not vary with loading. $\tau_{\alpha\beta_R}$ however is non-uniform, varies with loading, and plays an important role in the distribution of plastic strain.

Consider a thin band of uniform cross-section shown in Fig. 1 experiencing uniform plastic strain e^p_{xy}. The length of the band is much larger than its thickness ($2w$) and the thickness $2w$ is much smaller than the width $2d$. The strain distribution in the major central portion of the band away from the two ends is taken to be plane.

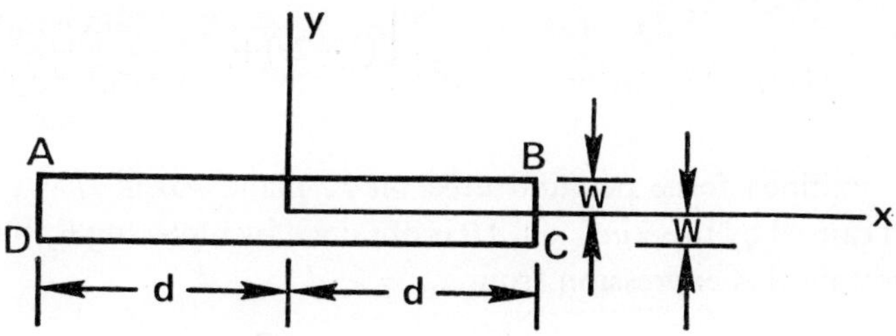

Fig. 1. Thin slice cross-section.

Imagine that this thin slice *ABCD* is cut out of the polycrystal, then the polycrystal has a long hole of uniform rectangular cross-section. This slice, after cut, is stress free but exposed to uniform plastic strain e^p_{xy}. Imagine that a uniform shear stress $-2\mu e^p_{xy}$ is applied to the slice to restore it back to the original configuration before the occurrence of the plastic strain and this slice is then inserted back to the hole and welded to the polycrystal. Since actually there is no such force $-2\mu e^p_{xy}$ applied to the slice, this force is relaxed by applying an equal and opposite force $2\mu e^p_{xy}$ on the boundary of the slice as shown in Fig. 2. This force is here referred to as the fictitious surface force caused by plastic strain. The stress field caused by this fictitious force is denoted by τ_{ij_F}. The residual stress field

$$\begin{aligned} \tau_{ij_R} &= \tau_{ij_F} - 2\mu e^p_{xy} \quad \text{within the slice,} \\ \tau_{ij_R} &= \tau_{ij_F} \quad \text{outside the slice.} \end{aligned} \tag{3}$$

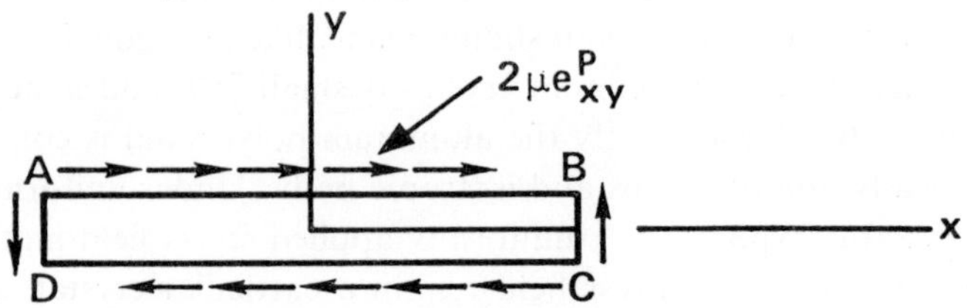

Fig. 2. Fictitious surface force caused by plastic strain.

This polycrystal is assumed to be of fine grain. The size of this sliding crystal is very small as compared to the polycrystal. Hence to calculate the stress field, this fictitious force may be considered to be applied in an infinite medium. The shear stress field τ_{xy} caused by a uniformly distributed force F_x along x-direction per unit length along z-axis at (x_0, y_0) is given [9] as

$$\tau_{xy} = \frac{-F_x(y-y_0)}{4\pi(1-\mu)[x-x_0)^2+(y-y_0)^2]}\left[(1-2\nu)+\frac{2(x-x_0)^2}{(x-x_0)^2+(y-y_0)^2}\right]. \tag{4}$$

Since the fictitious force per unit area on xz-plane AB is $2\mu e^p_{xy}$, the shear stress field caused by this $2\mu e^p_{xy}$ on AB is obtained by replacing F_x by $2\mu e^p_{xy} dx_0$ and integrating this expression from $x_0 = -d$ to $+d$.

$$\begin{aligned} \tau_{xy_{AB}} = {} & \frac{2\mu e^p_{xy}}{2\pi}\left[\tan^{-1}\frac{x-d}{y-w} - \tan^{-1}\frac{x-d}{y-w}\right] + \\ & + \frac{2\mu e^p_{xy}}{4\pi(1-\nu)}\left[\frac{(y-w)(x+d)}{(x+d)^2+(y-w)^2} - \frac{(y-w)(x-d)}{(x-d)^2+(y-w)^2}\right]. \end{aligned} \tag{5}$$

Similarly the shear stress field caused by this fictitious force on DC, CB and

DA are

$$\tau_{xyDC} = -\frac{2\mu e^p_{xy}}{2\pi}\left[\tan^{-1}\frac{x-d}{y+w} - \tan^{-1}\frac{x+d}{y+w}\right] +$$

$$+\frac{2\mu e^p_{xy}}{4\pi(1-\nu)}\left[\frac{(y+w)(x-d)}{(x-d)^2+(y+w)^2} - \frac{(y+w)(x+d)}{(x+d)^2+(y+w)^2}\right], \tag{6}$$

$$\tau_{xyCB} = \frac{2\mu e^p_{xy}}{2\mu}\left[\tan^{-1}\frac{y-w}{x-d} - \tan^{-1}\frac{y+w}{x-d}\right] +$$

$$+\frac{2\mu e^p_{xy}}{4\pi(1-\nu)}\left[\frac{(x-d)(y+w)}{(x-d)^2+(y+w)^2} - \frac{(x-d)(y-w)}{(x-d)^2+(y-w)^2}\right], \tag{7}$$

$$\tau_{DA} = \frac{2\mu e^p_{xy}}{2\pi}\tan^{-1}\frac{y+w}{x+d} - \tan^{-1}\frac{y-w}{x+d}\Bigg] +$$

$$+\frac{2\mu e^p_{xy}}{4\pi(1-\nu)}\left[\frac{(x+d)(y-w)}{(x+d)^2+(y-w)^2} - \frac{(x+d)(y+w)}{(x+d)^2+(y+w)^2}\right]. \tag{8}$$

τ_{xy} is the resolved shear stress in the sliding system. Since we are mainly interested in this component of stress, the subscript xy of τ_{xy} and e_{xy} are dropped: i.e., $\tau = \tau_{xy}$ and $e = e_{xy}$. The shear stress field τ_F caused by the uniform plastic strain e^p in this slice is the sum of the expressions given by eqs. (5) to (8)

$$\tau_F = \tau_{AB} + \tau_{DC} + \tau_{CB} + \tau_{DA}, \tag{9}$$

$$\tau_R = \tau_F - 2\mu e^p \tag{10}$$

where e^p is constant within the slice and vanishes elsewhere.

$$\tau_R = \frac{\mu e^p}{\pi(1-\nu)}\left[\frac{(y-w)(x+d)}{(y-w)^2+(x+d)^2} - \frac{(y-w)(x-d)}{(y-w)^2+(x-d)^2} + \right.$$

$$\left. + \frac{(y+w)(x-d)}{(x-d)^2+(y+w)^2} - \frac{(y+w)(x+d)}{(x+d)^2+(y+w)^2}\right]. \tag{11}$$

At $x = 0$, $y = 0$, and taking $w \ll d$,

$$\tau_{R_0} = -\frac{2\mu e^p 2w}{\pi(1-\nu)d}$$

where the subscript '0' denotes the origin.

Along the x-axis ($y = 0$)

$$\tau_R = \frac{4\mu e^p wd}{\pi(1-\nu)} \frac{(x^2-d^2)-w^2}{[w^2+(x+d)^2][w^2+(x-d)^2]}. \tag{12}$$

This shows that the relief of resolved shear stress is directly proportional to the product of the plastic strain e^p and the band thickness $2w$. Since $2w$ is very small, extremely large plastic strain e^p must occur to give a finite value of τ_R.

Outside of the slice τ_R is positive along x-axis. Plastic strain in this slice causes the positive resolved shear stress to increase and consequently the width of this thin slice ($2d$) to increase. This seems to explain why in general a slip band rapidly widens to cover the whole crystal.

Along the y-axis ($x = 0$),

$$\tau_R = \frac{2\mu e^p d}{\pi(1-\nu)} \left[\frac{(y-w)}{(y-w)^2+d^2} - \frac{(y+w)}{(y+w)^2+d^2}\right]$$

$$= \frac{4\mu e^p wd}{\pi(1-\nu)} \frac{(y^2-w^2)-d^2}{[(y-w)^2+d^2][(y+w)^2+d^2]}. \tag{13}$$

Near the band $y \ll d$, $w \ll d$

$$\tau_R \cong -\frac{4\mu e^p w}{\pi(1-\nu)} \frac{1}{d}\left(1-\frac{y^2}{d^2}\right), \tag{14}$$

$$\frac{d\tau_R}{dy} \cong \frac{4\mu e^p w}{\pi(1-\nu)d} \frac{2y}{d} \cong 0.$$

This shows that in these regions above AB and below CD in Figs. 1 and 2, the relief of the resolved shear stress τ_R is practically constant and is the same in the slice. Since these regions did not slide when the slice first slid, the τ_I in those regions was less than that in the slice. After this slice slides, $\tau = \tau_I+\tau_R+\tau_A$ would be still less in these regions than that in the slice. Assuming no strain-hardening, the resolved shear stress in the slice equals the critical shear stress τ_C and that in those neighboring regions would be less than τ_C, hence no slip occurs in these neighboring regions.

When w is small, eq. 12 reduces to

$$\tau_R = \frac{4\mu w}{\pi(1-\nu)d} e^p \frac{1}{(x/d)^2-1}.$$

Consider a slip band divided into $2N$ – equal thin slices along its width. Each slice has a width $2d$. The distance x at the centroid of the mth slice is $(2m-1)d$. Let e_m^p be the plastic strain caused by slip in the mth slice. We have

$$\tau_{R_m} = \frac{4\mu w}{\pi(1-\nu)d} \sum_{n=-N}^{N} \frac{1}{4(m-n)^2-1} e_n^p. \tag{15}$$

The resolved shear stresses in all the slices must be equal to τ_C

$$\tau_A + \tau_{R_m} = \tau_C, \qquad m = 1 \text{ to } N. \tag{16}$$

Solving eqs. (15) and (16), we obtain the plastic strain distribution e_n^p's in the slip band to give the uniform resolved shear stress in the band. This distribution of e^p in the band is not uniform and is shown in Fig. 3.

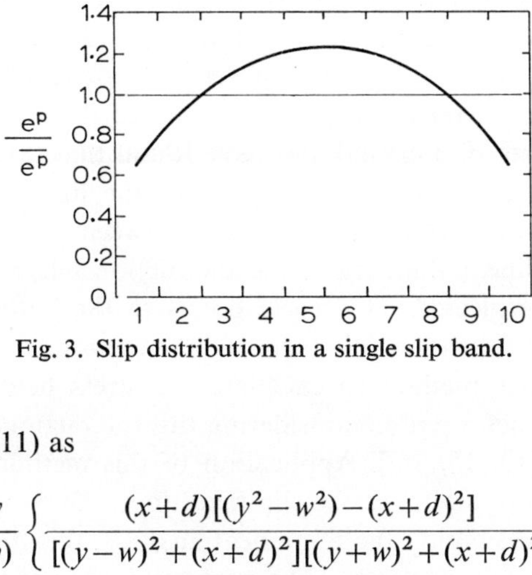

Fig. 3. Slip distribution in a single slip band.

Rewriting eq. (11) as

$$\tau_R = \frac{2\mu e^p w}{\pi(1-\nu)} \left\{ \frac{(x+d)[(y^2-w^2)-(x+d)^2]}{[(y-w)^2+(x+d)^2][(y+w)^2+(x+d)^2]} + \right.$$

$$\left. + \frac{(x-d)[(x-d)^2-(y^2-w^2)]}{[(x-d)^2+(y+w)^2][(x-d)^2+(y-w)^2]} \right\}. \tag{17}$$

Let $e^p w$ be kept constant, let e^p increase and let w decrease to zero, then

$$\tau_R = \frac{2\mu e^p w}{\pi(1-\nu)} \left\{ \frac{(x-d)[(x-d)^2-y^2]}{[(x-d)^2+y^2]^2} - \frac{(x+d)[(x+d)^2-y^2]}{[(x+d)^2+y^2]^2} \right\}. \tag{18}$$

The resolved shear stress due to the presence of an edge dislocation along the z-axis $(0, 0, z)$ with a Burger's vector pointing in the x-direction, is given [6] as

$$\tau = \frac{\mu b}{2\pi(1-\nu)} \frac{x(x^2-y^2)}{[x^2+y^2]^2}. \tag{19}$$

Letting $2we^p$ be the magnitude of the Burgers vector of the edge dislocation, it is seen that this resolved shear stress field caused by the movement of an edge dislocation having this Burger's vector orientated from $(-d, 0)$ to $(d, 0)$ is exactly the same as given by eq. (18). Hence, this plastic strain in the thin slice can be caused by the movement of dislocations [11, 12, 13].

Stress fields caused by slip in two and more slip bands

When a polycrystal is loaded, initially there is no plastic strain, $\tau_R = 0$, τ_A is maximum in the most favorably oriented crystal and τ_I is taken to be small. Slip bands start in this crystal and as loads increase, more crystals slide. The slip in one slid crystal causes a stress field in the aggregate. Lin, Uchiyama and Martin [14] have calculated the stress field caused by slip in an aggregate of a crystal and have found that this stress practically vanishes at a distance of about three times the linear dimension of the crystal. Hence, when the average distance between the two slid crystals is three times this linear dimension, the interaction effect between two slid crystals may be neglected. This corresponds to the initial stage of plastic deformation. Beyond the initial stage, the interaction effect should be considered. A rigorous method to calculate the stress field in a polycrystal caused by plastic deformation considering this interaction effect was shown by Lin and Ito [13, 15, 16,]. Application of this method to calculate the microstress field varying from one slip band to another would be extremely lengthy. Hence, to simplify the numerical analysis, this interaction between neighboring crystals is neglected. The problem is then reduced to the study of microstress and strain in a crystal of lower critical shear stress embedded in a matrix of much higher critical shear stress like a pure aluminum crystal embedded in an aluminum alloy.

Slip varies along the band width for a single slip band as shown in Fig. 3 and also varies along the width for two or more slip bands. However, in order to simplify the calculation, the relief of the resolved shear stress due to plastic strain in slip bands is assumed to be given by the relief at the centroid of the band caused by uniform plastic strain in each band. The relief of the shear stress in each band is then given by letting $x = 0$ in eq. (18) and may be rewritten as

$$\tau_R = \frac{4\mu e^p w}{\pi(1-\nu)d}\,\frac{d^2(y^2-d^2)}{(y^2+d^2)^2} = Ae^p\,\frac{(y^2-d^2)d^2}{(y^2+d^2)^2} \tag{20}$$

where

$$A = \frac{4\mu w}{\pi(1-\nu)d}.$$

In the region where y/d is small, the residual stress at $(0, y)$ caused by e^p may be expressed as

$$\tau_{R_y} = -Ae^p\,\frac{1-(y^2/d)}{[1+(y^2/d)]^2} \cong -Ae^p\left[1-3\left(\frac{y}{d}\right)^2\right] = \tau_{R_0}\left[1-3\left(\frac{y}{d}\right)^2\right] \tag{21}$$

where

$$\tau_{R_0} = -Ae^p. \tag{22}$$

Since this band continues to slide,

$$\tau_{I_0}+\tau_{R_0}+\tau_A = \tau_C$$

and under the assumption of no strain-hardening $\tau_{C_0} = \tau_{C_y} = \tau_C$.

Hence, from eqs. (21) and (22),

$$\tau_y = \tau_{I_y}+\tau_{R_0}\left[1-3\left(\frac{y}{d}\right)^2\right]+\tau_A = (\tau_{I_y}-\tau_{I_0})+3Ae^p\left(\frac{y}{d}\right)^2+\tau_C. \tag{23}$$

The first slip band occurs at $y = 0$, hence

$$\tau_{I_0}-\tau_{I_y} > 0.$$

When e^p reaches the value at which some nearby region $y = s$ first satisfies the condition

$$\tau_{I_0}-\tau_{I_y} = +3Ae^p\left(\frac{s}{d}\right)^2 \tag{24}$$

a second slip band starts to slide at $y = s$.

As the load increases further, the incremental shear stresses in these two bands must satisfy

$$\Delta\tau_{R_0} = \Delta\tau_{R_s} = -\Delta\tau_A \tag{25}$$

and the incremental plastic strain in both bands must be the same, i.e.

$$\Delta e_0^p = \Delta e_s^p = \Delta e^p,$$

$$\Delta \tau_{R_0} = \Delta \tau_{R_s} = -A\Delta e^p \left[2-3\left(\frac{s}{d}\right)^2\right], \tag{26}$$

$$\Delta \tau_{R_y} = -A\Delta e^p \left[1-3\left(\frac{y}{d}\right)^2\right] - A\Delta e^p \left[1-3\left(\frac{s-y}{d}\right)^2\right],$$

$$\Delta \tau_{R_y} - \Delta \tau_{R_0} = -3A\Delta e^p \frac{2y(s-y)}{d^2}. \tag{27}$$

Hence in the region between the two bands $0 < y < s$

$$\Delta \tau_R - \Delta \tau_{R_0} < 0.$$

This shows that relief of the resolved shear stress, caused by equal incremental plastic strain Δe^p in two closely located slip bands, is more in regions between the two bands than that in the bands. Hence, the resolved shear stress in regions between bands is always less than in the bands. Since the resolved shear stress in the bands is equal to the critical shear stress τ_c, the resolved shear stress in the region between the bands is always less than τ_c. Therefore, regions between two sliding bands do not slide while Δe^p increases in the two bands.

For these regions $y < 0$ and $y > s$, it follows from eq. (27) that

$$\Delta \tau_{R_y} - \Delta \tau_{R_0} > 0.$$

When Δe^p reaches the value, at which the resolved shear stress τ_y equals the critical shear stress τ_c, another slip band starts. Then the two exterior bands may continue to slip with no further sliding in the interior band, since the resolved shear stress in the latter is decreasing. Similarly additional slip bands may start one by one as the load increases. This explains why plastic deformation may concentrate in closely located slip bands.

In general, slip bands are not restricted to being closely located. Consider a fine-grain, aluminum polycrystal composed of individual crystals of cubic shape and 40μ in linear dimension. The crystal is assumed to develop slip bands 0.1μ thick and 40μ wide. The grain size of a fine-grain aluminum polycrystal is about 50μ (micron). The relief of the resolved shear stress is now given by eq. (20) as

$$\tau_{R_y} = -Ae^p \frac{1-(y/d)^2}{[1+(y/d)^2]^2} \tag{28}$$

in which y varies from 0 to $\pm 20\mu$. Dividing this 40μ into 80 equal intervals, the relief τ_{R_m} of the stress at $y = (m/2)\ \mu$ may be expressed as

$$\tau_{R_m} = -A \sum_{n=-40}^{40} \frac{1-\left(\dfrac{m-n}{80}\right)^2}{\left[1+\left(\dfrac{m-n}{80}\right)^2\right]^2} \tag{29}$$

where e_n^p denotes the plastic strain e^p at the nth band. If subscripts p, q denote the active slip bands and Q the total number of active slip bands,

$$\tau_p = \tau_{I_p} + \tau_A + \tau_{R_p} = \tau_c$$

$$\Delta\tau_p = \Delta\tau_A + \Delta\tau_{R_p} = 0$$

$$A \sum_{q=1}^{Q} \frac{1-\left(\dfrac{p-q}{80}\right)^2}{\left[1+\left(\dfrac{p-q}{80}\right)^2\right]^2} \Delta e_q^p = \Delta\tau_A. \tag{30}$$

This gives Q equations to solve for Q unknown Δe_q^{p}'s in terms of $\Delta\tau_A$.

Let subscript m denote any inactive band, then

$$\Delta\tau_m = -A \sum_{q=1}^{Q} \frac{1-\left(\dfrac{m-q}{80}\right)^2}{\left[1+\left(\dfrac{m-q}{80}\right)^2\right]^2} \Delta e_q^p \tag{31}$$

and since Δe_q^p's are given in terms of $\Delta\tau_A$, $\Delta\tau_m$ may be expressed by

$$\Delta\tau_m = K_m \Delta\tau_A. \tag{32}$$

At this rate of changing τ_m per unit $\Delta\tau_A$, the amount of $\Delta\tau_A$ required to bring τ_m to the critical shear stress is expressed as $(\tau_C - \tau_m)/K_m$. This is calculated for all inactive grids and the one with minimum value is becoming potentially active next. This slip band is added to the active slip bands and included in eq. (30). All the e_q^{p}'s obtained from eq. (30) must be in the same sense as τ_q; i.e., the incremental plastic work done by Δe_q^p must be positive. If some of the Δe_q^{p}'s do not satisfy this condition, some of these active grids must be deleted until all active grids will have Δe_q^p in the same sense as τ_q and all the non-active grids have $|\tau_m| < \tau_c$.

Consider a numerical example with $\tau_{I_0} = 0.01\ \tau_c$, $\tau_c = 50A \times 10^{-4}$ and all the other τ_{I_m}'s vanish. The subscript 0 denotes $y = 0$. As τ_A increases e_0^p increases until both τ_{40} and τ_{-40} reach τ_c and join in sliding. As τ_A increases further, τ_{24} and τ_{-24} reach τ_c and bands 24 and −24 join bands 1, 40 and −40 to slide. The sequence of active slip bands is listed in Table I.

TABLE I

Sequence of active slip bands

τ_A/τ_c	active slip band	τ_A/τ_c	active slip band
0.98	0	51.05	±16, ±17, ±34, ±40
1.00	0, ±40	54.27	0, ±17, ±34, ±40
1.08	0, ±23, ±40	57.48	0, ±17, ±18, ±34, ±40
1.09	0, ±22, ±23, ±40	62.50	0, ±18, ±19, ±34, ±40
1.10	0, ±21, ±22, ±40	68.72	0, ±19, ±20, ±34, ±40
1.15	0, ±20, ±21, ±40	74.95	0, ±20, ±34, ±35, ±40
1.20	0, ±19, ±20, ±40	77.43	0, ±20, ±21, ±35, ±40
1.25	0, ±18, ±19, ±40	85.68	0, ±21, ±22, ±45, ±40
1.34	0, ±17, ±18, ±40	97.37	0, ±22, ±23, ±35, ±40
1.47	0, ±16, ±17, ±40	112.37	0, ±23, ±24, ±35, ±40
1.68	0, ±15, ±16, ±40	131.20	0, ±24, ±35, ±36, ±40
2.01	0, ±14, ±15, ±40	131.85	0, ±24, ±25, ±36, ±40
2.57	0, ±13, ±14, ±40	156.04	0, ±25, ±26, ±35, ±40
3.63	0, ±12, ±13, ±40	188.34	0, ±26, ±27, ±36, ±40
5.90	0, ±11, ±12, ±40	201.02	0, ±14, ±27, ±36, ±40
11.98	0, ±10, ±11, ±40	201.64	0, ±14, ±15, ±27, ±36, ±40
33.85	±10, ±32, ±40	250.86	0, ±15, ±27, ±28, ±36, ±40
34.46	±11, ±32, ±33, ±40	252.47	0, ±15, ±16, ±28, ±36, ±40
34.64	±11, ±12, ±33, ±40		
35.80	±12, ±13, ±33, ±40		
37.57	±13, ±14, ±33, ±40		
40.29	±14, ±15, ±33, ±40		
44.51	±15, ±16, ±33, ±40		
48.69	±16, ±33, ±34, ±40		

Note: The numbers shown in the table denote the locations of the active slip band in terms of $y/40d$ from the center of the sliding crystal.

The resolved shear stress distribution across the crystal at different stages of loading are shown in Figs. 4 and 5. The amount of slip in different bands

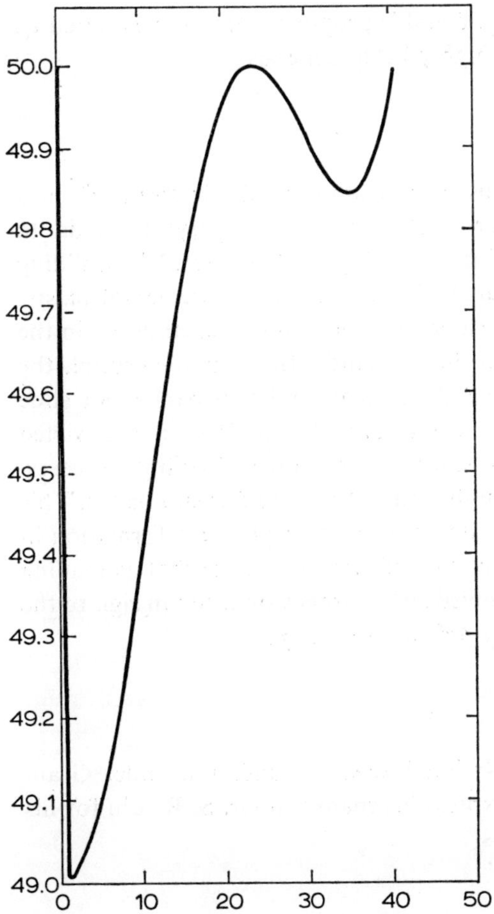

Fig. 4. Resolved shear stress distribution at $\tau_A/\tau_C = 1.08$.

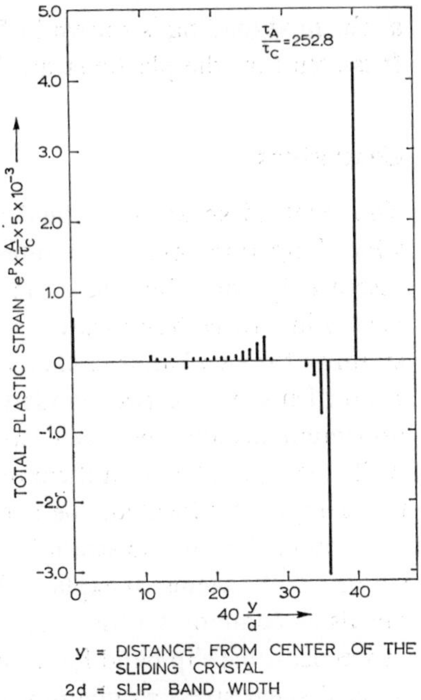

Fig. 6. Distribution of plastic strain across the crystal.

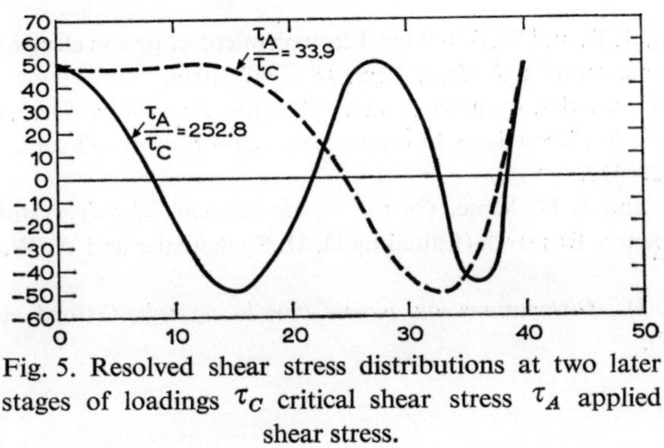

Fig. 5. Resolved shear stress distributions at two later stages of loadings τ_C critical shear stress τ_A applied shear stress.

at the final loading is shown in Fig. 6 and is proportional to the assumed τ_c. It is seen that the plastic strain is highly heterogeneous.

Conclusi ons

The resolved shear stress field caused by a uniform plastic strain e_{xy}^p in a wide, long thin layer in a interior crystal within a polycrystal tends to increase the width but tends to maintain the original thickness. When sliding occurs in *two closely* located parallel slip bands, some incremental plastic strain in the two bands causes a greater relief of resolved shear stress in the region between the two bands than in the bands. In a narrow crystal, the maximum distance between two bands is small and two bands can slide while the region between them remains unslid. If slip bands are not restricted to being closely located, the active slip bands change as loading increases. It is found that plastic strain is mainly concentrated in few slip bands. This simple model seems to explain the inhomogeneity of plastic deformation in metals as commonly observed. It may be of interest to note that as loading proceeds, some slip band has a resolved shear stress τ opposite in sign to the applied shear stress τ_A giving a slip Δe^p opposite to τ_A.

Acknowledgements

This research was sponsored by National Science Fundation under Grant GK 15349. The author wishes to express his thanks to Dr. S. R. Lin for his help in the numerical computations.

References

[1] Rosenhain, W., Further observations on slip bands. *Proc. Roy. Soc., 74* (1905) 557–562.
[2] Heidenriech, R. D. and W. Shockley, Electron microscope and electron diffusion study of slip in metal crystals. *J. Appl. Phys. 18* (1947) 1029.
[3] Nye, J. F., Plastic deformation of silver chloride. *Proc. Roy. Soc., A198* (1949) 190,
[4] Brown, A. F., Surface effects in plastic deformation of metals. *Advances in Physics. 14* (1952) 427–477.
[5] Dorn, J. F. and J. D. Mote, *On the plastic behavior of polycrystalline aggregates.* Material Sciences Research (Edited by H. H. Stadelmaier and W. W. Austin), Vol. I, 11–56.
[6] Cottrell, A. H., *Dislocations and plastic flow in crystals.* Oxford at the Clarendon Press, (1957).

[7] Taylor, C. I., Plastic strain in metals. *J. Inst. of Metals, 62* (1938) 307–324.
[8] Eshelby, J. D., The determination of the elastic field of an ellipsoidal inclusion and related problems. *Proc. Roy. Soc. London, A241* (1957) 398.
[9] Lin, T. H., *Theory of inelastic structures.* John Wiley and Sons, New York, Chapter 2, (1968).
[10] Barnett, C. S., *Structures of Metals.* McGraw Hill Book Co., (1952) 533.
[11] Read, T. A., *Dislocations in Crystals.* McGraw Hill Book Co., (1953) 36.
[12] Mura, T., *Continuum theory of dislocations and plasticity.* Proc. IUTAM Symp. on the Generalized Cosserat Continuum and the Continuum Theory of Dislocations as its Applications, Springer Verlag, (1967).
[13] Lin, T. H., *Physical theory of plasticity.* Advances in Applied Mechanics, Vol. 11, Academic Press, New York 1971.
[14] Lin, T. H., S. Uchiyama, and D. Martin, Stress fields in metals at initial stage of plastic deformation. *J. Mech. Phys. Solids, 9* (1961) 200–209.
[15] Lin, T. H. and Y. M. Ito, Theoretical plastic stress-strain relationship of a polycrystal and the comparison with the Von Mises and Tresca plasticity theories. *International Journal of Engineering Sciences, 4* (1966) 543–561.
[16] Ito, Y. M. and T. H. Lin, Theoretical plastic strain of a face-centered cubic polycrystal. *Recent Advances in Engineering Sciences, 5* (1970), 169–188, (Gordon & Breach, New York).

An analytical model of large deformation effects in crystalline aggregates

Kerry S. Havner

North Carolina State University, Raleigh, N.C., U.S.A.

A microcontinuum model of (rate-independent) deformation in crystalline aggregates under macroscopically uniform, quasi-static loading is presented, corresponding to the general problem of instantaneous response from a deformed state. A rational discretization, which includes all convection and rotation effects at large strain, is fully developed, and questions of existence and uniqueness of solution are investigated through consideration of alternative quadratic programming problems equivalent to the governing equations.

1. Introduction

As seen by an observer who can resolve distances to 10^{-3} mm, the deformation of crystal grains (of typical dimension 10^{-1} mm, say) within permanently 'strained' metal polycrystals is relatively smooth. At this level of observation (called 'microscopic' herein), one can just distinguish between slip lines on crystal surfaces after extensive plastic distortion of an aggregate specimen. Since these 'lines' are in actuality glide bands of order 10^{-4} mm width, containing numerous glide lamellae, a *sub*microscopic observer resolving distances to 10^{-5} mm (the order of 100 atomic spacings) is aware of grossly inhomogeneous and discontinuous deformation within each grain. Hence, a continuum point-of-view at this second level would seem untenable. (This important physical constraint on any micromechanics approach to macroscopic metal plasticity, first discussed clearly by Hill [1], has been often overlooked.) Here we adopt a microscopic continuum model consistent with the above, wherein a material 'point'[1]) has dimensions of order 10^{-3} mm (*i.e.*, $> 10^3$ lattice spacings). As in [3, 4], the mechanical behavior is taken

[1]) Krumhansl [2] has suggested the label 'elementary physical unit' for continuum mechanics studies.

to be representable via two *kinematically* independent mechanisms of deformation (which are phenomenological averages of complex processes occurring within the lattice volume defined by the 'point'). These mechanisms are: (a) elastic (mechanically recoverable) *infinitesimal* strain of the lattice, and (b) simple glide, on well-defined crystallographic slip systems, which translates material 'lines' of points (*i.e.*, glide packets) relative to one another but leaves the crystal structure unchanged.

In the sequel we present an *ad hoc* crystalline aggregate model, based upon the foregoing considerations and chosen with the objective of simulating quasi-static, large strain behavior of thin-walled tubes and flat sheet specimens. We investigate certain general relationships as regard volume averages of continuum microfields, but focus primary attention on the analytical development and formal solution of a consistent, discrete counterpart of the aggregate continuum problem. This new discrete model, which includes all convection and rotation effects, is an extension of the author's polycrystalline model at small strain presented in [5, 6].

2. Governing microfield equations

Let X represent a crystal material 'point' with position $\boldsymbol{x}$ and velocity $\boldsymbol{v}$ relative to a Cartesian reference frame at time t. We denote Cauchy (true) stress by $\boldsymbol{T}$, symmetric Kirchhoff stress by $\boldsymbol{\tau}$, and nominal stress by $\boldsymbol{s}$. Referred to the instantaneous deformed configuration as reference configuration (such that $\boldsymbol{T} = \boldsymbol{\tau} = \boldsymbol{s}$), material time derivatives of these stress tensors are related by (see [7], for example):

$$\dot{\boldsymbol{s}} = \dot{\boldsymbol{\tau}} - \boldsymbol{L}\boldsymbol{\tau} = \dot{\boldsymbol{T}} + D_1\boldsymbol{T} - \boldsymbol{L}\boldsymbol{T}. \tag{1}$$

$\boldsymbol{L} = \partial\boldsymbol{v}/\partial\boldsymbol{x}$ is the velocity gradient and D_1 is the trace of the rate of deformation tensor $\boldsymbol{D}$. In terms of the nominal stress rate $\dot{\boldsymbol{s}}$ within a crystal volume V_c and nominal traction rate $\dot{\boldsymbol{t}}$ on its surface S_c, the equations of changing motion are

$$\mathrm{div}_{\boldsymbol{x}}\, \dot{\boldsymbol{s}} + \rho \frac{\mathrm{d}}{\mathrm{d}t}(\boldsymbol{f} - \dot{\boldsymbol{v}}) = \boldsymbol{0} \quad \text{in} \quad V_c \tag{2}$$

and

$$\dot{\boldsymbol{s}}^T \boldsymbol{\nu} = \dot{\boldsymbol{t}} \quad \text{on} \quad S_c. \tag{3}$$

$\boldsymbol{f}$ is body force per unit mass, ρ is mass density and $\boldsymbol{\nu}$ is the unit outward normal on S_c. Alternatively, ın terms of $\dot{\boldsymbol{T}}$, (2) becomes

$$\mathrm{div}_x \dot{\boldsymbol{T}}+D_1 \,\mathrm{div}_x \boldsymbol{T}-\frac{\partial \boldsymbol{T}}{\partial \boldsymbol{x}}\cdot\boldsymbol{L}+\rho\frac{\mathrm{d}}{\mathrm{d}t}(\boldsymbol{f}-\dot{\boldsymbol{v}})=\boldsymbol{0}. \tag{2a}$$

(In component form referred to any spatial coordinate system, $(\partial \boldsymbol{T}/\partial \boldsymbol{x})$. $\boldsymbol{L}$ means $T^{ji}_{,m}v^m_{,i}$ with the commas denoting covariant differentiation.)

Based upon the physical considerations in sect. 1, the rate of deformation and rate of rotation tensors can be written

$$\boldsymbol{D}=\tfrac{1}{2}(\boldsymbol{L}+\boldsymbol{L}^T)=\boldsymbol{\varepsilon}+\dot{\gamma}_k\boldsymbol{N}^k, \tag{4}$$

$$\boldsymbol{W}=\tfrac{1}{2}(\boldsymbol{L}-\boldsymbol{L}^T)=\boldsymbol{\omega}+\dot{\gamma}_k\boldsymbol{\Omega}^k \tag{5}$$

(with summation on k over the instantaneously critical slip systems). $\boldsymbol{\varepsilon}=\dot{\boldsymbol{e}}$ is the lattice infinitesimal strain rate (velocity strain), $\boldsymbol{\omega}$ is the spin tensor of the lattice relative to the x_i reference frame, and $\dot{\gamma}_k$ is the plastic shear rate in the k-th crystallographic slip system (that is, the contribution of instantaneous glide in the k-th system to the velocity gradient at X, referred to the system axes). $\boldsymbol{N}^k$, $\boldsymbol{\Omega}^k$ are the symmetric and antisymmetric parts of the tensor (dyadic) product $\boldsymbol{b}^k\otimes\boldsymbol{n}^k$, with $\boldsymbol{n}^k$, $\boldsymbol{b}^k$ denoting unit vectors in the normal and glide directions, respectively, of the k-th slip system. Referred to a rigid, rectangular coordinate frame ξ_J rotating with the lattice at X, these vectors are expressed

$$\boldsymbol{n}^k_{(L)}=l^k_J\boldsymbol{h}^J,\quad \boldsymbol{b}^k_{(L)}=m^k_J\boldsymbol{h}^J \tag{6}$$

wherein the subscript (L) refers to the lattice and the $\boldsymbol{h}^J$ are unit base vectors of the Cartesian system ξ_J. To the order of approximation of infinitesimal lattice strains, l^k_J and m^k_J may be taken as constant and equal to the normalized Miller indices of the k-th slip system.

Introducing the fourth order elastic compliance tensor $\mathscr{C}(\boldsymbol{R})$, with $\boldsymbol{R}$ denoting the instantaneous (orthogonal) lattice orientation matrix $R_{Ji}=\cos(\xi_J, x_i)$, and defining the *lattice* co-rotational stress rate [3, 4]

$$\hat{\boldsymbol{\tau}}=\dot{\boldsymbol{\tau}}-\boldsymbol{\omega}\boldsymbol{\tau}+\boldsymbol{\tau}\boldsymbol{\omega}, \tag{7}$$

the elastic stress-strain rate equations are equivalently expressed:

$$\boldsymbol{\varepsilon}=\mathscr{C}\cdot\hat{\boldsymbol{\tau}},\quad \boldsymbol{\varepsilon}_{(L)}=\mathscr{C}_{(L)}\cdot\dot{\boldsymbol{\tau}}_{(L)}. \tag{8a, b}$$

($\mathscr{C}_{(L)}$ is the orthotropic crystal compliance of infinitesimal elastic strains referred to the lattice axes and taken as constant herein.) Kirchhoff stress-rate

$\dot{\tau}$ is adopted in (7), (8,a b) rather than Cauchy stress-rate $\dot{\boldsymbol{T}}$ (as chosen in [3, 4]) for convenience in subsequent analysis. Within the present context the difference is insignificant since, from (4), $\operatorname{tr}\boldsymbol{D} = D_1 = \dot{e}_1$, the infinitesimal elastic volumetric strain rate. Hence, from (1), $\dot{\tau}$ and $\dot{\boldsymbol{T}}$ are indistinguishable for practical purposes. Correspondingly, we incorporate $\dot{\tau}$ rather than $\dot{\boldsymbol{T}}$ in the general crystal hardening law introduced by Hill [3] and Mandel [8] and so obtain, for a critical (*i.e.*, potentially active) slip system:

$$\tau^k = \boldsymbol{N}^k \cdot \boldsymbol{\tau}, \tag{9}$$

$$\dot{\tau}^k = h^{kj}(\gamma)\dot{\gamma}_j \geqq \boldsymbol{N}^k \cdot \hat{\boldsymbol{\tau}}, \quad \dot{\gamma}_k \geqq 0, \tag{10a, b}$$

$$\dot{\gamma}_k(h^{kj}\dot{\gamma}_j - \boldsymbol{N}^k \cdot \hat{\boldsymbol{\tau}}) = 0 \qquad \text{(for each } k\text{)}. \tag{11}$$

In the above τ^k is the instantaneous crystal shear strength in the k-th system, h^{kj} are hardening moduli (see [3]), γ denotes the vector of plastic shears γ_k at X, and $\boldsymbol{N}^k \cdot \boldsymbol{\tau}$ (or $\hat{\boldsymbol{\tau}}$) denotes the trace of the matrix product of second order tensors. Opposite senses of slip in the same crystallographic slip system are denoted by distinct k's so that $\dot{\gamma}_k$ is always non-negative. Equations (1) through (11) complete the constitutive and continuum field equations governing the rate-independent crystal deformation at X.

3. The aggregate model and some general relations between macro- and micro-fields

Consider a thin-walled tube subjected to, say, axial load and internal pressure. The wall thickness of specimens studied experimentally in combined loading tests is often in the range 1–2 mm, with from 10–30 grains through the thickness (see [9, 10], for example). Thus, as an idealization of the physical situation, we assume a thickness of 1 mm (in the undeformed state) and define a unit cube $V = 1\ \text{mm}^3$ containing on the order 1000 crystal grains in the corresponding 'flat sheet' representation (*i.e.*, a *macroscopic* plane stress problem). Further, we wish the mathematical model to correspond to a macroscopically homogeneous physical specimen (that is, one for which strain gage readings, over distances of at least 1 mm on the surface, are essentially uniform from one location to another). We therefore require all the 'unit cubes' to deform identically and take the longitudinal and transverse faces to be planes of symmetry under biaxial macroscopic straining. (For additional discussion see [5].) Thus, for quasi-static deformation (neglecting inertia and body forces), but without restriction as to *total* strain magnitudes, we adopt as a model for analysis an initially unit cube (of generally anisotropic crystals) on each of whose faces either (1) a particular velocity component is prescribed, to give the appropriate *macroscopic* velocity strain, or (2) the associated traction is zero.

In a deformed configuration (in which the 'unit cube' is, essentially, a rectangular parallelopiped) it is easily shown (as in [5]) that the aggregate model satisfies exactly the well-known averaging theorem ([1], [11–13])

$$\bar{\boldsymbol{\tau}}\cdot\bar{\boldsymbol{D}} = \overline{\boldsymbol{\tau}\cdot\boldsymbol{D}} \equiv \overline{\boldsymbol{\tau}\cdot\boldsymbol{L}} \tag{12}$$

for any statically admissible stress field $\boldsymbol{\tau}$ and any continuous, piecewise-differentiable velocity field $\boldsymbol{v}$. (The bar above a tensor or scalar field denotes the aggregate volume average.) Consequently, macro-stress and -velocity strain are appropriately determined as

$$\boldsymbol{\Sigma}_M = \bar{\boldsymbol{\tau}}, \quad \boldsymbol{E}_M = \bar{\boldsymbol{D}}. \tag{13}$$

We admit the possibility of a family of macrostrain rates $\boldsymbol{E}_M^*$ producing purely elastic response of the aggregate from its present deformed configuration (such that $\dot{\gamma}_k^* \equiv 0$ in every slip system). Rotation and convection effects are then negligible for a particular $\boldsymbol{E}_M^*$, whence (to this approximation) $\hat{\boldsymbol{\tau}}^* = \dot{\boldsymbol{\tau}}^* = \dot{\boldsymbol{s}}^*$ and (neglecting body and inertia forces in (2)) $\mathrm{div}_x\, \dot{\boldsymbol{s}}^* = \boldsymbol{0} = \mathrm{div}_x\, \dot{\boldsymbol{\tau}}^*$. Moreover,

$$\boldsymbol{\varepsilon}^* = \mathscr{C}\cdot\dot{\boldsymbol{\tau}}^* = \psi\cdot\boldsymbol{E}_M^* \tag{14}$$

in which ψ is a tensor function of position within V corresponding to the instantaneous orientations of the crystal grains (*i.e.*, a weighting function of the elastic inhomogeneity). Thus

$$\overline{\dot{\boldsymbol{\tau}}^*\cdot\boldsymbol{\varepsilon}^*} = \boldsymbol{E}_M^*\cdot\overline{(\psi^T\cdot\mathscr{C}^{-1}\cdot\psi)}\cdot\boldsymbol{E}_M^* \tag{15}$$

and a symmetric, inverse elastic compliance for the overall aggregate can be calculated as

$$\mathscr{C}_M^{-1} = \overline{\psi^T\cdot\mathscr{C}^{-1}\cdot\psi}. \tag{16}$$

This equation reflects a dependence of instantaneous, macroscopic elastic response on the history of deformation, since the present configuration of grain orientations and shapes depends upon that history. However, this effect is probably not great in cubic crystalline aggregates. (Note that, for the boundary value problem defined herein, only two non-zero velocity strain components, E_{11} and E_{22}, say, are arbitrarily prescribed, the third being computed from (13). Thus, only a reduced elastic compliance corresponding to the macroscopic plane stress state can be determined from (16). See Sect. 7 of [5].)

From (10a) and the above simplification ($\hat{\boldsymbol{\tau}}^* = \dot{\boldsymbol{\tau}}^*$),

$$0 \geqq \boldsymbol{N}^k \cdot \dot{\boldsymbol{\tau}}^* = \boldsymbol{N}^k \cdot \mathscr{C}^{-1} \cdot \psi \cdot \boldsymbol{E}_M^* \tag{17}$$

in every critical system. Hence, pairing with $\boldsymbol{E}_M^*$ a particular macrostrain-rate $\boldsymbol{E}_M$ which produces instantaneous plastic shears, we can write (from (4), (8a), (10b), (14) and (17))

$$0 \geqq \overline{\dot{\gamma}_k \boldsymbol{N}^k \cdot \mathscr{C}^{-1} \cdot \psi \cdot \boldsymbol{E}_M^*} = \overline{\dot{\boldsymbol{\tau}}^* \cdot \boldsymbol{D}} - \overline{\hat{\boldsymbol{\tau}} \cdot \boldsymbol{\varepsilon}^*} \tag{18}$$

in which $\hat{\boldsymbol{\tau}}$ is the lattice co-rotational stress-rate and $\boldsymbol{D}$ the velocity strain corresponding to $\boldsymbol{E}_M$. Thus, from (15), (16) and (18),

$$0 \geqq \dot{\boldsymbol{\Sigma}}_M^* \cdot (\boldsymbol{E}_M - \mathscr{C}_M \cdot \overline{\psi^T \cdot \hat{\boldsymbol{\tau}}}) \tag{19}$$

wherein $\dot{\boldsymbol{\Sigma}}_M^* = \overline{\dot{\boldsymbol{\tau}}^*}$. Equation (19) is a generalization of the orthogonality condition derived by Hill [13] neglecting all convection and rotation effects. It can be interpreted as a plastic strain-rate normality condition in macro-stress space Σ_{11}, Σ_{22} if the following convention is adopted. We define macrostress rate

$$\hat{\boldsymbol{\Sigma}}_M = \overline{\psi^T \cdot \hat{\boldsymbol{\tau}}} \tag{20}$$

(the volume average of the weighted lattice co-rotational stress-rate) and 'elastic' and 'plastic' macrostrain rates

$$\boldsymbol{E}_M^e = \mathscr{C}_M \cdot \hat{\boldsymbol{\Sigma}}_M, \quad \boldsymbol{E}_M^p = \boldsymbol{E}_M - \boldsymbol{E}_M^e. \tag{21a, b}$$

Then, from (19)

$$\dot{\boldsymbol{\Sigma}}_M^* \cdot \boldsymbol{E}_M^p \leqq 0 \tag{22}$$

and the domains of all possible aggregate plastic strain-rates and elastic stress-rates at a yield point are orthogonal. In contrast, if we define *nominal* 'elastic' macrostrain rate $\mathscr{C}_M \cdot \dot{\boldsymbol{S}}_M$, with $\dot{\boldsymbol{S}}_M = \bar{\dot{\boldsymbol{s}}}$, the corresponding nominal 'plastic' macrostrain rate may not be strictly orthogonal since, from (1), (5), (7) and (19), we find

$$\overline{[\boldsymbol{\tau}\boldsymbol{W} + \boldsymbol{D}\boldsymbol{\tau} + \dot{\gamma}_k(\boldsymbol{\Omega}^k\boldsymbol{\tau} - \boldsymbol{\tau}\boldsymbol{\Omega}^k)] \cdot \boldsymbol{D}^*} \geqq \dot{\boldsymbol{\Sigma}}_M^* \cdot (\boldsymbol{E}_M - \mathscr{C}_M \cdot \dot{\boldsymbol{S}}_M) \tag{23}$$

with the left-hand side not obviously non-positive. It is of course $\dot{S}_M$, rather than the macrostress rate from (20), which can be readily determined from experimental data. Hence, the matter of (apparent) plastic strain rate normality in actual experiment would seem an open question at sufficiently large strains. However, for *significant* (measureable) additional yielding, as opposed to the precise yielding in any sub-volume considered here, elastic strain increments may as well be neglected (to within ordinary experimental error) in comparison with plastic strain increments. Generalized normality then follows from the principle of maximum plastic work established by Bishop and Hill [11].

4. Model discretization and matrix symbolism

We now turn attention to a rational discretization of the boundary value problem of instantaneous response from a deformed state, adopting the idea of constant strain crystallites introduced in [5] (*i.e.*, a 'finite element' approximation). The instantaneous velocity field is continuous and piecewise linear (hence, kinematically admissible) so that

$$\boldsymbol{v}(\boldsymbol{x}) = \sum_{M(q)} \phi_q^M(\boldsymbol{x})\boldsymbol{v}^M \tag{24}$$

within a tetrahedral crystallite q having nodal points M, where $\boldsymbol{v}^M = (v_1^M, v_2^M, v_3^M)^T$ and

$$\phi_q^M(\boldsymbol{x}) = \alpha_q^M + \beta_q^{Mj} x_j. \tag{25}$$

The repeated index indicates summation and the constants α_q^M, β_q^{Mj} are determined from the instantaneous nodal coordinates of q through the equations $\phi_q^M(\boldsymbol{x}^J) = \delta^{MJ}$, $J, M = 1, \ldots, 4$ (with δ^{MJ} the Kronecker delta). The local velocity gradient is

$$\boldsymbol{L}_{(q)} = \sum_{M(q)} \boldsymbol{v}^M \otimes \boldsymbol{\beta}_q^M \tag{26}$$

in which

$$\boldsymbol{\beta}_q^M = \frac{\partial \phi_q^M}{\partial \boldsymbol{x}}. \tag{27}$$

From virtual work considerations (since all prescribed tractions are zero) $\overline{\delta \boldsymbol{L} \cdot \dot{\boldsymbol{s}}} = 0$ for any kinematically admissible field $\delta \boldsymbol{v}$ satisfying prescribed velocities on the faces of the unit volume V. Hence, from (1) and (26), the field equation of continuing equilibrium $\mathrm{div}_x\, \dot{s} = \mathbf{0}$ is replaced by the set of discrete equations

$$\sum_{q(J)} (\beta^J \cdot \langle \dot{\tau} - L\tau \rangle)_{(q)} = \mathbf{0} \tag{28}$$

over the crystallite nodes J at which velocities are to be determined. ($\langle \ldots \rangle_{(q)}$ means $\int \ldots dV_q$.) Substituting (4), (5), (7), (8a), (26) and (27), the nodal continuing equilibrium equations can be expressed in final form as

$$\sum_{q(J)} \sum_{M(q)} (K^{JM}_{(q)} + \chi^{JM}_{(q)}) v^M - \sum_{q(J)} [\mathscr{D}\phi^J(\langle C^{-1}\rangle \bar{N}^k + \langle m^k \rangle)\dot{\gamma}_k]_{(q)} = \mathbf{0} \tag{29}$$

in which

$$K^{JM}_{(q)} = [\beta^J \cdot \langle \mathscr{C}^{-1} \rangle \beta^M]_{(q)} \equiv [\mathscr{D}\phi^J \langle C^{-1} \rangle \mathscr{D}^T \phi^M]_{(q)}, \tag{30}$$

$$\chi^{JM}_{(q)} = \tfrac{1}{2}[(\beta^J \cdot \langle \tau \rangle \beta^M) I - (\beta^J \cdot \beta^M)\langle \tau \rangle - \beta^M \otimes \beta^J \langle \tau \rangle - \langle \tau \rangle \beta^M \otimes \beta^J]_{(q)}, \tag{31}$$

$$(\mathscr{D}\phi^J \langle m^k \rangle)_{(q)} \equiv (\langle \Omega^k \tau - \tau \Omega^k \rangle \beta^J)_{(q)}. \tag{32}$$

The operator $\mathscr{D}$ is a 3×6 matrix representation of the spatial gradient, corresponding to 6×1 vector representations of stress- and strain-rates, and C is the 6×6 matrix representation of the local elastic compliance tensor $\mathscr{C}$ (see [5] for details). $\bar{N}_k$ is a row vector representation of the symmetric tensor N^k ($\bar{N}^k = \bar{N}_k^T$) and $\langle m^k \rangle$ is a vector representation of the symmetric tensor $\langle \Omega^k \tau - \tau \Omega^k \rangle$. The aggregate 'elastic stiffness' and 'geometric stiffness' matrices assembled from $K^{JM}_{(q)}$ and $\chi^{JM}_{(q)}$ obviously will be symmetric.

Having seen, once and for all, detailed results [(29)–(32)] of the discretization at a nodal point, henceforth we will write only general matrix equations for the overall aggregate, these being discrete counterparts of the various microfield equations in sect. 2. We introduce the necessary symbolism below and take up these equations in section 5. We denote $\hat{\Sigma} = (\ldots, \langle \hat{\sigma} \rangle^T_{(q)}, \ldots)^T$, $\mathring{\Sigma} = (\ldots, \langle \mathring{\sigma} \rangle^T_{(q)}, \ldots)^T$, and $\dot{E} = (\ldots, \dot{\xi}^T_{(q)}, \ldots)^T$, in which $\hat{\sigma}$ and $\mathring{\sigma}$ are vector representations of the symmetric stress-rates $\hat{\tau}$ and $\mathring{\tau}$, where (from [4])

$$\mathring{\tau} = \hat{\tau} - (\Omega^k \tau - \tau \Omega^k)\dot{\gamma}_k \tag{33}$$

is the 'material' co-rotational (Jaumann-Zaremba) stress-rate, and $\dot{\xi}$ is a vector representation of the velocity strain D. $\dot{\Gamma} = (\ldots, \dot{\gamma}^T_{(q)}, \ldots)^T$; $\dot{U} = [\dot{u}^T : (\dot{U}^0)^T]^T = (\ldots, (v^J)^T, \ldots)^T$, in which $\dot{u}$, $\dot{U}^0$ are the overall vectors of unknown and prescribed velocities, respectively; $\boldsymbol{S} = \lceil \langle \boldsymbol{C}^{-1} \rangle_{(q)} \rfloor$ (*i.e.*, diagonal in sub-matrices); $\bar{\boldsymbol{H}} = \lceil \boldsymbol{H}_{(q)} \rfloor$ (where $\boldsymbol{H}_{(q)}$ denotes the $N \times N$ matrix of instantaneous hardening moduli h^{kj} of q); $\bar{\boldsymbol{N}} = \lceil \boldsymbol{N}_{(q)} \rfloor$ (where $\boldsymbol{N}_{(q)}$ is an $N \times 6$ transformation matrix of row vectors $\bar{\boldsymbol{N}}_k$); and $\bar{\boldsymbol{M}} = \lceil \boldsymbol{M}_{(q)} \rfloor$, with $\boldsymbol{M}_{(q)}$ a $6 \times N$ matrix of column vectors $\langle \boldsymbol{m}^k \rangle$. Moreover, we denote $\bar{\boldsymbol{B}} = [\boldsymbol{B}_i \vdots \boldsymbol{B}_0]$, a matrix composed of 6×3 elements $\boldsymbol{B}_{qJ} = \mathscr{D}^T \phi_q^J$ (**0** if J is not a node of q), with $\boldsymbol{B}_i$, $\boldsymbol{B}_0$ corresponding to nodes of unknown and prescribed velocities, respectively. (For an exterior J at which only one or two velocity components are unknown, the super-column of elements $\boldsymbol{B}_{qJ}$ is divided between $\boldsymbol{B}_i$ and $\boldsymbol{B}_0$.) Lastly,

$$\chi = \begin{bmatrix} \chi_i & \chi_o \\ \chi_o^T & \chi_{oo} \end{bmatrix},$$

a symmetric matrix composed of 3×3 elements $\chi_{JM} = \sum_{q(J,M)} \chi_{(q)}^{JM}$ (**0** for nodal pairs J, M having no q in common). χ_i corresponds to pairs J, M of unknown nodal velocities; χ_o corresponds to pairs J unknown, M prescribed; χ_{oo} corresponds to pairs J, M prescribed; and we further denote $\bar{\chi} = [\chi_i \vdots \chi_o]$.

5. General matrix equations and quadratic programming

Based upon the foregoing definitions, we can set down the following general matrix equations which parallel (1) through (11) and (33):

$$\boldsymbol{B}_i^T \mathring{\boldsymbol{\Sigma}} + \bar{\chi} \dot{\boldsymbol{U}} = \boldsymbol{0}, \tag{34}$$

$$\dot{\boldsymbol{E}} = \bar{\boldsymbol{B}} \dot{\boldsymbol{U}} = \dot{\boldsymbol{E}}^e + \bar{\boldsymbol{N}}^T \dot{\boldsymbol{\Gamma}}, \tag{35}$$

$$\mathring{\boldsymbol{\Sigma}} = \hat{\boldsymbol{\Sigma}} - \bar{\boldsymbol{M}} \dot{\boldsymbol{\Gamma}}, \quad \hat{\boldsymbol{\Sigma}} = \boldsymbol{S} \dot{\boldsymbol{E}}^e, \tag{36a, b}$$

$$\bar{\boldsymbol{H}} \dot{\boldsymbol{\Gamma}} \geqq \bar{\boldsymbol{N}} \hat{\boldsymbol{\Sigma}}, \quad \dot{\boldsymbol{\Gamma}} \geqq \boldsymbol{0}, \tag{37a, b}$$

$$\dot{\boldsymbol{\Gamma}}^T (\bar{\boldsymbol{H}} \dot{\boldsymbol{\Gamma}} - \bar{\boldsymbol{N}} \hat{\boldsymbol{\Sigma}}) = 0. \tag{38}$$

Combining (34)–(36), we obtain the general matrix form of the discrete equations (29):

$$(\boldsymbol{B}_i^T \boldsymbol{S} \boldsymbol{B}_i + \chi_i) \dot{\boldsymbol{u}} - \boldsymbol{B}_i^T (\boldsymbol{S} \bar{\boldsymbol{N}}^T + \bar{\boldsymbol{M}}) \dot{\boldsymbol{\Gamma}} = -(\boldsymbol{B}_i^T \boldsymbol{S} \boldsymbol{B}_0 + \chi_0) \dot{\boldsymbol{U}}^0 \tag{39}$$

in which the aggregate elastic 'stiffness' matrix $\boldsymbol{K}_E = \boldsymbol{B}_i^T \boldsymbol{S} \boldsymbol{B}_i$, assembled from the $\boldsymbol{K}_{(q)}^{JM}$ over pairs J, M of unknown nodal velocities, is symmetric positive definite ($\mathscr{C}_{(L)}$ is positive-definite, hence $\boldsymbol{C}^{-1}$). $\boldsymbol{\chi}_i$ also is symmetric, and we denote

$$\boldsymbol{K} = \boldsymbol{B}_i^T \boldsymbol{S} \boldsymbol{B}_i + \boldsymbol{\chi}_i, \tag{40}$$

the overall stiffness matrix at large strains. From (29)–(32) and the general definitions of sect. 4, $\boldsymbol{K}$ and all other matrices appearing in (34)–(39) depend upon both the present state of the aggregate and its deformation history, but not upon instantaneous motion from that state. Hence, the problem defined by these equations is quasi-linear.

Consider the following condensed set of governing equations reduced from the above:

$$\boldsymbol{K}\dot{\boldsymbol{u}} - \boldsymbol{B}_i^T(\boldsymbol{S}\bar{\boldsymbol{N}}^T + \bar{\boldsymbol{M}})\dot{\boldsymbol{\Gamma}} = -(\boldsymbol{B}_i^T \boldsymbol{S} \boldsymbol{B}_0 + \boldsymbol{\chi}_0)\dot{\boldsymbol{U}}^0, \tag{41}$$

$$(\bar{\boldsymbol{H}} + \bar{\boldsymbol{N}}\boldsymbol{S}\bar{\boldsymbol{N}}^T)\dot{\boldsymbol{\Gamma}} - \bar{\boldsymbol{N}}\boldsymbol{S}\boldsymbol{B}_i\dot{\boldsymbol{u}} - \bar{\boldsymbol{N}}\boldsymbol{S}\boldsymbol{B}_0\dot{\boldsymbol{U}}^0 = \boldsymbol{\lambda}, \tag{42}$$

$$\boldsymbol{\lambda} \geqq \boldsymbol{0}, \quad \dot{\boldsymbol{\Gamma}} \geqq \boldsymbol{0}, \quad \dot{\boldsymbol{\Gamma}}^T\boldsymbol{\lambda} = 0, \tag{43a, b, c}$$

in which $\boldsymbol{\lambda}$ is simply a convenient auxiliary vector defined by (42). Since no assumptions have been made characterizing $\bar{\boldsymbol{H}}$, $\boldsymbol{K}$, or $\bar{\boldsymbol{M}}$. (41)–(43) represents the most general, compact form of the governing equations of the crystalline aggregate model at large strain. These equations bear a *formal* similarity to matrix equations given by Maier [14] as appropriate for the large displacement analysis of elastoplastic *structures*. The principal difference between the two sets of equations is the term containing $\bar{\boldsymbol{M}}$, which appears here but not in [14]. This term reflects the fact that, in the continuum physics of deforming crystals, elastic strain-rates referred to fixed axes must be related to *lattice* co-rotational rather than 'material' co-rotational stress-rates. [(8a, b), (36a, b). Also see [4].] In contrast, Maier's *macro*-constitutive assumptions, although defined in terms of so-called 'natural' generalized variables for 'finite elements' at the outset, essentially involve material co-rotational stress rates, so that a term like $\boldsymbol{B}_i^T\bar{\boldsymbol{M}}\dot{\boldsymbol{\Gamma}}$ does not occur. Nonetheless, general discussions in [14] of alternative representations of the equation sets as quadratic programming problems can guide us herein. Thus, we find that equations (41)–(43) define a 'linear complementarity problem' with which the following programming problem is associated. Minimize the quadratic functional

$$\Pi(\dot{\boldsymbol{u}}, \dot{\boldsymbol{\Gamma}}) = \dot{\boldsymbol{u}}^T\bar{\boldsymbol{\chi}}\dot{\boldsymbol{U}} + \dot{\boldsymbol{\Gamma}}^T\bar{\boldsymbol{H}}\dot{\boldsymbol{\Gamma}} - \dot{\boldsymbol{u}}^T\boldsymbol{B}_i^T\bar{\boldsymbol{M}}\dot{\boldsymbol{\Gamma}} + (\dot{\boldsymbol{u}}^T\boldsymbol{B}_i^T - \dot{\boldsymbol{\Gamma}}^T\bar{\boldsymbol{N}})\boldsymbol{S}(\bar{\boldsymbol{B}}\dot{\boldsymbol{U}} - \bar{\boldsymbol{N}}^T\dot{\boldsymbol{\Gamma}}) \tag{44}$$

subject to the linear constraints

$$(\boldsymbol{B}_i^T \boldsymbol{S}\bar{\boldsymbol{B}} + \bar{\chi})\dot{\boldsymbol{U}} - \boldsymbol{B}_i^T(\boldsymbol{S}\bar{\boldsymbol{N}}^T + \bar{\boldsymbol{M}})\dot{\boldsymbol{\Gamma}} = \boldsymbol{0}, \tag{45}$$

$$(\bar{\boldsymbol{H}} + \bar{\boldsymbol{N}}\boldsymbol{S}\bar{\boldsymbol{N}}^T)\dot{\boldsymbol{\Gamma}} - \bar{\boldsymbol{N}}\boldsymbol{S}\bar{\boldsymbol{B}}\dot{\boldsymbol{U}} \geqq \boldsymbol{0}, \quad \dot{\boldsymbol{\Gamma}} \geqq \boldsymbol{0}. \tag{46a, b}$$

If the minimum of Π is zero, a solution of the programming problem also satisfies (41)–(43). If not, no solution to (41)–(43) exists. That the minimum must be zero in order for the programming solution to satisfy the original problem follows trivially from (41)–(43). Forming scalar products of (41) with $\dot{\boldsymbol{u}}$ and (42) with $\dot{\boldsymbol{\Gamma}}$ and adding, $\Pi(\dot{\boldsymbol{u}}, \dot{\boldsymbol{\Gamma}}) \equiv 0$ is an immediate consequence of (43c). Conversely, for *any* $\dot{\boldsymbol{u}}^*$, $\dot{\boldsymbol{\Gamma}}^*$ satisfying (45), (46a, b) we find, upon forming scalar products and adding, $\Pi(\dot{\boldsymbol{u}}^*, \dot{\boldsymbol{\Gamma}}^*) \geqq 0$. Therefore, if particular vectors $\dot{\boldsymbol{u}}^*$, $\dot{\boldsymbol{\Gamma}}^*$ minimizing Π in the programming problem yield $\Pi(\dot{\boldsymbol{u}}^*, \dot{\boldsymbol{\Gamma}}^*) > 0$, there is no solution $\dot{\boldsymbol{u}}$, $\dot{\boldsymbol{\Gamma}}$ to (41)–(43) since its existence would require $\Pi(\dot{\boldsymbol{u}}, \dot{\boldsymbol{\Gamma}}) \equiv 0 \geqq \Pi(\dot{\boldsymbol{u}}^*, \dot{\boldsymbol{\Gamma}}^*)$, a contradiction. (Q.E.D.)

Henceforth, to direct attention to problems of the greatest physical interest (and to approach the form of the aggregate programming problem for small strains defined in [5]), we restrict consideration to invertible $\boldsymbol{K}$. Moreover, from (30), (31), and (40), we may as well take $\boldsymbol{K}$ positive-definite since $\boldsymbol{K}_E$ contains terms of order μ, a typical crystal shear modulus, whereas χ_i contains terms of order τ_{cr}, a typical critical shear strength $\ll \mu$ (from phenomenological crystal hardening laws). Thus, denoting

$$\boldsymbol{Q} = \boldsymbol{S}[\boldsymbol{I} - \boldsymbol{B}_i \boldsymbol{K}^{-1} \boldsymbol{B}_i^T \boldsymbol{S}], \tag{47}$$

we can eliminate $\dot{\boldsymbol{u}}$ from (41)–(43) and recast the equations in the form:

$$\boldsymbol{P}\dot{\boldsymbol{\Gamma}} - \bar{\boldsymbol{N}}\mathring{\boldsymbol{\Sigma}}^{(e)} \geqq \boldsymbol{0}, \quad \dot{\boldsymbol{\Gamma}} \geqq \boldsymbol{0}, \tag{48a, b}$$

$$\dot{\boldsymbol{\Gamma}}^T(\boldsymbol{P}\dot{\boldsymbol{\Gamma}} - \bar{\boldsymbol{N}}\mathring{\boldsymbol{\Sigma}}^{(e)}) = 0 \tag{49}$$

with

$$\boldsymbol{P} = \bar{\boldsymbol{H}} + \bar{\boldsymbol{N}}\boldsymbol{Q}\bar{\boldsymbol{N}}^T - \bar{\boldsymbol{N}}(\boldsymbol{I} - \boldsymbol{Q}\boldsymbol{S}^{-1})\bar{\boldsymbol{M}}, \tag{50}$$

$$\mathring{\boldsymbol{\Sigma}}^{(e)} = (\boldsymbol{Q}\boldsymbol{B}_0 - \boldsymbol{S}\boldsymbol{B}_i\boldsymbol{K}^{-1}\chi_0)\dot{\boldsymbol{U}}^0. \tag{51}$$

($\mathring{\boldsymbol{\Sigma}}^{(e)}$ is the overall vector of material co-rotational stress-rates corresponding to assumed elastic response of the aggregate, as defined for small strains in [6].) In the associated quadratic programming problem we minimize

$$\pi(\dot{\boldsymbol{\Gamma}}) = \dot{\boldsymbol{\Gamma}}^T\boldsymbol{P}\dot{\boldsymbol{\Gamma}} - \dot{\boldsymbol{\Gamma}}^T\bar{\boldsymbol{N}}\mathring{\boldsymbol{\Sigma}}^{(e)} \tag{52}$$

subject to the linear constraints (48a, b). As before, if min $\pi(\dot{\boldsymbol{\Gamma}}) = 0$, a solution of the programming problem satisfies (48)–(49). If min $\pi(\dot{\boldsymbol{\Gamma}}) > 0$, no solution to (48)–(49) exists (the proof following as before).

6. Uniqueness of solutions

Let $\Delta\dot{s}$, $\Delta\boldsymbol{v}$ denote differences between pairs of 'statically' admissible nominal stress-rate fields[1]) and pairs of kinematically admissible velocity fields, respectively. From the principle of virtual work,

$$\overline{\Delta\dot{s}\cdot\Delta\boldsymbol{L}} \equiv 0, \tag{53}$$

whence, a sufficient condition for uniqueness of solution is (see [15])

$$\overline{\Delta\hat{\boldsymbol{\tau}}\cdot\Delta\boldsymbol{D}} - \overline{(\boldsymbol{\tau}\Delta\boldsymbol{W}+\Delta\boldsymbol{D}\boldsymbol{\tau})\cdot\Delta\boldsymbol{L}} > 0. \tag{54}$$

For the crystalline aggregate, utilizing (1), (4), (5), (7), (8a), (10a, b) and the following inequality due to Hill [3],

$$\Delta\hat{\boldsymbol{\tau}}\cdot(\Delta\boldsymbol{D}-\Delta\boldsymbol{\varepsilon}) = \Delta\hat{\boldsymbol{\tau}}\cdot\Delta\boldsymbol{D}^p \geqq \Delta\dot{\gamma}_k h^{kj}\Delta\dot{\gamma}_j, \tag{55}$$

we can write [from (53)]

$$0 \geqq \overline{\Delta\hat{\boldsymbol{\tau}}\cdot\mathscr{C}\cdot\Delta\hat{\boldsymbol{\tau}}} + \overline{\Delta\dot{\gamma}_k h^{kj}\Delta\dot{\gamma}_j} - \overline{(\boldsymbol{\tau}\Delta\boldsymbol{W}+\Delta\boldsymbol{D}\boldsymbol{\tau})\cdot\Delta\boldsymbol{L}} - 2\overline{\Delta\dot{\gamma}_k\boldsymbol{\Omega}^k\boldsymbol{\tau}\cdot\Delta\boldsymbol{D}}. \tag{56}$$

In terms of the overall vectors and matrices of the discrete model, we find, corresponding to (53),

$$\sum_q \langle\Delta\dot{s}\cdot\Delta\boldsymbol{L}\rangle_{(q)} = \Delta\dot{\boldsymbol{u}}^T\boldsymbol{K}\Delta\dot{\boldsymbol{u}} - \Delta\dot{\boldsymbol{u}}^T\boldsymbol{B}_i^T(\boldsymbol{S}\bar{\boldsymbol{N}}^T+\bar{\boldsymbol{M}})\Delta\dot{\boldsymbol{\Gamma}} \equiv 0 \tag{57}$$

(since $\boldsymbol{K}\Delta\dot{\boldsymbol{u}} = \boldsymbol{B}_i^T(\boldsymbol{S}\bar{\boldsymbol{N}}^T+\bar{\boldsymbol{M}})\Delta\dot{\boldsymbol{\Gamma}}$ from (39), with $\Delta\dot{\boldsymbol{U}}^0 \equiv \boldsymbol{0}$). Moreover, from (34)–(39), corresponding to (56),

$$0 \geqq \Delta\hat{\boldsymbol{\Sigma}}\cdot\boldsymbol{S}^{-1}\Delta\hat{\boldsymbol{\Sigma}} + \Delta\dot{\boldsymbol{\Gamma}}^T\bar{\boldsymbol{H}}\Delta\dot{\boldsymbol{\Gamma}} + \Delta\dot{\boldsymbol{u}}^T\boldsymbol{\chi}_i\Delta\dot{\boldsymbol{u}} - \Delta\dot{\boldsymbol{u}}^T\boldsymbol{B}_i^T\bar{\boldsymbol{M}}\Delta\dot{\boldsymbol{\Gamma}}, \tag{58}$$

and, for $\boldsymbol{K}$ invertible,

$$0 \geqq \Delta\hat{\boldsymbol{\Sigma}}\cdot\boldsymbol{S}^{-1}\Delta\hat{\boldsymbol{\Sigma}} + \Delta\dot{\boldsymbol{\Gamma}}^T(\bar{\boldsymbol{H}}+\boldsymbol{G})\Delta\dot{\boldsymbol{\Gamma}} = \Delta\dot{\boldsymbol{\Gamma}}^T\boldsymbol{P}\Delta\dot{\boldsymbol{\Gamma}}, \tag{59}$$

wherein

$$\boldsymbol{G} = \bar{\boldsymbol{N}}_G(\boldsymbol{I}-\boldsymbol{Q}\boldsymbol{S}^{-1})(\boldsymbol{Q}\bar{\boldsymbol{N}}_G^T-\bar{\boldsymbol{M}}), \tag{60}$$

$$\bar{\boldsymbol{N}}_G = \bar{\boldsymbol{N}}+\bar{\boldsymbol{M}}^T\boldsymbol{S}^{-1}. \tag{61}$$

[1]) In the discrete model, stress-rate fields satisfying (28).

Obviously, for $\boldsymbol{P}$ positive-definite, if a solution to (48)–(49) exists it is unique. But for $\boldsymbol{P}$ merely positive semi-definite, $\pi(\dot{\boldsymbol{\Gamma}})$ is a convex functional, and the solution to the minimization problem (52), subject to (48a, b), also satisfies (49). Hence, positive-definiteness of $\boldsymbol{P}$ is sufficient to assure both existence and uniqueness of solution for the discrete aggregate model.

Recalling that $\boldsymbol{S} > 0$ (that is, positive-definite), we obtain from (59) the following continued inequality:

$$0 \geqq \Delta\dot{\boldsymbol{\Gamma}}^T \boldsymbol{P} \Delta\dot{\boldsymbol{\Gamma}} > \Delta\dot{\boldsymbol{\Gamma}}^T [\bar{\boldsymbol{H}} + \bar{\boldsymbol{N}}_G(\boldsymbol{I} - \boldsymbol{Q}\boldsymbol{S}^{-1})(\boldsymbol{Q}\bar{\boldsymbol{N}}_G^T - \bar{\boldsymbol{M}})] \Delta\dot{\boldsymbol{\Gamma}}. \tag{62}$$

At small strains this reduces to (from [5])

$$0 \geqq \Delta\dot{\boldsymbol{\Gamma}}^T \boldsymbol{P} \Delta\dot{\boldsymbol{\Gamma}} > \Delta\dot{\boldsymbol{\Gamma}}^T \bar{\boldsymbol{H}} \Delta\dot{\boldsymbol{\Gamma}}, \tag{63}$$

with $\boldsymbol{P} = \bar{\boldsymbol{H}} + \bar{\boldsymbol{N}}\boldsymbol{Q}\bar{\boldsymbol{N}}^T$, and the solution is unique for positive semi-definite $\boldsymbol{H}$ (aside from possible geometrically dependent slip systems within a crystallite producing the same $\boldsymbol{D}^p$). Equation (63) also holds, even at large strains, when convection and rotation effects can be *instantaneously* neglected during deformation from the current deformed state. As $\boldsymbol{P}$ is symmetric positive-definite in these special cases (taking $\boldsymbol{H}$ to be symmetric), two additional forms of the quadratic programming problem become available. In the first, minimize

$$I(\dot{\boldsymbol{\Gamma}}) = \tfrac{1}{2}\dot{\boldsymbol{\Gamma}}^T \boldsymbol{P} \dot{\boldsymbol{\Gamma}} - \dot{\boldsymbol{\Gamma}}^T \bar{\boldsymbol{N}} \mathring{\boldsymbol{\Sigma}}^{(e)} \tag{64}$$

subject only to $\dot{\boldsymbol{\Gamma}} \geqq \boldsymbol{0}$ [5, 6]. In the second, maximize $-\dot{\boldsymbol{\Gamma}}^T \boldsymbol{P} \dot{\boldsymbol{\Gamma}}$ subject to $\boldsymbol{P}\dot{\boldsymbol{\Gamma}} - \bar{\boldsymbol{N}} \mathring{\boldsymbol{\Sigma}}^{(e)} \geqq \boldsymbol{0}$ [14]. Neither of these (dual) forms can be adopted when convection and rotation effects must be included, however, since from (50) $\boldsymbol{P}$ will not be symmetric even though positive-definite.

In closing we note that, at small strains (with $\boldsymbol{H} \geqq 0$), an absolute minimum principle is available for the functional $2I_1 = \overline{\dot{\boldsymbol{\tau}} \cdot \boldsymbol{D}}$ [6] (also see [3]). By virtue of this principle, it is possible to prove strict convergence of the discrete stress- and strain-rate fields (of the aggregate model) to the corresponding micro-continuum fields as element sizes are reduced within crystal grains [6].

Acknowledgement

This work was supported by the Office of Naval Research and by the Advanced Research Projects Agency of the Department of Defense under Contract No. N00014-68-A-0187.

References

[1] Hill, R., *Surveys in Mechanics* (G. I. Taylor 70th Anniversary Volume), Cambridge University Press, (1956) 7–31.
[2] Krumhansl, J. A., *Mechanics of Generalized Continua* (IUTAM Symposium), ed., E. Kröner, Springer-Verlag, Berlin (1968) 298–311.
[3] Hill, R., *J. Mech. Phys. Solids 14* (1966) 95–102.
[4] Havner, K. S., Rept. 70–9, ONR Contract N00014–68–A–0187, North Carolina State University, December 1970, to appear in *Acta Mechanica.*
[5] Havner, K. S., *Int. J. Solids Struct. 7* (1971) 719–730.
[6] Havner, K. S., *Int. J. Solids Struct. 7* (1971) 1269–1275.
[7] Hill, R., *J. Mech. Phys. Solids 7* (1959) 209–225.
[8] Mandel, J., *Int. J. Solids Struct. 1* (1965) 273–295.
[9] Mair, W. M. and H. Ll. D. Pugh, *J. Mech. Engng. Sci. 6* (1964) 150–163.
[10] Ronay, M., *Int. J. Solids Struct. 4* (1968) 509–516.
[11] Bishop, J. F. W. and R. Hill, *Phil. Mag.*, ser. 7, *42* (1951) 414–427.
[12] Kocks, U. F., *Acta Metallurgica 6* (1958) 85–94.
[13] Hill, R., *J. Mech. Phys. Solids 15* (1967) 79–95.
[14] Maier, G., *Int. J. Solids Struct. 7* (1971) 345–372.
[15] Hill, R., *J. Mech. Phys. Solids 6* (1958) 236–249.

Composite and 'overlay' models in numerical analysis of elasto-plastic continua

O. C. Zienkiewicz

University of Wales, Swansea, U.K

G. C. Nayak

University of Roorkee (U.P.), India

D. R. J. Owen

University of Wales, Swansea, U.K

The formulation of physically correct constitutive relations for complex material behaviour presents many difficulties. The need for an explicit constitutive relation may, in numerical analysis, be avoided by using a composite or overlay model of the material in which the constituents behave in a relatively simple manner.

Here, using only elastic-ideally plastic constituents, more satisfactory behaviour patterns are achieved than those corresponding to isotropic or kinematic hardening laws. Numerically integrated isoparametric finite elements allow an easy implementation of the procedure which is of very wide applicability.

Some numerical examples illustrate the paper.

1. Introduction

Numerical methods and in particular the finite element procedures developed during recent years, permit the stress analysis of realistic configurations encountered in various branches of engineering and the incorporation of almost any material constitutive relationships [1]. Indeed at the present stage of their development the commonly heard complaint is that the lack of experimental data and of corresponding constitutive theory present the biggest stumbling block to further progress. Perhaps such criticism is too severe and the analyst is expecting too much when he demands mathematically neat and compact phenomenological laws to be presented to him by the experimentalists for physical problems of great complexity.

In this paper we shall attempt to indicate a new philosophy of approach which has proved successful in overcoming some difficulties of the problem in which the very characteristics of the numerical method are used to bypass the necessity for explicit constitutive relationship.

Two basic approaches seem to be present in the definition of physical behaviour of solids. One, strictly phenomenological, attempts to obtain material 'laws' by considering the behaviour of 'macro specimens' and formulating the theory by suitable description (with appropriate care taken not to violate overall thermodynamic principles). The second considers behaviour of the material on the 'micro' scale and attempts to underline the basic reasons behind the observable macro-relationships.

The conventional continuum plasticity theory developed by pioneers of this field [2]–[5] is based on the first approach while studies of slips, dislocations and other micro phenomena [6]–[10] give an example of the second.

Unfortunately, in general, the more fundamental micro-studies have not yet been generalized to a point where monocrystal predictions of overall constitutive relations can be obtained and on the other hand the phenomenological approaches often fail to decide adequately all phases of behaviour.

The numerical processes have a basic advantage which may allow us to bypass these difficulties. In the analysis the structure is in general divided into a series of 'elements'. Is it possible by assigning *simple* but different behaviour patterns to each element of finite size to obtain a composite behaviour which models reality? If the answer to this question is affirmative then some of the advantageous features of the 'micro' approach can be used directly in the analysis of the structure avoiding the need for an explicit formulation of a constitutive relation. In the next section we shall consider in more detail the methods by which this can be accomplished.

Throughout the paper we shall implicitly use a *displacement finite element* formulation and we shall assume that the reader is familiar with the basis of it. Summarising we have a displacement field defined by a series of nodal parameters (usually nodal displacements) of the form

$$\{u\} = \sum_{i=1}^{n} [N]_i\{\delta\}_i \tag{1}$$

where the matrices $[N]_i$ consist of appropriate shape functions defined in a piecewise manner throughout the domain. Thus the strains are given by

$$\{\varepsilon\} = \sum_{i=1}^{n} [B]_i\{\delta\}_i \qquad (2)$$

in which the matrices $[B]_i$ are again known functions of position.

The virtual work principle yields an approximate equilibrium condition as

$$\int_V [B]_i^T\{\sigma\}dV - \{F\}_i = 0. \qquad (3)$$

Here $\{F\}_i$ represents generalised forces corresponding to $\{\delta\}_i$, and $\{\sigma\}$ are the stresses at any point of the region.

If some constitutive law defines $\{\sigma\}$ in terms of $\{\varepsilon\}$ and its history, equations (3) give a non-linear set which, by solution, yields the full answer to the problem.

2. The general approach to simplified models

2.1 *A composite of elements*

The basic ideas of this section can best be illustrated on a rather complex problem perhaps falling outside the scope of simple 'plasticity' consideration. This concerns the behaviour of a rock mass whose description still eludes the effort of researchers in this field.

A simple model, however, exists for a particular case of a stratified, frictional, material as shown in Fig. 1. Here the behaviour is generally elastic provided (1) σ_n is negative (compressive), and (2) $\tau_{sn} \leqq \mu\sigma_n$.

If either of these limits is exceeded separation or slippage of the layers occur [11]. In Fig. 2(a) we show how a specimen (or a structure) of such a material responds to a simple compressive test.

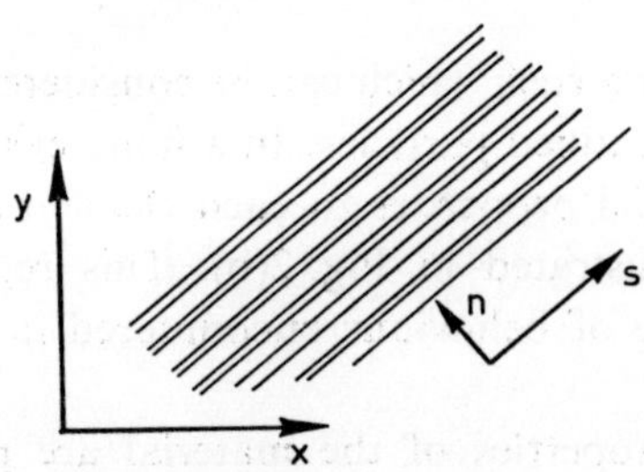

Fig. 1. A stratified continuum.

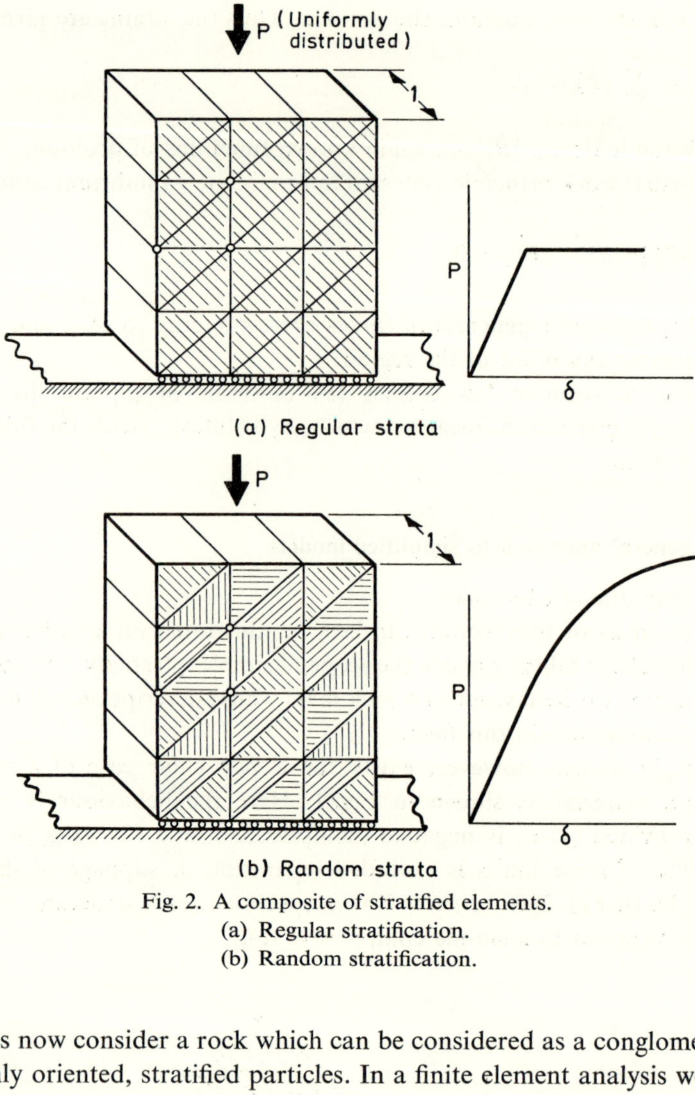

Fig. 2. A composite of stratified elements.
(a) Regular stratification.
(b) Random stratification.

Let us now consider a rock which can be considered as a conglomerate of randomly oriented, stratified particles. In a finite element analysis we could assign different material properties to each element and obtain the overall behaviour pattern illustrated in Fig. 2(b). This represents now (at least qualitatively) the type of behaviour encountered in a test of typical rock masses.

The 'directional' properties of the material are progressively lost with increasing number of elements and a randomness of their orientation. This concept was introduced into rock mechanics first by Zienkiewicz [12] and Dullage [13]. Indeed it follows a much older 'model' of describing a rock or concrete behaviour suggested by Brandtzaeg [14] in 1927. It is however only the finite element process that permits its practical utilisation. By the use of *simple* components we have in this case succeeded in representing a more complex material.

2.2 *A composite of integrating points*

The simple triangular element which is the 'original' element used in practice has in recent years proved to be uneconomic in linear analysis and is largely superseded by more complex, curved elements with numerous nodes as illustrated in Fig. 3.

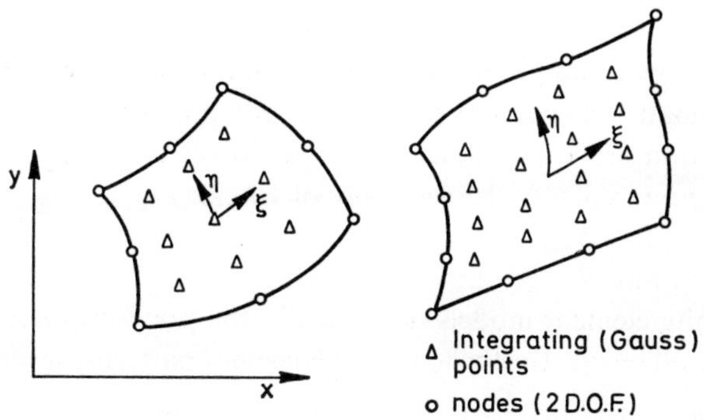

Fig. 3. Some 'modern' isoparametric elements in two dimensions.

Here exact integration of eq. (3) is always replaced by a *numerical integration* or summation at a selected number of integration points [1]). As a smaller number of elements are now invariably used in analysis the only apparent way is to introduce the randomness of material orientation by the variation of its properties at the selected integrating points. With the integral of eq. (3) replaced by a summation over l integration points, i.e.

$$\int_V [B]_i^T\{\sigma\}dV = \sum_{m=1}^{l} Am[B]_i^T\{\sigma\}_m \tag{4}$$

in which Am are the weighting coefficients.

[1]) These are known as Gauss points in the context of isoparametric elements since Gaussian integration rules are generally used.

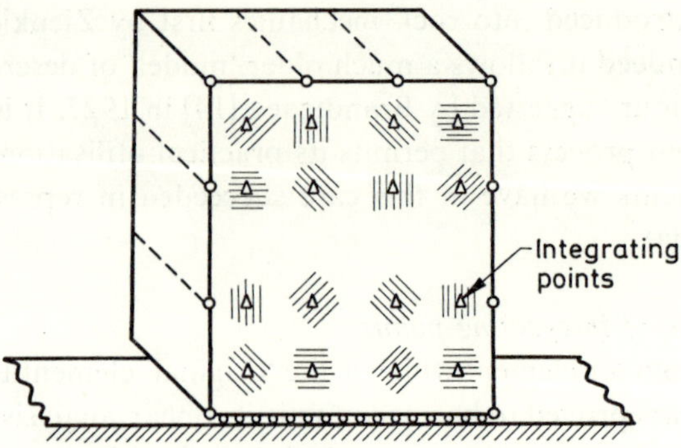

Fig. 4. A composite using integrating points of a complex element to model problem of Fig. 2b.

We apply now a simple but different material behaviour model at each integrating point achieving the overall effect implied in the conceptually simpler situation previously described. Fig. 4 illustrates the procedure where one element now replaces the assembly of Fig. 2.

2.3 *An overlay model*

While the finite element models described in the previous sections will yield excellent overall results for the load-displacement patterns, several difficulties exist:

(a) the number of elements or integration points must be sufficiently large to model the true behaviour pattern,

(b) the establishment of the relation between the experimentally observed, overall data and the behaviour of the composite is tedious,

(c) the stress pattern resulting from the analysis will obviously be random and only represent the behaviour *in the mean*. Thus the addition of some stress averaging procedure is essential.

These difficulties can either be overcome completely or at least alleviated by the introduction of the 'overlay' model. The name and its conceptual visualisation originate from the two dimensional example under discussion, though the principle is by no means limited to two dimensional problems.

Consider the problems of Fig. 2 and 4 now to be composed of κ-parallel layers of the material (not necessarily of equal thickness) as shown in Fig. 5. Each layer may now have different properties, (in the case under discussion a different stratification direction) and if a sufficient number of such layers

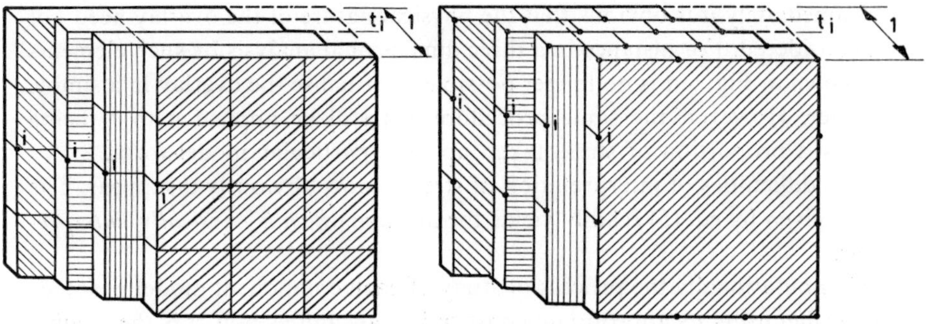

Fig. 5. The 'overlay' model with simple or complex element.

is included then, clearly, the same overall behaviour patterns as before can be achieved.

The advantage of the 'overlay' composite is obvious. The behaviour of integrating point or element is identical. With the nodes of each layer coincidental, identical strain is imposed in each layer. These in turn cause different stresses $\{\sigma\}_j$ to be developed in each layer, which contribute their appropriate portion to the integral of equation (3) or in the summation of equation (4). With the total 'thickness' of the layer being unity in the case considered the stress resultant $\{\sigma\}$ is equal to the sum of individual stresses suitably weighted i.e.

$$\{\sigma\} = \sum_{j=1}^{\kappa} \{\sigma\}_j t_j \quad \text{with} \quad \sum_{j=1}^{\kappa} t_j = 1. \tag{5}$$

If equally thick layers are considered

$$t_j = \frac{1}{\kappa} \tag{6}$$

but in general any subdivisions may be used in order to develop a desirable directional property.

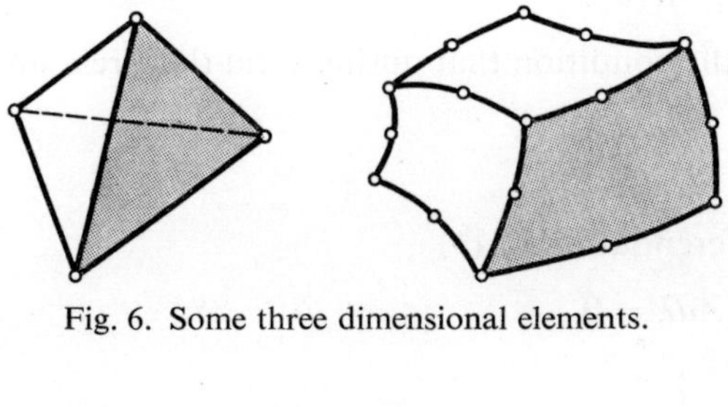

Fig. 6. Some three dimensional elements.

Immediately it is obvious that the physical nature of the overlay need not be considered. For instance in a three dimensional analysis by such elements as shown in Fig. 6 it is no longer possible to visualise the overlay in a simple manner. However, the same procedure can still be applied with t_j being simply a proportion of stress taken up by a particular model form.

The difficulties of interpretation of stresses now obviously disappear. The stress given by eq. (5) readily yields the *average which is physically interpretable.*

Computationally the process of study of each component model proceeds in parallel using its appropriate constitutive law and as now it is the *number of overlays* and not the number of elements (or integrating points) which determine the overall behaviour, a much reduced total number of elements need be employed in a given problem.

3. Applications in strain hardening plasticity

Formulation and practical solution of plasticity problem in the finite element field is now so well determined. A fairly detailed survey of the available solution processes is given in [1]. It will be found that in all processes the essential step consists of the introduction of the incremental elasto-plastic matrix.

Making use of the basic work by the processes of plasticity we note that plastic strain changes are related to gradients of the plastic potential Q. Thus

$$d\{\varepsilon\}_p = d\lambda\{a^*\}; \qquad Q \equiv Q(\{\sigma\}) \tag{7}$$

with

$$\{a^*\}^T = \left\{\frac{\partial Q}{\partial \sigma_x}, \ \ldots\right\} \tag{8}$$

and the elastic strain is given by

$$d\{\varepsilon\}_e = [D]^{-1} d\{\sigma\}.$$

Taking further the condition that during yield the stress must remain on the yield surface

$$F \equiv F(\{\sigma\}, \kappa) = 0. \tag{9}$$

We have by differentiation of (9)

$$\{a\}^T d\{\sigma\} - A d\lambda = 0 \tag{10}$$

where

$$\{a\}^T \equiv \left\{ \frac{\partial F}{\partial \sigma_x}, \; \dots \right\} \tag{11}$$

and

$$A = - \frac{1}{d\lambda} \frac{\partial F}{\partial \kappa} \, d\kappa. \tag{12}$$

As it is assumed that total strain changes are the sum of plastic and elastic components i.e.

$$d\{\varepsilon\} = d\{\varepsilon\}_p + d\{\varepsilon\}_e. \tag{13}$$

We find after elimination of the plastic multiplier $d\lambda$ that we can write

$$d\{\sigma\} = [D]_{ep} d\{\varepsilon\} \tag{14}$$

with

$$[D]_{ep} = [D] - [D]\{a^*\}[D]\{a\}^T/(A + \{a\}^T[D]\{a^*\}). \tag{15}$$

This explicit form was first introduced to the case of associated plasticity $F \equiv Q$ by Zienkiewicz, Valliappan and King [15] and generalised to non-associated situations recently [16].

For elastic-ideally plastic situations no difficulties arise. Indeed introduction

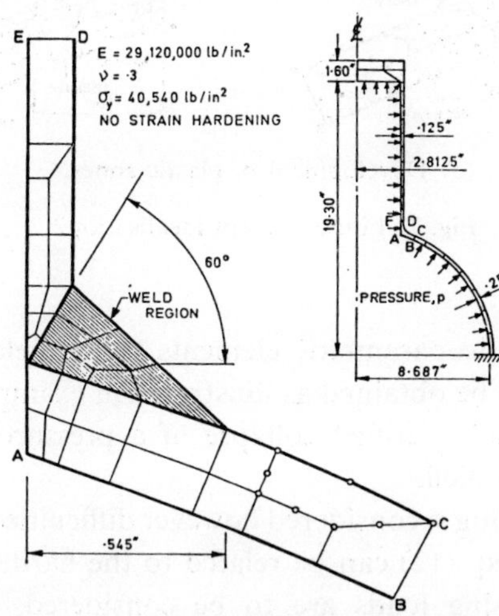

(a) of a spherical steel pressure vessel with flush nozzle junction.

Fig. 7a.

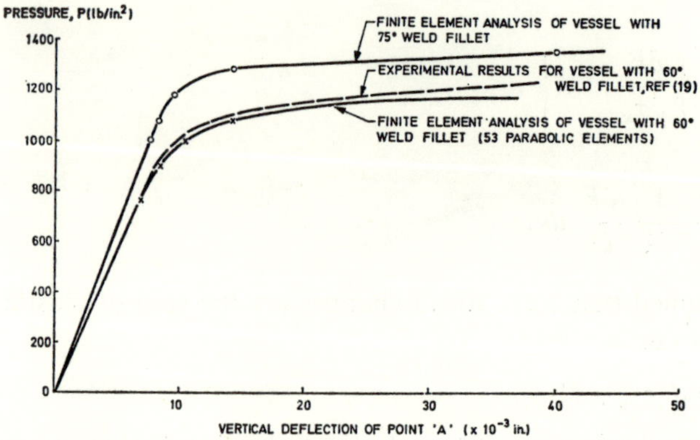

(b) Load displacement relationship for an internal point of the vessel.

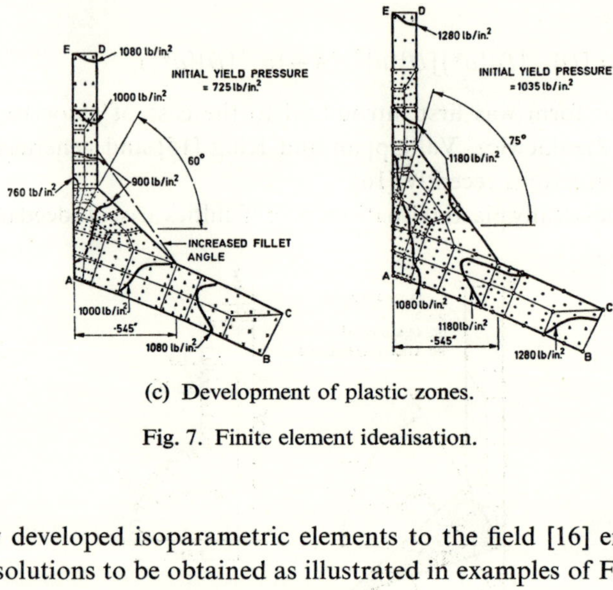

(c) Development of plastic zones.

Fig. 7. Finite element idealisation.

of recently developed isoparametric elements to the field [16] enables very economic solutions to be obtained as illustrated in examples of Fig. 7 and 8, where an experimentally verified collapse of a pressure vessel and an extrusion process are studied.

When strain hardening is considered however difficulties arise immediately. The parameter A of eq. (12) can be related to the hardening theory postulated. If only increasing loads are to be considered, so called *isotropic hardening* models the situation reasonably well and indeed in that case 'A' is simply related to the slope of the uniaxial stress strain curve of Fig. 9.

This type of situation has been well modelled in finite element analysis for some time but necessitating now an addition of the uniaxial curve characteristics as program data [16].

For unloading or load cycling the postulation of isotropic hardening law fails badly. If applied to the uniaxial case it results in completely fallacious behaviour on unloading indicated in Fig. 9.

To overcome these difficulties Prager [4] and later Ziegler [18] introduced the so called kinematic hardening hypothesis in which a shift of the yield surface is postulated. Whilst this improves the performance it introduces

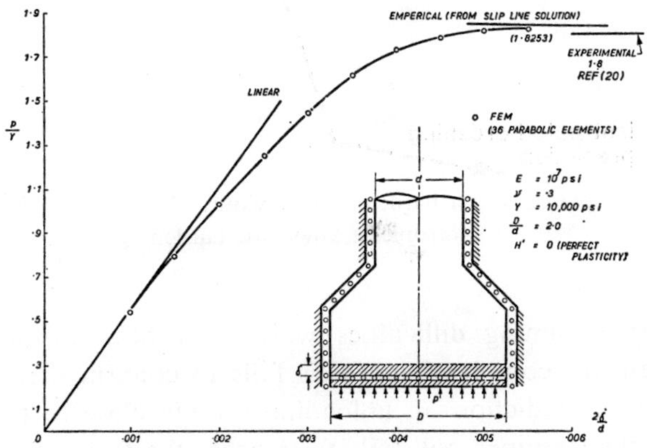

(a) Mean pressure/displacement curve.

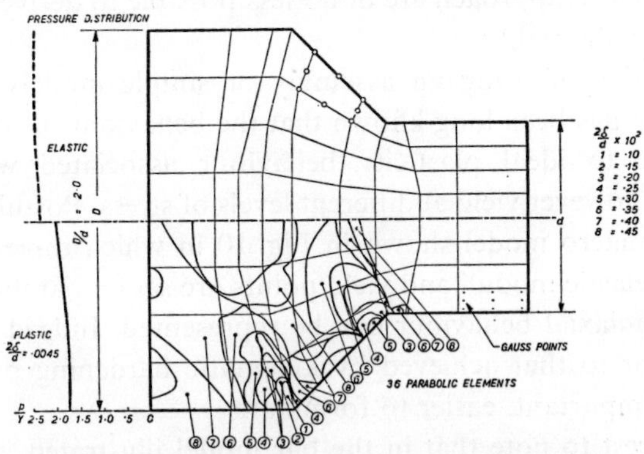

(b) Spread of plastic zones and pressure distribution

Fig. 8. An axisymmetric problem.

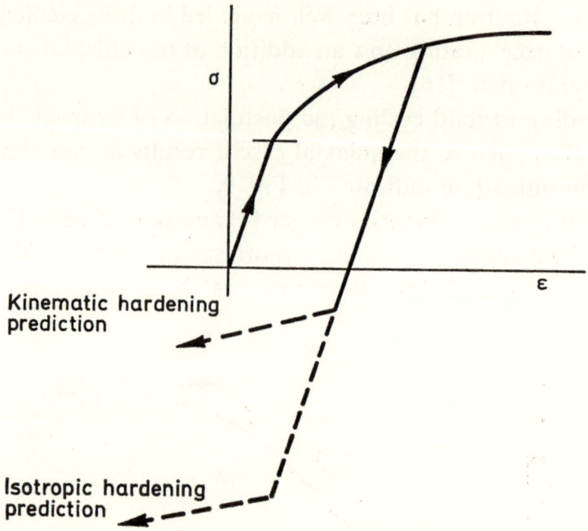

Fig. 9. Uniaxial plastic strain hardening behaviour. Unloading cycle prediction by isotropic or kinematic hardening.

additional programming difficulties which can be overcome [16], [18]. However even this complex hypothesis fails to confirm the reality. In the uniaxial case its prediction on unloading though close to the actual performance of the specimen still fails to indicate the form of 'reverse yield', particularly in the form of the slope of the curve. Further applications of the 'phenomenological' approach are doubtless possible to derive a model made in accordance with reality.

The alternative of using an assembly of simple models suggests itself immediately. It has been long known that the behaviour in strain hardening is largely due to ideal plasticity behaviour associated with individual crystals which however yield at different levels of stress. Postulating a simple, thin element, macro model shown in Fig. 10 in which three 'parallel bars' with different elastic moduli and yield points are set on parallel. We see here well the true uniaxial behaviour can be represented. Indeed the prediction is now superior to that achieved by kinematic hardening postulation and what is more important, easier to formulate.

It is of interest to note that in the bar model illustrated, the proportion t^A, t^B and t^C can be varied as well as the elastic moduli of the bars. In the present case we have taken the moduli as identical but always this is not necessary. With a suitable number of elastic-ideally plastic elements we can model uniaxial behaviour to a high degree of accuracy.

Immediately on proving the success of the one dimensional composite model, a two dimensional model may be similarly considered by using any of the three processes described in section 2. Thus for instance we can assign several elastic-ideally plastic proportions to individual elements; or integrating points or to a number of overlays.

The later model, for reasons already mentioned, forms the most advantageous possibility. In the example given below we achieve different patterns of realistic stress-strain behaviour of a 2×2 element assembly by a scattered distribution of simple properties.

Using four cubic type elements and assigning simply two different elastic-ideally plastic properties at a pattern of Gauss points shown in Fig. 11a, we

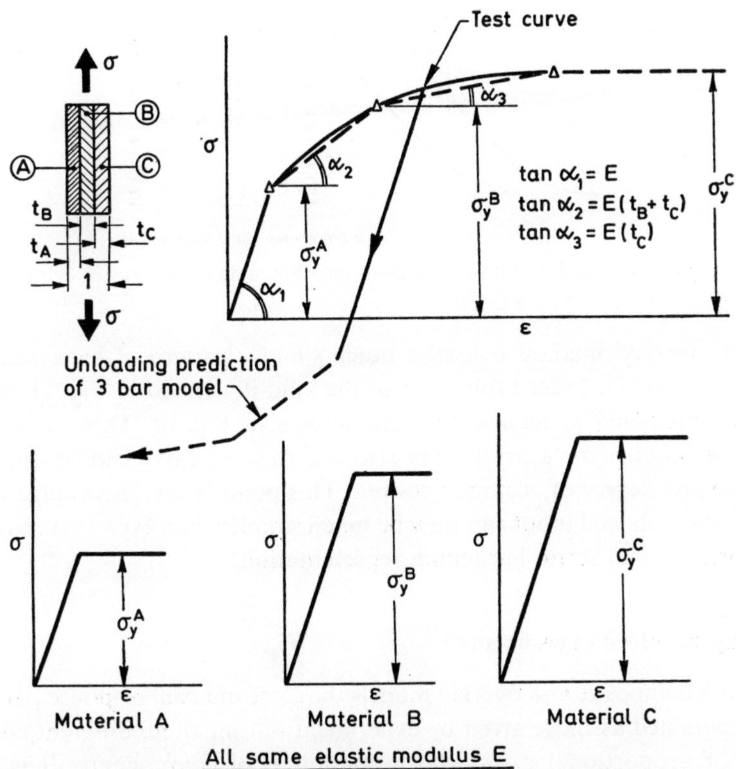

Fig. 10. Uniaxial plastic strain hardening behaviour using 3 different materials.

see that with different ratio of yield stress quite different stress-strain patterns are obtained as shown in Fig. 11b. Here smooth curves arise out of the distribution pattern and varying this can affect appreciably the result.

Clearly in a practical application a response study of a single element has to be made to check coincidence with specimen behaviour before. Such a representation can be incorporated in the analysis of this assembly. The wide range of possibilities offered is enormous.

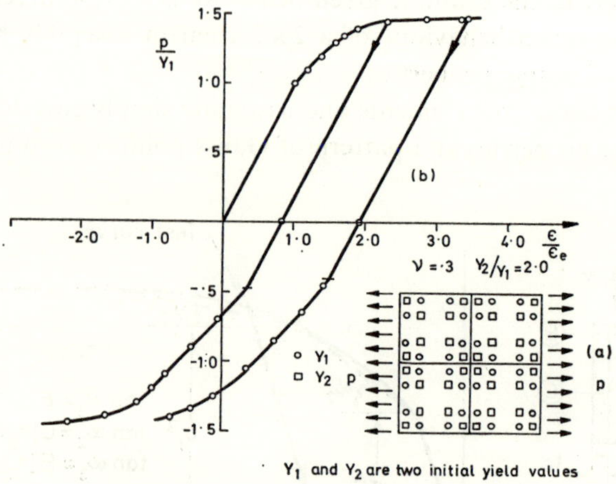

Fig. 11. A composite model with two materials distributed on cubic element Gauss points.

If the 'overlay' method is used a much simpler pattern of behaviour for one element arises. Indeed for a test of the kind illustrated in Fig. 11 results could be predicted as identical to that shown in Fig. 10. This has an important advantage as 'a priori' any strain hardening curve can be approximated to any degree of accuracy desired. This point is very advantageous in computing as indeed input can now be much simpler than even that required with conventional strain hardening representation.

4. Computer aided experiments

With both composite and overlay models the same uniaxial response patterns can be obtained as those given by experiment – using quite different combination of proportional t and elastic moduli E of components. It is now desirable to test the behaviour of the composite in other responses. Here it is predicted that feedback from numerical analysis will be of invaluable assistance in the laboratory in performing check experiments.

By satisfaction of a simple test with a limited number of spare parameters, numerical analysis can be carried out for other complex stress behaviour patterns. Laboratory tests used to verify these can in turn single out appropriate values of the parameters approximated to achieve agreements with predictions.

This feed back between laboratory and computer is in our opinion, one of the significant advances which are applicable beyond the realm of simple plasticity and can be generalised to much more complex behaviour patterns.

5. Concluding remarks

The methodology of the composite or overlay models illustrated in this paper has, obviously, very wide applications. Although it was illustrated here in the context of two materials only – and even for these much computer and experimental work still is in progress at our institution – it is not limited in any sense to these models.

It is in fact possible to overlay simple visco-elastic/plastic/visco-plastic/cracking models, to mention but a few, and obtain behaviour patterns of great complexity. Computationally considerable advantages immediately accrue.

Firstly general non-linear systems of computation such as the one originated from Swansea (*NOLIN*) which at present contain only a limited number of material choices (plastic, visco-elastic, creep of experimental type, Roscoe soil plasticity, concrete cracking etc.) can now be further extended by the *overlay structure* in which various models operate simultaneously with each load increment in specified proportion.

Secondly, experimental data needed to describe material behaviour can be reduced considerably.

Thirdly – this is an important point – programming effort is much reduced. Only simple behaviour patterns have been described with complex ones obtained by the overlay process.

Much work is in progress at the present to implement and verify a wide range of material patterns.

Authors' Postscript

It appears that the concept of 'sublayers' which play the same part as 'overlay models' has been introduced into literature earlier in the context of Von

Mises plastic material. The following references were brought to the authors' notice later but it appears that the idea of overlay models for any material behaviour has not been previously noted.

White, G. N., Application of the theory of perfectly plastic solids to stress analysis of stress hardening solids. Div. of Appl. Maths. Brown University, T.R.51, 1951.

Besseling, J. F., A theory of plastic flow for anisotropic hardening in plastic deformation of an initially isotropic material. Nat. Aero. Res. Inst. Amsterdam. Report S-410, 1953.

References

[1] Zienkiewicz, O. C., *The finite element method in engineering science*, McGraw-Hill (1971).

[2] Mises, R. Von, Mechanik der plastischen Formänderung von Kristallen. *Z. Angew, Math. Mech. 8* (1928) 161.

[3] Koiter, W. T., Stress-strain relations, uniqueness and variational theorems for elastic-plastic materials with sigular yield surface. *Quart. Appl. Math. 11* (1953) 350.

[4] Prager, W., The theory of plasticity – a survey of recent achievement. *Proc. Int. Mech. Engr. London, 119* (1955) 41.

[5] Hill, R., A variational principle of maximum plastic work in classical plasticity. *Quart. J. Mech. Appl. Math. 1* (1948) 18.

[6] Batdorf, S. B. and B. Budiansky, *A mathematical theory of plasticity based on the concept of slip*. NACA. T.N. No. 1971 (1949).

[7] Cottrell, A. H., *Dislocations and plastic flow in crystals*. Oxford University Press (1956).

[8] Kröner, E., Initial studies of a plasticity theory based upon statistical mechanics. In: *Inelastic Behaviour of Solids*, ed. M. F. Kannien et al. Battle Inst. Materials Science Colloquia, Sept. 1969 McGraw Hill (1970).

[9] Head, A. K., Dislocation theories of the plastic properties of metals and the transition to continuum mechanics. *In: Inelastic Behaviour of Solids* ed. M. F. Kannien, et al. Battle Inst. Materials Science Colloquia, Sept. 1969. McGraw Hill (1970).

[10] Owen, D. R. J., The application of dislocation theory to the determination of stress-strain relationships of composite materials. *Int. J. Non-Linear Mech. 6* (1971) 167.

[11] Zienkiewicz, O. C., B. Best, C. Dullage and K. G. Stagg, *Analysis of non-linear problems in rock mechanics with particular reference to jointed systems*. 2nd Int. Conf. on Rock Mechanics, Belgrade (1970).

[12] Zienkiewicz, O. C., *Numerical methods or art in rock and soil mechanics – nonlinear material models*. Proc. 13th meeting of the Ass. of Engg. Geologist. Oct. 22–23, 1970 Washington D.C.

[13] Dullage, C. R., Ph. D. Thesis, University of Wales, 1972.

[14] Brandtzaeg, A., *Failure of a material composed of non-isotropic elements.* Det Kgl. Norske Videnskabers Sels kabs Skrifter, No. 2 Trondheim, Norway (1927).

[15] Zienkiewicz, O. C., S. Valliappan and I. P. King, Elasto-plastic solution of engineering problems, initial-stress, finite element approach. *Int. J. Num. Meth. in Engg. 1* (1969) 75–100.

[16] Nayak, G. C. and O. C. Zienkiewicz, Elasto-plastic stress analysis with curved iso-parametric elements for various constitutive relations. To be published.

[17] Ziegler, H., A modification of Prager's hardening rule. *Quart. Appl. Math. 17* (1959) 55–65.

[18] Armeu, H. Jr., G. Isakson and A. Pifko, Discrete element methods for the plastic analysis of structures subjected to cyclic loading. *Int. J. Num. Meth. Engg. 2* (1970) 189–206.

[19] Dinno, K. S. and S. S. Gill, An experimental investigation into the plastic behaviour of flush nozzles in spherical pressure vessels. *Int. J. Mech. Sci. 7* (1965) 817.

[20] Johnson, W. and P. B. Mellor, *Plasticity for Mechanical Engineers.* Van Nostrand, London, (1962).

Matrix analysis of the large deformation of an elastic-plastic axially symmetric continuum

J. S. Gunasekera and J. M. Alexander

Imperial College, London, England

The finite element method is used to obtain a solution for the large deformation elastic-plastic problem of expanding a hole in a plate. Large deformations are taken into account by the use of the 'Initial Stress' matrix and the elastic-plastic material is represented by the Prandtl-Reuss relations together with the yield criterion of Von Mises. A computer program is developed and the analysis is carried out step-by-step using small linearised increments. The solution is compared with a previous theoretical solution obtained using a one variable theory.

Nomenclature

$\boldsymbol{B}$	Strain displacement matrix
$\boldsymbol{D}_e$	Elasticity matrix
$\boldsymbol{D}_{ep}$	Elastic-plastic stiffness matrix
E_p	Plastic modulus $= \mathrm{d}\bar{\sigma}/\mathrm{d}\bar{\varepsilon}_p$
G	Modulus of rigidity
$\boldsymbol{I}$	Unit matrix
$\boldsymbol{k}_s$	Small displacement, element stiffness matrix
$\boldsymbol{k}_\sigma$	Initial stress matrix (geometric matrix)
$\boldsymbol{k}_L$	Initial displacement matrix
$\boldsymbol{K}$	Overall stiffness matrix
r, z	Coordinates (cylindrical)
$\boldsymbol{r}$	Nodal displacements (whole structure)
$\boldsymbol{R}$	Nodal forces (whole structure)
u	Displacement in the r direction
v	Displacement in the z direction
V	Volume of the element

$\boldsymbol{\sigma}$	Stress
$\boldsymbol{\sigma}_D$	Deviatoric stress
$\boldsymbol{\varepsilon}$	Strain
$\bar{\sigma}$	Equivalent stress
σ_0	Initial yield stress
$\boldsymbol{\rho}$	Nodal displacements (element)
$\boldsymbol{P}$	Nodal forces (element)
$\boldsymbol{\phi}_e$	Elastic flexibility matrix
$\boldsymbol{\phi}_p$	Plastic flexibility matrix
ν	Poisson's ratio

Note: Matrices are denoted by bold letters.
Curly brackets { } denote transposed matrices, i.e. $\{A\} = \boldsymbol{A}^t$.

Introduction

Much progress has been made in the finite element method of analysis since the pioneering paper by Turner et al. [1]. Various types of elements have been developed and several kinds of nonlinearities are dealt with, notably for problems of nonlinear elasticity and plasticity. Relatively few papers deal with combined forms of nonlinearities.

Nonlinearities arise from two distinct sources, material nonlinearity and geometric nonlinearity. The most common material nonlinearity is represented by the idealised elastic-plastic material and several interesting problems of this type have been solved using incremental approaches and/or iterative techniques. The better known methods for elastic-plastic analysis are the so-called tangential stiffness method, the initial strain method and the initial stress method. The initial strain and stress methods are ideally suited for the solution of small displacement problems where the nonlinearity is only due to the nonlinear material properties (e.g. plasticity). In problems where combined forms of nonlinearities exist due to the presence of large displacements the tangential stiffness method may be the best method, as the analysis can then be carried out incrementally using small linearised steps. This not only takes into account the complete strain history at each point of the continuum, but also provides the complete deformation mode, giving valuable information about instability and collapse load conditions in structures, for example.

Geometric nonlinearities were first introduced by means of an incremental geometric matrix (2) (3) also known as the initial stress stiffness matrix. Two and three dimensional analysis of such problems has been carried out by many authors, notably by Argyris [4]. Later, derivation of the initial stress was made by the use of the Lagrangian or Green's strain tensor (Martin [5]). Recent analysis has established the importance of additional terms conveniently expressed as an initial displacement matrix and an initial load stiffness matrix (9). Direct methods of minimization to obtain solutions for geometrically nonlinear problems have been used by Bogner et al. [6]. Oden [7] and Oden and Kubitza [8] have solved problems of nonlinear elasticity with large strains and large displacements using the Newton-Raphson method for the solution of nonlinear simultaneous equations.

Hibbit, Marcal and Rice have used a Lagrangian approach for the problem of large strain and large displacement. It has been shown that this formulation is quite similar to a previous large rotation, small strain formulation (9) (10), except for the addition of the initial loads in the increment. Hibbit et al. have shown that for point loads with fixed direction such a matrix does not exist.

A series of papers by Nagamatsu et al. [11] [12] deal with 'Non uniform deformation of a block in plane-strain and axisymmetric compression caused by friction'. The material is assumed to be elastic-plastic, strain-hardening and deformation is pursued from elastic through partially plastic to totally plastic stages. The problem has been analysed using the calculus of variations and finite elements and the solution compared with results obtained experimentally.

Lee and Kobayashi [13] analysed axisymmetric upsetting and plane-strain side-pressing of solid cylinders by the finite element method. Solutions have been obtained up to 33 per cent reduction in height in axisymmetric upsetting and up to 19 percent in height in side-pressing, under conditions of complete sticking at the tool-workpiece interface.

In this paper an attempt is made to solve a problem involving material nonlinearity combined with large displacements, for an elastic-plastic material, using the tangential stiffness method. The nonlinearity due to large deformation is taken into account by the use of the initial stress matrix. The Prandtl-Reuss relations are used to represent the elastic-plastic material together with the yield criterion of Von Mises. A strain hardening law is assumed of the form $\bar{\sigma} = \sigma_0(1+B\bar{\varepsilon}_p)^n$, although any other form could easily be used in the computer program. The constants σ_0, B and n are determined by applying a least square regression analysis to the experimental values.

Anisotropy resulting from directional hardening is neglected in this analysis, although this too could be included with the proper knowledge of such data.

The model chosen for the finite element analysis is the problem of the expanding of a hole in a thin infinite plate having its initial thickness proportional to radial distance from the centre of the hole. This problem has been solved theoretically using a one variable theory (15). A finite element analysis was performed on a finite plate the hole being expanded from a small finite radius to obtain a sufficiently large elastic-plastic region. The results are compared with those obtained theoretically.

Theoretical considerations

The general theory of finite elements applied to problems of solids of revolution for the small displacement, elastic case is dealt with in numerous references (e.g. (14)) and is therefore not discussed here in any detail.

For the present analysis a triangular ring element with three nodal points is chosen and the whole body is assumed to be divided into such triangular elements (Fig. 1). A linear displacement function is assumed for the element, giving rise to a constant state of stress and strain inside every element. (This is not however strictly true for the axisymmetric case, where there is a slight variation of the circumferential strain over the element. The centroidal

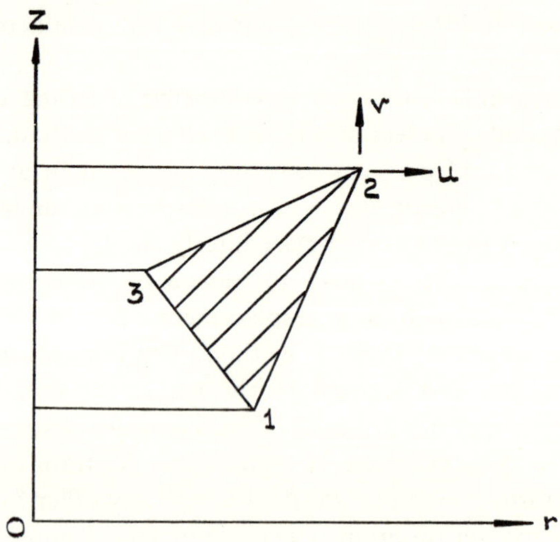

Fig. 1. Triangular ring element.

value is used for this analysis, for simplicity and is assumed constant inside the whole element.)

Following Argyris [3] and denoting the six nodal displacements for a typical element as $\boldsymbol{\rho}$ and the stresses and strains as

$$\boldsymbol{\sigma} = \{\bar{\sigma}_z \ \sigma_r \ \sigma_\theta \ \sqrt{2}\sigma_r\}, \tag{1}$$

$$\boldsymbol{\varepsilon} = \left\{\varepsilon_z \ \varepsilon_r \ \varepsilon_\theta \ \frac{1}{\sqrt{2}}\varepsilon_{rz}\right\} \tag{2}$$

and

$$\boldsymbol{\rho} = \{u_1 v_1 u_2 v_2 u_3 v_3\} \tag{3}$$

the strain-displacement relation can be established as

$$\boldsymbol{\varepsilon} = \boldsymbol{B\rho} \tag{4}$$

where $\boldsymbol{B}$ is the strain-displacement matrix (Appendix A). The stress-strain relationship for the elastic case can be written as,

$$\boldsymbol{\sigma} = \boldsymbol{D}_e\boldsymbol{\varepsilon} \tag{5}$$

and for the elastic-plastic case as

$$\mathrm{d}\boldsymbol{\sigma} = \boldsymbol{D}_{ep}\mathrm{d}\boldsymbol{\varepsilon} \tag{6}$$

where $\boldsymbol{D}_e$ and $\boldsymbol{D}_{ep}$ are the elastic and the elastic-plastic stiffness matrices respectively (Appendix B). Now, the element stiffness matrix can be obtained in the usual manner as,

$$\boldsymbol{k}_s = \int \boldsymbol{B}^t \boldsymbol{D}_{ep} \boldsymbol{B} \, \mathrm{d}V \tag{7}$$

the integration being over the volume of the triangular ring element.

The nonlinearity due to large deformations is taken into account by the use of the initial stress stiffness matrix (alternatively known as the Geometric matrix), $\boldsymbol{k}_\sigma$ and the initial displacement matrix, $\boldsymbol{k}_L$. Now the total stiffness matrix for a small increment can be shown to be

$$\boldsymbol{k} = \boldsymbol{k}_s + \boldsymbol{k}_L + \boldsymbol{k}_\sigma. \quad \text{(Appendices C, D).} \tag{8}$$

It has been shown (10) that, using a proper updating of the coordinates of the body, the initial displacement matrix $\boldsymbol{k}_L$ can be discarded conveniently. Therefore, we have,

$$\mathrm{d}\boldsymbol{P} = \boldsymbol{k}\mathrm{d}\boldsymbol{\rho} \tag{9}$$

where,

$$\boldsymbol{k} = \boldsymbol{k}_s' + \boldsymbol{k}_\sigma \tag{10}$$

and $\mathrm{d}\boldsymbol{P}$, $\mathrm{d}\boldsymbol{\rho}$ are the incremental nodal forces and the nodal displacements respectively.

The overall stiffness matrix relating all the nodal forces to the nodal displacements is obtained by assembling all the element stiffness matrices in the standard manner. Thus, we have,

$$\mathrm{d}\boldsymbol{R} = \boldsymbol{K}\mathrm{d}\boldsymbol{r} \tag{11}$$

where $\boldsymbol{K}$ is the overall stiffness matrix

$\mathrm{d}\boldsymbol{R}$, $\mathrm{d}\boldsymbol{r}$ — the incremental nodal forces and displacements respectively.

Method of solution

Since the stress-strain relationship for an elastic-plastic material is generally nonlinear, to obtain a realistic solution loading has to be carried out in small incremental steps. During the first load increment the material will at first behave elastically, than at a certain load condition the element (or a few elements) with the highest equivalent stress will yield and become plastic. When the loading is continued, more and more elements will yield and enter the plastic domain. All post yield elements are thereafter treated as plastic (until any such element starts to unload) and the correct element stiffness is evaluated using the current state of stress.

Ideally, of course, the load increment should be stopped as soon as one element becomes plastic, as this has to be treated as a post yield element, thereby changing the stiffness of this element and of the whole body. This would mean that each load increment will only make one element yield at a time, taking an enormous amount of computer time to solve any practical problem. The loading process can, however, be accelerated by the use of two methods, without impairing the accuracy of the solution greatly. In the first method all elements having an equivalent stress greater than $k\sigma_0$ (where k is a factor usually in the range 0.95–1.00) are assumed plastic for the next load increment. The alternative method is to continue the loading to make a prefixed number of elements yield, at one time. The first method is better suited for problems of fairly even stress distributions whereas the latter is suitable for problems with uneven stress distributions. Both methods are of course approximate in that the first assumes a few elements to be plastic when they are not yet plastic and the latter assumes a few elements to be elastic when they are in the transition period from elastic to plastic.

The operations carried out during a typical analysis on the computer can best be explained by the following sequence:

1. Perform an elastic analysis of the body in the standard manner. Determine the equivalent stress $\bar{\sigma}$ of all the elements using the relation

$$\bar{\sigma}^2 = \tfrac{3}{2}\sigma_D^t\sigma_D.$$

2. Find the lth maximum equivalent stress $\bar{\sigma}_l$ and scale up all quantities to make this element just yield. (This would of course make l elements plastic during the increment.) The scale factor for this case would be simply,

$$f = \frac{\bar{\sigma}_0}{\bar{\sigma}_l}$$ where $\bar{\sigma}_0$ is the initial yield stress of the material, at that point.

3. Evaluate the elastic-plastic stiffness matrix $\boldsymbol{D}_{ep}$ of all post yield elements and the elastic stiffness $\boldsymbol{D}_e$ of all elastic and unloading elements. ($\boldsymbol{D}_{ep}$'s are evaluated using the current state of stress.) Hence evaluate the element stiffness matrix $\boldsymbol{k}$.
4. Evaluate the initial stress matrix $\boldsymbol{k}_\sigma$ for all elements and hence the overall stiffness matrix of the body, $\boldsymbol{K}$.
5. Choose a suitable test increment (say $p \times \boldsymbol{R}$) and obtain a solution using the current stiffness matrix $\boldsymbol{K}$. Hence evaluate the increment of strain $\mathrm{d}\boldsymbol{\varepsilon}$ and the stress $\mathrm{d}\boldsymbol{\sigma}$. Choose $p = p_0$ to cause l elastic elements to become plastic.
6. Scale all incremental nodal displacements, forces, element stresses and strains using p_0 as the scale factor. Evaluate the total displacements, forces, stresses, strains, and the new coordinates of the nodal points.
7. Check the new equivalent stress of all the elements for yielding and unloading. A value of $LY(= 0, 1, 2)$ is assigned for elements which are elastic, plastic and unloading respectively.
8. Go back to (3) and proceed up to (7) to check whether the newly obtained stresses are within a prescribed tolerance to the previous stresses. If so, proceed to (9) and if not, repeat (3) to (7) until convergence is obtained.
9. Print all results. This completes the procedure for one load increment. For subsequent load increments go back to (3).

Finite element analysis

The model chosen for the finite element analysis is the expansion of a hole in a thin plate having the geometry shown in Fig. 2. This problem has been solved (for an infinite plate) using a one variable theory (15). For the finite element analysis, the plate was idealised into 286 elements and 189 nodal points, as shown in Fig. 3 where one quarter of a cross section of the plate is shown. The hole was of course expanded from a small finite radius. The

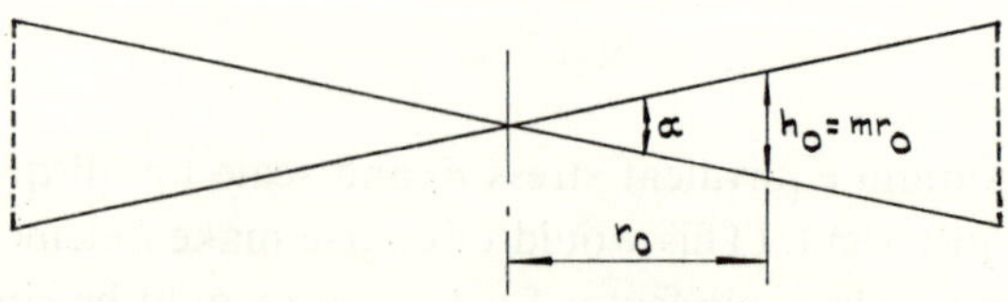

Fig. 2a. Initial profile of plate, $m = 2 \tan \alpha/2$.

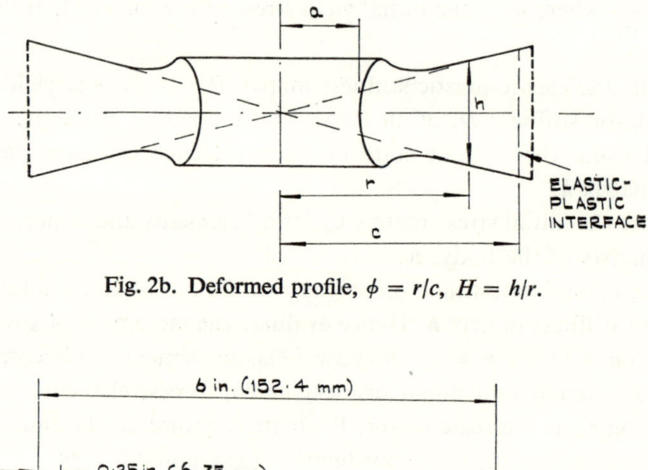

Fig. 2b. Deformed profile, $\phi = r/c$, $H = h/r$.

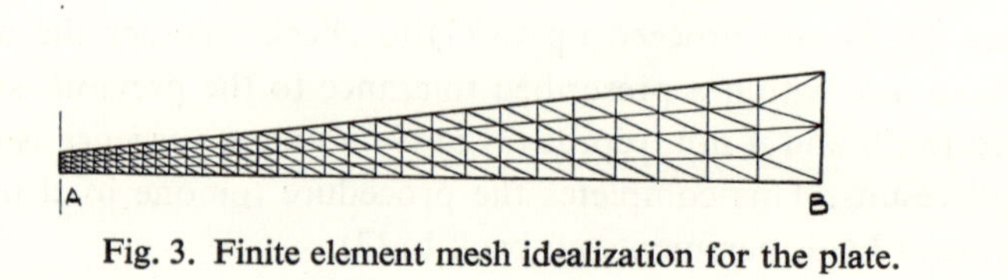

Fig. 3. Finite element mesh idealization for the plate.

plate was fixed at the edge and loading was carried out so that the radial displacements of the nodal points at the edge of the hole remained constant. The loading was carried out in small incremental steps until the hole expanded from a radius of 0.25 inches to 0.294 inches, in 44 incremental steps. In this analysis loading during each increment was carried out until one element reached the initial yield stress of the material and all elements with equivalent stresses greater than 0.95 times the initial yield stress were assumed plastic for the next load increment.

The solution was repeated using a second approach where 8 elements were made to yield during every incremental step. The results obtained are very similar to the previous method. The computing time was greatly reduced for the second method and this method seems to be better suited for most practical problems. The factors 0.95 and 8 in these two methods were arbitrarily chosen for the present problem to compromise between the accuracy of the solution obtained and the computing time consumed. In general, for other problems these two factors can be assumed as 0.9–1.0 for the first and from unity up to any number for the second, depending on the number of elements used in the mesh idealization, and the amount of accuracy required in the results.

Results and discussion

The finite element solutions obtained for this problem were compared with those obtained theoretically (15). The elastic solution shown in Fig. 4 is identical for the two cases. The results towards the edge of the plate (not shown here) revealed some discrepancy due to the fixed boundary condition at the edge of the finite plate. (The theoretical solution, of course, assumes an infinite plate.) This discrepancy (of very small order) was only localised at the edge of the plate and did not affect conditions towards the centre of the plate. A fixed boundary condition at the edge of the plate would give a higher value for σ_r and a lower value for σ_θ as compared with the infinite plate. (The alternative of course is to use a free boundary condition at the edge of the plate, which in turn would give a lower or zero value for σ_r and a higher one for σ_θ than the infinite plate.)

The theoretical solution obtained when the plate is partially plastic assumes plane stress conditions near the edge of the hole and the principal stresses to be in r, θ and z directions. This condition is obviously violated due to local large deformations and may have resulted in gross errors near the

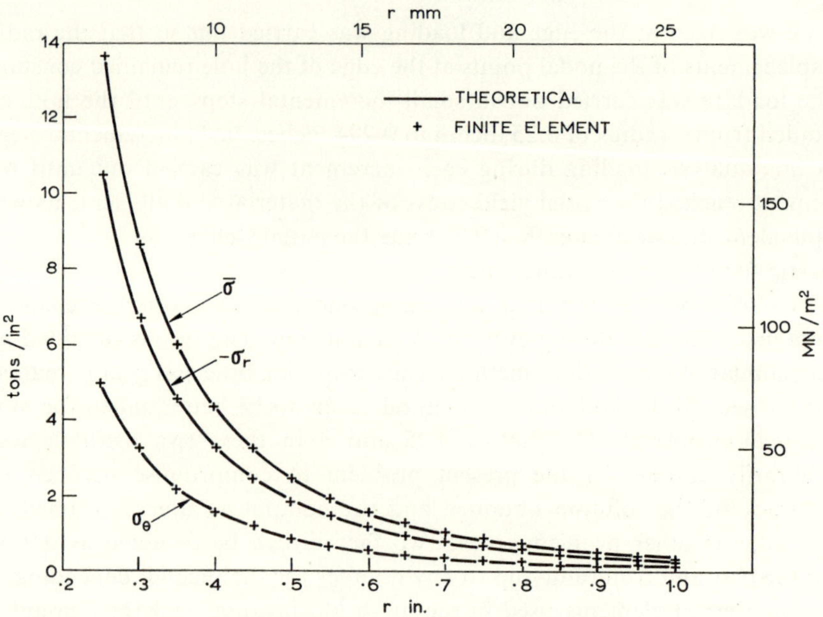

Fig. 4. Stress distribution (elastic).

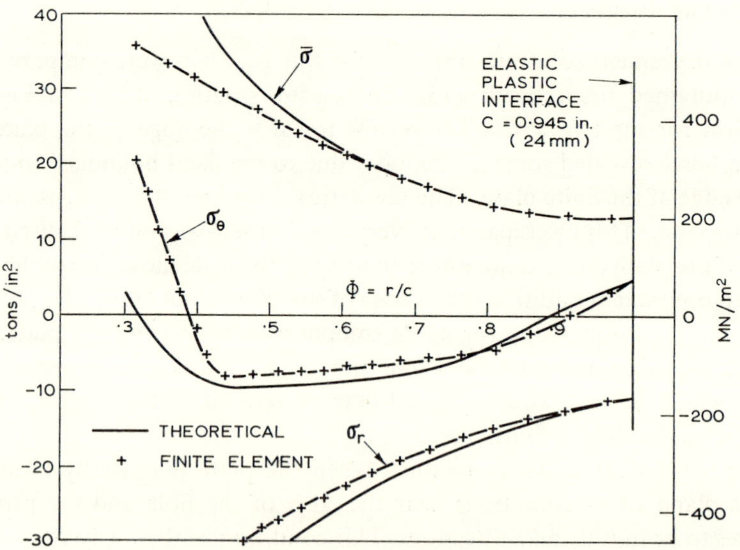

Fig. 5. Stress distribution (elastic-plastic).

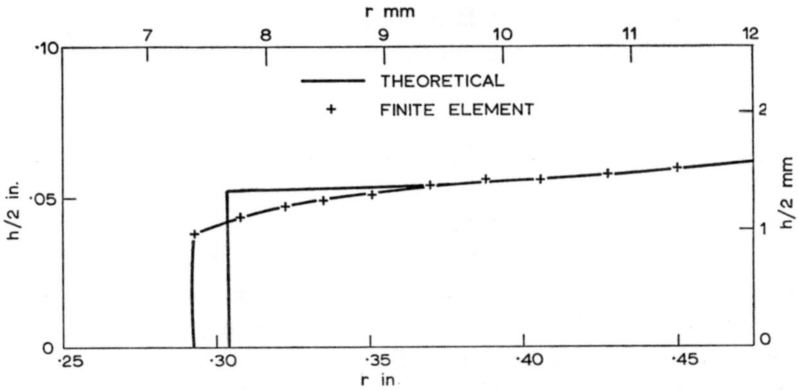

Fig. 6. Thickness profile (elastic-plastic).

edge of the hole. In fact axial and shear stresses of considerably large order were obtained with the finite element solution. (The axial and shear stresses were of course assumed to be zero in the theoretical analysis.) The stresses obtained from the finite element analysis were averaged over each section by considering all elements to be at the same radial distance. The stresses seem to agree fairly well close to the elastic-plastic interface but show some deviation near the edge of the hole (Fig. 5). The elastic part of the plate is not shown in Fig. 5 but the results of course agreed very well with the theoretical ones as before. The deformation in the z direction near the edge of the hole (Fig. 6) is smaller than that predicted by the theoretical analysis.

Conclusions

The computer program developed using the suggested method gives excellent comparison with the theory for the elastic case. The result obtained for the elastic-plastic large deformation case agrees very well near the elastic-plastic boundary but tends to deviate towards the region of very large deformation. The variation of σ_r, σ_θ and $\bar{\sigma}$ (Fig. 5) are however very similar and it is interesting to note the similarity in the variation of σ_θ with radius for the two solutions. The stresses σ_r and $\bar{\sigma}$ obtained using the finite element method are lower than those predicted by theory (15). None of the approximations used in the theory to simplify the analysis by converting it into a 'one-variable' case have been assumed for the finite element solution, and it would seem that the latter has given a more realistic solution to the problem. (In fact stresses of infinite or very large magnitude were obtained very near to the edge of the hole using the one variable theory, which are undoubtedly somewhat unrealistic.)

The finite element method can be used as a powerful tool for the solution of many nonlinear problems. The advantages of this method are that any type of nonlinearity can be handled and problems of any arbitrary shape can be solved. An approximation is at first made in the assumption of a displacement function for the particular element chosen and further approximations are often unnecessary. The disadvantages are the large amounts of computer storage and time required to solve nonlinear problems.

The method given may be suitable for elastic-plastic problems with small or large deformations. It is difficult to see whether the displacement approach can be used successfully to solve steady state problems of plastic flow, although a solution of this type may give an insight into the transient stages of such a process. A variational method using ψ, the stream function and its derivatives (the velocities) as the basic unknowns may be more suitable for the solution of such problems.

References

[1] Turner, M. J., R. W. Clough, H. C. Martin and L. J. Topp. Stiffness and deflection analysis of complex structures. *J. Aero. Sci. 23* (1956).

[2] Turner, M. J., E. H. Dill, H. C. Martin and R. J. Melosh. Large deflection analysis of complex structures subjected to heating and external loads. *J. Aero. Space Sci. 27* (1960) 97.

[3] Argyris, J. H., S. Kelsey and H. Kamel, *Matrix methods of structural analysis*. Pergamon Press (1964).

[4] Argyris, J. H., Matrix analysis of three dimensional elastic media. *AIAA 3* (1965).

[5] Martin, H. C., Derivation of stiffness matrices for the analysis of large deflection and stability problems. *Proc. 1st Conf. on Matrix Methods in Struct. Mech.* (1966) 697.

[6] Bogner, F. K., R. H. Mallet, M. D. Minich and L. A. Schmit, Development and evaluation of energy search methods of nonlinear structural analysis. *AFFDL-TR-65-113* (1965).

[7] Oden, J. T., Numerical formulation of nonlinear elasticity problems. *J. Struct. Div. ASCE 93* (1967) 235.

[8] Oden, J. T. and W. K. Kubitza, *Numerical analysis of nonlinear pneumatic structures*. Proc. 1st Int. Colloq. Pneumatic Structures. Stuttgart 1967.

[9] Hibbit, H. D., P. V. Marcal and J. R. Rice, A finite element formulation for problems of large strain and large displacements. *Int. J. Solids Struct. 6* (1970) 1069.

[10] Marcal, P. V., *Finite element analysis of combined problems of nonlinear material and geometric behaviour*. Tech. Report I Brown University (1969).

[11] Nagamatsu, A., T. Murota and T. Jimma, On the non-uniform deformation of block in plane strain compression caused by friction. Parts I to IV *Bull. JSME 13* (1970) and *14* (1971).
[12] Nagamatsu, A., T. Murota and T. Jimma, On the non-uniform deformation of material in axially symmetric compression caused by friction. Parts I and II *Bull. JSME 14* (1971).
[13] Lee, C. H. and S. Kobayashi, Analyses of axisymmetric upsetting and plane-strain side-pressing of solid cylinders by the finite element method. *J. Eng. for Industry, ASME 93* (1971).
[14] Zienkiewicz, O. C., *The finite element method in Engineering Science*. McGraw-Hill (1971).
[15] Alexander, J. M. and J. S. Gunasekera, On the geometrically similar expansion of a hole in a thin infinite plate. To be published in *Proc. Roy. Soc.*
[16] Hill, R., *Mathematical theory of plasticity*. Oxford (1950).

Appendix A

Derivation of *B* matrix

Assuming a linear displacement function for the displacement u and v we have

$$u = \alpha_1 + \alpha_2 r + \alpha_3 z, \qquad v = \alpha_4 + \alpha_5 r + \alpha_6 z \tag{12}$$

where

$$\boldsymbol{\alpha} = \{\alpha_1 \alpha_2 \ldots \alpha_6\} \tag{13}$$

are the generalised displacements.

Substituting for the displacements $u_i v_i$ at the nodal points $r_i z_i$ ($i = 1, 2, 3$) we get

$$\boldsymbol{\rho} = \boldsymbol{F\alpha} \tag{14}$$

where $\boldsymbol{F}$ is a 6×6 nonsingular matrix consisting of the coordinates $r_i z_i$.

Therefore we have

$$\boldsymbol{\alpha} = \boldsymbol{F}^{-1}\boldsymbol{\rho}. \tag{15}$$

The strain for the small displacement case can be expressed as

$$\boldsymbol{\varepsilon} = \left\{\varepsilon_z\, \varepsilon_r\, \varepsilon_\theta\, \frac{1}{\sqrt{2}}\varepsilon_{zr}\right\} = \left\{\frac{\partial v}{\partial z}\, \frac{\partial u}{\partial r}\, \frac{u}{r}\, \frac{1}{\sqrt{2}}\left(\frac{\partial u}{\partial z} + \frac{\partial v}{\partial r}\right)\right\}. \tag{16}$$

Substituting for u and v we get

$$\boldsymbol{\varepsilon} = \boldsymbol{\tau\alpha} \tag{17}$$

where

$$\boldsymbol{\tau} = \begin{bmatrix} 0 & 0 & 0 & 0 & 0 & 1 \\ 0 & 1 & 0 & 0 & 0 & 0 \\ \dfrac{1}{r} & 1 & \dfrac{z}{r} & 0 & 0 & 0 \\ 0 & 0 & 0 & \dfrac{1}{\sqrt{2}} & 0 & \dfrac{1}{\sqrt{2}} \end{bmatrix}. \tag{18}$$

Hence

$$\varepsilon = \tau F^{-1} \rho$$

i.e.

$$\varepsilon = B\rho \tag{19}$$

where

$$B = \tau F^{-1}. \tag{20}$$

Appendix B

Derivation of the stress-strain matrix

1. *Elastic case*

For an isotropic elastic material the stiffness matrix for the axisymmetric case can be expressed as

$$D_e = 2G\left[I_4 + \frac{\nu}{1-2\nu}\, e_3 e_3^t\right] \tag{21}$$

where

$$e_3 = \{1\;1\;1\;0\} \tag{22}$$

and the strain-stress flexibility matrix can be expressed as

$$\phi_e = \frac{1}{2G}\left[I_4 - \frac{\nu}{1+\nu}\, e_3 e_3^t\right]. \tag{23}$$

2. *Elastic-plastic case*

The Prandtl-Reuss equations in plasticity and the yield criterion of Von Mises are used to obtain the elastic-plastic stress-strain matrix for a work-hardening material.

Separating the stresses into their hydrostatic and deviatoric parts

$$\sigma = \sigma_D + \sigma_H \tag{24}$$

where

$$\boldsymbol{\sigma}_H = \tfrac{1}{3}\{I_1 I_1 I_1 0\} \tag{25}$$

and

$$I_1 = \sigma_z + \sigma_r + \sigma_\theta \tag{26}$$

is the first invariant of the stress tensor.

Eqn. (25) could also be written as

$$\boldsymbol{\sigma}_H = \tfrac{1}{3}\boldsymbol{e}_3\boldsymbol{e}_3^t\boldsymbol{\sigma}. \tag{27}$$

Therefore we have

$$\boldsymbol{\sigma}_D = \boldsymbol{\mu}\boldsymbol{\sigma} \tag{28}$$

where

$$\boldsymbol{\mu} = \boldsymbol{I}_{4\times 4} - \tfrac{1}{3}\boldsymbol{e}_3\boldsymbol{e}_3^t. \tag{29}$$

The following expressions also become evident.

$$\left.\begin{aligned} \boldsymbol{\mu}^t &= \boldsymbol{\mu} \\ \boldsymbol{\mu}\boldsymbol{\mu} &= \boldsymbol{\mu}^t\boldsymbol{\mu} = \boldsymbol{\mu} \\ \boldsymbol{\mu}\boldsymbol{D}_e &= \boldsymbol{D}_e\boldsymbol{\mu} = 2G\boldsymbol{\mu} \end{aligned}\right\} \tag{30}$$

The yield criterion of Von Mises can be expressed in matrix notation as

$$\bar{\sigma}^2 = \tfrac{3}{2}\boldsymbol{\sigma}_D^t\boldsymbol{\sigma}_D = \tfrac{3}{2}\boldsymbol{\sigma}^t\boldsymbol{\mu}\boldsymbol{\sigma}. \tag{31}$$

In a similar way the Prandtl-Reuss relations (15) can be written as,

$$\mathrm{d}\boldsymbol{\varepsilon} = \boldsymbol{\phi}_e\mathrm{d}\boldsymbol{\sigma} + \tfrac{3}{2}\,\frac{\boldsymbol{\sigma}_D}{\bar{\sigma}E_p}\,\mathrm{d}\bar{\sigma} \tag{32}$$

where

$$E_p = \frac{\mathrm{d}\bar{\sigma}}{\mathrm{d}\bar{\varepsilon}_p}. \tag{33}$$

The differential form of the Von Mises criterion can be expressed as

$$\bar{\sigma}\mathrm{d}\bar{\sigma} = \tfrac{3}{2}\boldsymbol{\sigma}_D^t\mathrm{d}\boldsymbol{\sigma}. \tag{34}$$

Eliminating $d\bar{\sigma}$ from eqn. (32)

$$d\boldsymbol{\varepsilon} = \boldsymbol{\phi}_e d\boldsymbol{\sigma} + \tfrac{9}{4}\frac{\boldsymbol{\sigma}_D \boldsymbol{\sigma}_D^t}{\bar{\sigma}^2 E_p} d\boldsymbol{\sigma} \tag{35}$$

or

$$d\boldsymbol{\varepsilon} = (\boldsymbol{\phi}_e + \boldsymbol{\phi}_p) d\boldsymbol{\sigma} \tag{35a}$$

where

$$\boldsymbol{\phi}_p = \tfrac{9}{4}\frac{\boldsymbol{\sigma}_D \boldsymbol{\sigma}_D^t}{\bar{\sigma}^2 E_p}. \tag{36}$$

Eqn. (35) can be inverted to yield

$$d\boldsymbol{\sigma} = \boldsymbol{D}_e \left[\boldsymbol{I}_4 + \tfrac{9}{4}\frac{2G}{E_p \bar{\sigma}^2} \boldsymbol{\sigma}_D \boldsymbol{\sigma}_D^t \right]^{-1} d\boldsymbol{\varepsilon}. \tag{37}$$

It can be proved that

$$[\boldsymbol{I} + \lambda \boldsymbol{a}\boldsymbol{a}^t]^{-1} = [\boldsymbol{I} - \gamma \boldsymbol{a}\boldsymbol{a}^t] \tag{38}$$

where $\boldsymbol{a}$ is a volumn vector and

$$\gamma = \frac{\lambda}{1 + \lambda \boldsymbol{a}^t \boldsymbol{a}}. \tag{39}$$

Using this transformation (with $\lambda = \tfrac{9}{4}(2G/H\bar{\sigma}^2)$ and $\boldsymbol{a} = \boldsymbol{\sigma}_D$) eqn. (37) becomes

$$d\boldsymbol{\sigma} = \boldsymbol{D}_e \left[\boldsymbol{I}_4 - \frac{3G}{E_p + 3G} \frac{\boldsymbol{\sigma}_D \boldsymbol{\sigma}_D^t}{\boldsymbol{\sigma}_D^t \boldsymbol{\sigma}_D} \right] d\boldsymbol{\varepsilon} \tag{40}$$

or it can be shown that

$$\boldsymbol{D}_{ep} = \boldsymbol{D}_e - \frac{2G}{(1 + E_p/3G)} \frac{\boldsymbol{\sigma}_D \boldsymbol{\sigma}_D^t}{\bar{\sigma}^2} \tag{41}$$

$\boldsymbol{\phi}_e$ and $\boldsymbol{\phi}_p$ are known as the elastic and the plastic flexibilities respectively. It is interesting to note that $\boldsymbol{\phi}_e$ can be inverted except for the trivial case when $\nu = \tfrac{1}{2}$ (incompressible material) and $\boldsymbol{\phi}_p$ is singular and cannot be inverted directly. This is quite obvious mathematically as any general matrix of the form $\boldsymbol{A} = \boldsymbol{a}\boldsymbol{a}^t$ where $\boldsymbol{a}$ is a column matrix has the lowest possible rank of 1 and is singular. Thus it would seem a direct displacement method of formulation of the stiffness matrix for the case of a rigid-plastic material is not feasible.

Appendix C

Large deformation theory – geometric nonlinearity

When the deformation in the continuum is large, the small displacement theory as described briefly earlier, is not applicable. The non-linearity due to the change of geometry of the body under stress has to be taken into account to predict accurately the deformations, the strains and the stresses.

The strain displacement matrix for the large deformation case would change considerably with the change of the geometry of the structure and thus would not remain constant as in the case of small displacement analysis. However, considering the current state and for a small load increment

$$\mathrm{d}\varepsilon = \bar{\boldsymbol{B}}\mathrm{d}\boldsymbol{\rho} \tag{42}$$

where

$$\bar{\boldsymbol{B}} = \boldsymbol{B}_0 + \boldsymbol{B}_L(\boldsymbol{\rho}). \tag{43}$$

$\boldsymbol{B}_0$ is the usual small displacement matrix, and $\boldsymbol{B}_L$ is due to the large deformation and is a function of the displacements $\boldsymbol{\rho}$. From the principle of virtual displacements, for an increment of displacements, it can be shown that

$$\boldsymbol{P} = \int \bar{\boldsymbol{B}}^t \boldsymbol{\sigma} \mathrm{d}V \tag{44}$$

[P.V.D. $\mathrm{d}\boldsymbol{\rho}^t\boldsymbol{P} = \int \mathrm{d}\varepsilon^t\boldsymbol{\sigma}\mathrm{d}V$].

Differentiating eqn. (44)

$$\mathrm{d}\boldsymbol{P} = \int \bar{\boldsymbol{B}}^t \mathrm{d}\boldsymbol{\sigma}\mathrm{d}V + \int \mathrm{d}\bar{\boldsymbol{B}}^t\boldsymbol{\sigma}\mathrm{d}V,$$

$$\mathrm{d}\boldsymbol{\sigma} = \boldsymbol{D}_{ep}\mathrm{d}\varepsilon = \boldsymbol{D}_{ep}\bar{\boldsymbol{B}}\mathrm{d}\boldsymbol{\rho}.$$

Hence

$$\mathrm{d}\boldsymbol{P} = \int \bar{\boldsymbol{B}}^t\boldsymbol{D}_{ep}\bar{\boldsymbol{B}}\mathrm{d}V\mathrm{d}\boldsymbol{\rho} + \int \mathrm{d}\bar{\boldsymbol{B}}^t\boldsymbol{\sigma}\mathrm{d}V, \tag{45}$$

$$\mathrm{d}\bar{\boldsymbol{B}}^t = \mathrm{d}\boldsymbol{B}_L^t. \tag{46}$$

Eqn. (46) becomes

$$\mathrm{d}\boldsymbol{P} = (\boldsymbol{k}_s+\boldsymbol{k}_L+\boldsymbol{k}_\sigma)\mathrm{d}\boldsymbol{\rho} \tag{47}$$

where

$$\boldsymbol{k}_s = \int \boldsymbol{B}_0\boldsymbol{D}_{ep}\boldsymbol{B}_0\mathrm{d}V, \tag{48}$$

$$\boldsymbol{k}_L = \int \boldsymbol{B}_0\boldsymbol{D}_{ep}\boldsymbol{B}_L\mathrm{d}V + \int \boldsymbol{B}_L\boldsymbol{D}_{ep}\boldsymbol{B}_0\mathrm{d}V + \int \boldsymbol{B}_L\boldsymbol{D}_{ep}\boldsymbol{B}_L\mathrm{d}V, \tag{49}$$

$$\boldsymbol{k}_\sigma\mathrm{d}\boldsymbol{\rho} = \int \mathrm{d}\bar{\boldsymbol{B}}^t\sigma\mathrm{d}V. \tag{50}$$

$\bar{\boldsymbol{k}} = \boldsymbol{k}_s+\boldsymbol{k}_L$ is known as the large displacement matrix. It may be seen that this is a matrix which could alternatively be obtained by using an infinitesimal strain but adjusting the element coordinates in the computation of the stiffness.

$\boldsymbol{k}_\sigma$ in equation (50) is a symmetric matrix dependent on the stress level. This matrix is known as the initial stress matrix or the geometric matrix. (See Appendix D)

$$\boldsymbol{k}_T = \boldsymbol{k}_s+\boldsymbol{k}_L+\boldsymbol{k}_\sigma \tag{51}$$

is known as the total tangential stiffness.

Appendix D

Large deformation formulation for axisymmetric finite elements

The most general definition of strain for large strains was introduced by Green and St. Venant and is known as Green's strain tensor. Referred to a fixed cartesian system of coordinates r, z and displacements u, v for the axisymmetric case this can be defined as,

$$\left.\begin{aligned}
\varepsilon_z &= \frac{\partial v}{\partial z} + \tfrac{1}{2}\left[\left(\frac{\partial u}{\partial z}\right)^2 + \left(\frac{\partial v}{\partial z}\right)^2\right] \\
\varepsilon_r &= \frac{\partial u}{\partial r} + \tfrac{1}{2}\left[\left(\frac{\partial u}{\partial r}\right)^2 + \left(\frac{\partial v}{\partial r}\right)^2\right] \\
\varepsilon_\theta &= \frac{u}{r} + \tfrac{1}{2}\frac{u^2}{r^2} \\
\frac{1}{\sqrt{2}}\varepsilon_{rz} &= \frac{1}{\sqrt{2}}\left(\frac{\partial u}{\partial z} + \frac{\partial v}{\partial r}\right) + \frac{1}{\sqrt{2}}\left(\frac{\partial u}{\partial r}\cdot\frac{\partial u}{\partial z} + \frac{\partial v}{\partial r}\cdot\frac{\partial v}{\partial z}\right)
\end{aligned}\right\} . \tag{52}$$

If the displacements are small the general first order linear strain approximation is obtained by neglecting the quadratic terms.

Derivation of the B_L matrix

The general Green's strain tensor for the axisymmetric case can be defined in terms of the small and large displacement components, i.e.

$$\varepsilon = \varepsilon_0 + \varepsilon_L \tag{53}$$

where

$$\begin{aligned}
\varepsilon_0 &= \left\{\varepsilon_z\ \varepsilon_r\ \varepsilon_\theta\ \frac{1}{\sqrt{2}}\varepsilon_{rz}\right\}, \\
&= \left\{\frac{\partial v}{\partial z}\ \frac{\partial u}{\partial r}\ \frac{u}{r}\ \frac{1}{\sqrt{2}}\left(\frac{\partial u}{\partial z} + \frac{\partial v}{\partial r}\right)\right\}.
\end{aligned} \tag{54}$$

Let

$$\psi = \{\psi_z \psi_r \psi_\theta\} \tag{55}$$

where

$$\left.\begin{aligned}
\psi_z &= \left\{\frac{\partial v}{\partial z}\ \frac{\partial u}{\partial z}\right\} \\
\psi_r &= \left\{\frac{\partial v}{\partial r}\ \frac{\partial u}{\partial r}\right\} \\
\psi_\theta &= \left(\frac{u}{r}\right)
\end{aligned}\right\} . \tag{56}$$

Therefore the quadratic terms of the Green's strain tensor could be written as,

$$\boldsymbol{\varepsilon}_L = \tfrac{1}{2}\begin{bmatrix} \boldsymbol{\psi}_z^t & \boldsymbol{O} & \boldsymbol{O} \\ \boldsymbol{O} & \boldsymbol{\psi}_r^t & \boldsymbol{O} \\ \boldsymbol{O} & \boldsymbol{O} & \boldsymbol{\psi}_\theta^t \\ \frac{1}{\sqrt{2}}\boldsymbol{\psi}_r^t & \frac{1}{\sqrt{2}}\boldsymbol{\psi}_z^t & \boldsymbol{O} \end{bmatrix} \begin{bmatrix} \boldsymbol{\psi}_z \\ \boldsymbol{\psi}_r \\ \boldsymbol{\psi}_\theta \end{bmatrix} \tag{57}$$

or

$$\boldsymbol{\varepsilon}_L = \tfrac{1}{2}\boldsymbol{A\psi},$$

$$\therefore\ \mathrm{d}\boldsymbol{\varepsilon}_L = \tfrac{1}{2}\boldsymbol{A}\,\mathrm{d}\boldsymbol{\psi} + \tfrac{1}{2}\mathrm{d}\boldsymbol{A\psi} = \boldsymbol{A}\,\mathrm{d}\boldsymbol{\psi}, \tag{58}$$

$$\boldsymbol{\psi} = \boldsymbol{S\alpha} \tag{59}$$

where

$$\boldsymbol{S} = \begin{bmatrix} 0 & 0 & 0 & 0 & 0 & 1 \\ 0 & 0 & 1 & 0 & 0 & 0 \\ 0 & 0 & 0 & 0 & 1 & 0 \\ 0 & 1 & 0 & 0 & 0 & 0 \\ \frac{1}{r} & 1 & \frac{z}{r} & 0 & 0 & 0 \end{bmatrix} \tag{60}$$

and $\boldsymbol{\alpha}$ is defined by equation (13).

Substituting for $\boldsymbol{\alpha}$ from (15)

$$\boldsymbol{\psi} = \boldsymbol{SF}^{-1}\boldsymbol{\rho},$$

$$= \boldsymbol{G\rho}, \tag{61}$$

$$\therefore\ \mathrm{d}\boldsymbol{\psi} = \boldsymbol{G}\,\mathrm{d}\boldsymbol{\rho} \tag{62}$$

if the variation of ε_θ inside the element is neglected assuming $r = \bar{r}$ and $z = \bar{z}$.

$$\therefore\ \mathrm{d}\boldsymbol{\varepsilon}_L = \boldsymbol{AG}\,\mathrm{d}\boldsymbol{\rho}.$$

Hence

$$\boldsymbol{B}_L = \boldsymbol{AG}. \tag{63}$$

Derivation of $\boldsymbol{k}_\sigma$ matrix
From eqn. (50) we have,

$$\boldsymbol{k}_\sigma \mathrm{d}\boldsymbol{\rho} = \int_V \mathrm{d}\boldsymbol{B}_L^t \boldsymbol{\sigma} \mathrm{d}V. \tag{64}$$

Substituting for $\boldsymbol{B}_L$

$$\boldsymbol{k}_\sigma \mathrm{d}\boldsymbol{\rho} = \int_V \boldsymbol{G}^t \mathrm{d}\boldsymbol{A}^t \boldsymbol{\sigma} \mathrm{d}V.$$

Expanding $\boldsymbol{A}$ and $\boldsymbol{\sigma}$ we get,

$$\begin{aligned}
\mathrm{d}\boldsymbol{A}^t\boldsymbol{\sigma} &= \begin{bmatrix} \boldsymbol{\psi}_z^t & \boldsymbol{O} & \boldsymbol{O} \\ \boldsymbol{O} & \boldsymbol{\psi}_r^t & \boldsymbol{O} \\ \boldsymbol{O} & \boldsymbol{O} & \boldsymbol{\psi}_\theta^t \\ \dfrac{1}{\sqrt{2}}\boldsymbol{\psi}_r^t & \dfrac{1}{\sqrt{2}}\boldsymbol{\psi}_z^t & \boldsymbol{O} \end{bmatrix} \begin{bmatrix} \sigma_z \\ \sigma_r \\ \sigma_\theta \\ \dfrac{1}{\sqrt{2}}\sigma_{rz} \end{bmatrix} \\
&= \begin{bmatrix} \sigma_z \boldsymbol{I} & \sigma_{rz}\boldsymbol{I} & \boldsymbol{O} \\ \sigma_r \boldsymbol{I} & \sigma_{rz}\boldsymbol{I} & \boldsymbol{O} \\ \boldsymbol{O} & \boldsymbol{O} & \sigma_\theta \end{bmatrix} \begin{bmatrix} \mathrm{d}\boldsymbol{\psi}_z \\ \mathrm{d}\boldsymbol{\psi}_r \\ \mathrm{d}\boldsymbol{\psi}_\theta \end{bmatrix} \\
&= \boldsymbol{M}\mathrm{d}\boldsymbol{\psi},
\end{aligned}$$

$$\begin{aligned}
\therefore\ \boldsymbol{k}_\sigma \mathrm{d}\boldsymbol{\rho} &= \int_V \boldsymbol{G}^t \boldsymbol{M} \mathrm{d}\boldsymbol{\psi} \mathrm{d}V, \\
&= \int_V \boldsymbol{G}^t \boldsymbol{M} \boldsymbol{G} \mathrm{d}\boldsymbol{\rho} \mathrm{d}V,
\end{aligned}$$

$$\therefore \quad \boldsymbol{k}_\sigma = \int_V \boldsymbol{G}^t \boldsymbol{M} \boldsymbol{G} \mathrm{d}V. \tag{66}$$

Static and dynamic elasto-plastic analysis by the method of finite elements in space and time

J. H. Argyris and A. S. L. Chan

Imperial College of Science and Technology, London, England

The paper initiates a major extension of the matrix displacement method concerning the analysis of dynamic phenomena in the presence of material and geometric non-linearities. In particular, elasto-plastic behaviour as well as large displacements are taken into account. An iterative procedure of solution of the non-linear matrix equations is discussed. The application of the theory is described in detail on two examples. The first considers the simple static problem of a rectangular flat strip in a tensile test. The iterative calculation may be carried out for deformations as large as required and shows clearly the necking effect. More ambitious is the second example which demonstrates the non-linear dynamic theory on a cylindrical deformable billet under the impact by a heavy rigid body. The momentum of the weight and the property of the billet are such that the latter will undergo large plastic deformations. If so required it is straight forward to incorporate damping and also allow for friction forces on the contacts. The direct applicability of the technique to forging problems is evident. The solution of the dynamic phenomenon is accomplished by extending the discretisation of space also into time. In particular, the inertia forces are taken to vary over a finite time element as a third order polynomial. Exceptional accuracy is achieved by this method.

1. Introduction

The finite element method, developed originally for the analysis of linearly elastic structures [1], has been successfully extended and refined beyond recognition in the course of less than twenty years. Apart from the introduction of a large variety of new elements, necessary for the realistic idealisation of different types of structures, the development of the geometric stiffness [2] enables the static analysis to deal with non-linearity arising from the change of geometry in a large deformation, thus breaking away from the usual small displacement restriction [3]. Non-linear stress-strain relationship has also been incorporated into the analysis by the application of the Prandtl-Reuss incremental strain theory, giving a plastic stiffness matrix in addition to the usual elastic stiffness [4, 5, 6], and the linear elastic assumption is no longer a necessity. With the help of the geometric stiffness matrix, the classical problem of elastic instability may be formulated and solved in the form of an eigenvalue analysis, from which the critical load and its associated buckling mode may be found. Progress has also been made in the field of structural dynamics, where the standard calculation of the normal modes and natural frequencies from an eigenvalue solution of the matrix equation of motion has recently been supplemented by the use of finite elements in time, as well as in space, enabling the complete time history of motion from a given initial state to be traced step-by-step. A comprehensive survey of these new innovations can be found in [7].

However, while many contributors continue to enlarge the compass of the method, and others engage in proving its validity and thus making it mathematically respectable, practical applications of these new developments are still confined, with a few exceptions [8, 9], to small demonstrative types of problems. The reason may be that any non-linear analysis would almost certainly require some sort of iterative approach, which could be extremely time consuming and thus prohibiting any large scale calculation. To a certain extent, this is inevitable but it can be alleviated somewhat by careful formulation, by good programming, and above all, by the use of an efficient iterative procedure. The following examples show how these may be achieved in statics as well as in dynamics.

The simple static problem considered here is a rectangular flat strip in a tensile test. The specimen is divided into a number of triangular elements, and displacement increments are imposed at the ends. The problem is non-linear from the effects of both plastic deformation and large displacements. For each incremental step, the displacement increments are calculated from the current value of the total stiffness, which consists of the geometric and plastic, as well as the elastic stiffness. The formulation of these matrices and the iterative procedure used in the calculation will be shown. The calculation may be continued to produce as large a deformation as required, but imperfections and dislocations would cause local deformation which leads eventually to failure. Such microscopic effects the finite element method is unfortunately unable to simulate.

The main purpose of this paper however is to show how the finite element method may be applied to non-linear dynamic problems. For illustration, the study of the behaviour of a cylindrical deformable billet under the impact by a heavy rigid body is chosen. The momentum of the weight and the property of the billet are taken to be such that the latter will have large plastic deformation. The stiffness matrix therefore involves once again the elastic, plastic and geometric components. A damping term may be incorporated if the damping property of the material is known, or if the material is visco-elastic of the Kelvin type, which amounts to the same thing. Furthermore, the friction forces on the contact surfaces of the billet with the weight, and with the rigid support may also be taken into account if they can be determined. Thus the problem is one encountered in the forging operation and may be taken as typical of any metal forming process.

The displacements in a dynamic problem are functions of both space and time. They may be obtained by solving the non-linear equation of motion incrementally, using the time element introduced in Ref. [10]. Here, however, a variation of this method is employed. Instead of prescribing the variation of the displacements in a given time interval, a variation of the inertia force is specified. This is equivalent to assuming a certain time function for the acceleration, and if, for example, this is taken to be cubic, the displacements will in effect be fifth order and should be superior to taking a cubic function of time for the displacements themselves directly. Another advantage is that the iterative procedure involves only matrix multiplication if the kinematic equivalent mass matrix can be assumed to remain sensibly constant. No matrix inversion is necessary for the incremental steps and hence the calculation can proceed extremely rapidly. This method was initiated by S. Kelsey, a former colleague of the authors, now Professor at the University of Notre Dame. It has been successfully applied in a number of non-linear elastic large displacement dynamic problems [11].

2. The Prandtl-Reuss incremental strain theory of plasticity

For deformation in the plastic range, the stress/strain relationship is assumed to be given by the Prandtl-Reuss incremental strain theory, in conjunction with the Von Mises criterion of yielding. This assumption is generally accepted in the static and quasi-static analysis, when the rate of loading is small. When this condition is not met, as in the case of the impact problem, its validity is less certain, but there is as yet no universally accepted alternative. It is therefore tentatively adopted for the dynamic analysis as well. The principle of the method is, at all events, unaffected by whatever theory one may choose to relate the stress to the strain; and indeed it may be used for the purpose of verifying a new theory. This is not the intention of this paper, however.

To formulate the Prandtl-Reuss theory in the matrix form, it is convenient to define the stresses and strains in a Cartesian coordinate system by the column matrices

$$\begin{aligned}\boldsymbol{\sigma} &= \{\sigma_{xx}\ \sigma_{yy}\ \sigma_{zz}\ \sqrt{2}\sigma_{xy}\ \sqrt{2}\sigma_{yz}\ \sqrt{2}\sigma_{zx}\},\\ \boldsymbol{\varepsilon} &= \{\varepsilon_{xx}\ \varepsilon_{yy}\ \varepsilon_{zz}\ \sqrt{2}\varepsilon_{xy}\ \sqrt{2}\varepsilon_{yz}\ \sqrt{2}\varepsilon_{zx}\}\end{aligned} \tag{1}$$

where

$$\varepsilon_{xy} = \frac{1}{2}\left(\frac{\partial u}{\partial y} + \frac{\partial v}{\partial x}\right) \quad \text{etc.} \tag{1a}$$

These may be separated into deviatoric and hydrostatic/volumetric parts

$$\begin{aligned}\boldsymbol{\sigma} &= \boldsymbol{\sigma}_D + \boldsymbol{\sigma}_H = \boldsymbol{\sigma}_D + \sigma_H \boldsymbol{e}_{3,3},\\ \boldsymbol{\varepsilon} &= \boldsymbol{\varepsilon}_D + \boldsymbol{\varepsilon}_H = \boldsymbol{\varepsilon}_D + \varepsilon_H \boldsymbol{e}_{3,3}\end{aligned} \tag{2}$$

where

$$\boldsymbol{e}_{3,3} = \{1\ \ 1\ \ 1\ \ 0\ \ 0\ \ 0\} \tag{2a}$$

and

$$\begin{aligned}\sigma_H &= \tfrac{1}{3}(\sigma_{xx}+\sigma_{yy}+\sigma_{zz}) = \tfrac{1}{3}\boldsymbol{e}^t_{3,3}\boldsymbol{\sigma},\\ \varepsilon_H &= \tfrac{1}{3}(\varepsilon_{xx}+\varepsilon_{yy}+\varepsilon_{zz}) = \tfrac{1}{3}\boldsymbol{e}^t_{3,3}\boldsymbol{\varepsilon}.\end{aligned} \tag{2b}$$

The deviatoric parts are thus given by

$$\begin{aligned}\boldsymbol{\sigma}_D &= (\boldsymbol{I}_6 - \tfrac{1}{3}\boldsymbol{e}_{3,3}\boldsymbol{e}^t_{3,3})\boldsymbol{\sigma},\\ \boldsymbol{\varepsilon}_D &= (\boldsymbol{I}_6 - \tfrac{1}{3}\boldsymbol{e}_{3,3}\boldsymbol{e}^t_{3,3})\boldsymbol{\varepsilon}.\end{aligned} \tag{3}$$

Notice that

$$\boldsymbol{e}^t_{3,3}\boldsymbol{\sigma}_D = \boldsymbol{e}^t_{3,3}\boldsymbol{\varepsilon}_D = 0. \tag{3a}$$

For isotropic elastic materials, their relationship is given by

$$\boldsymbol{\sigma}_D = 2G\boldsymbol{\varepsilon}_D = \frac{E}{1+\nu}\,\boldsymbol{\varepsilon}_D$$

and (4)

$$\sigma_H = 3K\varepsilon_H = \frac{E}{1-2\nu}\,\varepsilon_H$$

where E, ν, G and K are the Young's Modulus, Poisson's ratio, shear modulus and bulk modulus respectively. From these equations it may easily be shown that

$$\begin{aligned} \boldsymbol{\sigma} &= 2G\left(\boldsymbol{I}_6 + \frac{\nu}{1-2\nu}\, e_{3,3}e^t_{3,3}\right)\boldsymbol{\varepsilon} = \boldsymbol{\kappa}_e\boldsymbol{\varepsilon}, \\ \boldsymbol{\varepsilon} &= \frac{1}{2G}\left(\boldsymbol{I}_6 - \frac{\nu}{1+\nu}\, e_{3,3}e^t_{3,3}\right)\boldsymbol{\sigma} = \boldsymbol{\varphi}_e\boldsymbol{\sigma} \end{aligned} \tag{5}$$

where

$$\boldsymbol{\kappa}_e = \boldsymbol{\varphi}_e^{-1} \tag{5a}$$

is the elastic stiffness of the material.

To specify the behaviour of the material in the plastic range, a scalar quantity known as the equivalent stress $\bar{\sigma}$ is defined by

$$\bar{\sigma}^2 = \tfrac{3}{2}\boldsymbol{\sigma}^t_D\boldsymbol{\sigma}_D \tag{6}$$

which gives on differentiation

$$\bar{\sigma}\mathrm{d}\bar{\sigma} = \tfrac{3}{2}\boldsymbol{\sigma}^t_D\mathrm{d}\boldsymbol{\sigma}_D. \tag{6a}$$

The strain increment $\mathrm{d}\boldsymbol{\varepsilon}$ is split into elastic and plastic components, giving

$$\mathrm{d}\boldsymbol{\varepsilon} = \mathrm{d}\boldsymbol{\varepsilon}_e + \mathrm{d}\boldsymbol{\varepsilon}_p \tag{7}$$

where $\boldsymbol{\varepsilon}$ is assumed to have no volumetric complonent, i.e.

$$e^t_{3,3}\mathrm{d}\boldsymbol{\varepsilon}_p = 0. \tag{7a}$$

Likewise, the equivalent plastic strain increment is defined as

$$\mathrm{d}\bar{\varepsilon}^2_p = \tfrac{2}{3}\mathrm{d}\boldsymbol{\varepsilon}^t_p\mathrm{d}\boldsymbol{\varepsilon}_p. \tag{8}$$

The Von Mises yield criterion then asserts that yielding takes place when

$$\bar{\sigma} = Y\left(\int \mathrm{d}\bar{\varepsilon}_p\right) \tag{9}$$

which is a function of the past history of plastic deformation, represented by $\int \mathrm{d}\bar{\varepsilon}_p$. In the case of a uniaxial state of stress σ, $\bar{\sigma}$ is reduced to σ and $\mathrm{d}\bar{\varepsilon}_p$ to the plastic part of the corresponding strain, i.e.

$$\mathrm{d}\bar{\varepsilon}_p = \mathrm{d}\varepsilon_p = \mathrm{d}\varepsilon - \mathrm{d}\varepsilon_e. \tag{10}$$

Hence, the slope of the stress/strain diagram from a simple tensile test gives the tangent modulus

$$E_t = \frac{d\sigma}{d\varepsilon}. \tag{11}$$

Its inverse may be written as

$$\frac{1}{E_t} = \frac{d\varepsilon}{d\sigma} = \frac{d\varepsilon_e}{d\sigma} + \frac{d\varepsilon_o}{d\sigma} = \frac{1}{E} + \frac{1}{E_p} \tag{11a}$$

where

$$E = \frac{d\sigma}{d\varepsilon_e} = \text{Young's modulus}$$

and

$$E_p = \frac{d\sigma}{d\varepsilon_p} = \frac{d\bar{\sigma}}{d\bar{\varepsilon}_p} \tag{12}$$

may be called the plastic modulus.

The Prandtl-Reuss theory stipulates that the strain increments are proportional to the stress deviators by a constant $d\lambda$, providing that the material is yielding and the equivalent stress is increasing. This condition may be specified by (see Fig. 1)

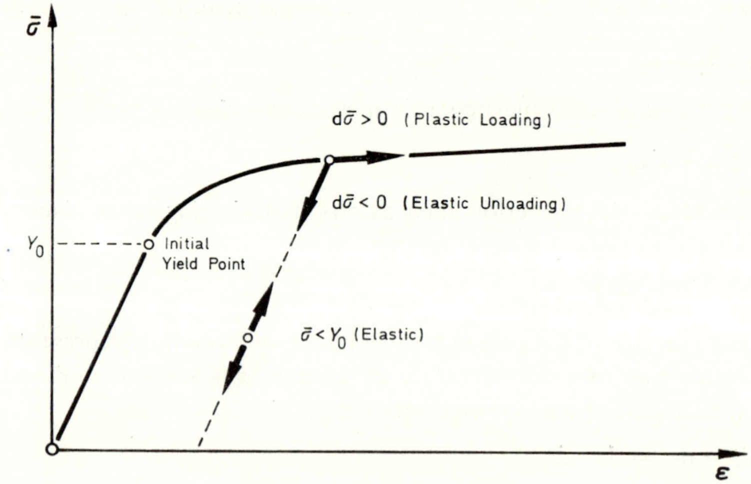

Fig. 1. Material behaviour.

$$
\begin{aligned}
\mathrm{d}\boldsymbol{\varepsilon} &= \mathrm{d}\lambda\boldsymbol{\sigma} && \text{for} \quad \bar{\sigma} = Y \quad \text{and} && \mathrm{d}\bar{\sigma} > 0, \\
&= \mathbf{0} && \text{for} \quad \bar{\sigma} = Y \quad \text{but} && \mathrm{d}\bar{\sigma} \leqq 0, \\
& && \text{or} \quad \bar{\sigma} < Y \quad \text{(elastic).}
\end{aligned} \tag{13}
$$

Substituting (13) and (6) into (8), the constant of proportionality $\mathrm{d}\lambda$ may be deduced. Since

$$
\begin{aligned}
\mathrm{d}\bar{\varepsilon}_p &= (\tfrac{2}{3}\mathrm{d}\boldsymbol{\varepsilon}_p \mathrm{d}\boldsymbol{\varepsilon}_p)^{\frac{1}{2}} = (\tfrac{2}{3}\mathrm{d}\lambda^2 \boldsymbol{\sigma}_D^t \boldsymbol{\sigma}_D)^{\frac{1}{2}} \\
&= \tfrac{2}{3}\mathrm{d}\lambda\bar{\sigma}
\end{aligned} \tag{14}
$$

giving

$$
\mathrm{d}\lambda = \frac{3}{2}\,\frac{\mathrm{d}\bar{\varepsilon}_p}{\bar{\sigma}} = \frac{3}{2E_p}\,\frac{\mathrm{d}\bar{\sigma}}{\bar{\sigma}} \tag{14a}
$$

after using the definition of E_p from (12).

It may be seen that the increment of plastic work, that is, the work done by the stress on the plastic strain increments, is given by

$$
\begin{aligned}
\mathrm{d}W_p &= \boldsymbol{\sigma}^t \mathrm{d}\boldsymbol{\varepsilon}_p = \boldsymbol{\sigma}_D^t \mathrm{d}\boldsymbol{\varepsilon}_p = \frac{3}{2E_p}\,\frac{\mathrm{d}\bar{\sigma}}{\bar{\sigma}}\,\boldsymbol{\sigma}_D^t \boldsymbol{\sigma}_D \\
&= \frac{1}{E_p}\,\bar{\sigma}\mathrm{d}\bar{\sigma} = \bar{\sigma}\mathrm{d}\bar{\varepsilon}_p.
\end{aligned} \tag{15}
$$

Hence $\mathrm{d}\bar{\sigma} > 0$ implies $\mathrm{d}W_p > 0$ and represents the condition of loading. Since $\mathrm{d}W_p$ cannot be negative, $\mathrm{d}\bar{\sigma} < 0$ must indicate unloading, for which $\mathrm{d}\boldsymbol{\varepsilon}_p = \mathbf{0}$ and $\mathrm{d}W_p = 0$. The condition $\mathrm{d}\bar{\sigma} = 0$ is a neutral case, corresponding to which no plastic work is done either.

Using (7), (4) and (6a) in (15) gives

$$
\begin{aligned}
\frac{1}{E_p}\,\bar{\sigma}\mathrm{d}\bar{\sigma} &= \boldsymbol{\sigma}_D^t \mathrm{d}\boldsymbol{\varepsilon}_p = \boldsymbol{\sigma}_D^t \mathrm{d}\boldsymbol{\varepsilon} - \boldsymbol{\sigma}_D^t \mathrm{d}\boldsymbol{\varepsilon}_e \\
&= \boldsymbol{\sigma}_D^t \mathrm{d}\boldsymbol{\varepsilon} - \frac{1}{2G}\,\boldsymbol{\sigma}_D^t \mathrm{d}\boldsymbol{\sigma}_D = \boldsymbol{\sigma}_D^t \mathrm{d}\boldsymbol{\varepsilon} - \frac{1}{3G}\,\bar{\sigma}\mathrm{d}\bar{\sigma}
\end{aligned} \tag{16}
$$

from which

$$
\left(\frac{1}{E_p} + \frac{1}{3G}\right)\bar{\sigma}\mathrm{d}\bar{\sigma} = \boldsymbol{\sigma}_D^t \mathrm{d}\boldsymbol{\varepsilon}. \tag{16a}
$$

The relationship between the total strain increment and the stress in the case of plastic loading is then

$$d\boldsymbol{\varepsilon} = d\boldsymbol{\varepsilon}_e + d\boldsymbol{\varepsilon}_p = \boldsymbol{\varphi}_e d\boldsymbol{\sigma} + \frac{3}{2E_p}\,\frac{d\bar{\sigma}}{\bar{\sigma}}\,\boldsymbol{\sigma}_D. \tag{17}$$

Hence the inverse relationship

$$\begin{aligned} d\boldsymbol{\sigma} &= \boldsymbol{\kappa}_e d\boldsymbol{\varepsilon} - \frac{3}{2E_p}\,\frac{d\bar{\sigma}}{\bar{\sigma}}\,\boldsymbol{\kappa}_e\boldsymbol{\sigma}_D = \boldsymbol{\kappa}_e d\boldsymbol{\varepsilon} - \frac{3G}{E_p}\,\frac{d\bar{\sigma}}{\bar{\sigma}}\,\boldsymbol{\sigma}_D \\ &= \left(\boldsymbol{\kappa}_e - \frac{3G}{(1+E_p/3G)\bar{\sigma}^2}\,\boldsymbol{\sigma}_D\boldsymbol{\sigma}_D^t\right) d\boldsymbol{\varepsilon} \\ &= (\boldsymbol{\kappa}_e - \boldsymbol{\kappa}_p)\,d\boldsymbol{\varepsilon} = \boldsymbol{\kappa}_t\,d\boldsymbol{\varepsilon} \end{aligned} \tag{18}$$

after using equations (5a), (3a) and (16a).

In the case of unloading, the plastic stiffness $\boldsymbol{\kappa}_p$, which is derived from the plastic strain increment $d\boldsymbol{\varepsilon}_p$, must be set to zero. The tangent stiffness $\boldsymbol{\kappa}_t$ is then reduced to the elastic stiffness $\boldsymbol{\kappa}_e$. This is the basic stress/strain equation for the matrix displacement method of elasto-plastic analysis. The analysis is essentially a non-linear one since the plastic stiffness $\boldsymbol{\kappa}_p$ as defined in (18) changes with the state of stress $\boldsymbol{\sigma}$. However, for a small incremental step, $\boldsymbol{\kappa}_p$ may be calculated according to the stress at the beginning of the step, and (18) may be regarded as a linearised relationship for a particular step.

3. Non-linear static analysis by the incremental approach

The general procedure for a small displacement linear analysis by the matrix displacement method has been well established. It is sufficient here to outline the steps of the linearised incremental approach, leading to the formulation of the geometric stiffness matrix.

As usual, the structure is idealised into an assembly of finite elements with a selected number of nodal points. In the small displacement case, the displacement $\boldsymbol{u}$ within an element is assumed to vary as a prescribed function of the displacements $\boldsymbol{\rho}$ at the nodes of the elements in the form of

$$\boldsymbol{u} = \{u\ v\ w\} = \boldsymbol{\omega\rho}\ ^{1)}. \tag{19}$$

The strain $\boldsymbol{\varepsilon}$ is then calculated from the displacements by

$$\boldsymbol{\varepsilon} = \boldsymbol{\delta}^t\boldsymbol{u} = \boldsymbol{\delta}^t\boldsymbol{\omega\rho} = \boldsymbol{\pi\rho} \tag{20}$$

[1]) { } denotes column matrix.

where, in the Cartesian coordinate system, the operator matrix

$$\boldsymbol{\delta} = \begin{bmatrix} \dfrac{\partial}{\partial x} & 0 & 0 & \dfrac{1}{\sqrt{2}}\dfrac{\partial}{\partial y} & 0 & \dfrac{1}{\sqrt{2}}\dfrac{\partial}{\partial z} \\ 0 & \dfrac{\partial}{\partial y} & 0 & \dfrac{1}{\sqrt{2}}\dfrac{\partial}{\partial x} & \dfrac{1}{\sqrt{2}}\dfrac{\partial}{\partial z} & 0 \\ 0 & 0 & \dfrac{\partial}{\partial z} & 0 & \dfrac{1}{\sqrt{2}}\dfrac{\partial}{\partial y} & \dfrac{1}{\sqrt{2}}\dfrac{\partial}{\partial x} \end{bmatrix} \tag{20a}$$

The matrix $\boldsymbol{\pi}$ is evaluated according to the undeformed geometry of the element.

Equilibrium of the corresponding nodal forces $\boldsymbol{P}$ and the stress $\boldsymbol{\sigma}$ within the element is then ensured by the application of the principle of virtual work, giving

$$\boldsymbol{P} = \int_V \boldsymbol{\pi}^t \boldsymbol{\sigma} \mathrm{d}V \tag{21}$$

where the integration is performed over the volume of the element material.

When the displacements are large, the relationship between strain and displacements becomes non-linear and (20) is no longer valid. However, they may be considered to increase by small steps, each sufficiently small, so that the strain increment $\mathrm{d}\boldsymbol{\varepsilon}$ may be related to the displacement increment $\mathrm{d}\boldsymbol{\rho}$ in a modified form of (20), giving

$$\mathrm{d}\boldsymbol{\varepsilon} = \boldsymbol{\pi}_0 \mathrm{d}\boldsymbol{\rho} \tag{22}$$

where $\boldsymbol{\pi}_0$ is based on the geometry at the beginning of the step. Equilibrium between the nodal force and the stress at the beginning of the step is according to (21)

$$\boldsymbol{P}_0 = \int_V \boldsymbol{\pi}_0^t \boldsymbol{\sigma}_0 \mathrm{d}V. \tag{23}$$

At the end of the step, the strain changes to $\boldsymbol{\varepsilon}_0 + \mathrm{d}\boldsymbol{\varepsilon}$, the stress $\boldsymbol{\sigma}_0 + \mathrm{d}\boldsymbol{\sigma}$, to nodal displacement $\boldsymbol{\rho}_0 + \mathrm{d}\boldsymbol{\rho}$, and the corresponding nodal force $\boldsymbol{P}_0 + \mathrm{d}\boldsymbol{P}$. The relationship between the strain and the displacement increments also change from (22) to

$$\mathrm{d}\boldsymbol{\varepsilon} = (\boldsymbol{\pi}_0 + \mathrm{d}\boldsymbol{\pi})\mathrm{d}\boldsymbol{\rho} \tag{24}$$

since the structure has deformed further during the step. The principle of virtual displacement now reads

$$\delta\boldsymbol{\rho}^t(\boldsymbol{P}_0+\mathrm{d}\boldsymbol{P}) = \int \delta\boldsymbol{\varepsilon}^t(\boldsymbol{\sigma}_0+\mathrm{d}\boldsymbol{\sigma})\mathrm{d}V$$

$$= \int \delta\boldsymbol{\rho}^t(\boldsymbol{\pi}_0+\mathrm{d}\boldsymbol{\pi})(\boldsymbol{\sigma}_0+\mathrm{d}\boldsymbol{\sigma})\mathrm{d}V \tag{25}$$

giving the equilibrium equation as

$$\mathrm{d}\boldsymbol{P} = \int (\boldsymbol{\pi}_0^t\mathrm{d}\boldsymbol{\sigma}+\mathrm{d}\boldsymbol{\pi}^t\boldsymbol{\sigma}_0)\mathrm{d}V = \mathrm{d}\boldsymbol{P}_m+\mathrm{d}\boldsymbol{P}_g \tag{25a}$$

after taking into account (23) and ignoring the second order term for the time being.

Substituting the stress/strain relationship (18) and neglecting again second order terms, the first part of $\mathrm{d}\boldsymbol{P}$

$$\mathrm{d}\boldsymbol{P}_m = \int \boldsymbol{\pi}^t\mathrm{d}\boldsymbol{\sigma}\mathrm{d}V = \left(\int \boldsymbol{\pi}_0^t\boldsymbol{\kappa}_t\boldsymbol{\pi}_0\mathrm{d}V\right)\mathrm{d}\boldsymbol{\rho} = \boldsymbol{k}_m\mathrm{d}\boldsymbol{\rho} \tag{26}$$

where

$$\boldsymbol{k}_m = \int \boldsymbol{\pi}_0^t\boldsymbol{\kappa}_t\boldsymbol{\pi}_0\mathrm{d}V \tag{26a}$$

is the material stiffness matrix of the element. The second part of $\mathrm{d}\boldsymbol{P}$ obviously arises from the effect of the existing stress $\boldsymbol{\sigma}_0$ on the change of geometry $\mathrm{d}\boldsymbol{\pi}$, which is a consequence of the deformation during the step. If σ_i is the ith element of the column $\boldsymbol{\sigma}_0$, corresponding to the ith column $(\mathrm{d}\boldsymbol{\pi}_i)$ of the $\mathrm{d}\boldsymbol{\pi}$ matrix, the second term may be re-written into the form

$$\mathrm{d}\boldsymbol{P}_g = \int \mathrm{d}\boldsymbol{\pi}^t\boldsymbol{\sigma}_0\mathrm{d}V = \int \sum_i \sigma_i\mathrm{d}\boldsymbol{\pi}_i\mathrm{d}V$$

$$= \sum_i\left(\int \sigma_i\frac{\mathrm{d}\boldsymbol{\pi}_i}{\mathrm{d}\boldsymbol{\rho}}\,\mathrm{d}V\right)\mathrm{d}\boldsymbol{\rho} = \boldsymbol{k}_g\mathrm{d}\boldsymbol{\rho} \tag{27}$$

where $\mathrm{d}\boldsymbol{\pi}_i/\mathrm{d}\boldsymbol{\rho}$ is a square matrix. Its jth column is the change of $\mathrm{d}\boldsymbol{\pi}_i$ due to the change of the jth element in $\mathrm{d}\boldsymbol{\rho}$. The matrix $\boldsymbol{k}_g$ is generally known as the geometric stiffness matrix. Eqn. (25a) now becomes

$$\mathrm{d}\boldsymbol{P} = (\boldsymbol{k}_m+\boldsymbol{k}_g)\mathrm{d}\boldsymbol{\rho} = \boldsymbol{k}_0\mathrm{d}\boldsymbol{\rho} \tag{28}$$

where $\boldsymbol{k}_0$ is the total stiffness matrix relating the increments of the nodal displacement and force, and is clearly a function of the state at the beginning of the step.

This linearised relationship can easily be improved by taking the second order terms in the strain/displacement equation and in the equilibrium condition into account. Thus, after the application of the load increment $\mathrm{d}\boldsymbol{P}$, let the force at the end of the step $\boldsymbol{P}_0+\mathrm{d}\boldsymbol{P}$ be equilibrated by a stress $\boldsymbol{\sigma}_0+\mathrm{d}\boldsymbol{\sigma}+\frac{1}{2}\mathrm{d}^2\boldsymbol{\sigma}$, where the second order term of the Taylor series expansion is included. Likewise, the incremental strain/displacement relationship at the end of the step it made more accurate by adding the second order change of $\boldsymbol{\pi}$ to (24), which becomes

$$\mathrm{d}\boldsymbol{\varepsilon} = (\boldsymbol{\pi}_0+\mathrm{d}\boldsymbol{\pi}+\tfrac{1}{2}\mathrm{d}^2\boldsymbol{\pi})\mathrm{d}\boldsymbol{\rho}. \tag{29}$$

The virtual work principle now reads

$$\delta\boldsymbol{\rho}^t(\boldsymbol{P}_0+\mathrm{d}\boldsymbol{P}) = \int \delta\boldsymbol{\varepsilon}^t(\boldsymbol{\sigma}_0+\mathrm{d}\boldsymbol{\sigma}+\tfrac{1}{2}\mathrm{d}^2\boldsymbol{\sigma})\mathrm{d}V \tag{30}$$

which gives

$$\mathrm{d}\boldsymbol{P} = \int [(\boldsymbol{\pi}_0^t\mathrm{d}\boldsymbol{\sigma}+\mathrm{d}\boldsymbol{\pi}^t\boldsymbol{\sigma}_0)+(\mathrm{d}\boldsymbol{\pi}^t\mathrm{d}\boldsymbol{\sigma}+\tfrac{1}{2}\mathrm{d}^2\boldsymbol{\pi}^t\boldsymbol{\sigma}_0+\tfrac{1}{2}\boldsymbol{\pi}_0^t\mathrm{d}^2\boldsymbol{\sigma})]\mathrm{d}V \tag{31}$$

after substituting (29), eliminating (23) and retaining only terms up to the second order. The first bracket on the right hand side contains the same terms as those in (25a) which gives $\boldsymbol{k}_0\mathrm{d}\boldsymbol{\rho}$. The second bracket is merely equal to half of the first order differential of the first bracket, and eqn. (31) is therefore

$$\mathrm{d}\boldsymbol{P} = (\boldsymbol{k}_0+\tfrac{1}{2}\mathrm{d}\boldsymbol{k})\mathrm{d}\boldsymbol{\rho} \tag{31a}$$

where $\mathrm{d}\boldsymbol{k}$ is the change of the stiffness during the step. Denoting the total stiffness at the end of the step by $\boldsymbol{k}_1$, then

$$\mathrm{d}\boldsymbol{k} = \boldsymbol{k}_1-\boldsymbol{k}_0 \tag{32}$$

and (31a) finally becomes

$$\mathrm{d}\boldsymbol{P} = \tfrac{1}{2}(\boldsymbol{k}_0+\boldsymbol{k}_1)\mathrm{d}\boldsymbol{\rho} = \boldsymbol{k}\mathrm{d}\boldsymbol{\rho}. \tag{31b}$$

This amounts to relating the force and displacement increments of a step by the average stiffness. It is itself a non-linear relationship since $\boldsymbol{k}_1$ is dependent on the increment $\mathrm{d}\boldsymbol{\rho}$. For its solution, a simple iteration within the step is therefore necessary. The first cycle assumes $\boldsymbol{k}_1 = \boldsymbol{k}_0$ and $\mathrm{d}\boldsymbol{\rho}$ is found from this linearised approximation. The calculation is repeated with the ensuing $\boldsymbol{k}_1$ until the result converges. This simple procedure, as has been seen, is entirely equivalent to the inclusion of second order terms in the incremental equations of equilibrium and between the strains and displacements.

The nodal displacement $\mathrm{d}\boldsymbol{\rho}$ of an element can be identified from the displacement vector $\mathrm{d}\boldsymbol{r}$ for the complete structure by the Boolean selection matrix $\boldsymbol{a}$ in the form of

$$\mathrm{d}\boldsymbol{\rho} = \boldsymbol{a}\mathrm{d}\boldsymbol{r}. \tag{33}$$

The static applied force $\mathrm{d}\boldsymbol{R}_s$ on the structure is then given by

$$\mathrm{d}\boldsymbol{R}_s = \sum \boldsymbol{a}^t\mathrm{d}\boldsymbol{P} = (\sum \boldsymbol{a}^t\boldsymbol{k}\boldsymbol{a})\mathrm{d}\boldsymbol{r} = \boldsymbol{K}\mathrm{d}\boldsymbol{r} \tag{34}$$

where the summation is over all the elements of the structure and the element stiffness $\boldsymbol{k}$ may be the average $\boldsymbol{k}$ of (31b), or the initial $\boldsymbol{k}_0$ of (28), as the degree of accuracy required. The matrix $\boldsymbol{K}$ is the assembled stiffness for the whole structure, and varies as a function of the displacement $\boldsymbol{r}$. Solution of (34) in the usual way gives a step-by-step analysis of the non-linear problem.

4. Formulation of the matrix equation of motion for a dynamic analysis with damping

In the cases when the dynamic response of the structure is significant but the deformation is small, the nodal force $\boldsymbol{P}$ induces not only displacement $\boldsymbol{u}$, but also velocity $\dot{\boldsymbol{u}}$ and acceleration $\ddot{\boldsymbol{u}}$. As a result there exists in the element the stress $\boldsymbol{\sigma}$, damping force $c\dot{\boldsymbol{u}}$ and inertia force $\rho\ddot{\boldsymbol{u}}$. The damping force is assumed to be proportional to the velocity by the coefficient c, and ρ is the mass per unit volume of material. Equilibrium between $\boldsymbol{P}$ and the internal forces is ensured once again by the principle of virtual work. Thus, for a system of displacement $\boldsymbol{u}$ and strain $\boldsymbol{\varepsilon}$ within the element, compatible with nodal displacement $\boldsymbol{\rho}$, the virtual work is

$$\boldsymbol{\rho}^t\boldsymbol{P} = \int [\boldsymbol{\varepsilon}^t\boldsymbol{\sigma} + \boldsymbol{u}^t(c\dot{\boldsymbol{u}} + \rho\ddot{\boldsymbol{u}})]\mathrm{d}V. \tag{35}$$

Substitution of (5), (19) and (20) gives

$$\boldsymbol{P} = \boldsymbol{k}_e\dot{\boldsymbol{\rho}} + \boldsymbol{c}\dot{\boldsymbol{\rho}} + \boldsymbol{m}\ddot{\boldsymbol{\rho}} \tag{36}$$

where

$$\boldsymbol{k}_e = \int \boldsymbol{\pi}^t\boldsymbol{\kappa}_t\boldsymbol{\pi}\mathrm{d}V$$

$$\boldsymbol{c} = c\int \boldsymbol{\omega}^t\boldsymbol{\omega}\mathrm{d}V \tag{37}$$

and

$$\boldsymbol{m} = \rho\int \boldsymbol{\omega}^t\boldsymbol{\omega}\mathrm{d}V.$$

Assembly of all elements by using the $\boldsymbol{a}$ matrix of (33) to identify each $\boldsymbol{\rho}$ with the appropriate displacement in $\boldsymbol{r}$ yields the equation of motion

$$\boldsymbol{R}_a = \boldsymbol{K}_e\boldsymbol{r}+\boldsymbol{C}\dot{\boldsymbol{r}}+\boldsymbol{M}\ddot{\boldsymbol{r}} \tag{38}$$

where $\boldsymbol{R}_a$ is the applied force, and

$$\begin{aligned}
\boldsymbol{K}_e &= \textstyle\sum \boldsymbol{a}^t\boldsymbol{k}_e\boldsymbol{a} && \text{the elastic stiffness matrix,}\\
\boldsymbol{C} &= \textstyle\sum \boldsymbol{a}^t\boldsymbol{c}\boldsymbol{a} && \text{the damping matrix,}\\
\boldsymbol{M} &= \textstyle\sum \boldsymbol{a}^t\boldsymbol{m}\boldsymbol{a} && \text{the mass matrix.}
\end{aligned} \tag{38a}$$

In this formulation, $\boldsymbol{C}$ is seen to be proportional to $\boldsymbol{M}$ as a result of the simple hypothesis that damping is proportional to the velocity. A more refined formulation may be obtained if the property of the material is assumed to be visco-elastic of the Kelvin type. Thus, the stress strain relationship, instead of the simple elastic one as given in (5), becomes

$$\boldsymbol{\sigma} = \boldsymbol{\kappa}_e\boldsymbol{\varepsilon}+\boldsymbol{\eta}\dot{\boldsymbol{\varepsilon}} \tag{39}$$

where

$$\boldsymbol{\eta} = 2\Gamma\left(\boldsymbol{I}_6+\frac{\mu}{1-2\mu}\,\boldsymbol{e}_{3,3}\boldsymbol{e}_{3,3}^t\right) \tag{40a}$$

and Γ and μ are the viscous equivalent of G and ν. In cases when the viscous part is assumed incompressible, i.e. if the volumetric strain is purely elastic, the matrix $\boldsymbol{\eta}$ takes instead of (40a) the form

$$\boldsymbol{\eta} = 2\Gamma(\boldsymbol{I}_6-\tfrac{1}{3}\boldsymbol{e}_{3,3}\boldsymbol{e}_{3,3}^t). \tag{40b}$$

The principle of virtual work expressing equilibrium between the applied force and the internal forces now reads

$$\boldsymbol{\rho}^t\boldsymbol{P} = \int (\boldsymbol{\varepsilon}^t\boldsymbol{\sigma}+\boldsymbol{u}^t\rho\ddot{\boldsymbol{u}})\mathrm{d}V \tag{41}$$

which leads back to eqn. (36) but with the matrix $\boldsymbol{c}$ given by

$$\boldsymbol{c} = \int \boldsymbol{\pi}^t \boldsymbol{\eta} \boldsymbol{\pi} \mathrm{d}V \tag{42}$$

the damping matrix is now similar to the stiffness matrix instead of the mass matrix. In fact, if $\mu = \nu$, $\boldsymbol{C}$ will be proportional to $\boldsymbol{K}_e$.

When the displacement is large and the structure undergoing plastic deformation, eqn. (38) changes to

$$\boldsymbol{R}_a = \boldsymbol{K}\boldsymbol{r} + \boldsymbol{C}\dot{\boldsymbol{r}} + \boldsymbol{M}\ddot{\boldsymbol{r}} \tag{43}$$

where the stiffness matrix $\boldsymbol{K}$ is not only the elastic stiffness, but consists also of the plastic stiffness and the geometric stiffness, as in the static analysis. This equation is valid at any given time, but is non-linear since $\boldsymbol{K}$ is a function of the displacement and hence also a function of the time t. To be more precise, it can be seen that $\boldsymbol{R}_a$ is made up of three components, representing the static force $\boldsymbol{K}\boldsymbol{r}$, the damping force $\boldsymbol{C}\dot{\boldsymbol{r}}$, and the inertia force $\boldsymbol{M}\ddot{\boldsymbol{r}}$. Suppose that, at the beginning of a preceding very short time interval, over which the stiffness matrix $\boldsymbol{K}$ remains substantially the same, the static force is given by

$$\boldsymbol{R}_{s0} = \boldsymbol{K}\boldsymbol{r}_0. \tag{44}$$

Subtracting (44) from (43) one obtains

$$\boldsymbol{R}_a - \boldsymbol{R}_{s0} = \boldsymbol{K}(\boldsymbol{r} - \boldsymbol{r}_0) + \boldsymbol{C}\dot{\boldsymbol{r}} + \boldsymbol{M}\ddot{\boldsymbol{r}}, \tag{45}$$

$$\boldsymbol{R}_a = (\boldsymbol{R}_{s0} + \boldsymbol{K}\mathrm{d}\boldsymbol{r}) + \boldsymbol{C}\dot{\boldsymbol{r}} + \boldsymbol{M}\ddot{\boldsymbol{r}}$$

or

$$= \boldsymbol{R}_s + \boldsymbol{C}\dot{\boldsymbol{r}} + \boldsymbol{M}\ddot{\boldsymbol{r}}. \tag{46}$$

This shows that the static force $\boldsymbol{R}_s$ must be accumulated from small increments of $\mathrm{d}\boldsymbol{R}_s = \boldsymbol{K}\mathrm{d}\boldsymbol{r}$ where $\boldsymbol{K}$ is a function of $\boldsymbol{r}$, and can be calculated in exactly the same way as described in the previous sections for the static analysis.

5. Solution of the non-linear dynamic problem by the finite element method

It is possible to solve this non-linear equation step-by-step, using the finite element method in time as proposed in ref. [9]. In this paper, an alternative approach is chosen. Instead of the displacement $\boldsymbol{r}$, the inertia force $\boldsymbol{M}\ddot{\boldsymbol{r}}$ is assumed to behave as a prescribed function of time within a given small time interval. Thus, from (46), the inertia force

$$\boldsymbol{R} = \boldsymbol{M}\ddot{\boldsymbol{r}} = \boldsymbol{R}_a - \sum \mathrm{d}\boldsymbol{R}_s - \boldsymbol{C}\dot{\boldsymbol{r}}. \tag{47}$$

Then, during the small time interval of duration τ, the force $\boldsymbol{R}$ is assumed to vary, for the purpose of the present calculations, as a cubic function of time. The four coefficients of the cubic function which defines this variation are determined by the conditions at the beginning and the end of this time interval. The time function can therefore be written into the form

$$\boldsymbol{R} = \varphi_1\boldsymbol{R}_0 + \varphi_2\boldsymbol{R}'_0 + \varphi_3\boldsymbol{R}_1 + \varphi_4\boldsymbol{R}'_1 \tag{48}$$

where φ_i are cubic functions of the non-dimensionalised time variable $\xi = t/\tau$, (giving $\mathrm{d}t = \tau\mathrm{d}\xi$) and

$$\boldsymbol{R}' = \mathrm{d}\boldsymbol{R}/\mathrm{d}\xi = \tau\mathrm{d}\boldsymbol{R}/\mathrm{d}t = \tau\dot{\boldsymbol{R}} \tag{48a}$$

is the rate of change of $\boldsymbol{R}$ with respect to ξ. The subscripts 0 and 1 denote values at the beginning and the end of the time interval when $\xi = 0$ and $\xi = 1$ respectively. The functions φ_i are therefore Hermitian polynomials of the second order, given by

$$\begin{aligned} \varphi_1 &= 1 - 3\xi^2 + 2\xi^3, \\ \varphi_2 &= \xi - 2\xi^2 + \xi^3, \\ \varphi_3 &= 3\xi^2 - 2\xi^3, \\ \varphi_4 &= -\xi^2 + \xi^3 \end{aligned} \tag{49}$$

each giving a unit value for its associated coefficient and zero for the others. The time interval τ may therefore be considered as a finite time element for the inertia force, and the entire time sequence is made up of such small time elements in series.

Having defined $\boldsymbol{R}$, eqn. (47) can now be integrated to obtain:

$$\begin{aligned} \boldsymbol{M}\dot{\boldsymbol{r}} = \boldsymbol{M}\dot{\boldsymbol{r}}_0 &+ \tau(\xi - \xi^3 + \tfrac{1}{2}\xi^4)\boldsymbol{R}_0 + \\ &+ \tau(\tfrac{1}{2}\xi^2 - \tfrac{2}{3}\xi^3 + \tfrac{1}{4}\xi^4)\boldsymbol{R}'_0 + \\ &+ \tau(\xi^3 - \tfrac{1}{2}\xi^4)\boldsymbol{R}_1 + \tau(-\tfrac{1}{3}\xi^3 + \tfrac{1}{4}\xi^4)\boldsymbol{R}'_1 \end{aligned} \tag{50a}$$

and

$$\begin{aligned} \boldsymbol{M}\boldsymbol{r} = \boldsymbol{M}\boldsymbol{r}_0 + \tau\xi\boldsymbol{M}\dot{\boldsymbol{r}}_0 &+ \tau^2(\tfrac{1}{2}\xi^2 - \tfrac{1}{4}\xi^4 + \tfrac{1}{10}\xi^5)\boldsymbol{R}_0 + \\ &+ \tau^2(\tfrac{1}{6}\xi^3 - \tfrac{1}{6}\xi^4 + \tfrac{1}{20}\xi^5)\boldsymbol{R}'_0 + \\ &+ \tau^2(\tfrac{1}{4}\xi^4 - \tfrac{1}{10}\xi^5)\boldsymbol{R}_1 + \tau^2(-\tfrac{1}{12}\xi^4 + \tfrac{1}{20}\xi^5)\boldsymbol{R}'_1. \end{aligned} \tag{50b}$$

The displacement $\boldsymbol{r}_1$ and velocity $\dot{\boldsymbol{r}}_1$ at the end of the time interval can be obtained from these two equations by simply putting $\xi = 1$, which using (48a) results in

$$\begin{aligned}\dot{\boldsymbol{r}}_1 &= \dot{\boldsymbol{r}}_0 + \tfrac{1}{12}\tau \boldsymbol{M}^{-1}(6\boldsymbol{R}_0 + \tau\dot{\boldsymbol{R}}_0 + 6\boldsymbol{R}_1 - \tau\dot{\boldsymbol{R}}_1)\\ &= \dot{\boldsymbol{r}}_0 + \mathrm{d}\dot{\boldsymbol{r}}\end{aligned} \tag{51a}$$

$$\begin{aligned}\boldsymbol{r}_1 &= \boldsymbol{r}_0 + [\tau\dot{\boldsymbol{r}}_0 + \tfrac{1}{60}\tau^2(21\boldsymbol{R}_0 + 3\tau\dot{\boldsymbol{R}}_0 + 9\boldsymbol{R}_1 - 2\tau\dot{\boldsymbol{R}}_1)]\\ &= \boldsymbol{r}_0 + \mathrm{d}\boldsymbol{r}.\end{aligned} \tag{51b}$$

The acceleration is given directly from (47) as

$$\ddot{\boldsymbol{r}}_1 = \boldsymbol{M}^{-1}\boldsymbol{R}_1. \tag{51c}$$

The two equations (51a and 51b) calculate the displacement and velocity at the end of the time step by adding increments to their initial values. However, they cannot be solved explicitly since they are dependent on the force and its time derivative at time $\xi = 1$, which depends on $\boldsymbol{K}$, $\mathrm{d}\boldsymbol{r}$ and $\mathrm{d}\dot{\boldsymbol{r}}$ according to the definitions in (46–47). Hence, an iterative solution for each time step is necessary. Starting from a given state at $\xi = 0$, the simple assumptions that $\dot{\boldsymbol{R}}_1 = \dot{\boldsymbol{R}}_0$ and $\boldsymbol{R}_1 = \boldsymbol{R}_0 + \tau\dot{\boldsymbol{R}}_0$ are made as a first approximation; this enables $\mathrm{d}\boldsymbol{r}$, $\mathrm{d}\dot{\boldsymbol{r}}$ and then $\boldsymbol{r}$ and $\dot{\boldsymbol{r}}$ to be determined by (51a and 51b), from which a second estimation of $\dot{\boldsymbol{R}}_1$ and $\boldsymbol{R}_1$ can now be calculated. The procedure is repeated until the results converge. The end values then become the initial values for the next time increment.

Strictly speaking, the kinematically equivalent mass matrix $\boldsymbol{M}$ (and also the damping matrix $\boldsymbol{C}$) are also functions of the geometry. If the displacements are large, they too will change with time. In practice, however, the influence of these changes are found to be small. The mass matrix can be assumed substantially constant during the complete time sequence, in which case the entire calculation involves only matrix multiplications after the inversion of the mass matrix at the very beginning. No further matrix inversion or solution is required in the iterative loops. The numerical calculation can therefore be performed extremely rapidly by a digital computer.

The assumption of force varying as a cubic function of time as given above can easily be improved if so desired. For example, the second time derivative of the force at the beginning and the end of the time interval may be taken as two more parameters to define the variation, which may now be taken as a function of the fifth order. The development is entirely analogous to the normal finite element theory in space (see e.g. ref. [10]), with the time t replacing the space coordinates as the independent variable.

6. Examples

6.1 *Rectangular thin strip in a tensile test*

This simple case is chosen as a representative example of the static large displacement plastic analysis. A quarter of the specimen is divided into a number of triangular elements with 3 nodal points (the constant strain TRIM 3 element in the ASKA library [12]). Axial displacement increments are imposed at the end of the strip to simulate the effect of the grip of the testing machine. The elastic stiffness and the geometric stiffness of this element has been given in refs. [2, 13] using the concept of the natural deformation modes.

The material is assumed in this calculation to be an aluminium alloy with a stress-strain relationship expressible by a modified Ramberg-Osgood formula giving the plastic strain

$$\varepsilon_p = \frac{1.1Y_0}{mE}\left[\left(\frac{\sigma}{1.1Y_0}\right)^m - \left(\frac{1}{1.1}\right)^m\right] \tag{52}$$

in which the initial yield stress Y_0 is taken to be 34,500 lbf/in^2, and the exponent $m = 10$. The inverse of the plastic modulus is then

$$\frac{1}{E_p} = \frac{d\varepsilon_p}{d\sigma} = \frac{1}{E}\left(\frac{\sigma}{1.1Y_0}\right)^{m-1}. \tag{53}$$

The calculation was performed firstly using the linearised incremental theory, i.e. each increment is based on the stiffness at the beginning of the step. The procedure for this is straightforward. Then the more refined parabolic incremental theory in which the average stiffness was used. The flow diagram for this procedure, involving an iteration for each step, is given in Fig. 2. The first step takes the equivalent stress near the point of initial yielding, then, for the 1st calculation, some very small increments are necessary to get around the elbow of the stress-strain curve, after that, relatively large increments can be imposed at each step. The calculation was terminated after the length had increased more than 50%. The shape of deformation of the specimen is shown in Fig. 3. Larger deformation could easily be calculated but most likely would be meaningless as the imperfections would have caused local deformation. As expected, the calculated average stress and strain follows closely the assumed curve as shown in Fig. 4. Generally speaking, the result of an elastic analysis using the constant strain triangular

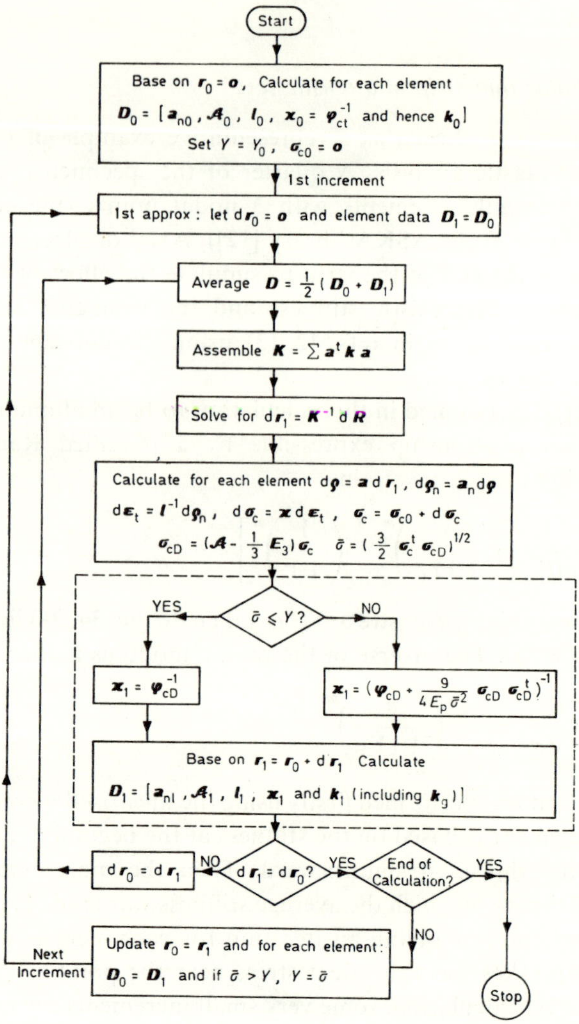

Fig. 2. Flow diagram for static analysis using triangular elements.

membrane element depends on the arrangement of the elements in the structural idealisation. In this plastic analysis, however, it is found that, although the detail stress distribution changes with different subdivision of the rectangular strip, the overall effect hardly changes at all. Moreover, even a very coarse grid (Fig. 3) gives very good agreement as shown in Fig. 4, refinement of the grid seems to be quite unnecessary.

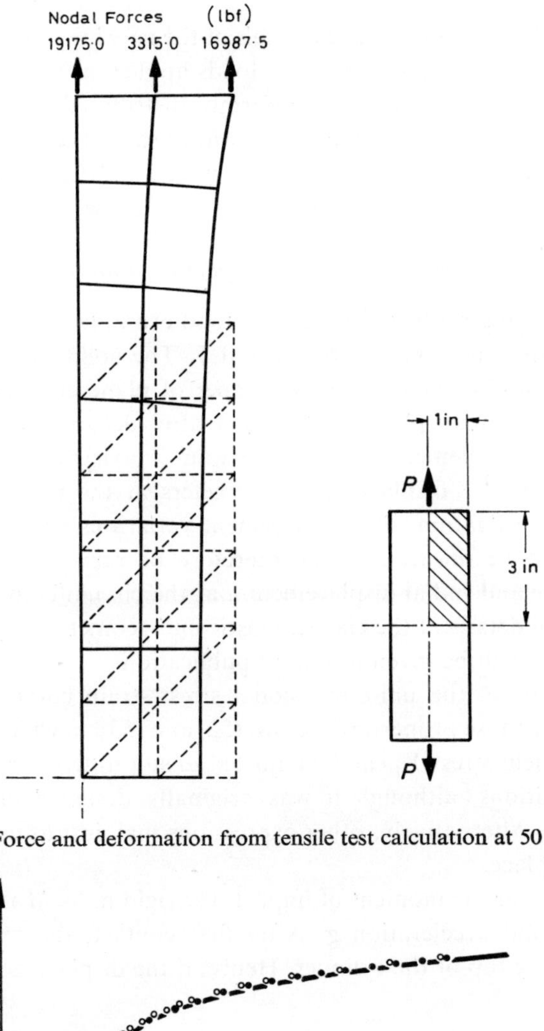

Fig. 3. Force and deformation from tensile test calculation at 50% strain.

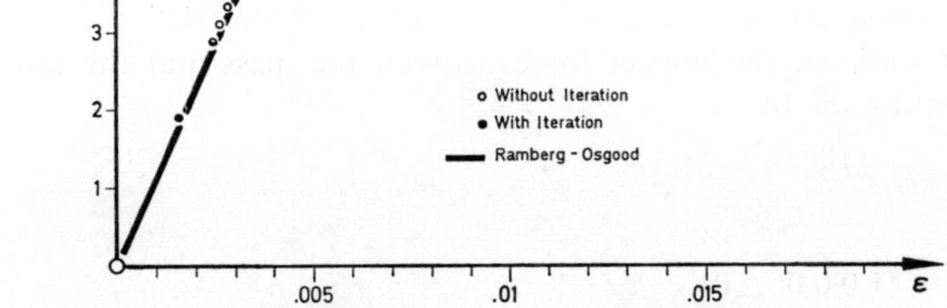

Fig. 4. Stress-strain relationship from tensile test calculation.

It is interesting to note that, as a result of the necking of the specimen due to plastic deformation, the equivalent loads applied at the gripping end do not any longer correspond to a uniformly distributed pressure, but with the centre of pressure (over half the total width) shifted considerably towards the centre. An analysis by load increments would not therefore give the same results if their proportions are kept constant as in the elastic case.

6.2 *Short circular cylinder under axial impact by a large rigid mass*

This typical forging problem serves as a good example of application of the dynamic method given in sections 4 and 5. The solid cylinder is divided axially into thin circular discs by a number of plane cross-sections, which are assumed to remain plane and circular after deformation. Each disc is treated as a finite element whose behaviour is axisymmetric and can be specified in terms of displacement parameters at the nodal planes. As a preliminary investigation, the displacement variations within each element are assumed to be linear, and can therefore be represented by the axial displacement w and radial displacement u at the circumference of the nodal circles. The full details of the elastic, plastic and geometric stiffness matrices for this element will be given in a later publication.

For convenience, the uni-dimensional stress-strain curve is once again assumed to give those properties of eqns. (52) and (53), in which the exponent m and initial yield stress Y_0 can be adjusted to suit different materials under different conditions, although it was originally designed chiefly for aluminium alloys. Alternatively, other expressions may just as easily be substituted in their place.

Suppose that, at the moment of impact, the rigid mass M is free falling at a velocity v_0 and acceleration g. After first contact, the mass will move together with the top of the cylinder. Hence, if the displacement vector is so arranged that

$$\boldsymbol{r} = \{w_1 w_2 \ldots w_n u_1 u_2 \ldots u_n\} \tag{54}$$

and it is assumed for the time being that there is no friction force on the contact surfaces, the impact force between the mass and the top of the cylinder is given by

$$F = -M\ddot{w}_1 = -M\boldsymbol{e}_1^t\ddot{\boldsymbol{r}} \tag{55}$$

where

$$\boldsymbol{e}_1 = \{1\,0\,0\,0 \ldots 0\}. \tag{55a}$$

The applied force is then

$$R_a = \{F\,0\,0\ldots0\} = Fe_1 = -Me_1e_1^t\ddot{r}. \tag{56}$$

For this particular problem under these assumptions, it is more convenient to modify eqn. (47) and define the force R, not by the inertia force of the cylinder alone, but by

$$R = M\ddot{r} - R_a = -\sum dR_s - C\dot{r}. \tag{57}$$

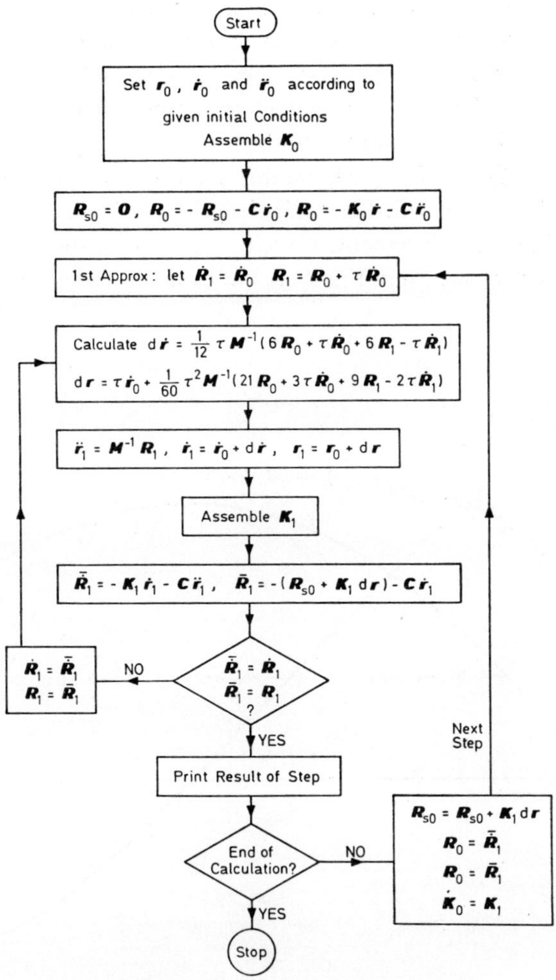

Fig. 5. Flow diagram for dynamic analysis.

After substitution of (56), it becomes

$$\boldsymbol{R} = (\boldsymbol{M}+M\boldsymbol{e}_1\boldsymbol{e}_1^t)\ddot{\boldsymbol{r}} = \boldsymbol{M}'\ddot{\boldsymbol{r}} = -\sum \mathrm{d}\boldsymbol{R}_s - \boldsymbol{C}\dot{\boldsymbol{r}} \tag{58}$$

where the $\boldsymbol{M}$ matrix has the mass M added to the leading element of the equivalent mass matrix $\boldsymbol{M}$ for the cylinder. The problem is then solved as described before in section 5, with eqn. (47) replaced by eqn. (58). The flow diagram for this calculation can be found in Fig. 5.

The example considered here is a circular cylinder of aluminium alloy of 1 in. both in height and in diameter, under the impact of a rigid mass M equal to 500 times that of the cylinder. No viscous damping has been allowed in these calculations. In the first case, M is travelling at $v_0 = 100$ in/sec. at the moment of impact, equivalent to a free fall from a height of 12.93 inches. The cylinder is taken to be at normal temperature, with $E = 11.4 \times 10^6$ lbf/in^2, $Y_0 = 34{,}500$ lbf/in^2, and an exponent $m = 10$ for eqn. (52). It is firstly divided along its length into 5 equal elements, for which the variations of displacement, velocity and acceleration at the top surface of the cylinder are shown in Fig. 6. A later calculation with 10 elements gives

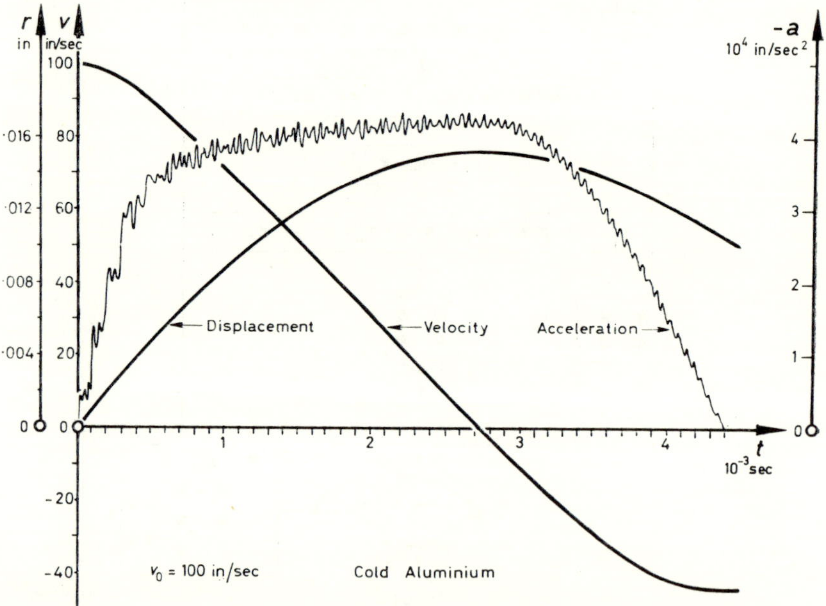

Fig. 6. Displacement velocity and acceleration at the top of the cylinder.

no noticeable difference to these results, apart from a more accurate stress variation. The deformation reaches a maximum after 2.7 msec. at which the average axial stress is about 54,500 lbf/in² and the height of the cylinder is reduced to 0.985 in., giving an average strain of 1.5%. Not surprisingly, the elastic resiliance of the cylinder causes the solid mass to bounce off. This happens when the acceleration changes sign from negative to positive 4.398 msec. after first contact, at a velocity $v = -44.6$ in/sec. (i.e. the top surface is on its way up). The height of the cylinder at this point is 0.989 in.

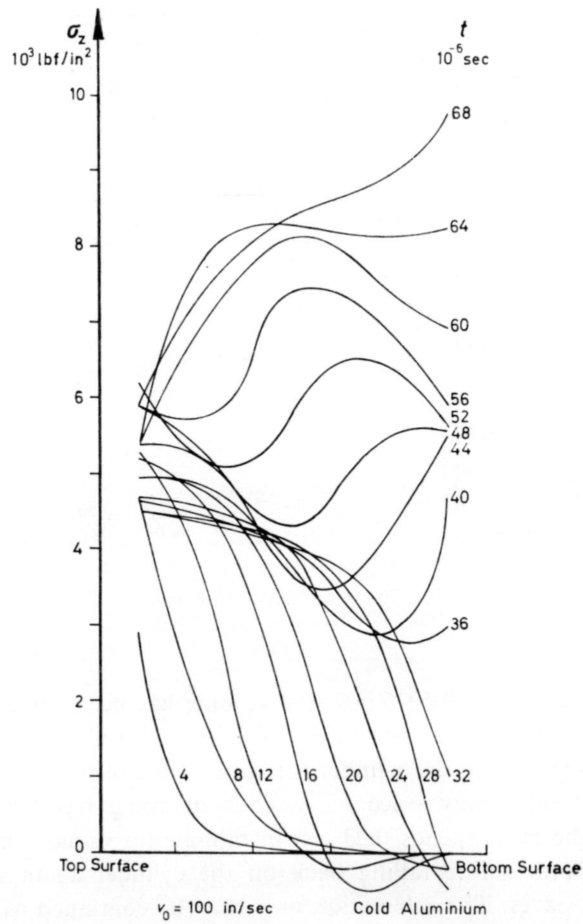

Fig. 7a. Axial stress distribution in axial direction of cylinder.

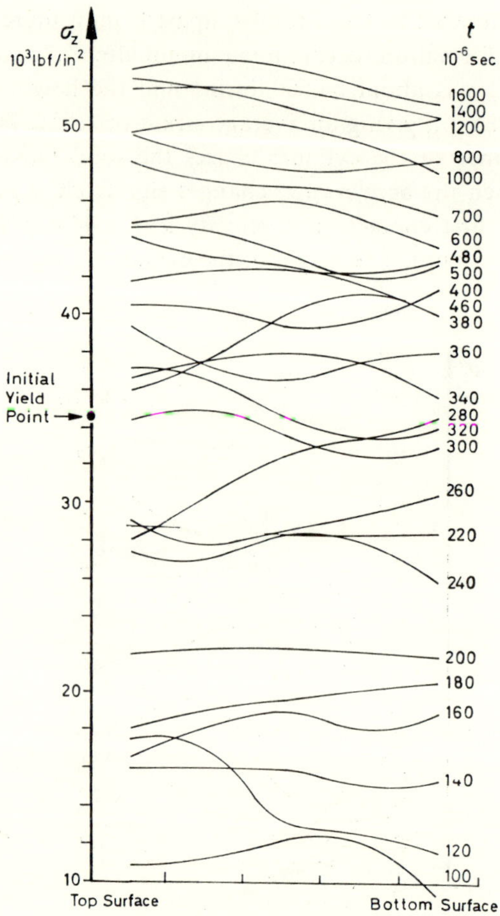

Fig. 7b. (Continued from 7a).

and the average radius 0.5026 ins. (the volume has been preserved). Thus, the elastic recovery agrees accurately with that given by the stress-strain curve. After separation, the cylinder is left in a state of free vibration which continues without decay since no viscous damping has been included. Meanwhile, the mass spends 0.23 sec. in mid-air (in vacuo) and reaches a height of 2.59 in. before falling back on the cylinder again at the same velocity as it leaves. The calculation may then be continued as if from the beginning if so desired. The entire calculation up to separation point is achieved by 2199 time elements of 2 μsec. each in duration. For each of these steps, the number of iterations varies consistently between 2 and 3, although the programme provides a facility which halfs the step time interval automatically should the result fail to converge after 10 iterations. The total time for the calculation on the CDC 6600 is 62 min., most of which is spent in calculating the plastic stiffness matrix after testing against elastic unloading.

Fig. 7a, b, and c, show the axial stress variation inside the cylinder from

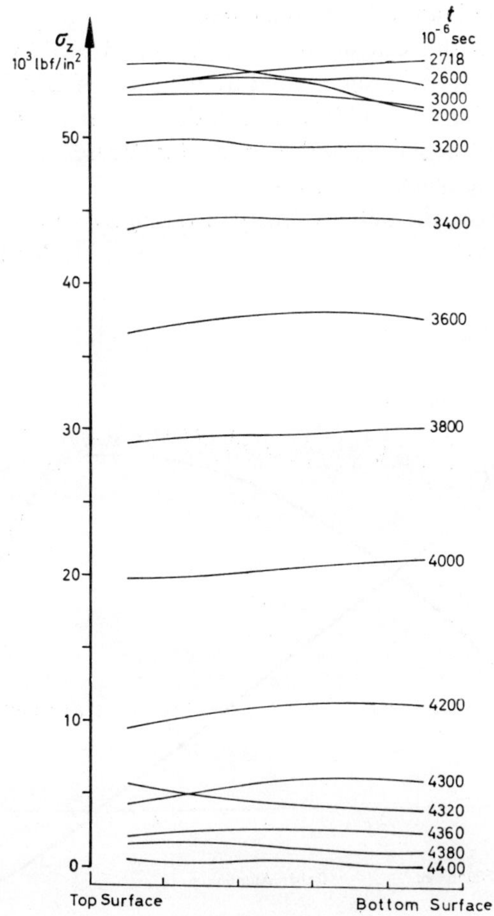

Fig. 7c. (Continued from 7b).

first impact. They picture clearly the propagation of the stress wave, its reflection from the support surface, and the subsequent combination which causes pockets of elastic unloading after the equivalent stress has entered the plastic range. The stress propagation is responsible for the wiggles in the acceleration curve in Fig. 6. The high frequency change is due to elastic stress wave. Here the calculation agrees accurately with the theoretical wave velocity of $\sqrt{E/\rho} = 2.09 \times 10^5$ in/sec. The presence of the slower plastic wave, travelling at a velocity of $\sqrt{E_p/\rho}$, is also descernible from the pulse-like effect modulating the amplitude of the curve. In this case, the plastic wave is roughly 3.5 to 4 times slower most of the time during plastic deformation. This fact is reflected faithfully by the curve.

The reliability of this calculation is further confirmed by a check on the energy level of the system, represented by the ratio of $\frac{1}{2}\dot{\boldsymbol{r}}^t\boldsymbol{M}'\dot{\boldsymbol{r}} + \frac{1}{2}\boldsymbol{r}^t\boldsymbol{K}\boldsymbol{r}$ to the initial energy of the mass $\frac{1}{2}Mv_0^2$. This ratio is 1 at all stages before the deformation becomes plastic, and a gradual decrease thereafter, recording the non-recoverable plastic dissipation.

In the second case, the initial velocity of the mass is raised to 200 in/sec., equivalent to a drop from 51.8 in. The cylinder is idealised by 5 elements

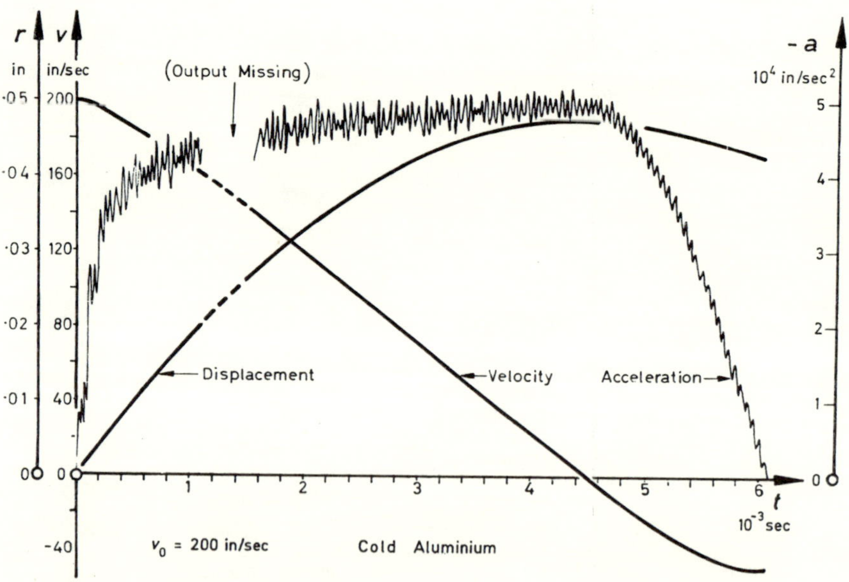

Fig. 8. Displacement, velocity and acceleration at the top of the cylinder.

and the results are shown in Fig. 8. Maximum deformation occurs after 4.44 msec., at which the height of the cylinder is 0.952 in., or an average strain of 4.8 %. The corresponding axial stress is approximately 63,000 lbf/in^2. While the elastic wave velocity remains the same, the theoretical plastic wave velocity is roughly around $0.3 - 0.27 \times 10^5$ in/sec., which is in agreement with the variations shown in the acceleration curve. The mass bounces off after 6.08 msec. with a velocity of -50 in/sec. (to a height of only 3.235 in.), leaving the cylinder with a permanent strain of 4.25 %. The calculation was performed again by 2 μsec time elements and takes 97 minutes.

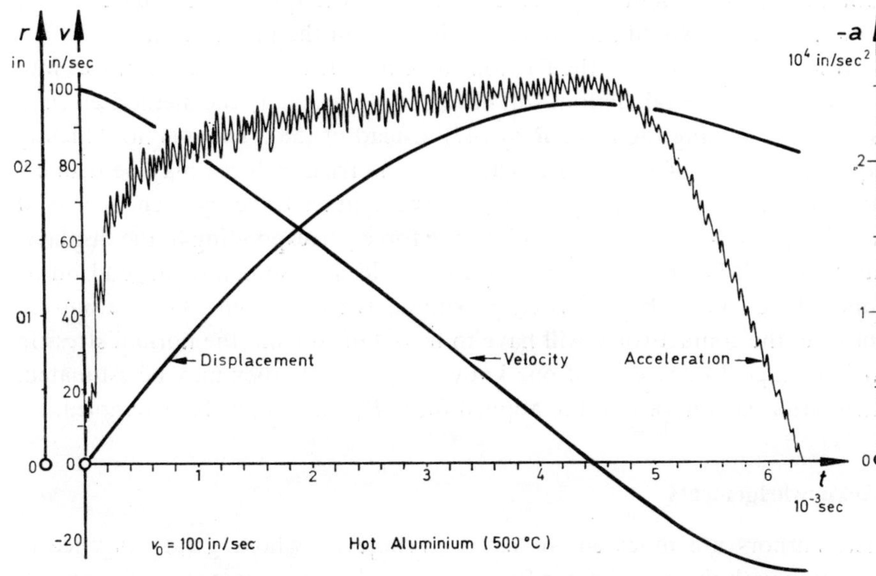

Fig. 9. Displacement, velocity and acceleration at the top of the cylinder.

For the third calculation, the initial velocity of the mass is again taken to be 100 in/sec. The material, however, has $E = 9.3 \times 10^6$ lbf/in^2., $Y_0 =$ $= 11{,}500$ lbf/in^2, and an exponent $m = 10$ for eqn. (52). The property is softer and is purported to represent aluminium alloy at 500°C. The calculation is performed by 4 μsec time elements, with the number of iterations for each step varying between 3 to 6. Results are shown in Fig. 9. Maximum deformation is reached after 4.416 msec at an average stress of 32,000 lbf/in^2 and strain of 2.41 %. Separation of the rigid mass occurs after 6.28 msec with a velocity of only -29.2 in/sec to a height of 1.1 in., leaving the cylinder with a permanent strain of 2.07%. All these are very much as expected. The calculation takes 66 mins. to complete.

It is perhaps worth mentioning that in all these cases, the effect of the body force in the geometric stiffness matrix is unimportant. The inclusion or otherwise of these terms makes no difference to the results.

Work is in progress to proceed to the next stage of the investigation, where a more refined element, with not only nodal displacements but also their derivatives as parameters, will be used in the calculation. This will enable the strain to vary within the element and to be continuous across element boundaries. It should improve even further the quality of the results or alternatively, it would require fewer elements in the idealisation.

If it is desired to take the friction force between the contact surfaces into account, then the calculation will have to follow exactly the method given in section 5, defining the force $\boldsymbol{R}$ by (47) instead of (58). There is no difficulty apart from that of knowing exactly how the friction force may be defined. Presumably it will be proportional to the contact force by a coefficient of friction μ, so that on the top surface, the force corresponding to the displacement u_1 will be equal to $-\mu F$. But it would also somehow depend on u_1 since there cannot be any friction without the movement. On the bottom surface, the contact force will have to be obtained from the normal stress in the cylinder. In any case, if one knows exactly how they may be estimated, then after each iteration, the applied force $\boldsymbol{R}_a$ can always be assembled.

Acknowledgements

The authors are indebted to Mr. V. Trbojevic, whose great diligence in assisting with the calculations for this research programme is much appreciated.

References

[1] Argyris, J. H., Energy theorems and structural analysis. Part I: General theory. *Aircraft Engineering*, 26 (1954) Oct. pp. 347–356, Nov. pp. 383–387, 394, 27 (1955), Feb. pp. 42–58, March pp. 80–94, April pp. 125–134, May pp. 145–158. Also as book: Butterworths Scientific Publications, London, Fourth Edition 1971.

[2] Argyris, J. H., *Recent advances in matrix methods of structural analysis*. Progress in Aeronautical Sciences, Pergamon Press, 4 (1964).

[3] Argyris, J. H., *Continua and discontinua, opening address to the international confer-*

ence on matrix methods of structural mechanics. Dayton, Ohio, Wright-Patterson U.S.A.F. Base, October 26th, 1965. Published in the Proceedings of the Conference by U. S. Government, January (1967) 1–198.

[4] Argyris, J. H., D. W. Scharpf and J. B. Spooner, *The elasto-plastic calculation of general structures and continua.* Third Conference on Dimensioning and Strength Calculations, Budapest (1968) 345–384.

[5] Argyris, J. H., D. W. Scharpf and J. B. Spooner, Die elasto-plastische Berechnung von allgemeinen Tragwerken und Kontinua. *Ingenieur Archiv*, 37 (1969) 326–352.

[6] Argyris, J. H. and D. W. Scharpf, *Methods of elasto-plastic analysis.* Proceedings of the Symposium on Finite Element Techniques, Committee Nine, Fourth International Ship Structures Congress, Stuttgart (1969) 381–416.

[7] Argyris, J. H., The impact of the digital computer on engineering sciences. Twelfth Lanchester Memorial Lecture, May 14th, 1969, *The Aeron. J. of the Roy. Aeron. Soc. 74* (1970) 13–41 and 111–127.

[8] Argyris, J. H. and D. W. Scharpf, *Berechnung vorgespannter Netzwerke.* Bayerische Akademie der Wissenschaften, Sonderdruck 4 aus den Sitzungsberichten, (1970) 25–58.

[9] Chan, A. S. L. and A. Firmin, The analysis of cooling towers by the matrix finite element technique. *The Aeron. J. of the Roy. Aeron. Soc.*, Part I, Small Displacements, 74 (1970) 826–835, Part II, Large Displacements, *74* (1970) 971–982.

[10] Argyris, J. H. and D. W. Scharpf, Finite elements in time and space. *The Aeron. J. of the Roy. Aeron. Soc.*, 73 (1969) 1041–1044. Also: *Nuclear Engineering and Design, 10* (1969) 456–464.

[11] Kelsey, S. and A. S. L. Chan, *Dynamic analysis of large displacements non-linear elastic problems.* Unpublished Report of Subdepartment of Aeronautical Structures, 1966.

[12] Schrem, E. and J. R. Roy, *An automatic system for kinematic analysis – ASKA*, Part I, ISD Report No. 98, Stuttgart, 1971.

[13] Argyris, J. H., Matrix analysis of three-dimensional anisotropic and inhomogeneous elastic media. Small and Large Displacements, *Ingenieur Archiv, 34* (1965) 33–55.

[14] Argyris, J. H. and D. W. Scharpf, Some general considerations on the natural mode technique. Part II: Large Displacements, *The Aeron. J. of the Roy. Aeron. Soc. 73* (1969) 361–368.

Computer aided visioplasticity solution of some deformation problems

A. H. Shabaik and E. G. Thomsen

University of California, Berkeley, U.S.A.

Procedures for obtaining complete computer solutions of steady state and non-steady state forming processes are discussed using the visioplasticity method as follows:

1) Steady State Axisymmetric Extrusion

Distribution fields of velocity, strain rate total effective strain and stress were obtained during the cold extrusion of a commercially pure aluminum through a conical die of 45° half-cone angle with an extrusion ratio of 3 : 1 at a speed of 1/8 ipm and flourocarbon as a lubricant.

Total effective strain and axial stress distribution during the hot extrusion of SAE 1018 steel at 1800°F were also obtained.

2) Non-steady State Forging

For the plane strain forging, the instantaneous grid patterns were obtained in a continuous upsetting process with and without lubricant. The test was carried out on a superplastic alloy of the eutectic lead-tin type at a speed of 0.015 in/min. The instantaneous distribution fields of velocity, total effective-strain rate, total effective strain and shear stress were obtained.

Introduction

The ultimate objective of an analysis of plastic deformation problems is the attainment of an all-inclusive method which would enable one to predict local values of strain rates, strain, temperature, and stresses during the course of deformation. Such information is essential in providing an insight into the mechanics of a forming process and to determine the final mechanical properties of the product, as for example, texture, anisotropy, and the state of residual stresses. Such information may aid also in the proper design of forming dies to achieve a sound product of specific quality.

Most methods of solution at present focus mainly on load prediction and it appears that only a few among these are available for the calculation of ideal rather than real stress distributions since they are still limited in their ability to treat other than idealized situations. Moreover, such solutions lack uniqueness and completeness in the mathematical sense due to approximations required in carrying out the necessary calculations. Hence the results predicted from these methods are insufficient and an analysis cannot be fully exploited to give information on the quality of the product or the design of the die. Results from these methods apparently have been utilized for predictions of load and it appears that an understanding of the interaction of process parameters, local values of stress and strain, and a determination of the final quality of the product have not been achieved. Thus, the requirement of a complete theory which can describe the mechanics of an actual deformation process and can make a prediction of actual velocities, strain rates, strain, and stresses at every point in the deformation region is not fulfilled.

The visioplasticity method, developed by Thomsen et al [1, 2] and recently computerized by Shabaik et al [3] apparently is the only method which can supply complete information on the detailed mechanics during deformation since it is based on an actual flow field obtained from an experiment. The method has been extensively utilized in obtaining complete solutions of stress and strain in steady-state forming processes through the use of the flow function concept introduced by Shabaik [4]. In addition, however, non-steady state forming processes, as for example, plain strain compression have also been analyzed by this method.

It is the purpose of this paper to show the application of the visioplasticity method aided by computer programs in the solution of both steady and non-steady state deformation processes. The wealth of information obtainable through this method as to quality of the product and die design will be described.

Basic equations

In a general problem of plastic deformation the calculation of stress and strain involves a solution of the basic equations which describe the mechanics of the deformation process. The basic equations for attainment of the complete solution by the visioplasticity method are outlined in this section for both steady and non-steady state forming processes.

a) *Steady-state forming process*

The axisymmetric steady state process is selected for this case.

From an experimentally determined steady state flow pattern, to be described later, the radial and axial velocity components (u, v) can be calculated at any point in the deformation zone from the known values of the flow function ψ as follows:

$$u = \frac{1}{2\pi r}\frac{\partial\psi}{\partial z},$$
$$v = -\frac{1}{2\pi r}\frac{\partial\psi}{\partial r}. \tag{1}$$

The strain rate components, in cylindrical coordinates can then be determined from:

$$\dot{\varepsilon}_r = \frac{\partial u}{\partial r} = -\frac{1}{2\pi r^2}\frac{\partial\psi}{\partial z} + \frac{1}{2\pi r}\frac{\partial^2\psi}{\partial r\partial z},$$
$$\dot{\varepsilon}_z = \frac{\partial v}{\partial z} = -\frac{1}{2\pi r}\frac{\partial^2\psi}{\partial r\partial z},$$
$$\dot{\varepsilon}_\theta = \frac{u}{r} = \frac{1}{2\pi r^2}\frac{\partial\psi}{\partial z}, \quad \text{and} \tag{2}$$
$$\dot{\gamma}_{rz} = \frac{\partial u}{\partial z} + \frac{\partial v}{\partial r} = \frac{1}{2\pi r}\left[\frac{\partial^2\psi}{\partial z^2} - \frac{\partial^2\psi}{\partial r^2} + \frac{1}{r}\frac{\partial\psi}{\partial r}\right].$$

Inspection will show that the continuity condition ($\dot{\varepsilon}_r + \dot{\varepsilon}_z + \dot{\varepsilon}_\theta = 0$) is automatically satisfied in equation (2).

The effective strain rate $\dot{\bar{\varepsilon}}$ is then calculated from its definition

$$\dot{\bar{\varepsilon}} = [\tfrac{2}{3}(\dot{\varepsilon}_r^2 + \dot{\varepsilon}_z^2 + \dot{\varepsilon}^2 + \tfrac{1}{2}\dot{\gamma}_{rz}^2)]^{\frac{1}{2}} \tag{3}$$

and the total effective strain $\bar{\varepsilon}$ can be calculated from the integration of $\dot{\bar{\varepsilon}}$ along the flow lines with respect to time

$$\bar{\varepsilon} = \int_0^t \dot{\bar{\varepsilon}} dt. \tag{4}$$

The stress components at any point in the deformation region can be determined from the strain rate components by considering the equilibrium and the plasticity equations as well as the mechanical properties of the materials [3].

$$\sigma_z(r, z) = \int_z^a \left[\frac{\partial}{\partial r}\left(\frac{\dot{\gamma}_{rz}}{2\dot{\lambda}}\right) + \frac{\dot{\gamma}_{rz}}{2r\dot{\lambda}} \right] dz + \\ + \int_0^r \left[\frac{\partial}{\partial r}\left(\frac{\dot{\varepsilon}_z - \dot{\varepsilon}_r}{\dot{\lambda}}\right) - \frac{\partial}{\partial z}\left(\frac{\dot{\gamma}_{rz}}{2\dot{\lambda}}\right) - \frac{\dot{\varepsilon}_r - \dot{\varepsilon}_\theta}{r\dot{\lambda}} \right]_{z=a} dr + \sigma_z(0, a), \qquad (5)$$

$$\sigma_r(r, z) = \sigma_z(r, z) + \frac{\dot{\varepsilon}_r - \dot{\varepsilon}_z}{\dot{\lambda}},$$

$$\sigma_\theta(r, z) = \sigma_z(r, z) + \frac{\dot{\varepsilon}_\theta - \dot{\varepsilon}_z}{\dot{\lambda}}, \qquad (6)$$

$$\tau_{rz}(r, z) = \frac{\dot{\gamma}_{rz}}{2\dot{\lambda}}$$

where $\dot{\lambda} = \frac{3}{2}(\dot{\bar{\varepsilon}}/\bar{\sigma})$, $\bar{\sigma} = f(\bar{\varepsilon}, \dot{\bar{\varepsilon}})$ is the flow stress of the material, and $\sigma_z(0, a)$ is axial stress at a reference point $(0, a)$ in the deformation zone.

b) *Non-steady state processes*

The procedures outlined in this section are for the attainment of a complete solution in non-steady state plane-strain forming processes. In this case, instantaneous photographs of the grid pattern during deformation must be obtained and the grid distortions must be analyzed as functions of time. From the two consecutive grid-patterns, the instantaneous velocity components can be determined assuming that a previous grid point moves to its new position with an average velocity during an incremental deformation step. The relative displacements of grid points between each consecutive step are calculated from the measurements of the coordinates of the grid points. The horizontal and vertical velocity components u and v in the x and y directions, respectively, at any time t are derived for the unit ram speed as follows:

$$u = \frac{\Delta x}{\Delta t}, \quad v = \frac{\Delta y}{\Delta t} \qquad (7)$$

where Δx and Δy are the horizontal and vertical displacements during an increment of time Δt.

The instantaneous strain rate components can be calculated for the known instantaneous velocity components.

$$\dot{\varepsilon}_x = \frac{\partial u}{\partial x}, \quad \dot{\varepsilon}_y = \frac{\partial v}{\partial y}, \quad \dot{\gamma}_{xy} = \frac{\partial u}{\partial y} + \frac{\partial v}{\partial x}. \tag{8}$$

The total effective strain rate is then calculated from

$$\dot{\bar{\varepsilon}} = \tfrac{2}{3}(3\dot{\varepsilon}_x^2 + \tfrac{3}{4}\dot{\gamma}_{xy})^{\frac{1}{2}}.$$

The total effective strain $\bar{\varepsilon}$ can be calculated from the integration of $\dot{\bar{\varepsilon}}$ along the instantaneous flow lines with respect to time.

$$\bar{\varepsilon} = \int_0^t \dot{\bar{\varepsilon}} dt.$$

The instantaneous stress components at any point in the deformation region can then be determined from the known instantaneous strain rate components by considering the equilibrium and plasticity equations as well as the mechanical properties of the material.

$$\begin{aligned} \sigma_y &= \sigma_y(0, a) - \int_a^y \frac{\partial \tau_{xy}}{\partial x} dy - \int_0^x \left[\frac{\partial}{\partial x}\left(\frac{\dot{\varepsilon}_x - \dot{\varepsilon}_y}{\dot{\lambda}}\right) + \frac{\partial \tau_{xy}}{\partial y} \right]_{y=a} dx, \\ \sigma_x &= \sigma_y + \frac{\dot{\varepsilon}_x - \dot{\varepsilon}_y}{\dot{\lambda}}, \\ \tau_{xy} &= \dot{\gamma}_{xy}/2\dot{\lambda} \end{aligned} \tag{9}$$

where $\dot{\lambda} = \frac{3}{2}(\dot{\bar{\varepsilon}}/\bar{\sigma})$, and $\sigma_y(0, a)$ is the stress component σ_y at a reference point $(0, a)$ which can be evaluated from the condition that the load normal to the free boundary is zero.

In both cases of steady state and non-steady state processes, computer programs were developed to carry out complete analyses of the deformation processes.

Complete analysis

a) *Steady state forming processes*

The axisymmetric extrusion process was selected for this case and the complete analysis of stress and strain was obtained as previously described.

The steady state flow lines in an axisymmetric forward extrusion process of a commercially pure aluminum were obtained experimentally. Split specimen of 1 inch diameter and 3 inches long with grid lines of 0.002″ thick and 0.020″ spacing were printed on the interface, using Kodak photo resist method, before extrusion. The two halves of the billet were assembled in the extrusion chamber and were extruded from a diameter of 1 inch to 0.577 inch. (extrusion ratio = 3.0). A conical die of 90° cone angle was used. The extrusion process was carried until approximately half of the original length

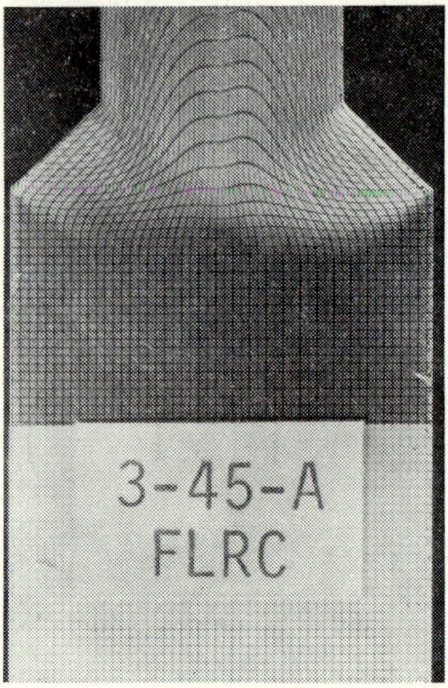

Fig. 1. Distorted grid pattern during steady state flow of an axisymmetric extrusion of commercially pure aluminum. Ext. ratio = 3.0, half-cone angle = 45°. Lubricant: carbon.

was extruded, to assure the establishment of the steady state condition [3], with a ram speed of 0.125 in/min., using fluorocarbon as lubricant. A photograph of the steady state flow pattern is shown in Fig. 1. The coordinates of different points along the flow lines at equally spaced sections were tracked using a scanning machine and were used as input data to a computer program developed for the complete analysis. Fig. 2, plotted by a computer, shows these points along the 25 flow lines at different k-sections selected during tracking.

In steady state axisymmetric extrusion, the values of the flow function ψ at any k-section can be obtained from the known r-coordinates at the entry

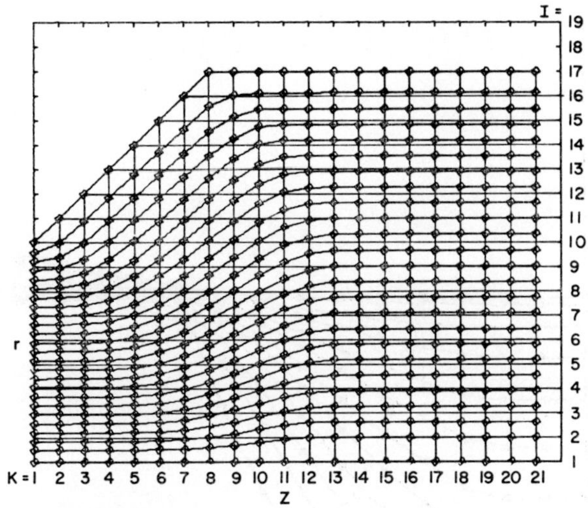

Fig. 2. Grid system of equally-spaced points and steady state flow lines of commercially pure aluminum in axisymmetric extrusion process. Ext. ratio = 3.0, half-cone angle = 45°. Lubricant: fluorocarbon.

section ($k = 21$) as follows:

$$\psi_i = \pi R_i^2 |V_0| \tag{10}$$

where V_0 is the ram speed, R_i is the radius of the ith flow line at $K = 21$, and ψ_i is the value of the flow function along the ith flow line at any K-section.

In the evaluation of the magnitudes of the velocity components [eq. (1)] the first derivative of the flow function must be determined. The results obtained are not usually smooth [3] because of the limited accuracy with which the experimental data are determined.

Smoothing of the flow function ψ was carried out in the r and z directions. The differences of values of ψ up to the fourth order were evaluated and smoothed with respect to r and z, respectively, using a numerical smoothing process [3]. A computer program was developed to carry out the smoothing of ψ and the entire calculations for stress and strain. The difference of the values of ψ before and after smoothing was found to be very small.

The velocity components (u, v) are then obtained by numerical differentiation of ψ with respect to, z and r and substitution in eq. (1). A computer plot of the axial and radial velocities as a function of r at different k-sections are shown in Fig. 3 and 4, respectively.

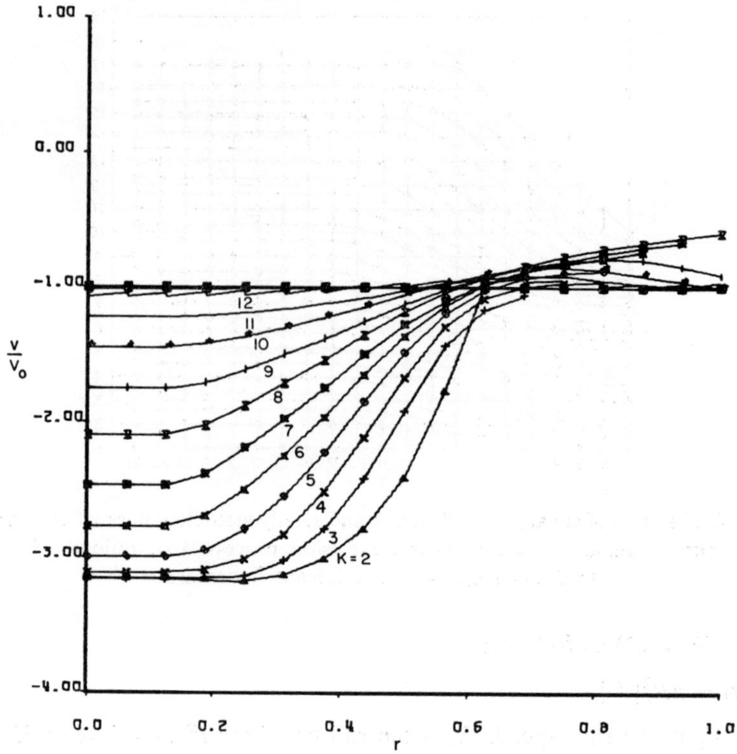

Fig. 3. Axial velocity distribution $V_0 = 0.125''/\text{min}$.

The strain rate components were calculated, from equation (2), at the mesh points of square grids, while the effective strain rate $\dot{\bar{\varepsilon}}$ is calculated from eq. (3). The total effective strain $\bar{\varepsilon}$ was evaluated by numerical integration of the effective strain rates along a particular flow line with respect to time.

Fig. 5 shows $\bar{\varepsilon}$ as a function of r at different k-sections, respectively. The evaluation of stress components can be obtained from eqs. (5) and (6) and the known values of the strain rates. Inasmuch as the flow stress of commercially pure aluminum can be given by the equation

$$\bar{\sigma} = C\bar{\varepsilon}^n$$

where, $C = 31000$ and $n = 0.25$, it is possible to evaluate the axial stress components at each mesh point by numerical integration of eq. (5) and by

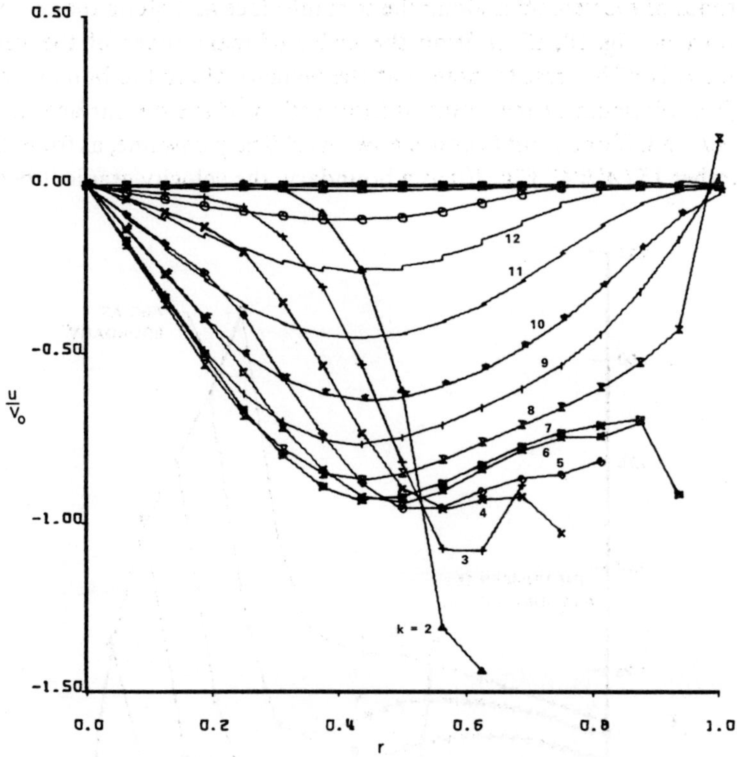

Fig. 4. Radial velocity distribution $V_0 = 0.125''/\text{min}$.

substituting $\frac{3}{2}(\dot{\bar{\varepsilon}}/C\bar{\varepsilon}^n)$ for $\dot{\lambda}$. The axial stress $\sigma_z(0, a)$ is evaluated such that

$$\int \sigma_z(r, z) \cdot 2\pi r\, dr$$

at the exit section is zero.

The axial stress component as a function of r at different k-sections is shown in Fig. 6. The previous results demonstrate the potentiality of the visioplasticity method in attaining the complete analysis of stress and strain in an axisymmetric extrusion problem. Several conclusions and useful information can be drawn from these results. If, for example, the velocity components, shown in Fig. 3 and 4 are added vectorially then the resultant velocities along the center line and along the boundary can be obtained. Both magnitudes of the velocities along the tool interface and along the center line are shown in Fig. 10, illustrating the order of magnitudes of the velocity gradient. It is of interest to note that the velocity along the boundary goes through a minimum at the point of intersection of the die surface with the container wall. If one would choose now an arbitrary flow line, as for example line number 15 ($A'B'C'$ Fig. 10) as a boundary, the velocity gradient between

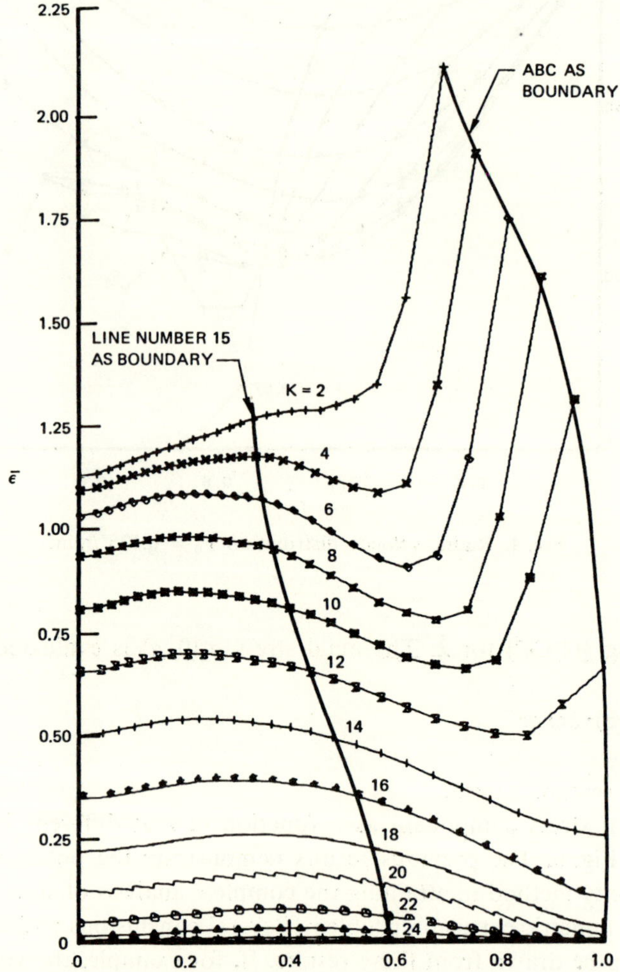

Fig. 5. Total effective strain distribution.

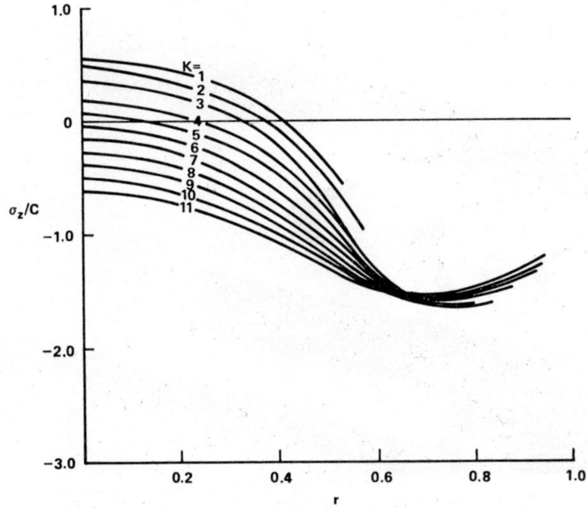

Fig. 6. Axial stress distribution $C \simeq 31000$ psi.

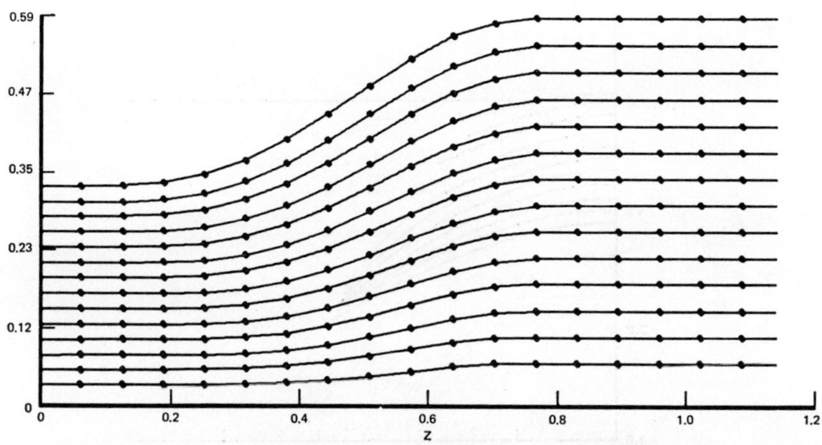

Fig. 7. Flow lines with line number (15) as boundary.

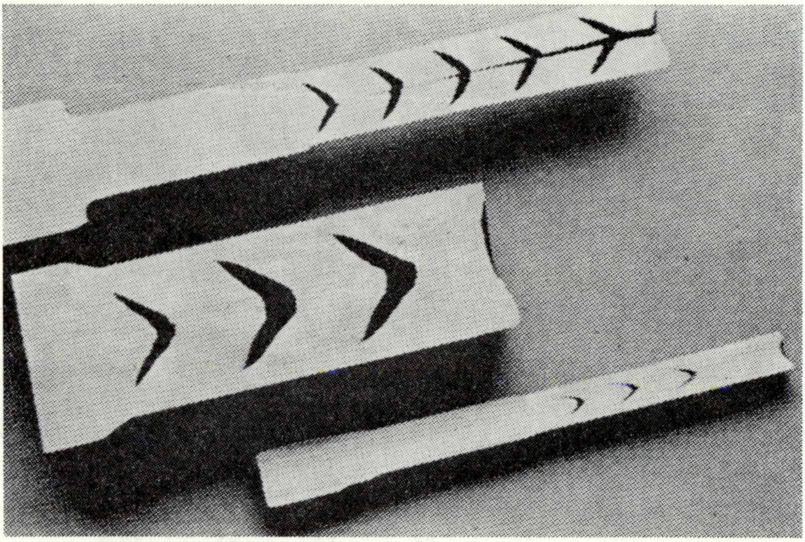

Fig. 8. Center bursts in forward extension of steel.

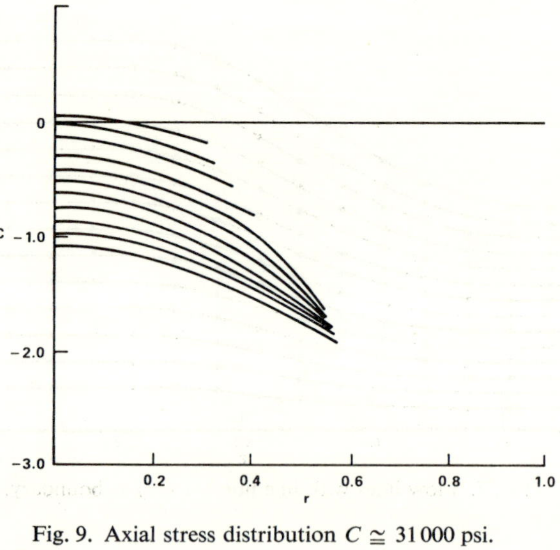

Fig. 9. Axial stress distribution $C \cong 31\,000$ psi.

the center line and this particular boundary would be very much reduced and, therefore, the velocity field would approach that of uniform deformation. Consequently, if uniform properties in the product is an objective, then a boundary having the shape similar to that of line 15 should be selected. This choice would result in flow lines similar to those in Fig. 7 assuming no change in the frictional drag along the boundary.

Fig. 5 also shows that the total effective strain $\bar{\varepsilon}$ in the product is higher near the tool boundary than it is at the center. A comparison of the $\bar{\varepsilon}$ distribution, for line 15 as the boundary with that of the real boundary *ABC*, is also shown in Fig. 5. It is seen that the total effective strain distribution for the case when line 15 is the boundary is reasonably uniform and the sudden rises in $\bar{\varepsilon}$ are eliminated. This part demonstrates the superiority of a die shape similar to that of line 15 as compared with that of a conical die in attaining uniform properties in the final product.

Fig. 6 shows that tensile axial stresses exist in the central portion of the deformation region during extrusion. This means that the material will have

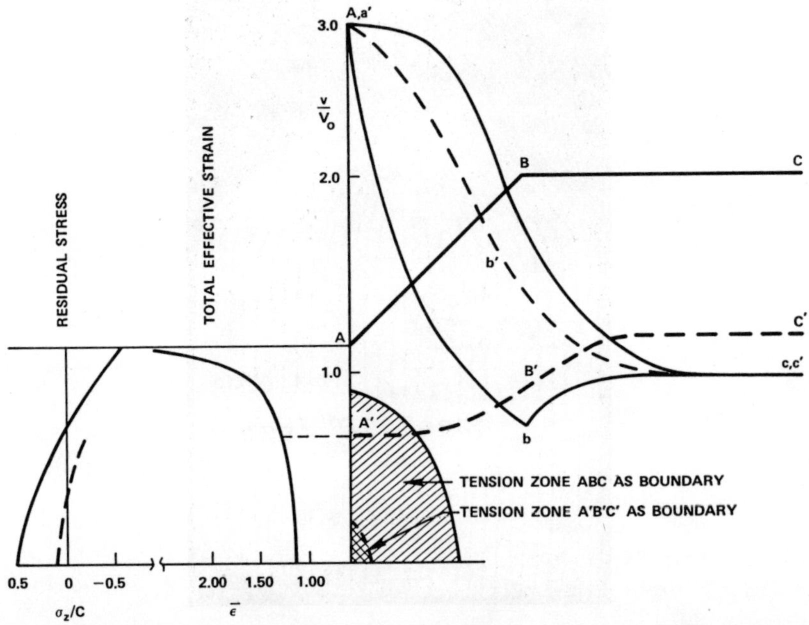

Fig. 10. Effect of boundary shape on velocity magnitude along the boundary and center line, tension zone, total effective strain, and residual stresses.

a tendency to form cavities in the center, especially when defects exist or when the material is weak in tension. Such cavity formation is referred to as center bursts which can sometimes be observed during cold extrusion [6] as shown in Fig. 8. It is also of interest to note that the axial stress distribution at the exit section ($K = 1$) can serve as a measure of the residual stress distribution in the product.

If a die shape similar to line 15 ($A'B'C'$ Fig. 10) is used the axial stress distribution will be as shown in Fig. 9. As a consequence the tension zone in the central part of the extrusion is almost completely eliminated and the axial stress distribution in the product becomes nearly uniform. Hence, the product will be substantially free of defects and of residual stresses. This again is a demonstration of the superiority of a die shape similar to the flow line number 15 over that of a conical die in obtaining uniform properties, and defect and residual stress free products.

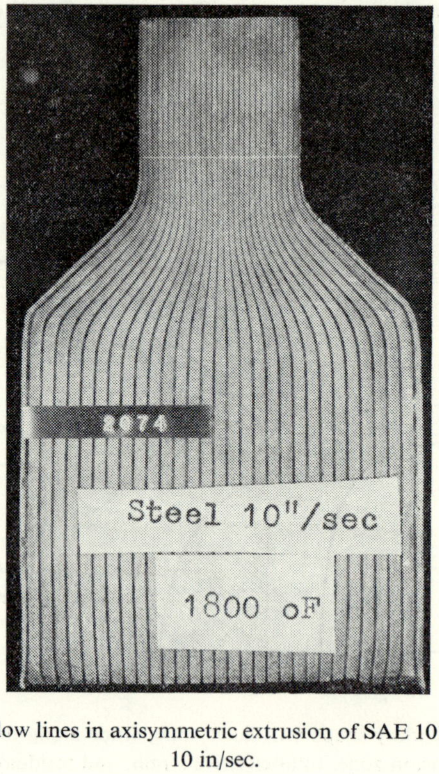

Fig. 11. Steady state flow lines in axisymmetric extrusion of SAE 1018 steel at 1800°F and 10 in/sec.

The foregoing conclusions are summarized in Fig. 10.

Recently [8] a computer program was developed for the analysis of the extrusion process through curved dies. In this case the normalized coordinate $r_n(r_n = r/R_b$ and R_b is the radius along the boundary) was used to transform the curved boundary to a straight line parallel to the extrusion axis in the mapped plane.

An analytical form in r and z of the same trend as that of the flow function ψ was selected and the least mean square fit was used to determine the values of its coefficients. The components of velocity, strain rates, and stresses were then evaluated. The foregoing method was applied to an analysis of the hot extrusion of SAE 1018 steel at 1800°F and at an extrusion speed of 10 in/sec. The flow lines, shown in Fig. 11, given in [7] were used and the complete solution was obtained [9]. Figures 12 and 13 show a computer plot of the

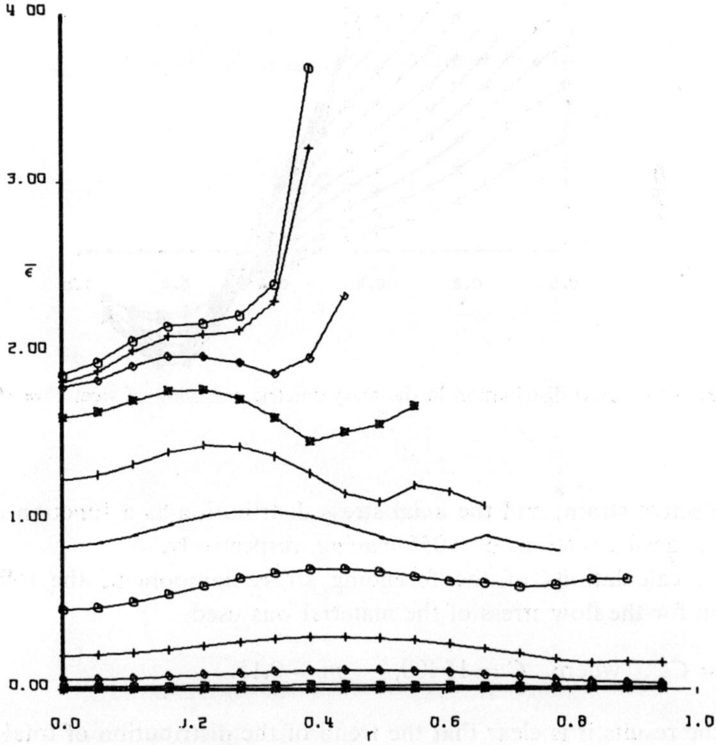

Fig. 12. Total effective strain distribution.

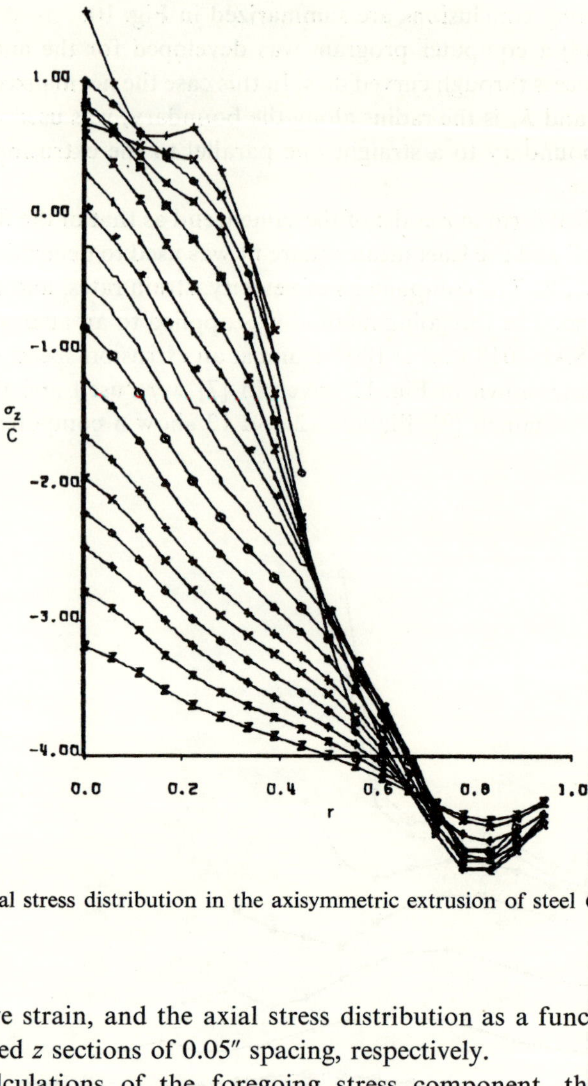

Fig. 13. Axial stress distribution in the axisymmetric extrusion of steel $C = 15400$.

total effective strain, and the axial stress distribution as a function of r at equally spaced z sections of 0.05″ spacing, respectively.

In the calculations of the foregoing stress component, the following equation for the flow stress of the material was used

$$\bar{\sigma} = C\dot{\bar{\varepsilon}}^m \quad \text{where} \quad C = 15400, \qquad m = 0.15.$$

From the results it is clear that the trend of the distribution of total strain and axial stress for hot extrusion is the same as that for cold extrusion.

b) *Non-steady state forming processes*

Plane strain compression was selected for this case and the analysis previously described for the non-steady state processes was applied using a computer program. Specimens 1 inch high, 1.25 inch wide, and 0.59 inch thick were prepared from a superplastic alloy of lead and tin. Square grids of 0.040″ were printed on one face of the specimen using the Kodak-photo-resist technique. The tests were carried out behind a glass plate in a Kudo-Type Apparatus which provided the condition of plain strain so that the instantaneous grid pattern can be photographed without stopping the process. The simple compression test was performed for two different conditions of friction between the anvils and the specimen, in one case fluorocarbon was used as the lubricant while in the other no lubricant was used at all.

Instantaneous photographs of the distorted grid lines were obtained in a continuous upsetting process for the two cases of friction. The tests were carried out at a speed of approximately 0.015 in./min., and photographs and load readings were taken every two minutes. Figure 14 shows some selected photographs of the instantaneous grid patterns with and without lubricant.

The instantaneous coordinates of each point in these photographs were punched on IBM cards, using a scanning machine. A computer program was developed and the coordinates of these points were used as input data. A computer plot of the instantaneous grid patterns is shown in Fig. 15. The instantaneous flow lines and the free boundary changes were obtained and compared with those of uniform deformation as shown in Fig. 16. The results show that when an effective lubricant is applied, the flow field becomes almost identical with that of uniform deformation.

Instantaneous velocity components at each grid point were obtained using equation (7) and two consecutive grid-patterns.

A numerical smoothing process was used to smooth the values of u and v. The strain rate components were thus obtained from a numerical differentiation. The continuity equation was checked and the maximum deviation was approximately 15%. The total effective strain was obtained from the integration of the effective strain rate along the instantaneous flow lines shown in Fig. 16.

Finally, from the known values of the shear strain rate along the interface the shear stress τ was obtained from:

$$\tau = \frac{\dot{\gamma}}{3\dot{\bar{\varepsilon}}}\,\bar{\sigma} \tag{11}$$

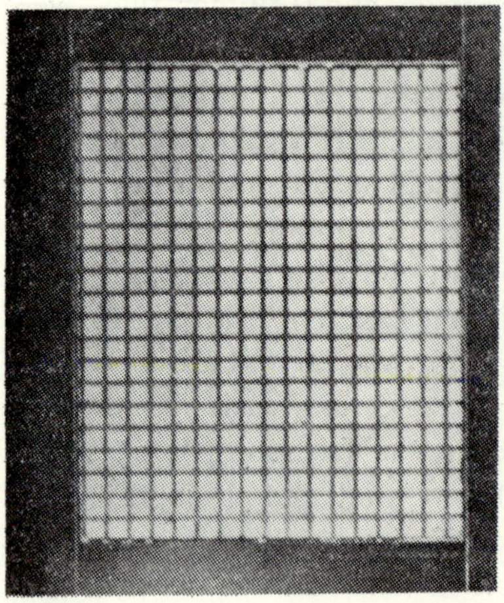

$\Delta h = 0$

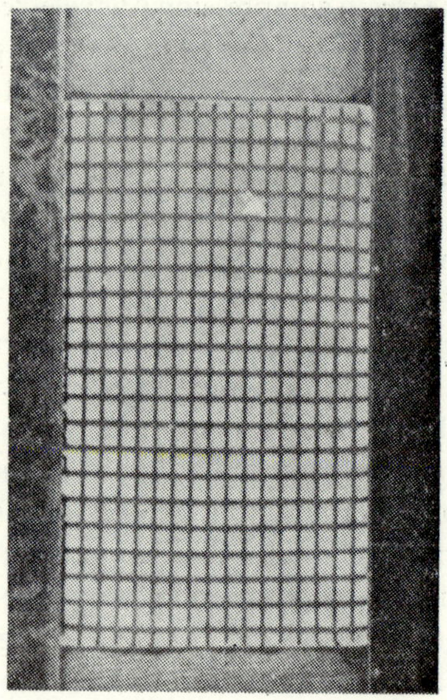

$\Delta h = 0.165''$

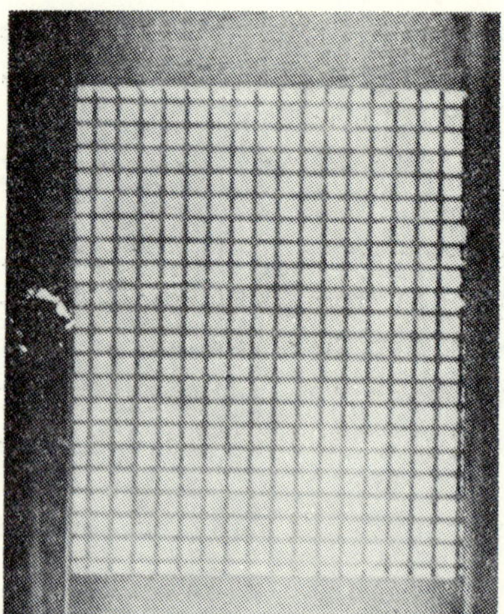

$\Delta h = 0$

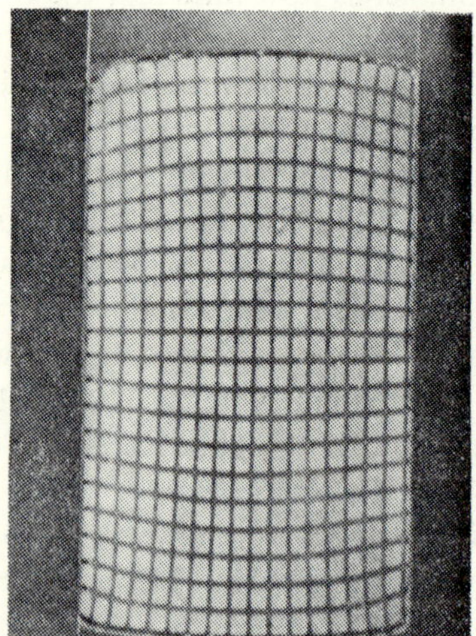

$\Delta h = 0.165''$

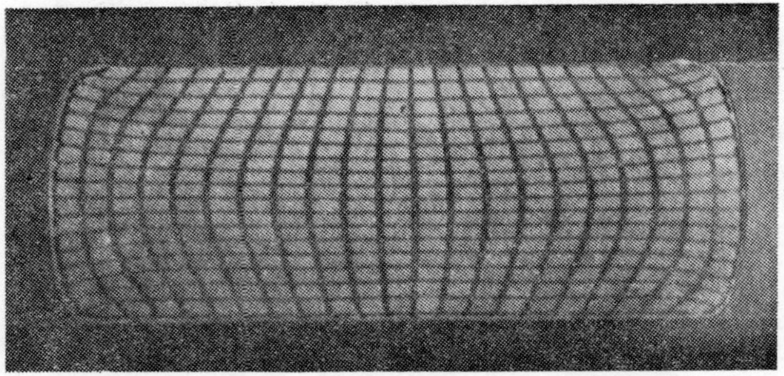

$\Delta h = 0.315''$

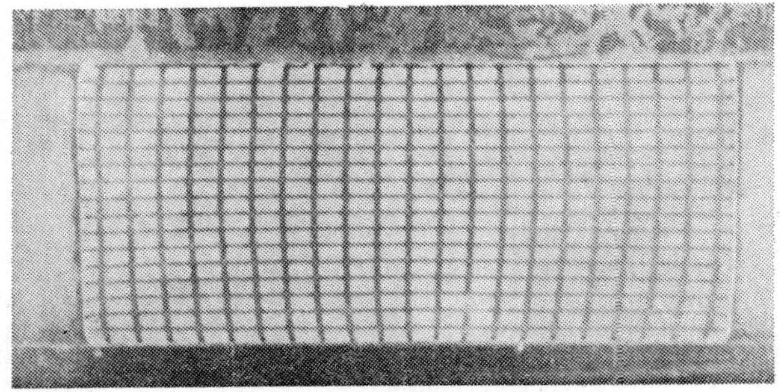

$\Delta h = 0.314''$

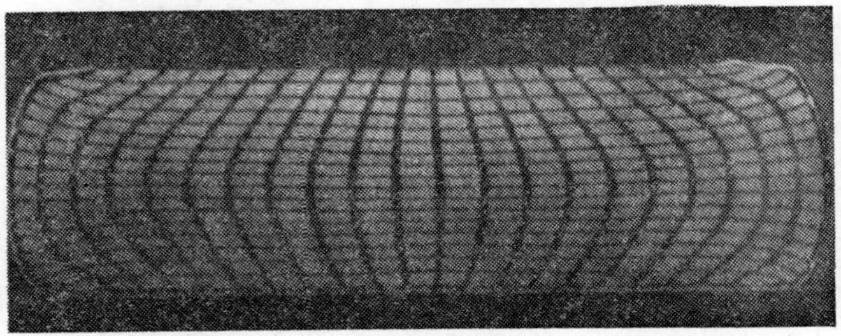

$\Delta h = 0.409''$

(a) unlubricated

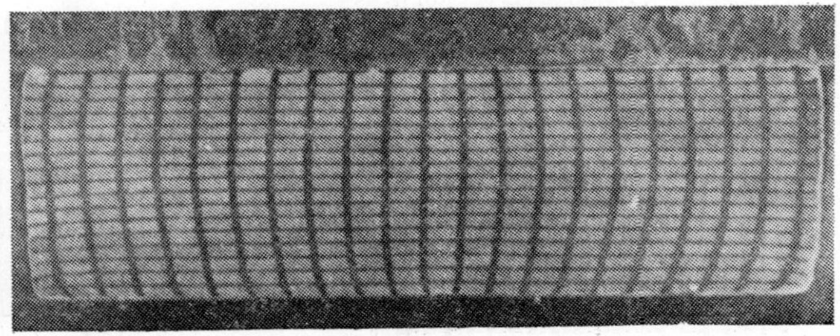

$\Delta h = 0.401''$

(b) lubricated

Fig. 14. Flow patterns of lead-tin superplastic alloy in plane strain compression.

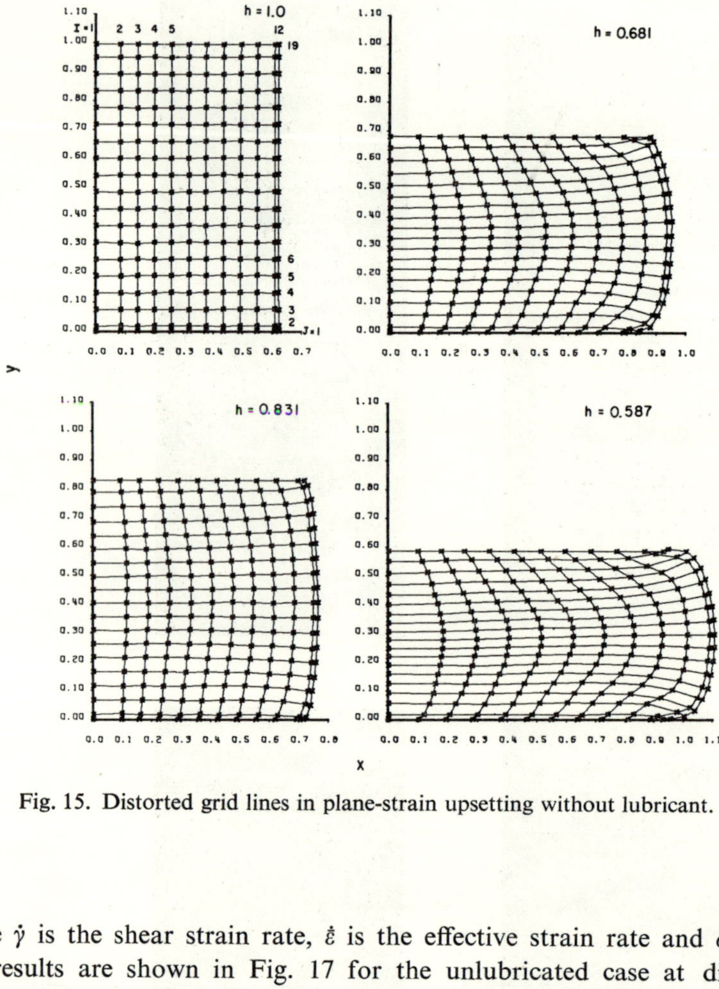

Fig. 15. Distorted grid lines in plane-strain upsetting without lubricant.

where $\dot{\gamma}$ is the shear strain rate, $\dot{\bar{\varepsilon}}$ is the effective strain rate and $\bar{\sigma} = C$. The results are shown in Fig. 17 for the unlubricated case at different degrees of deformation. It is evident that the shear stress varies from zero at the center and increases towards the free edge of the specimen. These results show that if sticking were to take place it would be more likely that it would start at the free edge of the specimen rather than at the center. These results confirm the observations of a similar trend of the shear stress along the boundary reported by Pearsall and Backofen [10] during compression tests of cylindrical specimens. These investigators used pressure sensitive pins to determine experimentally local values of normal pressure and shear stress along the interface. It is also clear now that these observations do not

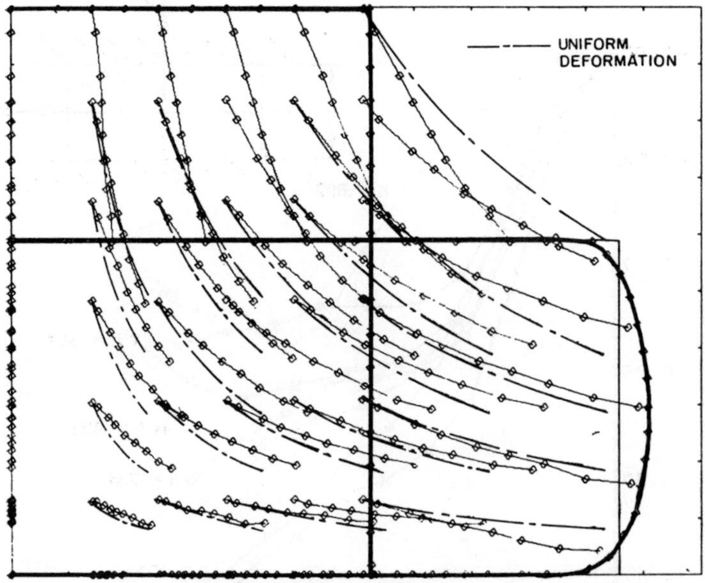

Fig. 16(a) Comparison of flow lines.

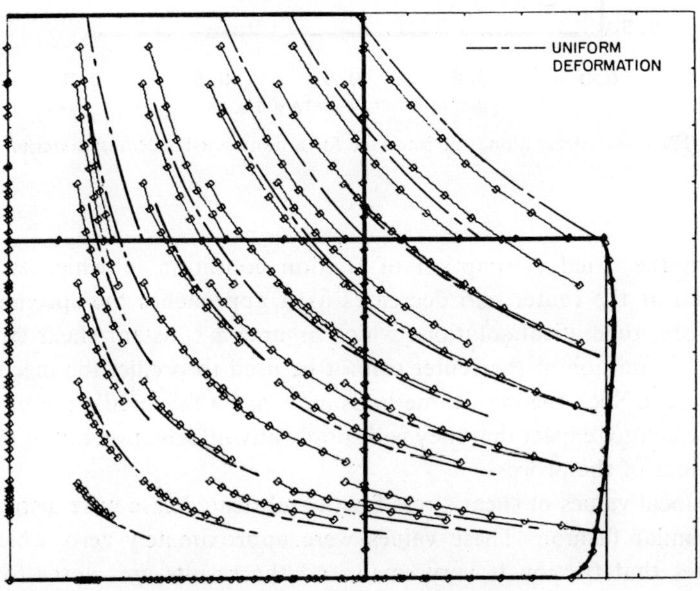

Fig. 16(b). Comparison of flow lines.

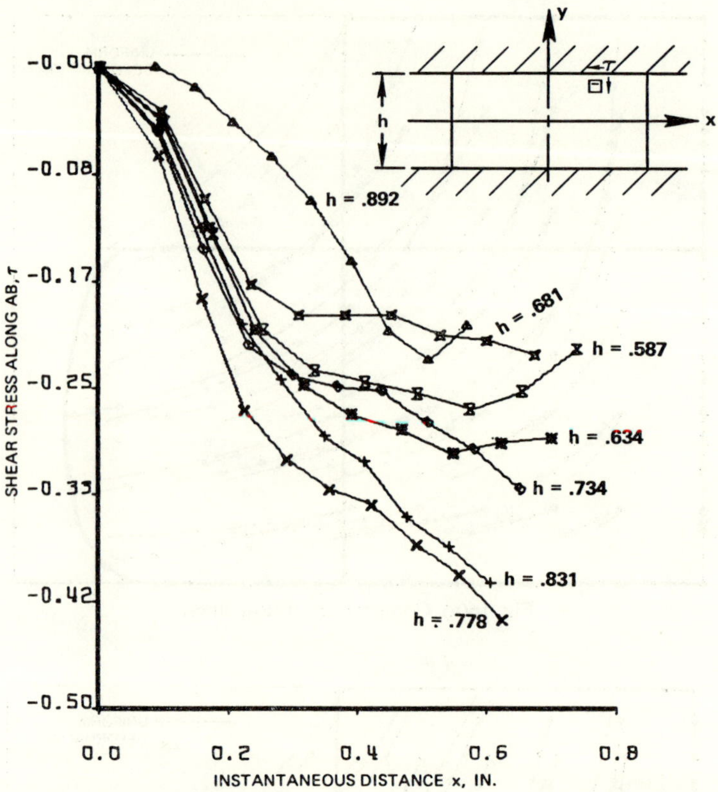

Fig. 17. Shear stress along the interface for a nonwork-hardening material $\bar{\sigma} = c$.

confirm the usual assumption of friction condition in which sticking is assumed at the center and decreases as it approaches the specimen edge. Therefore, theoretical solutions which assume a constant shear stress or a sticking condition at the center cannot be used to predict the mechanics of the process. Such theoretical methods may give a fair prediction of the load but one cannot expect that they will supply any information about the actual mechanics of the process.

The local values of shear stress for the lubricated case were also obtained in a similar fashion. These values were approximately zero which again indicates that friction is very small and the results are closer to that of uniform deformation.

Acknowledgements

The authors wish to thank Sharon DuPre for typing the manuscript. They also wish to thank the UCLA Campus Computing Network for the use of their computer.

References

[1] Thomsen, E. G. and J. T. Lapsley, Jr., Experimental stress determination within a metal during plastic flow. *Proceedings of the Society of Experimental Stress Analysis, 11* (1954) 59–68.

[2] Thomsen, E. G., C. T. Yang and J. B. Bierbower, *An experimental investigation of the mechanics of plastic deformation of metals*. University of California Press, Berkeley and Los Angeles, California, (1954) 89–144.

[3] Shabaik, A. H. and S. Kobayashi, Computer application to the visioplasticity method. *ASME Trans., J. of Eng. for Ind.* (1967) 339–346.

[4] Shabaik, A. H., *Theoretical methods for analysis of metal deformation problems*. Ph. D. dissertation, University of California, May (1966).

[5] Shabaik, A. H. and E. G. Thomsen, Flow studies in extrusions. *Annals of the C.I.R.P. XVII* (1969) 149–156.

[6] Blickwede, D. J., *Cold extrusion of steel*. Metal Progress, May (1970) 78–82.

[7] Altan, T., S. Kobayashi, V. DePierre and C. M. Pierce, Plastic deformation of copper and steel in hot extrusion through conical die. *ASME, J. Eng. for Ind.* (1968) 371–380.

[8] Shabaik, A. H., *Computer aided visioplasticity solution to axisymmetric extrusion through curved boundaries*. Submitted for presentation at the 1971 ASME Winter Annual Meeting.

[9] Shabaik, A. H. and A. M. Virani, *The use of superplastic alloy of the eutectic of lead-tin for the analysis of hot extrusion process*. Submitted for presentation at the 1971 ASME Winter Meeting.

10] Pearsall, G. W. and W. A. Backofen, Frictional boundary conditions in plastic compression. *J. of Eng. for Ind., Ser. B, 85* (1963) 68–76.

Recent advances in optimal plastic design

G. I. N. Rozvany and S. R. Adidam
Monash University, Clayton, Victoria, Australia

Optimality conditions are summarised for various classes of convex specific cost functions and for multi-load and multi-component systems. Extensions of these conditions to non-convex specific cost functions and to problems involving partially preassigned geometry are presented. Examples of optimal beams, grillages, shells and fibre-reinforced plates are given. An extension of Foulkes's theorem to two-dimensional systems is introduced.

1. Terminology

The aim of optimal design is to minimize a quantity termed *cost* (Φ) within specified design constraints. The cost per unit length or unit area of a structure is called *specific cost* ψ. Using Prager's notation [1], [2], the *generalised stresses* $Q_i(\xi)$ $(i = 1, 2, \ldots, n)$ define a local state of stress at an arbitrary point ξ of the structure and the corresponding *generalised strains* are $q_i(\xi)$ $(i = 1, 2, \ldots, n)$.

A load $\boldsymbol{p}(\xi)$ is said to be *associated* with a stress field $\boldsymbol{Q}(\xi)$ if it equilibrates the latter and a load is said to be associated with a displacement field if the latter furnishes the power of dissipation [1] which equals the power of the load [1]. A total cost Φ associated with a specified stress field $\boldsymbol{Q}^*(\xi)$ is given by

$$\Phi = \int_D \psi(\boldsymbol{Q}^*)d\xi \tag{1}$$

where D is the domain of the structure. The total cost associated with a displacement field $\boldsymbol{u}(\xi)$ is defined as

$$\Phi = \int_D p_i u_i d\xi. \tag{2}$$

If the *specific cost function* $\psi(\boldsymbol{Q})$ is differentiable on the interior of a 'regime'

R_i, then the *cost gradient vector* for all $\boldsymbol{Q}^* \in \operatorname{int} R$ is defined as

$$\boldsymbol{G} = (\mathbf{grad}\ \psi)_i = \left(\frac{\partial\psi}{\partial Q_1}, \ldots, \frac{\partial\psi}{\partial Q_n}\right)\bigg|_{\boldsymbol{Q}=\boldsymbol{Q}^*}. \tag{3}$$

If **grad** ψ is discontinuous at $\boldsymbol{Q}^{**}$, then $\boldsymbol{G}$ is defined as any convex combination of the gradient vectors for the adjacent regimes R_i:

$$\boldsymbol{G} = \sum_{i=1}^{m} \lambda_i\,(\mathbf{grad}\ \psi)_i\bigg|_{\boldsymbol{Q}=\boldsymbol{Q}^{**}} \tag{4}$$

with

$$\lambda_i \geqq 0 \quad \text{and} \quad \sum_{i=1}^{m} \lambda_i = 1.$$

A specific cost function is called *conical* if

$$\psi(\lambda \boldsymbol{Q}) = \lambda\psi(\boldsymbol{Q}) \quad \text{for all} \quad \lambda > 0. \tag{5}$$

The graph of such functions in R^{n+1} is a cone having its vertex at $\boldsymbol{Q} = 0$. The *cost gradient surface* is the locus of all cost gradient vectors associated with non-zero stress vectors.

Example (1): Consider a rectangular cross-section of variable width b but preassigned depth d = const. Since the yield moment of such section in pure flexure is $M_0 = bd^2\sigma_0/4$ and the yield force in simple tension is $N_0 = bd\sigma_0$ where σ_0 is the yield stress, the quantity $\beta = N_0/M_0 = 4/d$ = const for all admissible cross-sections. The yield condition (Fig. 1a, continuous line)

$$\frac{|M|}{M_0} + \left(\frac{N}{N_0}\right)^2 = 1 \tag{6}$$

and flow rule (Fig. 1a) for this problem were described by Prager [1]. Adopting the specific cost function $\psi = 2\sigma_0 A$ where $A = bd$ is the cross-sectional area and substituting $M_0 = N_0/\beta$ into eq. (6),

$$\psi = \beta|M| + \sqrt{\beta^2M^2 + 4N^2} \tag{7}$$

which, by eq. (5), is clearly a 'conical' function. Two contour lines of this function representing $\psi = 2N_0$ (continuous line) and $\psi = N_0$ (broken line) are given in Fig. (1a). The components of the cost gradient vector are

$$\frac{\partial\psi}{\partial N} = \frac{4N}{\sqrt{\beta^2M^2+4N^2}}, \quad \frac{1}{\beta}\,\frac{\partial\psi}{\partial M} = \operatorname{sgn} M + \frac{\beta M}{\sqrt{\beta^2M^2+4N^2}}. \tag{8}$$

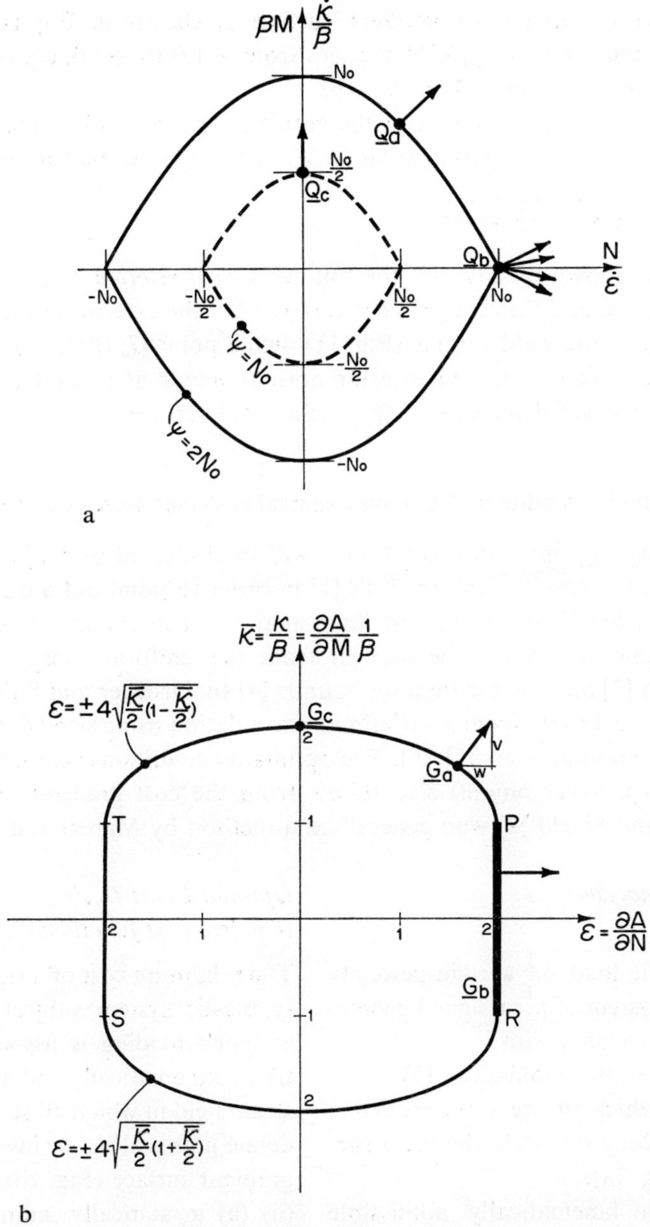
βM
κ̇/β
N₀
Qa
N₀/2
Qc
N
ε̇
Qb
-N₀
-N₀/2
N₀/2
N₀
ψ = N₀
-N₀/2
ψ = 2N₀
-N₀
a
κ̄ = κ/β = ∂A/∂M 1/β
ε = ±4√(κ̄/2 (1 - κ̄/2))
Gc
2
v
Ga
w
T
1
P
2
1
1
2
ε = ∂A/∂N
Gb
S
1
R
2
ε = ±4√(-κ̄/2 (1 + κ̄/2))
b

Fig. 1

The corresponding cost gradient surface is shown in Fig. 1b. Since at $M = 0$, the value of $\partial\psi/\beta\partial M$ changes from -1.0 to $+1.0$, eq. (4) gives the line segments PR and ST in Fig. 1b.

Denoting the components of the gradient vector ε and $\bar{\kappa}$, and expressing both in terms of the parameter $\alpha = M/N$, eqs. (8) can be combined to give

$$\varepsilon = \pm 4\sqrt{\pm\bar{\kappa}(\tfrac{1}{2}\mp\bar{\kappa}/4)}. \tag{9}$$

It can be shown readily that the following *dual relation* exists between the yield surface and the cost gradient surface: Let the exterior normal vector at point $\boldsymbol{Q}_a$ of the yield surface (Fig. 1a) define point $\boldsymbol{G}_a$ (Fig. 1b) on the cost gradient surface. Then an exterior normal vector at point $\boldsymbol{G}_a$ of the cost gradient surface defines point $\boldsymbol{Q}_a$ on the yield surface.

2. Optimality conditions for convex conical cost functions - single load system

The foregoing optimality conditions will be discussed in conjunction with various theorems on limit analysis [1] in order to point out a dual relationship between limit design and limit analysis. The statements on optimal design that follow can be derived from the uniform energy dissipation principle [3] and the theorem on bounds [4] by Drucker and Shield or, in a more general form, from a variational formulation using slack functions and Lagrangian multipliers [5], [6]. The optimality conditions (but not the statement on a lower bound) also follow from the cost gradient principle by Prager and Shield [2] who generalised a method by Marcal and Prager [7].

Limit Analysis	*Optimal Limit Design (Conical cost functions)*
The limit load on a rigid-perfectly plastic system of preassigned geometry is associated with	The minimum cost of a rigid perfectly plastic system subject to a preassigned loading is associated with
(i) a statically admissible [1] stress field in which all stress vectors define points along or inside the yield surface (Fig. 1a).	(i) a kinematically admissible [1] strain field in which all strain vectors define points along or inside the cost gradient surface (Fig. 1b).
(ii) (a) a kinematically admissible [1] strain field in which any strain vector is	(ii) (a) a statically admissible [1] stress field in which any stress vector is

(b) *exterior normal* to the yield surface (Fig. 1a) at the point defined by the corresponding stress vector if such point is on the yield surface and

(c) *zero* if such point is inside the yield surface (Fig. 1a).

(b) *exterior normal* to the cost gradient surface (Fig. 1b) at the point defined by the corresponding strain vector if such point is on the cost gradient surface and

(c) *zero* if such point is inside the cost gradient surface (Fig. 1b).

If only condition (i) is satisfied, a *lower bound* on the limit load (minimum volume) is obtained.

If only condition (ii) is satisfied, an *upper bound* on the limit load (minimum volume) is obtained.

Parts (b) and (c) of condition (ii) can be disregarded if an upper bound on the minimum volume is calculated from the stress components using the primal formulation (eq. 1). However, they give the correct strain field if the upper bound is calculated using the dual formulation (eq. 2).

The dual relationship between the yield surface and cost gradient surface was explained in section (1).

The foregoing conditions have been used for deriving bounds on the minimum volume of rectangular Tresca sandwich plates and rectangular plates reinforced with straight fibres [5], [6], [8].

3. Optimality conditions for convex conical cost functions - alternate load systems

The foregoing conditions have been extended by Rozvany [5], [6] to multiple load conditions. Let the specific cost in an optimal design be $\psi = \psi_0$ at some point ξ of the structure. The corresponding contour line $\psi(\boldsymbol{Q}) = \psi_0$ of the specific cost function shall be called the *design cost surface*. Fig. 1a shows such a surface in continuous line for the cost value of $\psi_0 = 2N_0$. For the class of cost functions considered, the design cost surface may also be interpreted as the yield surface at a point ξ of an optimized structure.

Let the stress vector for loading condition 'k' be $\boldsymbol{Q}_k$ and the corresponding strain vector $\boldsymbol{q}_k$. Then the conditions for optimality may be stated as follows:

The minimum cost of a rigid-perfectly plastic system subjected to several alternative preassigned load systems is associated with

(i) statically admissible stress fields $\boldsymbol{Q}_k(\xi)$ for each loading condition and

(ii) (a) kinematically admissible strain fields $\boldsymbol{q}_k(\xi)$ such that

(b) $\boldsymbol{q}_k = \nu_k \boldsymbol{G}_k$ where $\boldsymbol{G}_k$ is the gradient vector associated with $\boldsymbol{Q}_k$ (see eqs. 3 and 4)

(c) $\nu_k \geqq 0$ if $\boldsymbol{Q}_k$ defines a point on the design cost surface and $\nu_k = 0$ if $\boldsymbol{Q}_k$ defines a point inside the design cost surface and

(d) $\sum_{k=1}^{z} \nu_k = 1$ where z is the total number of loading conditions.

For example, let points $\boldsymbol{Q}_a$, $\boldsymbol{Q}_b$ and $\boldsymbol{Q}_c$ in Fig. 1a, represent statically admissible stress vectors for three alternative load systems. The corresponding cost gradient vectors are given by points $\boldsymbol{G}_a$ and $\boldsymbol{G}_c$ and line segment $\boldsymbol{G}_b$ (i.e. PR) in Fig. 1b. The admissible set of strain values are given by

$$\left.\begin{array}{l} \boldsymbol{q}_a = \lambda_1 \boldsymbol{G}_a, \quad \boldsymbol{q}_b = \lambda_2 \boldsymbol{G}_b, \quad \boldsymbol{q}_c = 0 \\ \text{with} \quad \lambda_1 + \lambda_2 = 1 \quad \text{and} \quad \lambda_i \geqq 0 \quad (i = 1, 2) \end{array}\right\} \tag{11}$$

A *lower bound* on the minimum cost is given by the integral

$$\Phi = \int_D \sum_{k=1}^{z} \boldsymbol{p}_k \cdot \boldsymbol{u}_k d\xi \tag{12}$$

where the strain vectors $\boldsymbol{q}_k$ associated with the displacement fields $\boldsymbol{u}_k$ must observe the following restrictions:

$$\boldsymbol{q}_k = \nu_k \boldsymbol{G}_k^* \tag{13}$$

such that

(a) $\boldsymbol{G}_k^*$ is any strain vector defining a point on or inside the cost gradient surface (Fig. 1b),

(b) $\nu_k \geqq 0 \qquad (k = 1, \ldots, z)$ and

(c) $\sum_{k=1}^{z} \nu_k = 1.$

An *upper bound* can be obtained by finding statically admissible stress fields $\boldsymbol{Q}_k(\xi)$ for each loading condition and adopting the highest cost value for each point:

$$\psi = \max_k [\psi(\boldsymbol{Q}_k), \qquad k = 1, \ldots, z]. \tag{14}$$

4. Optimality conditions for convex non-conical cost functions

For this class of problems, optimality conditions can be obtained from the Prager-Shield cost gradient principle [2] or by variational methods [5], [6].

For a *single load system*, it is necessary to find

(i) a statically admissible stress field $\boldsymbol{Q}(\xi)$ and

(ii) a kinematically admissible strain field $\boldsymbol{q}(\xi)$ such that

(iii) $\boldsymbol{Q}(\xi)$ and $\boldsymbol{q}(\xi)$ satisfy the relations (3) and (4) [i.e. $\boldsymbol{q}(\xi)$ is the gradient vector of $\psi(\boldsymbol{Q})$].

For *alternate load systems*,t he optimality conditions given for conical cost functions (section 3) are also valid for non-conical convex cost functions. However, the statements on lower bounds in sections (2) and (3) cannot be extended in that form to non-conical cost functions.

5. Optimality conditions for convex cost functions associated with multi-component systems and alternate load systems

In order to visualise this class of problems, consider a fibre-reinforced plate or shell with fibres running in *two specified directions* and *along both faces* of the plate or shell. The cross-sectional areas of these sets of fibres constitute four independent cost components at any point of the middle surface and each cost component at a given point may be governed by a different loading condition.

Let the specific cost function ψ consist of the following components:

$$\psi = \sum_{l=1}^{m} \psi_l. \tag{15}$$

The optimality conditions, for m cost components and w alternate load conditions, as derived by Rozvany [6], can be summarised as follows:

At a given point ξ, let the design value of the cost components ψ_l $(l = 1, 2, \ldots, m)$ be $\psi_{l0}(l = 1, 2, \ldots, m)$.

Then the minimum cost is associated with:

(i) a statically admissible stress field $\boldsymbol{Q}_k(\xi)$ for each loading condition 'k' and

(ii) a kinematically admissible strain field $\boldsymbol{q}_k(\xi)$ for each loading condition 'k' such that

(a) $q_k = \sum_{l=1}^{m} v_{lk} G_{lk}$ where G_{lk} is the gradient vector (equations 3 and 4) of the cost component function $\psi_l(Q)$ at $Q = Q_k$,

(b) $v_{lk} \geqq 0$ if $\psi_l = \psi_{l0}$
(ie. if the loading condition k is critical in relation to the cost component ψ_l at the point ξ considered),

(c) $v_{lk} = 0$ if $\psi_l \leqq \psi_{l0}$,

(d) $\sum_{k=1}^{w} v_{lk} = 1$ for each cost component $l = 1, 2, \ldots, m$.

Optimal solutions involving alternate load systems on fibre-reinforced plates were obtained by Charrett [6], [8].

6. Plastic optimal design problems involving non-convex specific cost functions

If the specific cost function is non-convex, then the kinematic conditions derived by a variational method [5], [6] are the same as those given in sections (4) and (5) but they constitute only a necessary condition for a local minimum and not a sufficient condition for a global minimum. However, in a number of plastic optimization problems, these kinematic conditions admit a sufficiently small number of solutions such that it is possible to find the global optimum. In the two examples that follow, these conditions lead to a global optimum directly.

Example (1). *Optimal design of clamped beam of variable depth.* Consider the optimal design of a clamped beam subjected to a uniformly distributed load p (Fig. 2a) and having a preassigned width b but variable depth $d(x)$. The material of the beam is assumed to have a rigid perfectly plastic stress-strain relation, but the optimal solutions derived will be shown to be valid for any linearly or non-linearly elastic stress-strain relation.

It can be shown readily that the cross-sectional area of the beam is

$$\psi = k\sqrt{|M(x)|} \tag{16}$$

where $M(x)$ is the bending moment and k is a constant (Fig. 2b). The corresponding kinematic condition is (Fig. 2c)

$$\kappa(x) = \frac{dA(x)}{dM} = \frac{k}{2\sqrt{|M(x)|}} \operatorname{sgn} M(x) \tag{17}$$

where κ is the curvature.

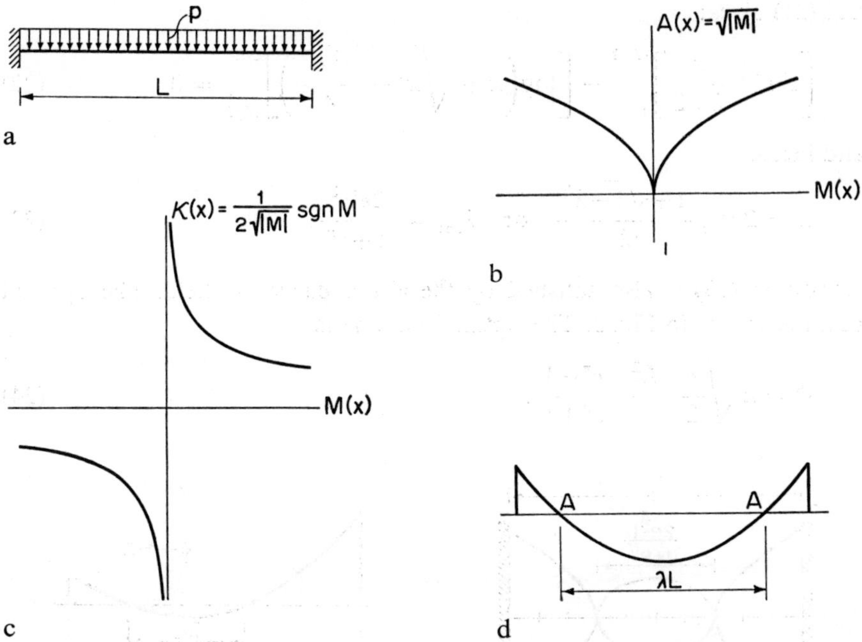

Fig. 2

Since the beam is clamped at both ends, the following conditions must be met by the curvature field $\kappa(x)$:

$$\int_{-L/2}^{L/2} \kappa(x)dx = 0, \tag{18}$$

$$\int_{-L/2}^{L/2} \int_{-L/2}^{L/2} \kappa(x)dx = 0. \tag{19}$$

Using eqs. (17), (18) and (19), the optimal solution can be derived easily. Adopt the moment field (Fig. 2d)

$$M = p\left(\frac{\lambda^2 L^2}{8} - \frac{x^2}{2}\right) \tag{20}$$

in which λ is an unknown constant. By eqs. (17), (18) and (20),

$$k\sqrt{\frac{2}{p}}\left[\int_{0}^{\lambda L/2} \frac{dx}{\sqrt{(\lambda^2 L^2/4) - x^2}} - \int_{\lambda L/2}^{L/2} \frac{dx}{\sqrt{x^2 - (\lambda^2 L^2/4)}}\right] = 0. \tag{21}$$

Eq. (21) gives:

$$\left[\sin^{-1}\frac{x}{\lambda L/2}\right]_0^{\lambda L/2} - \left[\log\left(x+\sqrt{x^2-\frac{\lambda^2L^2}{4}}\right)\right]_{\lambda L/2}^{L/2} = 0 \tag{22}$$

and hence

$$\pi = 2\log\frac{1+\sqrt{1-\lambda^2}}{\lambda} \quad \text{or} \quad \lambda_{\text{opt}} = \frac{2e^{\pi/2}}{1+e^{\pi}}. \tag{23}$$

Condition (19) is also satisfied by the above curvature field. The optimal beam is shown in Fig. 3. The optimal volume is

$$\Phi = k\sqrt{\frac{p}{2}}\,\frac{L^2}{4}\,\frac{e^{\pi}-1}{e^{\pi}+1}. \tag{24}$$

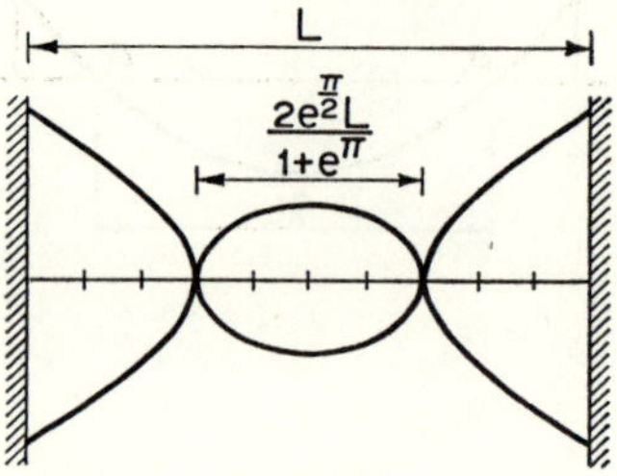

Fig. 3

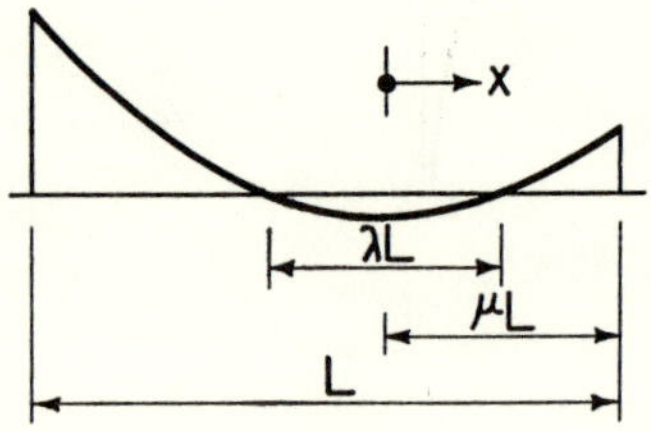

Fig. 4

Check by differentiation

In the case of non-convex cost functions, optimal solutions are not necessarily symmetric for symmetric load and boundary conditions. Introducing the indeterminate parameters λ and μ, (Fig. 4) represents all statically admissible moment diagrams. The moment values are given by

$$M = p\left(\frac{\lambda^2L^2}{8} - \frac{x^2}{2}\right) \tag{25}$$

and the total volume is:

$$\Phi = k\sqrt{p/2}\left(\int_{-\lambda L/2}^{\lambda L/2}\sqrt{\frac{\lambda^2L^2}{4}-x^2}\,dx + \right.$$
$$\left. + \int_{\lambda L/2}^{\mu L}\sqrt{x^2-\frac{\lambda^2L^2}{4}}\,dx + \int_{-L(1-\mu)}^{-\lambda L/2}\sqrt{x^2-\frac{\lambda^2L^2}{4}}\,dx\right). \tag{26}$$

After integrating, the condition $\partial\Phi/\partial\lambda = 0$ gives $\mu = \frac{1}{2}$ which means that the optimal solution is symmetric. Then the condition $\partial\Phi/\partial\lambda = 0$ gives

$$\pi\lambda - 2\lambda \log \frac{1+\sqrt{1-\lambda^2}}{\lambda} = 0. \tag{27}$$

Eq. (27) yields the same solution as the one in eq. (23). The foregoing solution coincides with the minimum of a curve representing the volume for all symmetric solutions for this problem in ref. [9].

Remark: If the cross-sectional area-moment relation takes the form

$$\psi = k|M|^{\alpha} \tag{28}$$

where k and α are known constants, then the above method gives a rate of energy dissipation

$$D = M\kappa = |M|k\alpha|M|^{\alpha-1} = \alpha\psi \tag{29}$$

which is everywhere proportional to the cross-sectional area A. Similar results were obtained by Drucker and Shield [3] for solid structures of variable depth.

Example (2): *Plastic optimal design of anisotropic cylindrical shells.* Consider a class of cylindrical shells (Fig. 5a) in which (i) the width (W) and spacing (S) of the ribs have preassigned[1]) values but the depth (d) is a variable quantity; (ii) the shell between ribs is very thin and hence the entire load can be assumed to be resisted by ribs; (iii) there is no interaction between the yield capacity of longitudinal and circumferential ribs and (iv) the spacing (S) of the ribs is small in comparison with the radius (R) of the shell, (v) both the loading and the boundary conditions are rotationally symmetric.

The equilibrium equation for this problem is

$$\frac{dM}{dx} + \frac{N}{R} = p_M(x) + p_N(x) = p(x) \tag{30}$$

where M is the longitudinal bending moment per unit shell width, N is the hoop force per unit width, R is the radius of the middle surface of the shell, x is the longitudinal coordinate, $p(x)$ is the internal pressure and $p_M(x)$ and

[1]) In problems in which W and S are not preassigned, the optimal solution reduces to densely spaced, infinitely deep ribs having a theoretical total volume of zero [13].

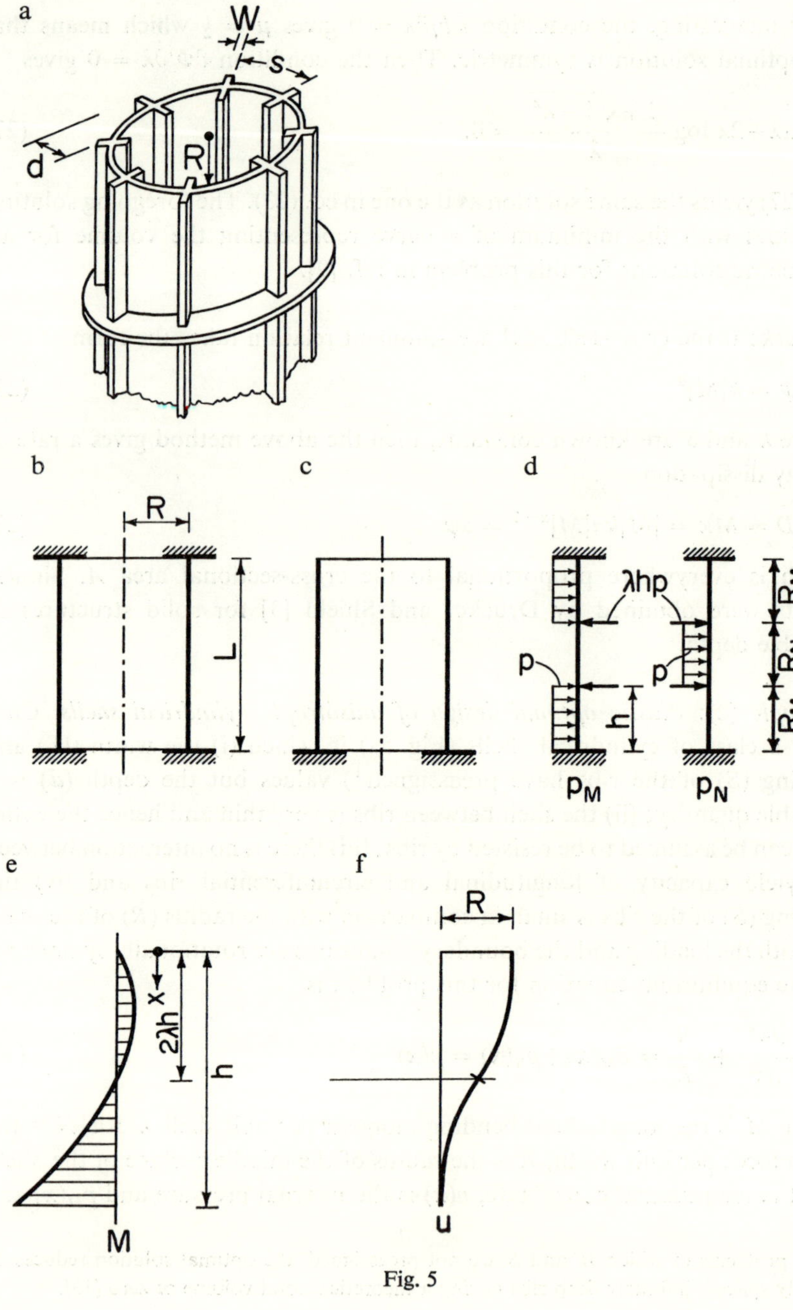

Fig. 5

$p_N(x)$ are loads resisted by bending moments and hoop forces, respectively.
Introducing the parameters

$$\Phi = \frac{V\sigma_0}{2\pi R} \quad \text{and} \quad k = 2\sqrt{b\sigma_0} \tag{31}$$

where V is the total volume of the ribs, σ_0 is the yield stress and b is the aggregate width of ribs per unit shell width, the following relation can be obtained [10]

$$\Phi = \int_0^L (|N| + k\sqrt{|M|})dx. \tag{32}$$

The specific cost function in eq. (32) gives the following kinematic conditions for optimality:

$$\varepsilon = \operatorname{sgn} N \qquad (N \neq 0), \tag{33}$$

$$-1 \leqq \varepsilon \leqq 1 \qquad (N = 0), \tag{34}$$

$$\kappa = \frac{k}{2\sqrt{|M|}} \operatorname{sgn} M. \tag{35}$$

Considering cylindrical shells with either both ends clamped (Fig. 5b) or one end clamped and the other end free (Fig. 5c), the type of optimal solution to be tried consists of the following regions:

$$\left.\begin{array}{lll} R_M\text{-regions:} & N = 0, & M \neq 0 \\ R_N\text{-regions:} & M = 0, & N \neq 0 \end{array}\right\}. \tag{36}$$

It will be assumed that R_M-regions occur adjacent to the support (s) and R_N-regions occur over the central part of the shell, with a concentrated hoop force along the region boundaries (Fig. 5d). In a typical R_M-region, the moment field is given by (Fig. 5e):

$$M = -p(x^2 - 2\lambda hx)/2. \tag{37}$$

The second derivative of the radial deflection u'' is given by eqs. (35) and (37):

$$u'' = \frac{k}{2\sqrt{|M|}} \operatorname{sgn} M = \begin{cases} \dfrac{k}{2\sqrt{p/2}\sqrt{x^2 - 2\lambda hx}} & \text{for} \quad 2\lambda h < x < h, \\ \dfrac{k}{2\sqrt{p/2}\sqrt{2\lambda hx - x^2}} & \text{for} \quad 0 \leqq x \leqq 2\lambda h. \end{cases} \tag{38}$$

Eq. (33) implies that the radial displacement in R_N is $y=R$. Hence the boundary conditions for $u(x)$ in R_M (Fig. 5f) are:

$$u(0) = R, \quad u'(0) = 0, \quad u(h) = 0, \quad u'(h) = 0. \tag{39}$$

Eqs. (38) and (39) give

$$\lambda = 2e^{\pi}/(e^{\pi}+1)^2, \quad h = R(e^{\pi}+1)/2(e^{\pi}-1) \tag{40}$$

where $v = 2\sqrt{2p/k}$.

Independent proof of global optimality
It can be shown [10] using theorems of concave programming [11], [12] that the solution must consist of the type of regions given in eq. (36). Further, by considering various combinations of these regions, it can be shown [10] that the configuration given in 5d gives the smallest volume over a certain L/vR ratio. Then the only set of statically admissible moment fields is given by eq. (37), and, from eq. (32), the total volume for shells clamped at one end (Fig. 5c) becomes:

$$\Phi = [L-(1-\lambda)h]qR+k\left[\int_0^{2\lambda h}\sqrt{\lambda hpx-\frac{px^2}{2}}\,dx+\right.$$
$$\left.+\int_{2\lambda h}^{h}\sqrt{\frac{px^2}{2}-\lambda hpx}\,dx\right] = pR\left\{L-(1-\lambda)h+\right.$$
$$\left.+\frac{h^2}{vR}\left[\lambda^2\pi-\lambda^2\log\frac{1-\lambda+\sqrt{1-2\lambda}}{\lambda}+(1-\lambda)\sqrt{1-2\lambda}\right]\right\}. \tag{41}$$

Conditions $\partial\Phi/\partial h = 0$ and $\partial\Phi/\partial\lambda = 0$ then give the results in eq. (40).

7. Optimal design involving partially preassigned strength of distribution

Extension of Foulkes's theorem [14] to two dimensional systems. In this section, the particular problem of torsionless prismatic grillages is discussed. However, the formulation used can be extended to any one, two or three dimensional system and any preassigned strength distribution.

Consider an orthogonal grillage (Fig. 6a) consisting of densely spaced straight beams such that the cross section is constant for any one beam. Let the cross sectional area ψ_i of a beam be given by

$$\psi_i = k|\overline{M}_i| \qquad (i = x, y) \tag{42}$$

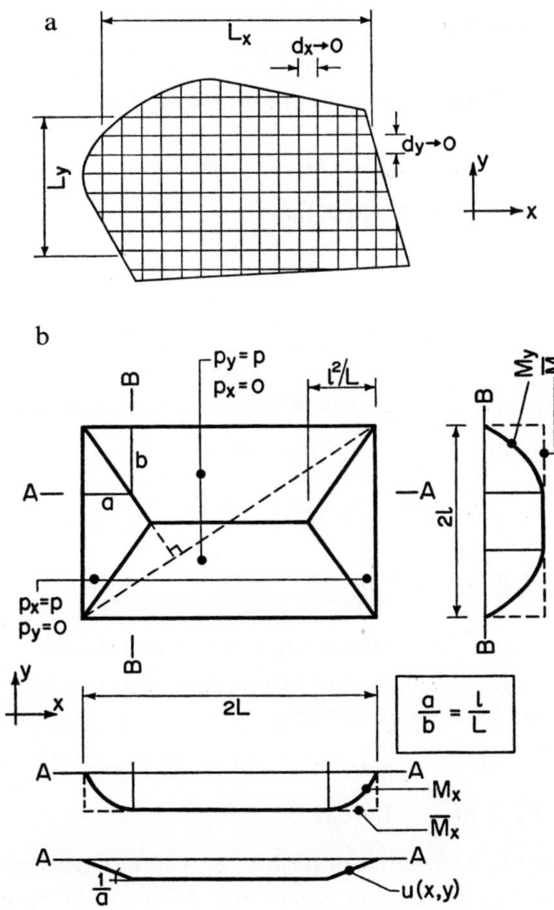

Fig. 6

where $\overline{M}_i$ is the design yield moment capacity of a beam and k is a constant; and let the equilibrium equation, due to neglecting the torsion, be

$$\frac{\partial^2 M_x}{\partial x^2} + \frac{\partial^2 M_y}{\partial y^2} = p_x(x, y) + p_y(x, y) = p(x, y) \tag{43}$$

where M_x and M_y are bending moments in x and y directions and p_x and p_y are loads carried by the x-beams and y-beams respectively.

Using Lagrangian multipliers η_i and u and slack variables s_i [15], the problem can be formulated as

$$\min \iint_D [\psi_x+\psi_y+\eta_1(kM_x+s_1-\psi_x)+\eta_2(-kM_x+s_2-\psi_x)+$$
$$+\eta_3(kM_y+s_3-\psi_y)+\eta_4(-kM_y+s_4-\psi_y)+$$
$$+u(p-\partial^2M_x/\partial x^2-\partial^2M_y/\partial y^2)]dxdy. \tag{44}$$

Considering that

$$\int_{L_x(y)} \psi_x(y)dx = L_x(y)\psi_x(y), \text{ etc.} \tag{45}$$

variation of M_x, M_y, s_i, ψ_x and ψ_y give the Euler-Lagrange equation [15]:

$$k(\eta_1-\eta_2) = \partial^2u/\partial x^2, \quad k(\eta_3-\eta_4) = \partial^2u/\partial y^2, \tag{46}$$

$$(\text{for } s_i = 0)\ \eta_i \geqq 0, \quad (\text{for } s_i \geqq 0)\ \eta_i = 0, \tag{47}$$

$$L_x = \int_{L_x} (\eta_1+\eta_2)dx, \quad L_y = \int_{L_y} (\eta_3+\eta_4)dy. \tag{48}$$

The kinematic meaning of eqs. (46) through (48) is as follows:

It is necessary to find a displacement field $u(x, y)$ such that the curvatures κ_x and κ_y satisfy the following conditions:

(i) $\kappa_i \leqq 0$ if $M_i = \overline{M}_i$, $\kappa_i \leqq 0$ if $-M_i = \overline{M}_i$ and
$\kappa_i = 0$ if $|M_i| \leqq \overline{M}_i \quad (i = x, y)$ (49)

(i.e. a curvature can be negative or positive only if the negative or positive moment in the same direction equals the design yield moment) and

(ii) the sum of the modulus of the rotations along any one beam is proportional to the length (L_x or L_y) of such beam.

Fig. 6b shows the optimal moment field and displacement field satisfying the foregoing requirements for a rectangular grillage, subjected to a uniform load p. This solution was originally obtained, somewhat intuitively, for elastic prestressed plates [16] and plates reinforced with elastic-plastic fibres [17].

References

[1] Prager, W., *Introduction to plasticity*. Addison Wesley, Reading, Mass. (1959).

[2] Prager, W., and R. T. Shield, A general theory of optimal plastic design. *J. appl. Mech. Trans. ASME, 34* (1967) 185–186.

[3] Drucker, D. C. and R. T. Shield, Design for minimum weight. *Proc. Int. Cong. appl. Mech.*, Brussels, (1956) Book 5, 212–222.

[4] Drucker, D. C. and R. T. Shield, Bounds on minimum weight design. *Quart. appl. Math. 15* (1957) 269–281.

[5] Rozvany, G. I. N. and S. R. Adidam, Dual formulation of variational problems in optimal design. *Proc. Int. Conf. Design Automation*, Toronto 1971, also: in press, *Trans. ASME.*

[6] Charrett, D. E. and G. I. N. Rozvany, Extensions of the Prager-Shield theory of optimal design. In press, *Int. J. Non-Linear Mech.*

[7] Marcal, P. V. and W. Prager, A method of optimal plastic design. *J. de Méc. 3* (1964) 509–530.

[8] Charrett, D. E., *Plastic analysis and optimal design of slabs, plates and frames*. Ph. D. Thesis, Monash University, (1970).

[9] Megarefs, G. J. and P. G. Hodge, Jnr., Singular cases in the optimal design of frames. *Quart. appl. Math. 21* (1963) 91–103.

[10] Rozvany, G. I. N. and S. R. Adidam, Optimal design of anisotropic cylindrical shells. *Monash University CE Reports 8* (1971).

[11] Rozvany, G. I. N., Concave programming in structural optimization. *Int. J. mech. Sci. 12* (1970) 131–142.

[12] Rozvany, G. I. N., Concave programming and piece-wise linear programming. *Int. J. Numer. Methods in Engrg. 3* (1971) 131–144.

[13] Kozlowski, W. and Z. Mróz, Optimal design of solid plates. *Int. J. Solids Structures 5* (1969) 781–794.

[14] Foulkes, J., The minimum-weight design of structural frames. *Proc. Roy. Soc.*, Ser. A. *223* (1954) 482–494.

[15] Elsgolc, L. E., *Calculus of variations*. Pergamon, Oxford, (1961).

[16] Rozvany, G. I. N., Optimum synthesis of prestressed structures. *J. Struct. Div. Proc. Amer. Soc. Civ. Engrs. 90* (1964) 189–211.

[17] Rozvany, G. I. N., Minimum volume of uncurtailed orthogonal reinforcement in freely supported slabs. *Concrete and Constructional Engineering 61* (1966) 281–286.

On the thermodynamic states of plastically deformed solids

Y. Horie

North Carolina State University, Raleigh, N.C., U.S.A.

An argument is presented that the existing difficulties in the thermodynamics of plasticity lie in the overemphasis placed on macroplastic strains and a particular history associated with it. It is proposed that any plastically deformed states can be adequately represented by an adjacent elastic state constrained with stationary dislocations. Some elementary, but critical, results are obtained including (1) the resolution of Drucker's dilemma concerning the temperature dependence of yield stress and (2) a proof of the path independence of adiabatic plastic deformations in the small. Also a conjecture is offered that the 'chemical potential' may be assigned so as to deal with elastic-plastic transitions.

1. Introduction

Thermodynamicists and continuum mechanists rarely agree with each other about the applicability of classical thermodynamics to plasticity. The former maintain the negative view simply because it is an irreversible process. The latter propose that a generalized system which is mathematically consistent can be constructed so as to handle the phenomenon by use of the invariance principle on the energy rate equation and the constitutive relations. Therein physical supports of generalized concepts appear to be of secondary interest. There are other notable approaches, but none seem to be accepted as the final answer to the problem.

The confrontation is little abated and is often avoided with indifference to the other side's arguments. Sadly, it is rarely settled by means of critical experimentation.

The author, stimulated by Bridgman's work [1], has conceived the idea that the problem arises from the fact that one has associated too much thermodynamical meaning with the current level of plastic strains. Once admitted, the abyss which lies between the two camps seems approachable. Therefore, a new survey is undertaken so as to demonstrate that one can approach plasticity by means of the classical thermodynamics. However, it will be limited to the simplest cases which are adequate to test the basic premises.

The theory first reexamines the concept of plastic strains in the framework of classical thermodynamics. The basis is the observation that dislocation movements are primarily responsible for plastic deformations in solids. It is then proposed that plastically deformed states may be represented by elastic state constrained with stationary dislocations. The study leads to an inevitable proposal of chemical potential for dislocations which enables one to interpret an elastic-plastic transition as a thermodynamical one.

Later as an application Drucker's dilemma concerning an increase of yield stress with temperature is resolved. Attempts will also be made to understand some of the basic premises of a theory in which entropy is given as a primitive concept.

2. Dislocations and their thermodynamic representation

A cross-section of a dislocation in a crystal is shown in Fig. 1. It is well established that the process of plastic deformations consist in a migration of

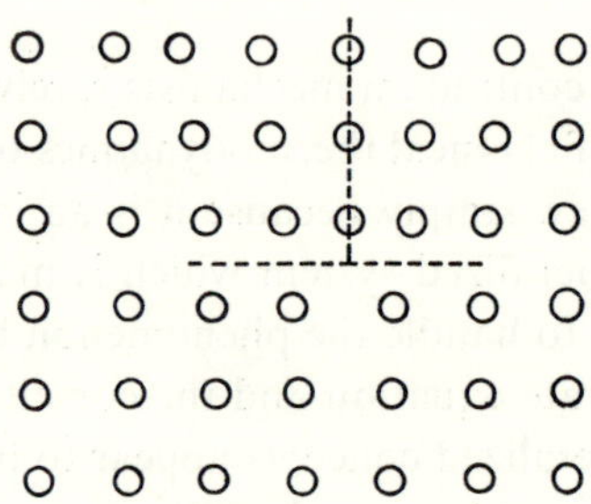

Fig. 1. A dislocation cross section in a lattice.

dislocations through a crystal. For example, if the dislocation in the figure slides to the right-hand edge due to shear stresses applied at the top and bottom surface, the result is equivalent to a plastic shear deformation of one lattice spacing (Fig. 2).

Although in thermodynamics, one does not consider the motion of a single dislocation; the number does not alter the essential features of the problem we want to consider. The problem is to coordinate processes such as those

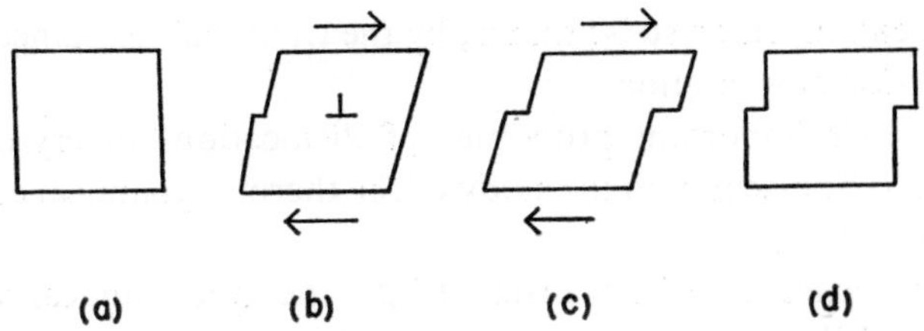

Fig. 2. Permanent deformation due to a dislocation movement.

described in the above paragraph into a logical system which is consistent with classical thermodynamics. The answer cannot be supplied, the author feels, by merely adding plastic strains into the argument of internal energy which is a state variable and which must be independent of processes. Hence, if one adheres strictly to the classical principles, the inclusion is an invalid procedure.

This practice, regardless whether or not it be justified, seems to have been established through the formulation of the energy conservation law in continuum mechanics which states it in the rates of energy transfer. Hence, irreversible heat transfer may now be expressed explicitly in terms of state variables. This is not true in classical thermodynamics where there exists only an inequality for irreversible heat transfer. Classical statements concern the energy balance between two equilibrium states and never mention natural time. Hence, the two formulations are not equivalent without some physical qualifications on the former. Hence, as we see presently, one needs to distinguish the thermodynamical meanings of plastic strains and strain rates. The former will be shown to be insignificant in thermostatics.

The basic properties of dislocations in crystals are quoted below from the standard references [2, 3].

(P-1) The energy of a stationary dislocation is mostly the strain energy caused by the dislocation.

(P-2) The total energy of a strained crystal with a stationary dislocation is the sum of the lattice strain energy and the dislocation energy. It is independent of the position of the dislocation.

(P-3) The elastic properties of the crystal are unaltered by the presence of stationary dislocations. Until the dislocations move, one cannot tell that it is there by measuring the response of the body to applied forces.

(P-4) Yield stresses are determined primarily by the stress required to move dislocations in crystals; that is, by the frictional resistance of a crystal to dislocation motion.

There are other important properties of dislocations in crystals, but the above seem to be essential in the study of the thermodynamical consideration of plasticity.

These properties may be translated into thermodynamical variables by adopting the following assumption:

(A-1) Crystal lattices are perfectly elastic, and the total internal energy is the sum of the lattice energy and the energy associated with the dislocations in the crystal.

(A-2) Dislocations in a crystal are adequately described by the density function n_{ik}:

$$n_{ik} = \sum_{l} l q_{ik}(l)$$

where q_{ik} is a volume density of dislocation segments of length l whose orientation and Burger vector are given by $\boldsymbol{t}_i$ and $\boldsymbol{b}_k$.

(A-3) Elastic properties of a crystal is independent of n_{ik}.

(A-4) The contribution of surface energy to internal energy is considered to be negligible for the sake of simplicity.

(A-5) The internal lattice energy is an additive quantity as in classical thermodynamics.

Suppose a crystal undergoes an isothermal plastic deformation due to the motion of a single dislocation illustrated in Fig. 2. Then by (A-4) the initial and the final states must be, thermodynamically speaking, identical. Otherwise, the additive property of internal lattice energy is violated. This can also be recognized by the fact that elastic strains are independent of the overall shape of the body. Thermodynamics does not dictate shapes. Hence, it may be concluded that in thermodynamics one needs to distinguish macro-plastic strains created by dislocation movements and the internal crystal distortion due to stationary dislocations in the crystal. The instantaneous values of plastic strains play little role in thermodynamics. The contribution if any is, as we will see in later sections, through the integrated work.

Next consider an incremental deformation of a crystal involving changes in the lattice energy and n_{ik}. Then due to (A-1), (A-3), and (A-4), the total incremental change of energy of a crystal may be given by the expression:

$$\mathrm{d}E = \mathrm{d}E^L(T, \varepsilon^L) + \mathrm{d}E^D = \mathrm{d}Q + \mathrm{d}W, \qquad (1)$$

where E^L is the specific lattice energy, E^D the specific dislocation energy,

T temperature, and ε^L lattice strain. The decomposition of internal energy is necessary because of the fact that elastic properties of lattices are assumed to be independent of stationary dislocations.

Then if the density change of dislocations whose orientation and Burger's vector are $\boldsymbol{t}_i$ and $\boldsymbol{b}_k$ is given by $\mathrm{d}n_{ik}$, $\mathrm{d}E^D$ may be expressed as

$$\mathrm{d}E^D = \mu_{ik}\mathrm{d}n_{ik}, \tag{2}$$

where μ_{ik} is the energy associated with the density change $\mathrm{d}n_{ik}$. It may be observed by rearranging the terms in Eqs. (1) and (2),

$$\mathrm{d}E^L = \mathrm{d}W + \mathrm{d}Q - \mu_{ik}\mathrm{d}n_{ik}. \tag{3}$$

Now since the crystal lattice is assumed to be perfect, one can associate with it the entropy in the classical sense. That is, one can conceive a reversible path between any two states of E^L

$$\mathrm{d}E^L = T\mathrm{d}S^L + \boldsymbol{\sigma}^L \cdot \mathrm{d}\boldsymbol{\varepsilon}^L. \tag{4}$$

On the contrary, one cannot always associate the entropy and other thermodynamical potentials with the overall internal energy E. The reason is that the change in n_{ik} is usually produced by the migration of dislocations which causes dissipation of mechanical work. The exception to this is the case of spontaneous generation and annihilation of dislocations which may be important at high temperatures. But since our interest is in plastic deformation due to dislocation movements this complication will be ignored in this treatise. That is, $\mathrm{d}n_{ik}$ always accompanies the movements of dislocations.

In classical thermodynamics as Milne exemplifies [4], 'catastrophic changes' are studied in their after effects as the results of processes under consideration. The applicability is indeed independent of whether or not real processes are reversible. Then from Eq. (3) one can infer that there exists an equivalent lattice state for any plastically deformed crystals. And this state can be determined by an infinitesimal unloading without a change in n_{ik}.

3. Stability analysis of elastic-plastic transition

That dislocations are of shadowy existence (see (P-3)) enables one to conceive an interesting view of Eq. (3) which clarifies the thermodynamic nature of plastic deformations and the motivation which led Bridgman to introduce the idea of the sea of irreversibility.

Equation (3) suggests in form that a crystal lattice interacts with three agencies: work, heat, and density number (Fig. 3). However, the fact that the lattice energy lacks the density number in its argument requires that it is a one-way interaction as shown in Fig. 3. The effects of the change in n_{ik} are always communicated through the other two agencies. But this is precisely the reason why plastic deformations are inherently irreversible. Let it be static or dynamic. What is important, however, is that the change can be registered through crystal lattice.

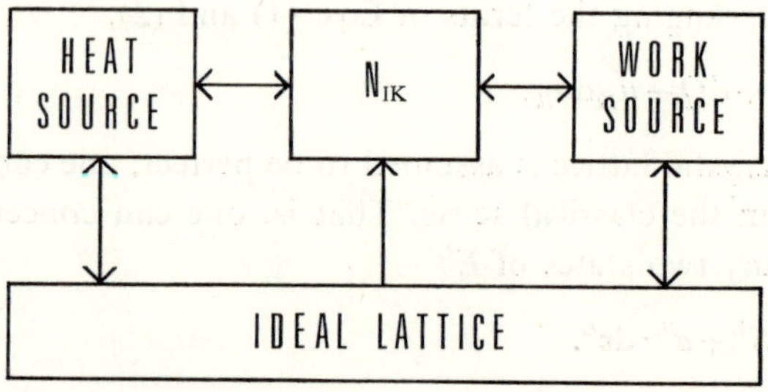

Fig. 3. Thermodynamic interactions of ideal lattices.

Once the above view is adopted as the basis of the thermodynamical consideration of plastic deformations, it may be appropriate to consider μ_{ik} as a 'chemical' potential for the change in the density of dislocation segments whose orientation and Burger's vector are $\boldsymbol{t}_i$ and $\boldsymbol{b}_k$. The use of

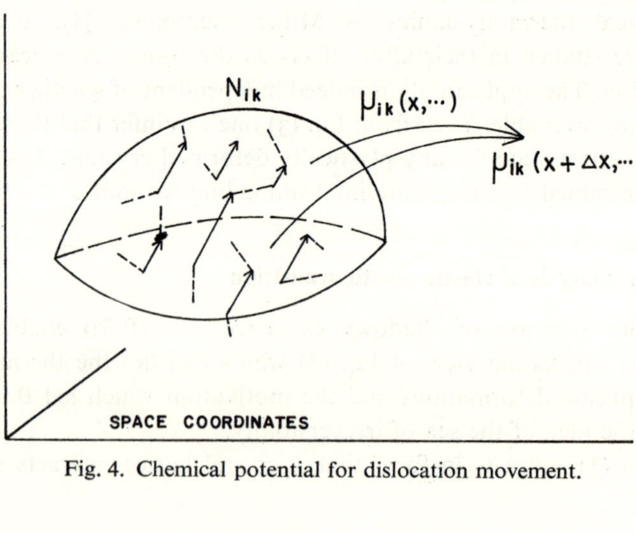

Fig. 4. Chemical potential for dislocation movement.

the word 'chemical' may also be justified from the fact that the free energy of distributed dislocations is mostly the strain energy. But contrary to the usual chemical potential, the partial derivative of μ_{ik} with respect to n_{ik} does not determine the density. Hence, the collection of a crystal lattice and dislocations in it is not an equilibrium mixture in the sense of classical thermodynamics. Dislocations interact with the lattices as a kind of movable constraint, and μ_{ik} is the potential for their movements as a whole.

In terms of this 'chemical' potential, the density n_{ik} of dislocation segments is subject to change whenever the following conditions are satisfied (Fig. 4)

$$\mu_{ik}(\boldsymbol{\sigma}, T, \boldsymbol{n}|\boldsymbol{x}) = \mu_{ik}(\boldsymbol{\sigma}, T, \boldsymbol{n}|\boldsymbol{x}+\Delta\boldsymbol{x}),$$

$$\mu_{ik}(\boldsymbol{\sigma}, T, \boldsymbol{n}|\boldsymbol{x}) = \mu_{ij}(\sigma, T, \boldsymbol{n}|\boldsymbol{x}+\Delta\boldsymbol{x})$$

where $\boldsymbol{x}$ is parametric.

The second condition involves the rotation of dislocation segments. If the movements are restricted to sliding, the first condition may be written as

$$\frac{\partial\mu_{ik}}{\partial\boldsymbol{x}} = 0. \tag{5}$$

This is the incipient yielding condition.

Now we are ready to test these conceptual bases in simple examples, and to find what the facts are in the 'new' domain.

4. Static analysis without dislocation energy

At room temperatures the dislocation energy E^D is relatively small in comparison with E^L. One may ignore as the first approximation the energy associated with dislocations. However, we must retain the interactions between the ideal lattice and the distributed dislocations in it. Hence, the stresses in the work term in Eq. (3) cannot be set equal to $(\partial E^L/\partial\varepsilon^L)$.

First, we examine how the preceding preparation enables one to handle the simplest thermodynamics of plastic deformations. As an example, we employ mostly a one-dimensional model.

(i) *Isothermal deformations*

A familiar stress-strain relation is illustrated in Fig. 5. The path *AB* shown in the figure is of course irreversible. But in the previous sections, it is proposed that for any state on the stress-strain curve there exists a thermo-

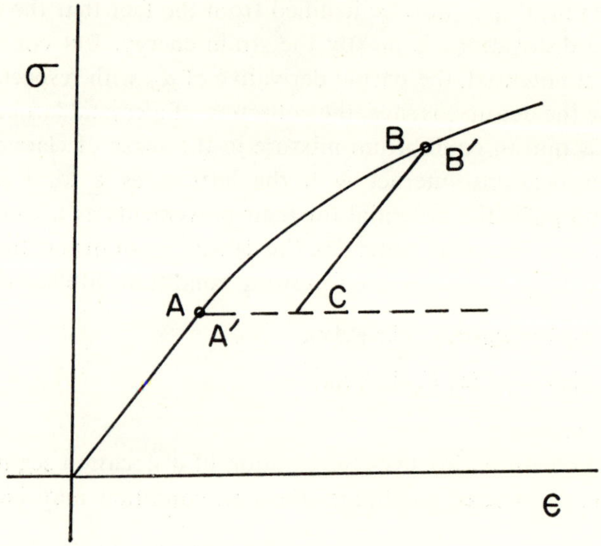

Fig. 5. An isothermal stress-strain relation.

dynamically equivalent elastic state which is infinitesimally close to it. For example, A' and B' are such ones in relation with A and B respectively. Any plastically deformed crystals can be represented by the lattice state.

In isothermal deformation without E^D, one can further conclude that A' is exactly A and that any state on the horizontal line emanating from the stress-strain curve is equivalent to the state at the intersection. This conclusion has been reached by Bridgman [1] also. The conclusion is consistent with the classical principles.

Hence, as far as the internal energy of the lattice is concerned, a process ABC shown in the figure is a thermodynamical cycle. Then applying the classical inequality relation for thermal conduction in any irreversible processes, one obtains

$$\oint_{ABC} \delta Q = \oint_{ABC} (\mathrm{d}E - \mathrm{d}W) = -\oint_{ABC} \mathrm{d}W < 0. \tag{6}$$

Or it may be written as

$$\oint_{ABC} \mathrm{d}W > 0.$$

In terms of stress strain relations, this becomes

$$\oint_{ABC} \sigma \mathrm{d}\varepsilon > 0.$$

The three dimensional analog is given by

$$\oint_{\mathrm{cyclic}} \sigma_{ij} \mathrm{d}\varepsilon_{ij} \geqq 0, \tag{7}$$

where the equality is included for elastic deformations. The inequality is applicable for all cyclic plastic deformations.

Now Drucker's work-hardening rule can be obtained by considering an infinitesimal cycle from the state $(\sigma_{ij}, \varepsilon_{ij})$ to $(\sigma_{ij}+\mathrm{d}\sigma_{ij}, \varepsilon_{ij}+\mathrm{d}\varepsilon_{ij})$. That is, on completing a cycle Eq. (7) produces a relation, which is often introduced in plasticity as a postulate,

$$\mathrm{d}\sigma_{ij}(\mathrm{d}\varepsilon_{ij}-\mathrm{d}\varepsilon_{ij}^{e}) = \mathrm{d}\sigma_{ij}\mathrm{d}\varepsilon_{ij}^{p} \geqq 0, \tag{8}$$

where unloading is of course elastic, and $\mathrm{d}\varepsilon_{ij}^{p}$ plastic strain increment. Furthermore, since B is equivalent to B', it may be easily seen that

$$\int_{AB} \sigma_{ij}\mathrm{d}\varepsilon_{ij} = \int_{AB} \sigma_{ij}(\mathrm{d}\varepsilon_{ij}^{L}+\mathrm{d}\varepsilon_{ij}^{p}) = \text{a function of } \sigma_{ij} \text{ in lattice.}$$

Hence, the equation yields in general

$$\mathrm{d}\varepsilon_{ij}^{p} = A_{ijkl}\mathrm{d}\sigma_{kl}. \tag{9}$$

The heat exchange along the irreversible, path AB can be determined from Eq. (6).

$$\oint_{ABC} \delta Q = \int_{AB} \delta Q = \int_{BC} \delta Q = -\int_{AB} \mathrm{d}W - \int_{BC} \mathrm{d}W.$$

Since the path BC is reversible, one can replace the heat by the lattice entropy.

$$\int_{AB} \delta Q = -\int_{BC} T\mathrm{d}S^{L} - \int_{AB} \mathrm{d}W - \int_{BC} \mathrm{d}W. \tag{10}$$

Under isothermal conditions, the entropy is given by

$$T\mathrm{d}S^{L} = T\left(\frac{\partial S^{L}}{\partial \sigma}\right)_{T} \mathrm{d}\sigma = T\alpha \mathrm{d}\sigma$$

where α is the thermal expansion coefficient. For an infinitesimal cycle, one finds from Eqs. (10) and (11)

$$\delta Q_{d\sigma} = -[T\alpha + d\varepsilon^p]d\sigma. \tag{12}$$

That is, the heat flow to the outside is greater than the plastic working. The excess is small due to the magnitude of α, but it should be there.

It appears that Eqs. (5), (8), and (9) may form the basis for isothermal stress-strain relations. But the difficulty must be registered:

1. dn_{ik} is the density change measured after the completion of a plastic flow and is not directly related to $d\varepsilon^p$ which is a measure of the area swept by the mobile dislocations, except for some idealized situation [5].

2. To be rigorous, the theory must provide mechanisms for mobile dislocations and separate them from stationary ones. This is anticipated from the fact that plasticity, as stated in (P-4), is a dynamic phenomenon.

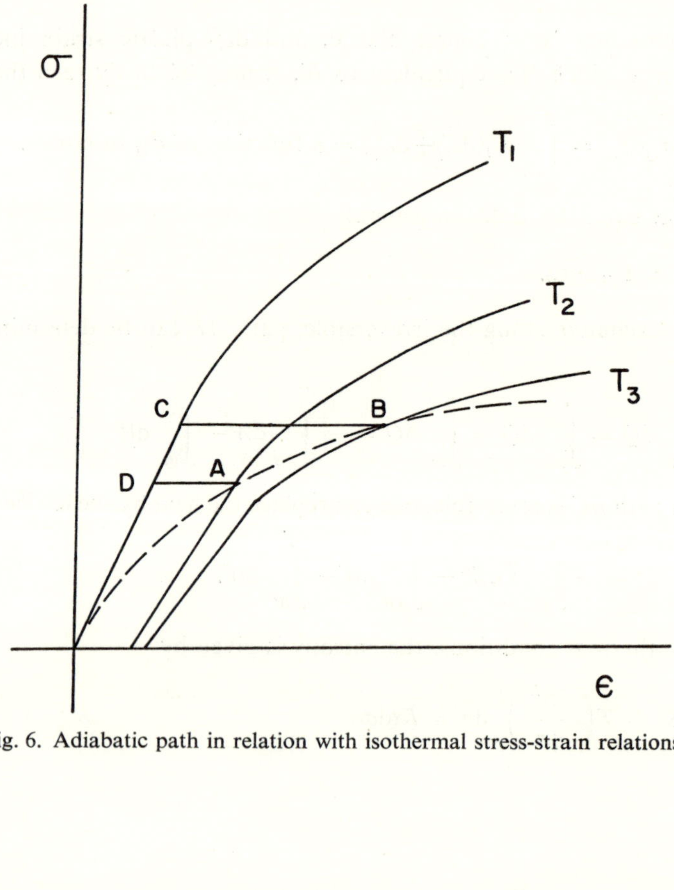

Fig. 6. Adiabatic path in relation with isothermal stress-strain relations.

However, the advantage of thermostatic approach to plasticity is so that one can circumvent internal mechanisms of energetics.

That is, the lack of any specific constitutive relations will be of no importance in studying the thermostatic relations of plastic deformations. The general prediction such as Eq. (12) can be experimentally tested without referring to any specific form of stress-strain relation.

(ii) *Adiabatic deformations*

Contrary to the rather ordinary results in the previous section, an interesting result may be obtained under adiabatic conditions purely by thermodynamical reasoning. It is the temperature dependence of yield stress. Mathematically speaking, it is in Eq. (5) or (10). If thermodynamics is applicable to plasticity, it must be explicitly related to other thermodynamical coefficients.

Suppose a series of isothermal stress-strain relations is given by the curves in Fig. 6. Then the adiabatic deformation traces a path shown by a dotted line. In order to consider the temperature dependency of yield stress, a set of four states, *A*, *B*, *C*, and *D* is chosen. *A* is the adiabatic yielding point. *B* is a plastic state such that there exists a corresponding yield point *C* at the same stress, but at a lower temperature T_1. *D* at the stress level of *A* lies on the isothermal curve T_1 of *C*. Then it may be seen from the preceding analysis that the internal energy difference between *A* and *D*, or *B* and *C* are given by $\int_{T_2}^{T_1} C_\sigma \mathrm{d}T$ and $\int_{T_3}^{T_1} C_\sigma \mathrm{d}T$ respectively. C_σ is the heat capacity at the constant stress. Then one finds

$$\oint_{ABCDA} \mathrm{d}E = \int_{AB} \mathrm{d}W + \int_{T_3}^{T_1} C_\sigma \mathrm{d}T + \int_{\sigma C}^{\sigma D} \left(\frac{\partial E^L}{\partial \sigma}\right)_T \mathrm{d}\sigma + \int_{T_1}^{T_2} C_\sigma \mathrm{d}T = 0.$$

This equality relies strongly on the fact that during the irreversible process $A-B$, there is no heat flow.

Assuming C_σ and $(\partial E^L/\partial\sigma)_T$ constant, one obtains

$$(\sigma_C - \sigma_D)\left(\frac{\partial E^L}{\partial \sigma}\right)_T = \int_{AB} \mathrm{d}W + C_\sigma(T_2 - T_3). \tag{13}$$

This leads to a result analogous to Drucker's [6]

$$\left|\frac{\mathrm{d}\sigma_y}{\mathrm{d}T}\right| \leqq \frac{C_\sigma E^p_{\tan}}{\sigma_y}, \tag{14}$$

where $E^p_{\tan}$ is a plastic tangent modulus. It is, however, derived by use of isothermal processes. Approximating the segment AB by a linear hardening rule with an adiabatic plastic modulus E^p_{ad}, one finds from (13) that

$$(\sigma_C-\sigma_D)\left(\frac{\partial E^L}{\partial \sigma}\right)_T = \frac{(\sigma_C-\sigma_D)(\sigma_C+\sigma_D)}{2E^p_{ad}} + C_\sigma(T_2-T_3). \tag{15}$$

Hence, it is found by letting σ_C approach σ_D that

$$\frac{d\sigma_y}{dT} = -\frac{C_\sigma}{[(\partial E^L/\partial \sigma)_T - \sigma_y/E^p_{ad}]}. \tag{16}$$

Comparing Eqs. (14) and (16), the dilemma that Drucker faced can be explained. If the term $(\partial E^L/\partial \sigma)_T$ is missing in Eq. (16), the yield stress increases with temperature. It will happen in our formula if the following condition is met

$$\left(\frac{\partial E^L}{\partial \sigma}\right)_T < \frac{y}{E^p_{ad}}$$

where it may be found that

$$\left(\frac{\partial E^L}{\partial \sigma}\right)_T = \frac{\sigma}{E^y} + \alpha T$$

and E^y Young's modulus. But the possibility is very remote.

A similar analysis can be carried out by use of isothermal processes, but it becomes a little complicated due to the fact that there is heat exchange during the irreversible process $A-B$.

The pressure dependency of yield stress can be derived from Eq. (16).

$$\frac{d\sigma_y}{dp} = -\frac{dT}{dp}\,\frac{C_\sigma}{[(\partial E^L/\partial \sigma)_T - \sigma_y/E^p_{ad}]}.$$

What is interesting is the classification of pressure dependency with respect to the sign of (dT/dp). But this classification may be more difficult to confirm experimentally than the measurements of the absolute magnitudes of $d\sigma_y/dT$.

5. Analysis with dislocation energy

(i) *Statics*

The basic relation is the local energy equation

$$dE^L = dQ + dW - \mu_{ij}dn_{ij}.$$

In the one-dimensional case, it may be written as

$$dQ = dE^L - \sigma d\varepsilon + \mu dn. \tag{17}$$

The density change will not take place unless a certain condition is met on σ, T, n. But thermostatically speaking, the mechanism is of no concern. We only examine the end products of a process.

The lattice states at A and C in Fig. 5 are still equivalent regardless whether or not there exist dislocations in actual crystals (P-3). Then one can show that the heat exchange during an isothermal plastic working is

$$(\delta Q)_{d\sigma} = -(T\alpha + d\varepsilon^p)d\sigma + \mu dn. \tag{18}$$

Since μdn is positive, the density change draws heat into a crystal. That is, dislocations can be considered metaphorically as heat sinks. Their energy is released into the lattice when annihilated. This is a more satisfying picture of dislocations in thermostatics than that of constraints.

Adiabatic consideration again leads to an interesting result which is a proof of Drucker's 'path independence in the small' [6]. Consider an infinitesimal cyclic path $ABCA$ similar to the one in Fig. 5. Now, due to the temperature change along ABC, the states A and C are not equivalent. The separation in internal energy is given by $\int_{T_C}^{T_A} T(\partial S^L/\partial T)_\sigma dT$. Hence, it may be found that

$$\oint_{ABCA} dE^L = \int_{ABC} dW - \int_{AB} \mu dn + \int_{T_C}^{T_A} T(\partial S^L/\partial T)_\sigma dT = 0.$$

When the cycle is infinitesimal, it becomes

$$d\sigma d\varepsilon^p = \mu dn + T(\partial S^L/\partial T)_\sigma dT,$$

where the coefficients on the right-hand side are constant. This shows that $d\sigma d\varepsilon^p$ is path-independent.

(ii) *Dynamics*

Combining Eqs. (3) and (4), one finds in one dimension

$$T dS^L = dQ + dW - \sigma^L d\varepsilon^L - \mu dn.$$

This appears to be differentiable with respect to time without adding any qualifications. This is not so as explained in the introduction. It is admissible only if the energy inputs in Eq. (1), heat or work or both, were transformed instantly into the other forms of energy on the left-hand side. This is nothing but the assumption of local equilibrium. Then with this qualification in mind, the above equation becomes

$$\dot{S}^L = -\partial(q/T)/\partial x - q\partial(lnT)/\partial x + (\sigma/T)\dot{\varepsilon}^p - (\mu/T)\dot{n}$$

where σ is $(\partial E^L/\partial S^L)_{\varepsilon L}$, q heat flux.

It is clear that the lattice entropy does register the irreversible effects in its surrounding. The formal application of a phenomenological theory yields

$$q = L_{00}\partial(lnT)/\partial x + \ldots,$$

$$\dot{\varepsilon}^p = L_{11}(\sigma - \sigma^*)/T + \ldots, \tag{19}$$

$$\dot{n} = L_{22}(\mu - \mu^*)/T + \ldots, \tag{20}$$

where σ^* and μ^* are introduced to take into account the fact that the flow will not be initiated until σ and μ reach certain critical values, and L_{ij} are constant.

These results are not particularly critical, but they convey the feeling that the view expressed by Eyring [8] needs to be explored further and that the model is accessible to critical testing. For example, if one relates $\dot{\varepsilon}^p$ to dislocation velocity v, Eqs. (19) and (20) become

$$v = L'_{11}(\sigma - \sigma^*)/T + \ldots,$$

$$\dot{n} = L_{22}(\mu - \mu^*)/T + \ldots.$$

Then it may be suggested that the crystal with dislocations may be considered as an open moving 'spin' system with a 'chemical' potential μ.

If this conjecture is correct, we should be able to study the specific effects of moving dislocations on heat flow, i.e., the dynamical version of Eq. (18).

6. Concluding remarks

As far as thermostatics is concerned, the idea of replacing plastically deformed states by equivalent elastic equilibrium states is substantiated by non-trivial results which are useful and can be experimentally tested very easily.

Little can be said about the dynamics without knowing the kinetics of dislocation movements. But the idea is not inconsistent with the theory of irreversible processes.

References

[1] Bridgman, P. W., *Rev. Mod. Phys. 22* (1950) 56.

[2] Cottrell, A. H., *Dislocations and plastic flow in crystals.* Oxford Univ. Press, London, 1953.

[3] Gilman, J. J., *Physical nature of plastic flow and fracture.* In: Plasticity, Proc. Second Symp. on Naval Structural Mechanics. Pergamon Press, New York, 1960.

[4] Milne, E. A., *Thermodynamics of the stars.* In: Selected Papers on the Transfer of Radiation. Ed. by D. H. Menzel. Dover, New York, 1946.

[5] Zarka, J., *Sur la viscoplasticité des métaux.* In: Behavior of Dense Media under High Dynamic Pressures. IUTAM, Symp. H.D.P., Paris. Gordon and Breach, New York, 1968.

[6] Drucker, D. C., *Extension of stability postulate with emphasis on temperature changes.* In: Plasticity, Proc. Second Symp. on Naval Structural Mechanics. Pergamon, New York, 1960.

[7] De Groot, S. R., *Thermodynamics of irreversible processes.* North-Holland, Amsterdam, 1951.

[8] Eyring, H., *J. Chem. Phys. 4* (1936) 283.

The author was informed by a referee of the following references, which, he was told, contain similar ideas and fundamental principles.

[9] Kestin, J., IUTAM Symposium, Vienna, Springer, (1967) 177.

[10] Kestin, J. and J. R. Rice, Int. Symp. on Thermodynamics, Pittsburgh, Mono Book Corp., 1970.

[11] Rice, J. R., *J. Appl. Mech.*, Sept. (1970) 728.

[12] Rice, J. R., Brown University Report NRO 64-424, No. 12.

Observed plastic behavior of metals vis-a-vis the endochronic theory of plasticity

K. C. Valanis

University of Iowa, Iowa City, U.S.A.

1. Introduction

The endochronic theory of viscoplasticity is applied to a number of significant problems which have previously been investigated experimentally by other authors. These involve the mechanical response of aluminum and copper under complex strain histories. It has been found that phenomena such as cross-hardening loading-unloading loops, cyclic hardening as well as behavior in tension in the presence of shear stress, can be described quantitatively and accurately by the theory.

In a previous paper [1], a theory of viscoplasticity was developed on the basis of the concept that the current state of stress is a functional of the entire *history* of deformation and temperature,

> but the history was defined with respect to a time scale which is in itself a property of the material at hand.

For this reason we have termed the theory, 'endochronic'. In particular for the case of strain-rate-independent materials under isothermal conditions, which is the topic of this work, it was shown that within the restriction of small strains

$$\sigma_{ij} = \delta_{ij} \int_0^z \lambda(z-z') \frac{\partial \varepsilon_{kk}}{\partial z'} \, dz' + 2 \int_0^z \mu(z-z') \frac{\partial \varepsilon_{ij}}{\partial z'} \, dz' \tag{1.1}$$

where

$$\lambda^{(z)} = \lambda_\infty + \sum_{r=1}^{n} \lambda_r \, e^{-\rho_r z} \tag{1.2}$$

$$\mu(z) = \mu_\infty + \sum_{r=1}^{n} \mu_r \, e^{-\alpha_r z} \tag{1.6}$$

where λ_∞, λ_r, μ_∞, ρ_r and α_r are positive constants and

$$z = z(\zeta); \qquad \frac{dz}{d\zeta} > 0, \qquad z \geqq 0. \tag{1.4a, b}$$

The symbol z denotes a positive monotonically increasing *time scale* with respect to a *time measure* $d\zeta$ such that

$$d\zeta^2 = p_{ijkl} d\varepsilon_{ij} d\varepsilon_{kl} \tag{1.5}$$

where p_{ijkl} is a material tensor, which is positive definite and which, for the isotropic materials envisioned in Eq. (1.1), has the form

$$p_{ijkl} = k_1 \delta_{ij} \delta_{kl} + k_2 \delta_{ik} \delta_{il} \tag{1.6}$$

where k_1 and k_2 are material parameters, such that $k_1 + (k_2/3) \geqq 0$, $k_2 \geqq 0$.

Materials described by Eq. (1.1) are strain history dependent but strain rate independent. The derivation of constitutive equation (1.1) was given in detail and its relation to classical theory of plasticity[1]) was examined, in [1].

In the present paper we shall be concerned with the real behavior of metals under conditions of room temperature and slow straining. By examining experimental data on copper and aluminum, we shall show that Eq. (1.1) does indeed have the capability of explaining *quantitatively* and with accuracy such diverse phenomena as cross-hardening i.e. hardening in tension due to torsion, loading-unloading *loops*, hysteresis loops during repetitive tension-compression histories, as well as behavior in tension in the presence of shear stress due to torsion. The question of material compressibility in the plastic range will also be discussed.

At the present time one cannot find sufficient experimental data in the literature to determine the functions $z(\zeta)$, $\lambda(z)$ and $\mu(z)$. Therefore, in order to use the theory at all, with the object of interpreting *available* experimental data, we have sought other avenues, essentially heuristic, to determine the form of the above functions.

It will be recalled that,

$$\zeta = \int_0^\zeta \{p_{ijkl} d\varepsilon_{ij} d\varepsilon_{kl}\}^{\frac{1}{2}} \tag{1.7}$$

[1]) By 'classical theory of plasticity' we imply what is commonly understood as the incremental theory with a cylindrical isotropically expanding yield surface of Huber-Mises type.

i.e. ζ is a FUNCTIONAL of the strain history. We set[1])

$$dz = \frac{d\zeta}{f(\zeta)}. \tag{1.8}$$

At this point various possibilities present themselves. The simplest is to take $f(\zeta)$ = constant, in which case,

$$z = \zeta_1 \zeta + \zeta_0 \tag{1.9}$$

where ζ_0 and ζ_1 are constants. However it can be shown (and will be demonstrated in later sections) that this choice eliminates 'cross-hardening' in the sense that a change in the uniaxial stress-strain behavior due to shear prestrain cannot be accommodated. This effect has been observed in practically all experiments on metals that have ever been reported. Therefore, though Eq. (1.9) is convenient, it is not very useful.

This observation applies to Ilyushin's [3] and Rivlin's [4] theory, where $z = \zeta + \zeta_0$ and

$$\zeta = \int_0^\zeta \{d\varepsilon_{ij} d\varepsilon_{ij}\}^{\frac{1}{2}}. \tag{1.10}$$

Note that Eq. (1.10) is a particular case of Eq. (1.7); also note the absence of the material parameters k_1 and k_2 from Eq. (1.10), which renders ζ independent of the material at hand. In other words their theories are not *endochronic*.

The next simplest choice is to consider $f(\zeta)$ to be a linear function of ζ i.e.

$$f(\zeta) = 1 + \beta\zeta \tag{1.11}$$

where β must be a positive constant if $dz/d\zeta$ is to remain positive for all ζ. As a result of Eq.'s (1.8) and (1.11),

$$z = \frac{1}{\beta} \log(1 + \beta\zeta) \tag{1.12}$$

an expression which has been found to give satisfactory agreement in cases of some significant experiments, to be presented in subsequent Sections.

In the absence of experimental data, the question of the form of the

[1]) This step is guided by considerations of entropy generation and was explained in detail in Ref. 2.

'relaxation' functions $\lambda(z)$ and $\mu(z)$ is equally difficult. There are two simplifying assumptions, however, which lead to a relation between $\lambda(z)$ and $\mu(z)$, so that one is left with the problem of finding the form of only one of these functions. One is that of an elastic hydrostatic response and the other is the assumption of constant Poisson's ratio.

In classical plasticity, elastic hydrostatic response is, of course, taken for granted and the physical plausibility of this assumption makes it particularly attractive. However, the assumption of constant Poisson's ratio (not necessarily equal to one half) also has analytical advantages. In effect then, under condition of plane stress, or uniaxial strain, the strain in the unstressed direction is related to the strains in the stressed directions by a multiplicative constant. Thus the strain increments in the direction of zero stress may be readily eliminated from the expression for $d\zeta$ so that the latter may be expressed solely in terms of the strain increments in the stressed directions. Thus in order to decide which of the two assumptions to adopt we resorted to experiments on copper and aluminum. The results of these experiments are discussed below.

2. Material compressibility under hydrostatic tension

The experiments, simple in nature, consisted in monitoring the axial stress as well as the axial strain and one transverse strain of a thin flat bar in uniform uniaxial tension[1]). This simple test has enjoyed a great deal of popularity but we could find only one instance [5] where 'transverse strain' measurements were made in the plastic range. This is surprising, since this test affords a very critical evaluation of the assumption of plastic incompressibility.

The classical plasticity theory yields the following result, assuming elastic compressibility and a Von-Mises yield surface:

$$-\frac{d\varepsilon_y}{d\varepsilon_x} = \tfrac{1}{2} - (\tfrac{1}{2} - \nu_0)\frac{E_t}{E_0} \tag{2.1}$$

where ε_x and ε_y are the longitudinal and transverse strains respectively, E_0 and ν_0 are the elastic Young's modulus and Poisson's ratio respectively, and E_t is the tangent modulus of the tensile stress-strain curve.

[1]) Detailed description of the apparatus and the experimental procedure is to be found in Ref. 10.

Eq. (2.1) can be integrated numerically to yield a relation between ε_y and ε_x. This has been done for aluminum and copper.

The aluminum specimens were sheared from a 1100–0 aluminum sheet and annealed at 600°F for two hours and then oven cooled to room temperature. The copper specimens were cut out from an electrolytic tough pitch copper 110(99.9+ % cu) bus bar and annealed at 750°F and then oven cooled to room temperature.

The results are shown in Figs. 1 and 2. It may be seen that the classical plasticity theory gives a poor prediction of transverse strain response particularly for aluminum.

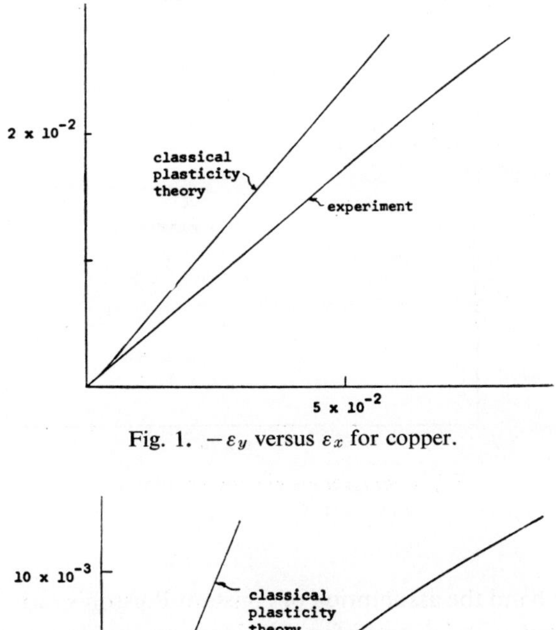

Fig. 1. $-\varepsilon_y$ versus ε_x for copper.

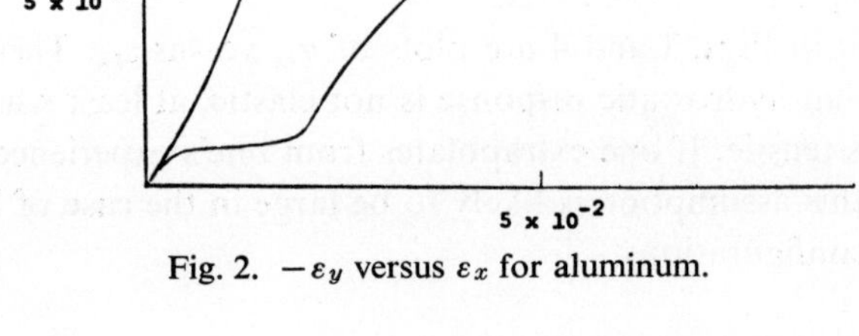

Fig. 2. $-\varepsilon_y$ versus ε_x for aluminum.

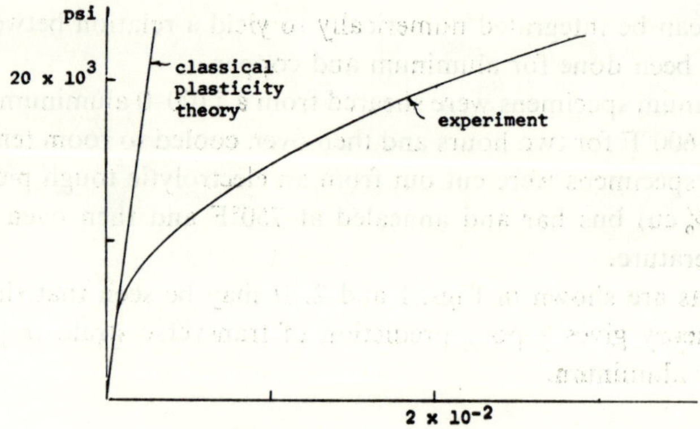

Fig. 3. σ_{kk} versus ε_{kk} for copper.

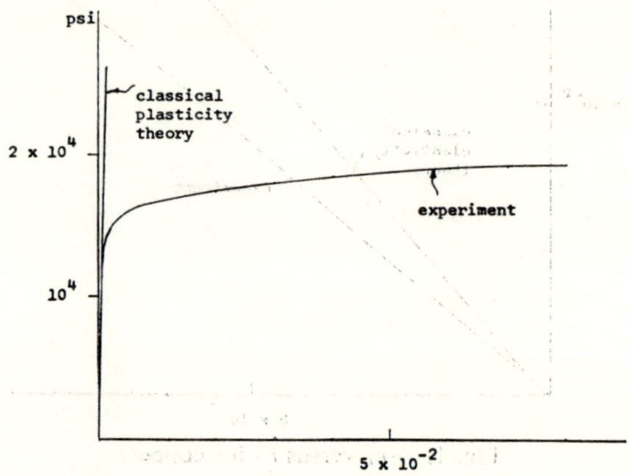

Fig. 4. σ_{kk} versus ε_{kk} for aluminum.

On the other hand the assumption of constant Poisson's ratio is quite good over a wide range of axial strain for aluminum and is almost exactly true in the case of copper. For this reason the assumption of constant Poisson's ratio was adopted in the analysis.

Also shown in Figs. 3 and 4 are plots of σ_{kk} versus ε_{kk}. These plots show very vividly that hydrostatic response is not elastic, at least when the hydrostatic stress is tensile. If one extrapolates from one's experience in elasticity, the effect of this assumption is likely to be large in the case of kinematically constrained configurations.

3. Crosshardening in tension-torsion

It has been observed in aluminum and copper as well as in other metals, that prestraining in torsion, well into the plastic range, has a significant hardening effect on the stress strain curve in tension.

In this Section we shall analyze data by Mair and Pugh [6], who have investigated this effect on annealed copper. Their experiments were performed on very thin circular cylinders which were twisted well into the plastic region, so that upon unloading there remained a permanent residual shear strain. The effect of initial shear prestrain on the tensile response was then obtained by loading the cylinders in tension.

The constitutive equations pertinent to the above situation are:

$$\sigma = \int_0^z E(z-z') \frac{de}{dz'} dz' \tag{3.1}$$

$$\tau = 2\int_0^z \mu(z-z') \frac{d\eta}{dz'} dz' \tag{3.2}$$

where σ and ε are the axial stress and strain, respectively, and τ and η are the respective shear stress and tensorial shear strain; the moduli $E(z)$ and $\mu(z)$ are interrelated through the bulk modulus $K(z)$. Their relation is best expressed through their Laplace transforms, i.e.,

$$\bar{E} = \frac{3\bar{\mu}}{1+(\bar{\mu}/3\bar{K})}. \tag{3.3}$$

A constant Poisson ratio leads to the result

$$E(z) = E_0 G(z), \qquad E_0 = \frac{3\mu_0}{1+(\mu_0/3K_0)} \tag{3.4, 5}$$

where $G(z)$ was taken simply as

$$G(z) = e^{-\alpha z}. \tag{3.6}$$

Comparison with experimental data

In the tension-torsion test the effect of constant Poisson ratio is to reduce $d\zeta^2$ to the form

$$d\zeta^2 = k_1 d\varepsilon^2 + k_2 d\eta^2 \tag{3.7}$$

where k_1 and k_2 are material constants, not the same as those in Eq. (1.6).

During torsion ($\varepsilon = 0$),

$$\zeta = k_2 \eta \tag{3.8}$$

whereas during tension ($\eta = \eta_0$) and after pretorsion

$$\zeta = k_2 \eta_0 + k_1 \varepsilon \tag{3.9}$$

where η_0 is the maximum shear prestrain.

Equation (3.1) may now be written in the form

$$\sigma = E_0 \int_{\zeta_0}^{\zeta} G[z(\zeta) - z(\zeta')] \frac{\partial \varepsilon}{\partial \zeta'} \, d\zeta' \tag{3.10}$$

(where $\zeta_0 = k_2 \eta_0$) since $\varepsilon = 0$ in the range $0 \leqq \zeta \leqq k_2 \eta_0$. Thus cross-hardening is taken fully into account by Eq. (3.10), through the shear prestrain parameter ζ_0, which appears as a lower limit on the integral on the right hand side of Eq. (3.10). If, in particular, we assume that $G(\zeta)$ is given by Eq. (3.6) and use of this is made in Eq. (3.10) the latter becomes

$$\sigma = E_0 \, e^{-\alpha z(\zeta)} \int_{\zeta_0}^{\zeta} e^{\alpha z(\zeta')} \frac{\partial \varepsilon}{\partial \zeta'} \, d\zeta'. \tag{3.11}$$

The integral in the right hand side of Eq. (3.11) can be evaluated explicitly by using Eq. (2.9) and noting that during monotonically increasing extension $d\varepsilon/d\zeta' = k_1$. Omitting the algebra,

$$\sigma = \frac{E_0(1+\beta\zeta)}{k_1 \beta n} \left\{ 1 - \left(\frac{1+\beta\zeta}{1+\beta\zeta_0} \right)^{-n} \right\} \tag{3.12}$$

where

$$n = 1 + \frac{\alpha}{\beta} \tag{3.16}$$

and

$$\zeta_0 \leqq \zeta = \zeta_0 + k_1 \varepsilon. \tag{3.14}$$

Equation (3.12) represents a family of stress strain waves in tension, in terms of the prestrain parameter ζ_0 and the 'cross-hardening' parameter[1]) β.

[1]) There is ample justification for calling β the cross-hardening parameter. Indeed in the limit of $\beta = 0$, and using Eq. (3.9) Eq. (3.12) becomes:

$$\sigma = \frac{E_0}{k_1 \alpha} (1 - e^{-\alpha \varepsilon})$$

which is independent of ζ_0; in other words cross-hardening cannot take place when $\beta = 0$, as pointed out earlier.

To determine the material parameters in Eq. (3.12) we note that in the absence of shear prestrain ($\zeta_0 = 0$),

$$\sigma = \frac{E_0(1+\beta_1\varepsilon)}{\beta_1 n}\{1-(1+\beta_1\varepsilon)^{-n}\} \tag{3.15}$$

where $\beta_1 = k_1\beta$.

It may be verified that as $\varepsilon \to 0$, $\sigma = E_0\varepsilon$ i.e. E_0 is the initial slope of the stress-strain curve. Also as ε increases, σ tends symptotically to the linear expression

$$\sigma = \frac{E_0}{\beta_1 n}(1+\beta_1\varepsilon). \tag{3.16}$$

If E_t is the slope (tangent modulus) of the asymptotic straight line, then

$$n = \frac{E_0}{E_t}. \tag{3.17}$$

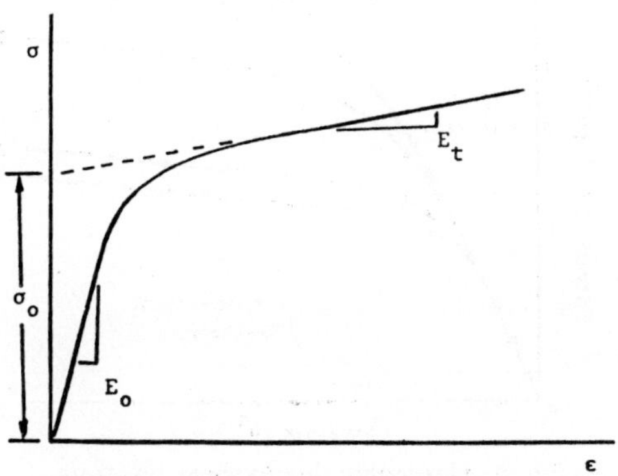

Fig. 5. Typical stress-strain curve.

Also, as shown in Fig. 5, if one extrapolates backwards the asymptotic straight line to intersect the stress axis, one obtains an intercept σ_0 from which β_1 is determined by the relation

$$\beta_1 = \frac{E_t}{\sigma_0}. \tag{3.18}$$

Similarly integration of Eq. (3.2) yields an equation analogous to Eq. (3.16);

this is

$$\tau = \frac{2\mu_0(1+\beta_2\eta)}{\beta_2 n}\{1-(1+\beta_2\eta)^{-n}\} \tag{3.19}$$

where $\beta_2 = k_2\beta$. Thus, β_2 and μ_0 may be determined from Eq. (3.19).

Finally we observe from Eq. (3.12) that the intercepts σ_0' in the presence of shear prestrain ζ_0 are given from the expression

$$\sigma_0' = \sigma_0(1+\beta\zeta_0) = \sigma_0(1+\beta_2\eta_0). \tag{3.20}$$

Equation (3.20) was used to confirm the self-consistency of the theory. However Eqs. (6.12), (3.1) and (3.19) can only yield the ratio (k_1/k_2) but the constants k_1 and k_2 cannot be evaluated. In this sense, and for these experiments one may choose k_2 arbitrarily; we chose $k_2 = 1$.

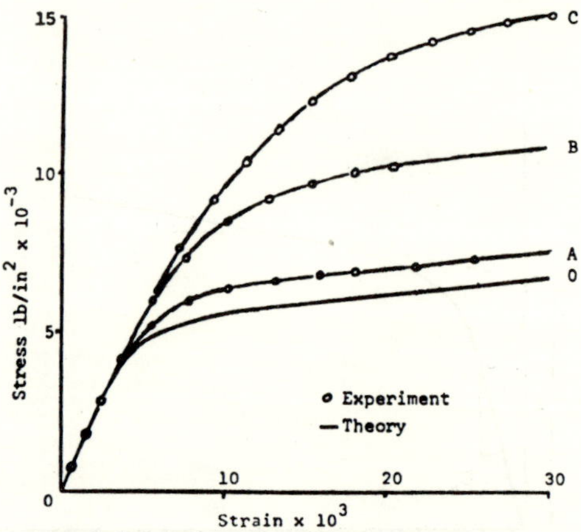

Fig. 6. Hardening due to shear prestrain.

Experimental data obtained by Mair and Pugh [6] that illustrate the effect of cross-hardening are given in Fig. 6.

Curve 0 is the virgin stress strain curve for the type of copper they used. The circles on the curves A, B and C are experimental points corresponding to initial shear prestrains of 0.25×10^{-2}, 1.5×10^{-2} and 6×10^{-2} respectively.

From curve 0, $E_0 = 14 \times 10^6$ lb/ln^2, $\beta_1 = 0.53 \times 10^2$, $n = 46$. With $k_2 = 1$, Eq. (3.20) was used to give $k = 1.00$. The curves A, B and C were then calculated and plotted as shown. The agreement between theory and experiment is remarkable.

4. Repetitive loading-unloading cycles

The tensile strain history $\varepsilon(\zeta)$ corresponding to a typical tensile straining-unstraining sequence is shown in Fig. 7. We use the terms 'straining' and 'unstraining' in their obvious sense. Points on the ζ-axis denoted by ζ_r $(r-1, 2\ldots)$ represent points of discontinuity in $d\varepsilon/d\zeta$, brought about by reverting from straining to unstraining histories, or vice-versa.

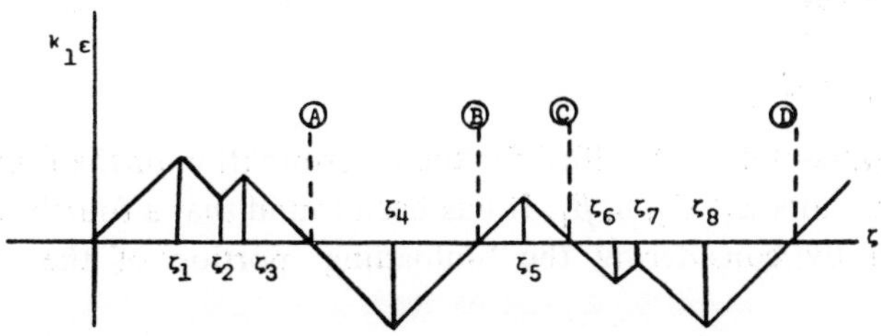

Fig. 7. Typical loading-unloading-loading sequence.

A perusal of experimental data on copper, shows that the constitutive equation of the metal varies depending on its previous history of manufacture and subsequent annealing. The single term form of $G(\zeta)$ that explained Mair and Pugh's data [6] so well was found inadequate to explain data by Lubahn [7] and by Wadsworht [8]. In this case we used,

$$G(z) = G_1 + G_2\,e^{-\alpha z}, \qquad E(z) = E_1 + E_2\,e^{-\alpha z}. \tag{4.1, 2}$$

Let ζ_n $(m = 1, 2\ldots)$ be the last point of discontinuity in $d\varepsilon/d\zeta$. Then using Eqs. (2.9), (3.11) and (4.1) and in the range $\zeta_m \leqq \zeta$

$$\sigma = (1+\beta\zeta)\sigma_0\left\{(-1)^m - \frac{1}{(1+\beta\zeta)^n} + 2\sum_{r=1}^{m}(-1)^{r+1}\left(\frac{(1+\beta\zeta_r)}{1+\beta\zeta}\right)^n\right\} + E_1\varepsilon. \tag{4.3}$$

The quantities ζ_r may be evaluated explicitly in terms of ε_r (the values of strain corresponding to ζ_r) by the formula

$$\zeta_r = 2k_1\sum_{s=1}^{r}(-1)^{s-1}\varepsilon_s + k_1(-1)^r\varepsilon_r. \tag{4.4}$$

The effect of E_1 on the unstraining characteristics is remarkable, especially since its effect on the shape of the straining part of the stress-strain curve is minimal. Let the history $\varepsilon(z)$ be one of continuous straining. Then Eq. (4.3) becomes:

$$\sigma = E_1\varepsilon + \frac{E_2(1+\beta_1\varepsilon)}{n\beta_1}\{1-(1+\beta_1\varepsilon)^{-n}\}. \tag{4.5}$$

From Eq. (4.5) we obtain the following relations in the notation of Section 3:

$$E_1+E_2 = E_0, \tag{4.6a}$$

$$\frac{E_2}{\beta_1 n} = \sigma_0, \tag{4.6b}$$

$$E_1+\sigma_{01} = E_t. \tag{4.6c}$$

Equations (4.6a-c) do not suffice for the determination of the four unknown material constants E_1, E_2, n, β_1. It has been found that a fourth relation can be obtained by considering the 'unloading' portion of the stress-strain history.

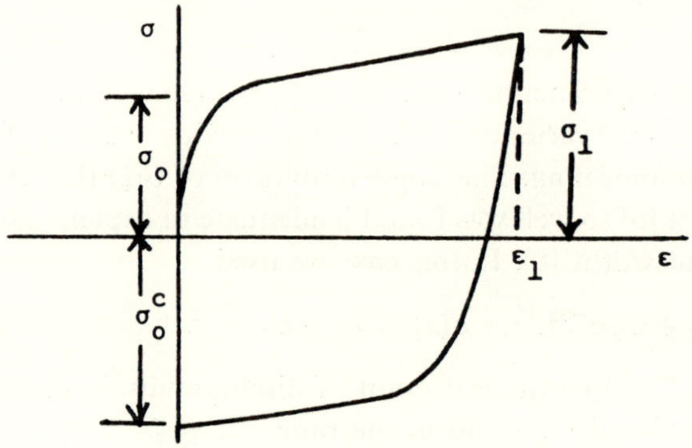

Fig. 8. Tension-compression test.

Fig. 8 shows the stress-strain relation for a uniaxial specimen which has been strained in tension to a strain value ε_1; whereupon it is unloaded and compressed until the final strain is zero. The strain-intrinsic time measure history $\varepsilon(\zeta)$ corresponding to the above stress-strain history is also shown in Fig. 9.

Equation (4.3) in conjunction with the above history yields the relation,

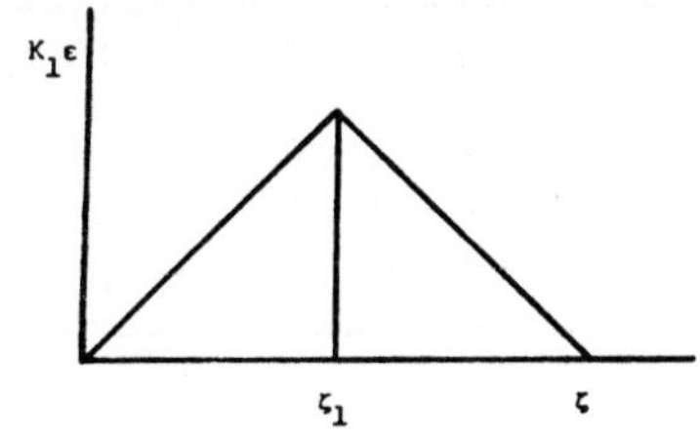

Fig. 9. Strain history of fig. 6.

at $\varepsilon = 0$:

$$\sigma = \sigma_0(1+\beta\zeta)\left\{-1+\frac{2(1+\beta\zeta_1)^n-1}{(1+\beta\zeta)^n}\right\}. \tag{4.7}$$

If the value of ε_1 is sufficiently large (in the case of copper this vlaue was found to be 50×10^{-3}, or so) then σ_0^c is given very nearly by the expression

$$\sigma_0^c = -\sigma_0(1+2\beta\zeta_1) = -\sigma_0(1+2\beta_1\varepsilon_1). \tag{4.8}$$

The constant β_1 can now be obtained from Eq. (4.8) and the constants E_1, E_2 and n can be found from Eqs. (4.6a–c).

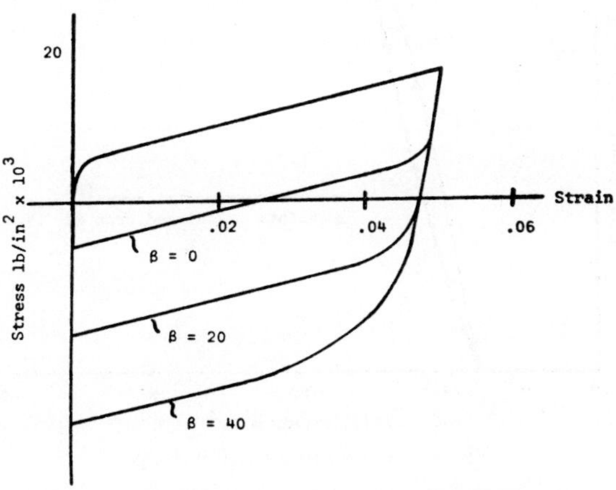

Fig. 10. Bauschinger effect and its relation to β.

We illustrate the points made in the above discussion in Fig. 10 where stress-strain curves for three different materials are given when these are subjected to the same strain history shown in Fig. 9.

The constants for these materials are given in the following table:

	E_1	E_2	β	n
1	0	6.14×10^6	0.4×10^2	25
2	0.12×10^6	6.02×10^6	0.2×10^2	50
3	0.24×10^6	5.9×10^6	0	∞

In Fig. 11 we illustrate an attempt to predict analytically the loading-unloading-loading response of copper in simple tension. The solid line is an experimental curve obtained by Lubahn [7] for a copper specimen which had already undergone similar strain cycles. We have assumed, however, that these have a negligible effect in the response shown because they occurred sufficiently far in the distant 'past'.

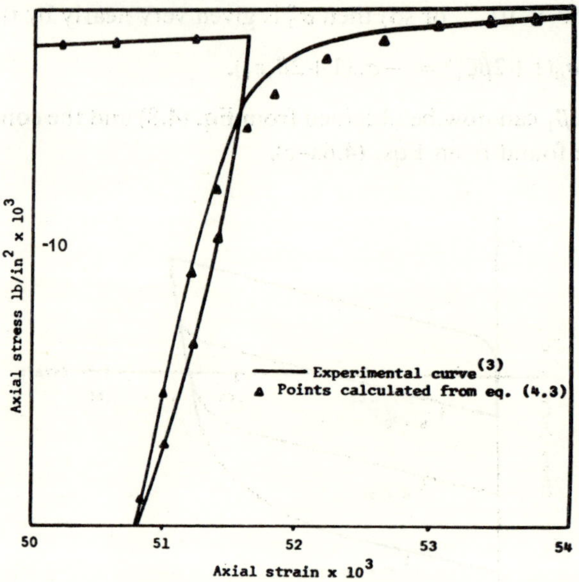

Fig. 11. Unloading-loading loop.

The triangular points shown, were obtained theoretically from Eq. (4.3) by assuming that the specimen was continually extended (without unstraining) until the strain $\varepsilon = 51.6 \times 10^{-3}$ was reached. The unstraining-straining cycle was then applied.

Despite the fact that $E(z)$ was approximated by two terms, as in Eq. (4.2) the agreement between theory and experiment is good. The constants employed were, $\sigma_0 = 3 \times 10^3$, $E_1 = 0.125 \times 10^6$, $\beta = 0.02 \times 10^3$, $n = 130$.

We conclude this Section by considering the effect of work hardening under cyclic straining. In particular we examine the work of Wadsworth's [8] and show that our theory again provides a sound analytical basis for his results. In this work single copper crystals were tested under conditions of uniaxial cyclic strain. The data was presented in terms of the resolved shear stress and strain in the plane of slip.

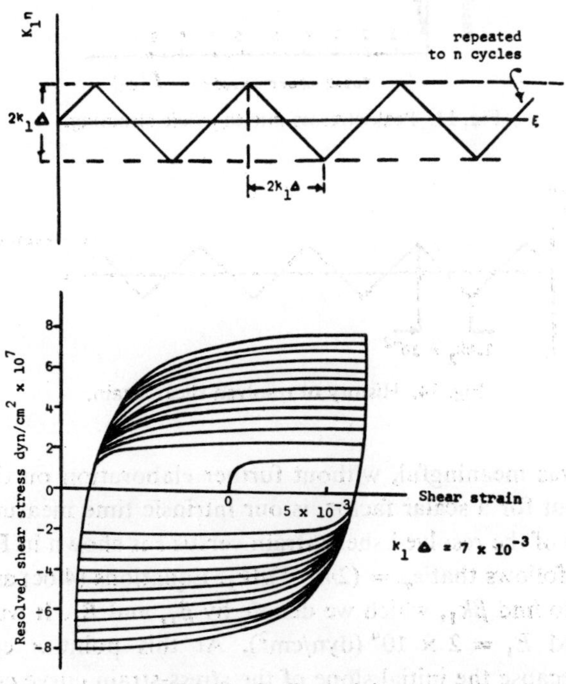

Fig. 12. Cyclic straining.

Fig. 12 gives the first few cycles of his straining program, in which a crystal was cycled under fixed limits of resolved shear strain of 7×10^{-3}. The 'peak stresses' corresponding to the extreme values of tensile and compressive strain increased monotonically with the number of cycles.

In Fig. 13 the values of peak tensile and compressive stresses have been plotted by Wadsworth against $|d\eta|$. It is rather interesting that he felt that

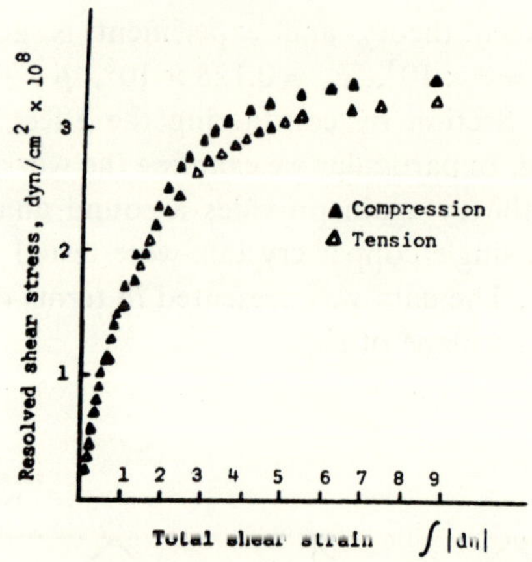

Fig. 13. Peak stresses due to cylcic straining.

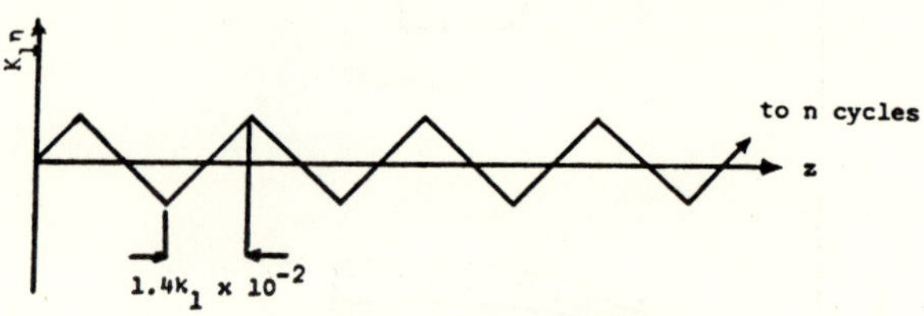

Fig. 14. History of resolved shear strain.

such a plot was meaningful, without further elaboration on this point. Of course $|d\eta|$, but for a scalar factor, is our intrinsic time measure.

The history of the resolved shear strain versus ε is shown in Fig. 14. From this Figure it follows that $\varepsilon_m = (2m-1)\Delta k_1$. Equations (4.6c) and (4.8) were now utilized to find βk_1, which we denote by β_1, and E_1. It was found that $\beta_1 = 12.3$ and $E_1 = 2 \times 10^9$ (dyn/cm^2). At this point n could not be determined because the initial slope of the stress-strain curve corresponding to $\xi < \xi_1$ could not be evaluated accurately.

However letting $\tau_m \equiv \tau|_{\xi=\xi,m}$ it was found that as $m \to \infty$, Eq. (4.3) yields the asymptotic expression:

$$\tau = n\beta\tau_0\Delta + E_1\Delta. \tag{4.9}$$

Hence, from the tensile experimental curve of Fig. 4 of Ref. 4, n could be determined explicitly and was found to be equal to 225. For this vlaue of n the term $1/(1+\beta\zeta_m)^n$ was found to be negligible for $m \geqq 1$.

Thus for the history in Fig. 14, Eq. (4.3) gives

$$\tau_{m+1} = \tau_0[1+\beta_1(2m+1)\Delta]\left\{(-1)^m + 2\sum_{r=1}^{m}\left[\frac{1+(2r-1)\beta_1\Delta}{1+(2m+1)\beta_1\Delta}\right]^n (-1)^{r+1}\right\} + E_1\varepsilon. \tag{4.10}$$

The above equation can be simplified further for large values of m. In particular for $m \geqq 50$, it was found that the series in the bracket on the right hand side of Eq. (4.10) degenerates into the geometric series

$$\sum_{r=1}^{m}(-u_m)^r \sim -\frac{u_m}{1+u_m}, \qquad u_m = \left[\frac{1+(2m-1)\beta_1\Delta}{1+(2m+1)\beta_1\Delta}\right]^n. \tag{4.11, 12}$$

Equation (4.10) may now be written in an symptotic form in terms of the absolute value of the shear stress as follows:

$$|\tau_m| = \{1+\beta(2m+1)\Delta\}\tau_0\left\{-1+\frac{2}{1+u_m}\right\} + E_1\Delta. \tag{4.13}$$

For very large values of $m(m \gg n)$ Eq. (4.13) simplifies further and becomes

$$|\tau_m| = \frac{n\beta_1\tau_0\Delta}{1-(n/m)} + E_1\Delta. \tag{4.14}$$

Thus,

$$\lim_{m\to\infty} |\tau_m| = n\beta_1\tau_0\Delta + E_1\Delta. \tag{4.15}$$

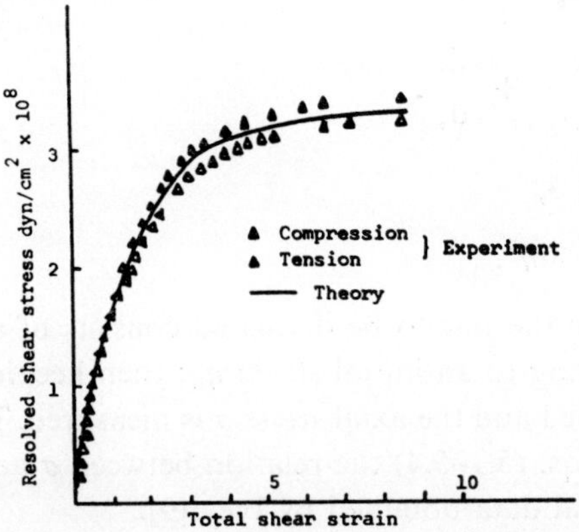

Fig. 15. Theoretical prediction of cylcic hardening.

In Fig. 15 a plot has been made of the theoretical relation between $|\tau_m|$ and $m\Delta$ obtained from Eq. (4.13). The experimental points obtained by Wadsworth are also shown. The following comments are in order. Though our theory does give values for τ_m which are different in tension from those in compression, the difference is not as great as the experimental data indicate, and is too small to be plotted on the scale shown. However, the theoretical curve lies very close to, and is in fact bounded by the experimental points, which indicate a deviation between the values of compressive stress and those of tensile stress which increases with m but is never greater than 5.5% per cent

This is the first time that a theory of plasticity has provided a rational explanation for the phenomena of cyclic hardening.

5. Tensile response in the presence of initial shear stress

In this Section we shall examine the effect of initial constant *prestress* in torsion on the stress-strain curve in tension. To do this, we have assumed, just as we did in Section 3, that $E(z)$ and $\mu(z)$ are proportional to the same relaxation function $G(z)$, and furthermore that $G(z)$ consists of a single exponential term i.e. it is given by Eq. (3.6). Thus

$$E(z) = E_0\, \mathrm{e}^{-\alpha z}. \tag{5.1}$$

In the light of Eq. (5.1) and bearing in mind Eq. (2.9), the constitutive Eqs. (3.1) and (3.2) can be reduced to the differential equations

$$E_0 \frac{\mathrm{d}\varepsilon}{\mathrm{d}\zeta} = \frac{\alpha\sigma}{1+\beta\zeta} + \frac{\mathrm{d}\sigma}{\mathrm{d}\zeta}, \tag{5.2}$$

$$2\mu_0 \frac{\mathrm{d}\eta}{\mathrm{d}\zeta} = \frac{\alpha\tau}{1+\beta\zeta} + \frac{\mathrm{d}\tau}{\mathrm{d}\zeta}, \tag{5.3}$$

where, as in Section 3,

$$\mathrm{d}\zeta^2 = k_1 \mathrm{d}\varepsilon^2 + k_2 \mathrm{d}\eta^2. \tag{5.4}$$

As mentioned above the test to be discussed consists of applying an initial stress τ^0 corresponding to an initial strain η_0; then keeping τ^0 constant, an axial strain ε is applied and the axial stress σ is measured. The object at hand is to deduce from Eqs. (5.1–5.4) the relation between σ and ε, and compare with the experimental data obtained by Ivey [9].

From Eq. (5.4) it follows that the axial straining process begins at $\zeta = \zeta_0$

where

$$\tau_0 = k_2\eta_0. \tag{5.5}$$

During this process $d\tau/d\zeta = 0$, so that from Eq. (5.3)

$$d\eta = \frac{\alpha\tau^0}{2\mu_0}\,\frac{d\zeta}{1+\beta\zeta}. \tag{5.6}$$

Equations (5.4) and (5.3) now combine to show that during the axial straining process,

$$d\zeta^2 = k_1^2 d\varepsilon^2 + k_2^2\left(\frac{\alpha\tau^0}{2\mu_0}\right)^2 \frac{d\zeta^2}{(1+\beta\zeta)^2}. \tag{5.7}$$

Let $(k_2/k_1) = k$, $k_1\beta = \beta_1$ and $c = (k\alpha\tau^0/2\mu_0)$.

Eqs. (5.2) and (5.7) then yield the following relation between σ and ε (true for large n, $n > 30$), in terms of the parameters ϕ and ϕ_0:

$$\sigma = \frac{E_0 c}{\beta_1 n}\left\{\sinh\phi - \left(\frac{\cosh\phi_0}{\cosh\phi}\right)^{n-1}\sinh\phi_0\right\}, \tag{5.8}$$

$$\varepsilon = \frac{c}{\beta_1}\{F(\phi) - F(\phi_0)\} \tag{5.9}$$

where

$$F(\phi) \equiv \sinh\phi - \tan^{-1}(\sinh\phi), \tag{5.10}$$

$$1+\beta z = c\cosh\phi, \qquad \phi_0 = \cosh^{-1}\left\{\frac{(1+\beta_1 k\eta_0)}{c}\right\}. \tag{5.11, 12}$$

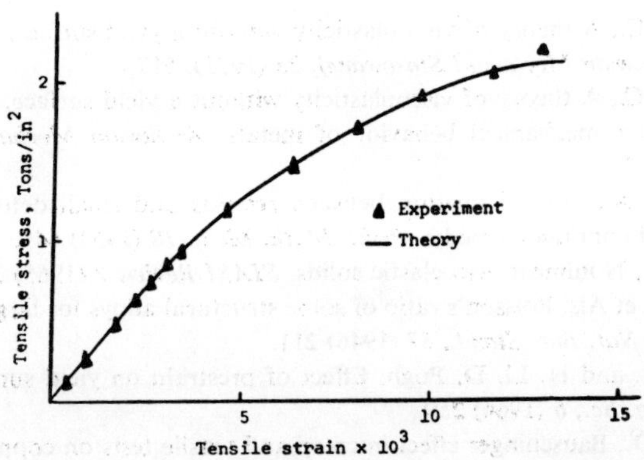

Fig. 16. Theoretical prediction of effect of shear prestress.

The relation between σ and ε has been calculated with the following values of the constants:

$$k = 1, \quad \sigma_0 = \frac{E_0}{\beta_1 n} = 17.4 \times 10^3 \, \text{lb/ln}^2, \quad \beta = 20, \quad n = 33.$$

The result was compared with one of Ivey's experiments in which $\tau^0 = 14 \times 10^3 \, \text{lb/m}^2$, $_0 = 2.35 \times 10^{-3}$. The predicted and experimental stress-strain responses compare very favorably. See Fig. 16.

6. Conclusion

It appears that the endochronic theory of plasticity, provides a rational analytical framework within which the mechanical response of copper and aluminum in the plastic range can be predicted in a number of significant cases quantitatively and accurately, with a single constitutive equation. The application of the theory to a broader class of phenomena will be the object of a future investigation.

7. Acknowledgement

This research was sponsored by the Air Force Office of Scientific Research, Office of Aerospace Research, United States Air Force, under AFSOR Grant Nr 70–1916.

References

[1] Valanis, K. C., A theory of viscoplasticity without a yield surface, Part I – General theory. *Archiwum Mechaniki Stosowanej*, *23* (1971), 517.

[2] Valanis, K. C., A theory of viscoplasticity without a yield surface, Part II – Application to the mechanical behavior of metals. *Archiwum Mechaniki Stosowanej*, *23* (1971), 535.

[3] Ilyushin, A. A., On the relation between stresses and small deformations in the mechanics of continuous media. *Prikl. Math. Meh.*, *18* (1954) 641.

[4] Rivlin, R. S., Nonlinear viscoelastic solids. *SIAM Review 7* (1965) 323.

[5] Stang, A. H. et Als, Poisson's ratio of some structural alloys for large strain. *Journal of Research*, *Nat. Bur. Stand.*, *37* (1946) 211.

[6] Mair, W. M. and H. Ll. D. Pugh, Effect of prestrain on yield surfaces in copper. *J. Mech. Eng. Sc.*, *6* (1964) 250.

[7] Lubahn, J. D., Bauschinger effect in creep and tensile tests on copper. *J. of Metals*, *205* (1955) 1031.

[8] Wadsworth, N. J., Work hardening of copper crystals under cyclic straining. *Acta Metallurgica, 11* (1963) 663.

[9] Ivey, H. J., Plastic stress-strain relations and yield surfaces for aluminum alloys. *J. Mech. Eng. Sc., 3* (1961) 15.

[10] Valanis, K. C. and Han-Chin Wu, *Material instability in the experimental study of the plastic compressibility of some important metals.* Mechanics and Hydraulics Report 1.03, University of Iowa, August, 1971.

Plane waves in elastic-plastic solids

Paul Germain

University of Paris, Paris, France

E. H. Lee

Stanford University, Stanford, U.S.A.

Analysis of the propagation of plane waves in elastic-plastic solids is considered, with particular emphasis on shock waves. It is shown that for satisfactory treatment of shock wave propagation, non-linear theory is needed with inclusion of thermodynamic characteristics. Finite deformation thermo-elastic-plastic theory is applied to this problem, which contrasts with the hydrodynamic theory commonly used for shock analysis in metals at extremely high pressure loading, where shear-stress effects can be neglected. For work-hardening materials, it is shown that shock calculations call for additional information beyond that required for the analysis of continuous loading problems, which is not so in hydrodynamic theory. An additional relation governing plastic work for discontinuous strains is suggested to overcome this difficulty, which can also be resolved in a more satisfactory manner by studying the shock structure in detail. The latter calls for rate dependent theory, and an analysis on this basis is also given.

Introduction

This paper is concerned with the propagation of plane waves in elastic-plastic materials, considering both continuous variations of stress and shock waves. Such waves can be conveniently generated experimentally by the impact of one flat plate on another or by explosive loading, and provide a means of measuring material characteristics [1][1], [2]. Plane waves also comprise a convenient configuration for application of constitutive theories of materials, and hence of testing their validity.

The discussion is formulated from the vantage point of shock-wave theory in fluid mechanics, since it is in that discipline that shock wave phenomena have been most fully studied, and the significance of inviscid analysis, and

[1]) Numbers in square brackets refer to the bibliography at the end of the paper.

the inclusion of dissipation effects, have been most completely understood. Moreover, hydrodynamic theory of shock waves in metals [1] has been shown to provide a satisfactory means of analysis of certain problems associated with loading at extremely high pressures. In these circumstances, shear stress effects become negligible compared with the energy transformations associated with dilatation, and so the stress tensor can be approximated by the isotropic tensor: hydrostatic pressure, and a fluid type theory results. In this paper we develop a theory for wave propagation in elastic-plastic solids which specializes naturally to the hydrodynamic model in the limit of high applied pressures. This property is likely to provide a convenient extension of hydrodynamic theory valid when shear effects are small but not negligible.

A system governed by conservation laws has continuous solutions of the corresponding partial differential equations, but there can exist surfaces across which the dependent variables are discontinuous. The conservation laws then yield jump conditions, and the associated discontinuous solutions represent shock waves. This phenomenon is in principle associated with non-linear behavior, where, for example, a non-linear convection effecs causes a wave front to steepen as in compressible fluid theory, where successive pressure increments are propagated with increasing speeds. This tendency of a wave front to steepen and eventually 'over shoot' or 'break' is inhibited by dissipation effects which become important for sufficiently steep gradients, even though the coefficients governing these influences are small enough to exert a negligible influence over most of the field. Thus the shock wave solution is the limit of rate independent analysis, but the effect of additional dissipation on the shock relations does appear as an inequality, usually connected with the second law of thermodynamics. It turns out, as discussed hereafter, that for plastic waves such a dissipation-reduced analysis cannot be deduced directly from rate-independent elastic-plastic theory. The term 'dissipation reduced' must be used in place of the common 'inviscid' adjective of fluid mechanics, since continuous plastic flow is dissipative, and a conservative assumption corresponding to inviscid fluid flow would be unsatisfactory. Thus, for plastic deformation, entropy increases in both continuous and discontinuous waves in contrast to the usual analysis of fluid problems.

For a more detailed study of shock phenomena, the discontinuity must be considered as a narrow region in which the dependent variables vary very rapidly. The space scale is expanded in this region to permit evaluation of the shock structure: the local variation in detail of stress and other state variables. For this study the dissipation characteristics are needed in contrast to the discontinuity approach.

From this general description of the development of shock wave theory, based mainly on developments in fluid mechanics, it is clear that a satisfactory analysis of plastic shock-waves must be based on a non-linear elastic-plastic theory, containing a complete thermodynamic description of the material characteristics. This should include both geometric and material nonlinearity. In this paper we use the constitutive relations presented in [3] and [4], and hence this application may provide a means of testing the validity of that approach by comparison with shock wave measurements. Space limitations prevent a complete development of this plastic shock-wave theory in the present paper, and a more complete treatment is expected to appear in due course [5].

Hydrodynamic theory and related analysis

As mentioned in the Introduction, for very high pressures applied on the surface of a metal, shear stress effects may become negligible, the stress tensor can then be replaced by hydrostatic pressure, and the constitutive equations reduce to those of a compressible fluid. Only two state variables are then needed, for example: density ρ and absolute temperature θ. Development of appropriate such laws governing deformation of metals is discussed in [1] and [6].

Although the resulting hydrodynamic theory is similar to classical gas dynamics, many of the usual results in the latter discipline arise from convexity properties of pressure volume relations, which can be violated even by the uniaxial motion of a non-linear elastic material, which is governed by a theory similar in mathematical structural to hydrodynamic theory. The usual geometric property that the curvature of the Hugoniot has constant sign can be disobeyed, leading to an essential change in interpreting the significance of the Hugoniot curve. For example, Fig. 1. shows the $(-T, \tau)$ plane for such a case taken from a discussion by Duvaut [7], where $-T$ is the compressive traction normal to the wave front, and $\tau = \rho^{-1}$ the specific volume. $(-T, \tau)$ corresponds to the pressure-volume plane in hydrodynamic theory. The curve $\mathscr{I}$ is the isentrope through the state A, having a single inflexion point at I. The Hugoniot curve $\mathscr{H}$ of A can be shown to intersect the isentrope $\mathscr{I}$ at a point, say F, and lie below $\mathscr{I}$ between A and F.

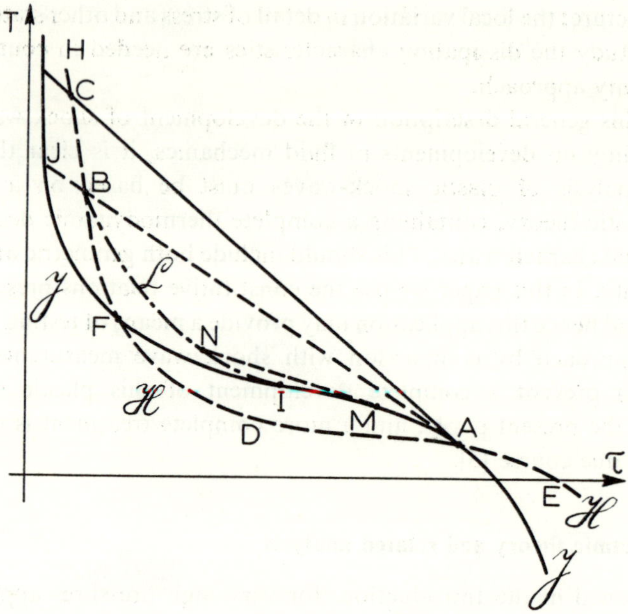

Fig. 1

Moreover, on any chord of $\mathscr{J}$ passing through A with a slope greater than that of the tangent ABJ to $\mathscr{J}$ at A, there is one and only one point of $\mathscr{H}$, for instance C. $\mathscr{H}$ is thus the curve $EADFBC$.

The part of the Hugoniot curve ADF lies below the isentrope $\mathscr{J}$, has a lower entropy, and hence cannot represent adiabatically compressed states for downstream tractions up to that corresponding to F. Nor can points on the Hugoniot segment FB represent downstream states for higher pressures, since the corresponding shock velocity, given by the gradient of the chord through A, would travel more slowly than small disturbances at A which travel with a velocity associated with the steeper gradient of the tangent to the isentrope at A. This implies that the simple discontinuity would be unstable. For down-stream tractions corresponding to an increase in pressure less than that corresponding to the inflexion point I, a continuous isentropic wave profile occurs. For down-stream pressures corresponding to tractions between those associated with points I and B, for example that corresponding to the point N, a continuous wave corresponding to the isentrope segment AM is followed by a discontinuity corresponding to MN moving with the velocity of the rear of the continuous part (MN is tangent to the isentrope at

M). For pressures above B and below A, the propagation of a single discontinuity yields states according to the Hugoniot curve $\mathscr{H}$. Thus for this circumstance, in contrast to classical gas dynamics, the locus of down-stream states associated with the upstream state A is given by the curve $EAMINBCH$, and not simply by the Hugoniot curve $\mathscr{H}$. The former locus, called the Hugoniot Contour or Hugoniot Indicatrix, comprises arcs of the Hugoniot curve joined by arcs of the isentrope, and Hugoniot curves for points on the isentrope other than A, for example M. Thus when a point of inflexion occurs in the isentrope, continuous waves as well as simple discontinuities must be studied to determine wave configurations corresponding to discontinuous applied pressure changes. For this reason, continuous wave solutions may need to be studied when considering waves generated by pressures applied discontinuously at the surface of a solid, as well as waves of discontinuity of traction. Both types are considered in the study of waves in elastic-plastic media presented in this paper.

The elastic-plastic law

The formulation presented in [3] and [4] is adopted here, since it provides a consistent nonlinear theory with a thermodynamic structure which includes energy dissipation in plastic flow, features required for a study of shock wave analysis as discussed in the Introduction.

We shall be concerned with the propagation of plane waves with traction vectors normal to the wave fronts, and hence can utilize the special case of Cartesian axes parallel to and normal to the wave fronts, which material isotropy and problem symmetry determine to be principal directions for stress and strain throughout the motion. As pointed out in [8], it is then convenient to use natural, or logarithmic, strain components. Elastic and plastic components are then additive at finite strains and effectively uncouple the elastic and plastic laws [9]. If λ_i ($i = 1, 2, 3$) are the stretch ratios (or components of the deformation gradient $\boldsymbol{F}$) along the axes x_i,

$$\varepsilon_i = \log \lambda_i, \tag{1}$$

$$\varepsilon_i = \varepsilon_i^e + \varepsilon_i^p \tag{2}$$

where superscripts e and p denote the elastic and plastic components, then for plane waves with motion in the x_1 direction normal to the wave fronts, inhibition of lateral strain by inertia effects determines:

$$\varepsilon_i = \varepsilon_i^e + \varepsilon^p = 0, \qquad i = 2, 3. \tag{3}$$

Equation (2) corresponds to the matrix product relation:

$$F = F^e F^p \tag{4}$$

where F^e and F^p are the deformation gradients associated with elastic and plastic strain respectively (in the present case, diagonal matrices with components λ_i^e and λ_i^p).

The theory presented in [3] and [4] for isotropic work hardening elastic-plastic behavior, prescribes the thermodynamic properties in terms of the free energy function per unit mass: $\Psi(F^e, \theta, w)$, where w is the total plastic work on an element per unit mass, and θ is the absolute temperature. The stress Σ and entropy s, are then given by the thermo-elastic relations

$$\Sigma = \rho F^e \frac{\partial \Psi}{\partial F^e}, \qquad s = -\frac{\partial \Psi}{\partial \theta}. \tag{5}$$

The yield locus is given by

$$\mathscr{F}(\rho_0 \rho^{-1} \Sigma, \theta) - \kappa(w) \leqq 0 \tag{6}$$

and no plastic volume change prescribes

$$\varepsilon_1^p + \varepsilon_2^p + \varepsilon_3^p = 0 \quad (\det F^p = 1). \tag{7}$$

For the one-dimensional motion case:

$$\dot{w} = \rho^{-1}(\sigma_1 - \sigma_2)\dot{\varepsilon}_1^p \tag{8}$$

where σ_i are the diagonal components of the stress tensor ($\sigma_2 = \sigma_3$ by symmetry), and (6) becomes

$$Y(w, \theta) \mp \rho_0 \rho^{-1}(\sigma_1 - \sigma_2) \geqq 0 \tag{9}$$

the upper sign referring to tensile plastic flow in the x_1 direction, the lower to compressive. Equations (8) and (9) yield:

$$\rho_0 \dot{w} = \pm Y(w, \theta)\dot{\varepsilon}_1^p \tag{10}$$

for plastic flow. Unloading, or neutral loading and elastic response give:

$$\dot{\varepsilon}_1^p = \dot{w} = 0. \tag{11}$$

For many metals over a moderate strain range, elastic properties are found to be insensitive to changes in plastic flow (see [9], p. 445) as long as the plastically deformed new natural state is used as the reference state at each instant, hence we can write:

$$\Psi = \bar{\Psi}(\boldsymbol{F}^e, \theta) + \hat{\Psi}(w) \tag{12}$$

and (5) then guarantees this property since $\boldsymbol{F}^e$ is the deformation gradient from this natural state.

The energy equation now becomes:

$$\rho\theta\dot{s} = \rho w\left(1 - \frac{\mathrm{d}\hat{\Psi}}{\mathrm{d}w}\right) - q_{i,i} \tag{13}$$

where q_i is the heat flux, and the second law of thermodynamics then takes the form:

$$\dot{w}\left(1 - \frac{\mathrm{d}\hat{\Psi}}{\mathrm{d}w}\right) \geqq 0 \tag{14}$$

which is satisfied for the selection adopted:

$$\hat{\Psi} = (1-\gamma)w, \qquad \gamma = 0.9 \tag{15}$$

which was shown by Farren and Taylor [10] to provide a good approximation for many metals. For continuous adiabatic motion

$$\theta\dot{s} = \gamma\dot{w} \tag{16}$$

so that $\gamma\dot{w}$ is the part of the plastic work which contributes to the increase in entropy, thus representing the irreversible character of plastic flow.

In selecting an appropriate form for $\bar{\Psi}$ in (12), we bear in mind the fact that hydrodynamic theory, which treats the material as a fluid, provides a satisfactory approximation at high pressures, and hence the elastic free energy function is expressed as the sum of a fluid-like term and a term depending on the stress deviator. Isotropic laws for both elastic and plastic deformation are assumed, hence the free energy must be a function of temperature θ, and of three invariants I_1, J_2 and J_3 defined by:

$$\begin{aligned} I_1 &= \varepsilon_1^e + \varepsilon_2^e + \varepsilon_3^e, \\ J_2 &= \tfrac{1}{2}[(e_1^e)^2 + (e_2^e)^2 + (e_3^e)^2], \\ J_3 &= \tfrac{1}{3}[(e_1^e)^3 + (e_2^e)^3 + (e_3^e)^3] \end{aligned} \tag{17}$$

where e_i^e are the principal values of the deviator of the strain tensor:

$$e_i^e = \varepsilon_i^e - \frac{I_1}{3}. \tag{18}$$

Thus we assume:

$$\bar{\Psi}(\varepsilon_1^e, \varepsilon_2^e, \varepsilon_3^e, \theta) = f(I_1, \theta) + h(J_2, J_3) \tag{19}$$

where we have dropped θ from the second term since thermo-mechanical coupling is associated with dilatation. Because of (2) and (7)

$$I_1 = \log\left(\frac{\rho_0}{\rho}\right), \tag{20}$$

ρ_0 being the uniform density in the reference configuration, and it is convenient to write the first term in (19) in the usual fluid form:

$$f(I, \theta) = \bar{f}(\rho, \theta). \tag{21}$$

Equation (5) yields:

$$\begin{aligned} \sigma_i &= \rho \frac{\partial \bar{\Psi}}{\partial \varepsilon_1^e} = \rho \frac{\partial f}{\partial I_1} + \rho \frac{\partial h}{\partial J_2} e_i^e + \rho \frac{\partial h}{\partial J_3} [(e_i^e)^2 - \tfrac{2}{3} J_2], \\ s &= \frac{\partial \bar{\Psi}}{\partial \theta} = -\frac{\partial f}{\partial \theta}. \end{aligned} \tag{22}$$

Thus the pressure, entropy, density and temperature correspond to those of an associated fluid defined by f, (19), and h determines the deviator components of the stress tensor. Finally we select h in (19) as:

$$\rho_0 h = 2\mu J_2 \tag{23}$$

which can be expected to provide a useful approximation [8]. Note, as discussed in [8], the single shear constant μ does not imply Hooke's law, since the geometrical nonlinearity associated with finite strain is included in (22).

We can now apply these elastic-plastic constitutive relations to the analysis of wave propagation.

Adiabatic continuous processes

For the plane waves under consideration, (3) and (7) and symmetry about the x_1 axis give the first of (22) in the form:

$$\begin{aligned} \sigma_1 &= -p + \frac{4\mu}{3} \frac{\rho}{\rho_0} \varepsilon_1 - \frac{2\mu\rho}{\rho_0} \varepsilon_1^p, \\ \sigma_2 = \sigma_3 &= -p - \frac{2\mu}{3} \frac{\rho}{\rho_0} \varepsilon_1 + \frac{\mu\rho}{\rho_0} \varepsilon_1^p \end{aligned} \tag{24}$$

where the first term of (22) gives

$$p = \rho \frac{\partial f}{\partial I_1} = p(s, \varepsilon_1); \tag{25}$$

using (20), (21) and the second of (22) to express

$$s = s(\varepsilon_1, \theta) \tag{25}$$

as an independent variable instead of θ, having used (2), (3) and (7) to eliminate the other strain components. This manipulation also yields:

$$\theta = \theta(s, \varepsilon_1) \tag{26}$$

and

$$\varepsilon_1 = \log \frac{\rho_0}{\rho}. \tag{27}$$

For states inside the yield surface, (11) and (16) require s and ε_1^p to remain constant, and (24) and (25) provide the stress strain relation governing the wave propagation.

When plastic flow occurs, (8), (10) and (16) combined with the yield condition (9), which according to (24) takes the form

$$\frac{3\varepsilon_1^p}{2} = \varepsilon_1 \mp \frac{Y}{2\mu}, \tag{28}$$

determine the differential relations:

$$\rho_0\theta \frac{\mathrm{d}s}{\mathrm{d}\varepsilon_1} = \rho_0 \quad \gamma \frac{\mathrm{d}w}{\mathrm{d}\varepsilon_1} = \pm Y\gamma \frac{\mathrm{d}\varepsilon_1^p}{\mathrm{d}\varepsilon_1}. \tag{29}$$

Using (26), (28) and $Y = Y(w, \theta)$, ε_1^p can be eliminated to give the differential equation system:

$$\rho_0\theta \frac{\mathrm{d}s}{\mathrm{d}\varepsilon_1} = \rho_0\gamma \frac{\mathrm{d}w}{\mathrm{d}\varepsilon_1} = \tfrac{2}{3}\gamma Y \frac{\pm 1 - \dfrac{1}{2\mu}\dfrac{\partial Y}{\partial \theta}\dfrac{\partial \theta}{\partial \varepsilon_1}}{1 + \dfrac{Y}{3\mu}\left(\dfrac{\gamma}{\rho_0\theta}\dfrac{\partial Y}{\partial \theta}\dfrac{\partial \theta}{\partial s} + \dfrac{1}{\rho_0}\dfrac{\partial Y}{\partial w}\right)} \tag{30}$$

for the determination of s and w as functions of ε_1. The stress components are then given by (24), with p from (25) and ε_1^p from (28).

This approach is greatly simplified for ideal plasticity, $Y = Y_0$, a constant, for (28) gives

$$\dot{\varepsilon}_1^p = \frac{2\dot{\varepsilon}_1}{3} \tag{31}$$

and (29) can be integrated directly with ε_1 as independent variable. Note that since the relations are rate independent, any variable changing monotonically with time, such as ε_1, can be used as a 'reduced time' independent variable,

If Y is a function of w only, independent of θ, then (10) determines ε_1^p by quadratures, and (28) determines ε_1.

A similar simplification arises for $Y = Y(\theta)$. Then ε_1^p can be eliminated from (28) and (29), utilizing (25a) to yield:

$$\frac{d\theta}{d\varepsilon_1}\left(\rho_0\theta\frac{\partial s}{\partial\theta}+\frac{\gamma}{3\mu}Y\frac{dY}{d\theta}\right) = \pm\gamma Y-\rho_0\theta\frac{\partial s}{\partial\varepsilon_1} \tag{32}$$

which is a differential equation to determine θ as a function of ε_1. The plastic strain component ε_1^p is then given by (28) and the stress components by (24).

In all these cases, when $\sigma_1(\varepsilon_1)$ has been determined, it can be combined with the equation of motion to permit evaluation of the wave profile for continuous waves.

Analysis for shock waves

The previous section gives a method of determining curves in the $(\sigma_1, \varepsilon_1)$ Hugoniot plane for continuous adiabatic processes. It was shown, by considering as an example the case of an elastic solid, that, in addition to these, the Hugoniot curves representing propagation of a discontinuity in stress are needed to give the Hugoniot contours, and hence wave systems for all choices of downstream pressure.

According to the assumptions (12), (15), (19), (21) and (23), the specific internal energy is given by:

$$e = \Psi+s\theta = \tilde{e}+\frac{2\mu J_2}{\rho_0}+(1-\gamma)w \tag{33}$$

where $\tilde{e}$ is the specific internal energy of the associated fluid.

The Hugoniot relation for energy jump across a shock wave takes the form

$$[e] = \bar{\sigma}_1\left[\frac{1}{\rho}\right] \tag{34}$$

where square brackets represent the discontinuous increase of a variable across the shock, and the bar the average value of the variable on the two sides.

If we assume that material on each side of the shock is at the plastic yield limit, as, for example, would be the case if a pre-cursor to the yield limit preceded the shock, σ_1 can be eliminated from (34), using the yield condition (9) in the form

$$\sigma_1 + p = \pm \frac{2\rho}{3\rho_0} Y \tag{35}$$

to give

$$\left[\tilde{e} + \frac{2\mu J_2}{\rho_0} + (1-\gamma)w\right] + \left(\bar{p} \mp \frac{2\overline{\rho Y}}{3\rho_0}\right)\left[\frac{1}{\rho}\right] = 0 \tag{36}$$

in which expression (33) has been used for e. Now

$$[\tilde{e}] + \bar{p}\left[\frac{1}{\rho}\right] = H(p, \rho) \tag{37}$$

is the Hugoniot function of the associated fluid, the contour $H = 0$ prescribing the corresponding Hugoniot curve. By expressing $[J_2]$ in terms of strain components and utilizing the yield condition, $[J_2]$ is given by:

$$\frac{2\mu}{\rho_0}[J_2] = \frac{1}{3\mu\rho_0}\bar{Y}[Y]$$

and hence (36) can be written:

$$H(p, \rho) + \frac{1}{3\mu\rho_0}\bar{Y}[Y] + (1-\gamma)[w] \mp \frac{2}{3\rho_0}\overline{\rho Y}\left[\frac{1}{\rho}\right] = 0. \tag{38}$$

In the special case

$$\gamma = 1, \qquad Y = Y_0 \quad \text{(constant)}$$

(38) reduces to

$$H(p, \rho) \mp \frac{2Y_0}{3\rho_0}\bar{\rho}\left[\frac{1}{\rho}\right] = 0 \tag{39}$$

which gives the Hugoniot curve for the elastic-plastic material directly. A similar result is also valid if

$$\gamma = 1, \qquad Y = Y(\theta)$$

but in general when $\gamma \neq 1$ and Y depends on w, (38) does not provide sufficient information to determine the Hugoniot curve. Such a deficiency was pointed out in [8]. Thus the global approach, replacing the shock by a discontinuity and not examining the shock structure, fails, since no law is available which defines $[w]$ analogous to (10) for the continuous case.

An approach which resolves this difficulty is mentioned in the next section. by analyzing the shock structure on the basis of a rate dependent plasticity law. If we wish to restrict the discussion to the global approach, an additional relation must be assumed, analogous to (10) for the continuous case, and a mathematically convenient form is the finite difference analogue of (10):

$$\rho_0[w] = \overline{Y}[\pm\varepsilon_1^p] \tag{40}$$

which, using (28), can be written in the form:

$$\rho_0[w] = 2\overline{Y}\frac{[\pm\varepsilon_1]}{3} - \frac{\overline{Y}[Y]}{3\mu}. \tag{41}$$

This can be considered as a relation defining w implicitly for any pair of values of the thermodynamic state of the associated fluid. Combining this relation with [38], permits the Hugoniot curve to be determined.

Shock structure, rate effects and conclusions

It has been shown that whenever the yield stress or free energy depend on the plastic work w, it is not possible to determine the Hugoniot curve without introducing a new phenomenological relation to prescribe the change in w across the shock. This predicament suggests that the rate independent laws utilized for plastic flow are unsuitable for investigating conditions within the shock layer, and that rate effect terms are needed to represent behavior in this extremely high strain-rate region. This provides the possibility of enhancing the dissipation as required to properly express the shock phenomenon.

An approach which is related to the plastic rate-effect law given by Malvern [11], is to express the plastic strain rate in terms of the variable:

$$z = (\tfrac{3}{2}\varepsilon_1^p - \varepsilon_1)^2 - \frac{Y^2}{4\mu^2} \tag{42}$$

which according to (28) would be zero at yield and negative inside the yield surface according to the rate-independent theory used so far. However, (28) will be discarded and the plastic strain rate expressed in terms of z so that it grows extremely rapidly for positive z, and decreases rapidly towards zero for negative z. The growth is such for positive z that plastic flow occurs to reduce z to zero in a short time interval. Thus the material is effectively elastic ($z < 0$) or effectively at the yield limit ($z \simeq 0$) throughout most of the solution, but carries higher stress in the shock layer according to (24) and thus encompasses the enhanced dissipation there.

The above requirements can be achieved by writing

$$\zeta = \frac{z}{\delta}, \tag{43}$$

δ being a positive non-dimensional relaxation parameter, which can be chosen to be small, and

$$\dot{\varepsilon}_1^p = (\varepsilon_1 - \tfrac{3}{2}\varepsilon_1^p)k(\zeta) \tag{44}$$

where $k(\zeta) \to 0$ as $\zeta \to -\infty$, and $k(\zeta) \to \infty$ as $\zeta \to \infty$.

The shock structure is studied by drawing on the experience of other shock wave analyses, and assuming the motion in the very thin shock layer to be stationary with respect to a frame moving with the shock. Hence, for one-dimensional strain motion, all state variables can be expressed as functions of:

$$\xi = x_1 - Wt \tag{45}$$

where W is the shock wave velocity. The integrals of the conservation laws of mass, momentum and energy, combined with (44) and (24) then yield a pair of differential equations with independent variable ξ, and a pair of algebraic relations for the dependent variables ε_1, θ, ε_1^p and w. Solution of these equations provides the shock structure:

$$\sigma_1 = \sigma_1(\xi)$$

and the form of the differential equations is such that for certain situations ξ can be eliminated, and a differential relation involving $\mathrm{d}w/\mathrm{d}\varepsilon_1^p$ occurs, the solution of which provides the required additional shock condition, without the necessity of evaluating the shock structure. Details of the procedure will be found in [5], where it is shown that in certain circumstances this additional shock condition does not depend on the precise form of the rate-law, but only on the limits for large and small ζ given above. Moreover, as in the classical theory given in previous sections, simplifications arise for $\gamma = 1$ and Y independent of w.

This paper has presented a consistent theory of shock waves in elastic-plastic materials for circumstances when the pressure is not sufficiently high to permit hydrodynamic analysis to be used. Qualitative differences arise in the structure of the theory, the inclusion of shear effects and plasticity causing complications in both algebra and principle. For many applications, simplifications of this theory will be adequate, for example when shear effects have a minor but non-negligible influence, and it is expected that specializations sufficiently simple for practical applications will arise.

References

[1] Rice, M. H., R. G. McQueen and J. M. Walsh, *Solid state physics*. Vol. 6 Eds. F. Seitz and D. Turnbull, Academic Press, 1–63 (1958).
[2] Herrmann, W., *Wave propagation in solids*. Ed. J. Miklowitz, ASME (1969) 129–183.
[3] Lee, E. H. and D. T. Liu, *Jour. Appl. Phys. 38* (1967) 19–27.
[4] Lee, E. H., *Jour. Appl. Mech. 36* (1969) 1–6.
[5] Germain, P. and E. H. Lee, On shock waves in elastic-plastic solids. To appear in *Jour. Mech. Phys. Solids*.
[6] Zel'dovich, Y. B. and Y. P. Raizer, *Physics of shock waves and high temperature phenomena*. Vol. II, Chap. XI, Academic Press, 1967.
[7] Duvaut, G., *Jour. de Mécanique*, *6* (1967) 371–404.
[8] Lee, E. H. and T. Wierzbicki, *Jour. Appl. Mech.*, *34* (1967) 931–936.
[9] Lee, E. H., *Inelastic behavior of solids*. Eds. M. F. Kanninen et. al., 423–446, McGraw-Hill, 1970.
[10] Farren, W. S. and G. I. Taylor, *Proc. Roy. Soc., London, 143* (1934) 307–326.
[11] Malvern, L. E., *Jour. Appl. Mech.*, *18* (1951) 203–208.

Superplasticity: Constitutive equations and forming problems

A. R. Ragab and J. L. Duncan

McMaster University,
Hamilton, Ontario, Canada

Superplasticity is the nonlinear viscous behaviour of a number of alloys which have a characteristic, ultra-fine grain microstructure. The plasticity theories and constitutive equations generally employed in the analysis of stationary creep in structures are shown to be applicable to these materials. In a tensile process, non-uniformities develop continuously and the characteristics of superplastic deformation in such a process are illustrated by the results of numerical analysis of initially non-uniform tensile bars and thin walled tubes expanded by internal pressure.

Nomenclature

σ, $\dot{\varepsilon}$	unaxial stress and natural strain-rate
σ_e, $\dot{\varepsilon}_e$	representative stress and strain-rate
σ_0, $\dot{\varepsilon}_0$	material parameters ($\dot{\varepsilon}_0 = 1$)
m	strain rate sensitivity index
n	strain hardening index
L	grain size
D	diameter of tensile testpiece
ΔD	current (Dmax $-$ Dmin)
h	wall thickness of tube
H	nonuniformity of tube wall thickness
r_m	mid-wall radius of tube
R	diametral expansion

Introduction

Superplasticity is a distinct deformation behaviour exhibited by many alloys. Like creep, it occurs at elevated temperatures and is a time-dependent phenomenon, although superplastic deformation rates are some orders of magnitude greater than creep rates and approach those in hot working processes such as press forging. The subject is a relatively new one, and perhaps not familiar to many plasticians; in this paper an attempt is made to present a general description of the behaviour of superplastic alloys and to examine plasticity theories and constitutive equations used in the analysis of superplastic forming processes.

Phenomenologically, superplasticity is characterized by an unusually high rate-dependence approaching that of viscous liquids such as molten glass and heated thermoplastics and, in comparison with hot working of other alloys, a very low resistance to deformation at low strain rates. Many superplastic alloys also exhibit extreme resistance to fracture and as a consequence of these characteristics, the elongation in a controlled tensile test generally exceeds 500% and may be as high as 2000% [1]. Metallurgically, superplasticity is associated with an ultra-fine microstructure and it appears to be a general deformation phenomenon in many alloys in which this microstructure can be induced by thermal or mechanical methods and retained in a stable form at superplastic working temperatures.

A brief history [2] traces superplasticity to the work of Rosenhain [3], Hargreaves [4] and Pearson [1] in Britain, followed by Russian work [5–7] published between 1944 and 1961. Recent interest stems from a review by Underwood [8] and work by Backofen [9–12] which established the modern foundations of the subject in the 1960's. The literature is now extensive but in only a handful of papers has any attempt been made to apply the methods of continuum mechanics to technological processes of forming superplastic alloys. Industrially, the interest is in exploiting the viscous nature of these materials in processes similar to glass blowing and vacuum forming of thermoplastics and also in utilizing the low flow resistance in forging processes which resemble the injection molding of plastics [13].

Metallurgical considerations

The grain size of superplastic alloys is typically between 1 and 10 microns (m^{-6}) and the microstructure is unaffected by large deformations; individual grains do not elongate in the direction of extension. Various mechanisms have been proposed to account for this behaviour – principally grain boundary sliding, atomic diffusion and slip followed by recrystallization. The resistance to flow reduces with decreasing grain size and this is the converse of the behaviour of conventional alloys at lower temperatures [14].

Various constitutive equations can be derived assuming particular deformation mechanisms. Backofen and Avery [9–11] suggested the relation,

$$\dot{\varepsilon}=A\sigma/L^2+B\sinh\beta\sigma \tag{1}$$

which was extended by Packer and Sherby [15–16] to,

$$\dot{\varepsilon}=\frac{K'\sigma}{L^2}+\frac{A'\sigma^2}{L^3}\,(\mathrm{e}^{-\eta/\sigma})+B'\sigma^2\sinh\beta'\sigma^{0.25} \tag{2}$$

where $\dot{\varepsilon}$ is the true strain rate and σ is the true stress. In these equations, L is the grain size and A, B, β etc., are material parameters. As controversy exists regarding deformation mechanisms, the elaboration of equation (2) may not be justified. Following the approach in creep, a simpler relation has been suggested, namely,

$$\dot{\varepsilon}=\frac{\sigma^b}{L^a}\exp(-Q/kT) \tag{3}$$

where Q is the activation energy, T the absolute temperature and k Boltzmann's constant. A regression analysis may be used to determine parameters in equation (3) from test data; one approach is given in ref. [17].

Phenomenological model

In discussing mathematical models in general, one can hardly improve on Drucker's suggestion [18] that 'the best model to use is the simplest one which will exhibit the characteristics of greatest interest'. On this basis, the most appropriate and widely used constitutive equation for superplastic alloys is,

$$\frac{\sigma_e}{\sigma_0}=\left(\frac{\dot{\varepsilon}_e}{\dot{\varepsilon}_0}\right)^m \tag{4}$$

where σ_e and $\dot{\varepsilon}_e$ are the generalized stress and strain-rate respectively and σ_0, $\dot{\varepsilon}_0$ and m are the material parameters. σ_0 indicates the strength of the material at a particular strain-rate, $\dot{\varepsilon}_0$ is a constant customarily taken as unity and m is frequently called the strain-rate sensitivity index. The limitations of this equation which neglects prior history and the dependence of stress on strain and time are indicated below.

Strain-rate sensitivity

Over an extended range of strain rates, the $\sigma_e/\dot{\varepsilon}_e$ curve in a logarithmic diagram is typically sigmoidal. In the truly superplastic range, which may extend over one or two decades of strain-rate in a stable alloy, equation (4) will provide a reasonable fit to the experimental data. The value of m for typical superplastic alloys is between 0.35 and 0.9; it is significantly greater than the strain-rate sensitivity in most conventional alloys under creep conditions but it is less than that applying to molten glass and heated thermoplastics. On either side of this superplastic range, the strain-rate sensitivity diminishes and equation (4) is not appropriate. It has become customary therefore to present curves showing the variation of m with $\dot{\varepsilon}_e$. These are

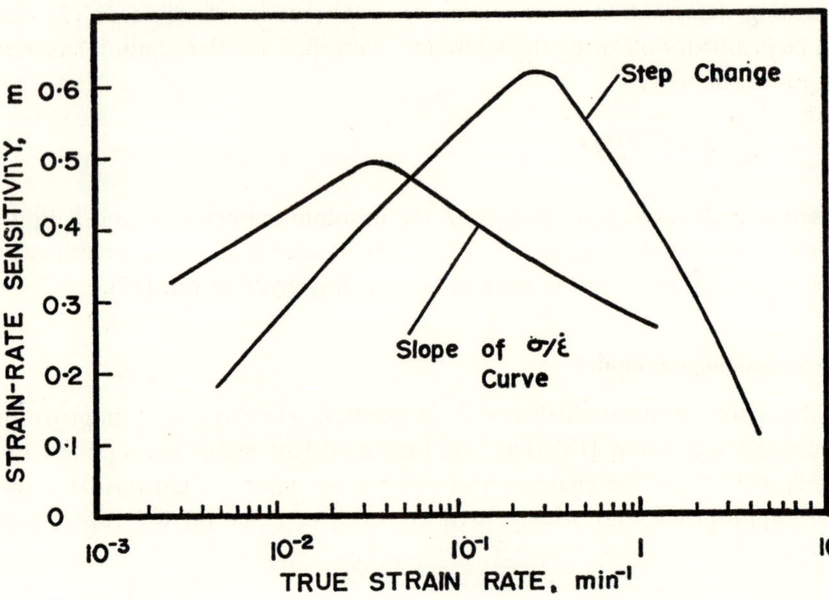

Fig. 1. Variation of strain-rate sensitivity with strain rate as determined by two different methods for Cu-9.5% Al-4% Fe bronze (from ref. 19).

derived either by differentiating a smooth curve drawn through σ_e, $\dot{\varepsilon}_e$ points calculated from many tests, or, alternatively, from a single test in which a number of step changes are made in the strain-rate [9]. In Fig. 1, $m/\dot{\varepsilon}_e$ curves from each method are compared and in general the results do not agree [19]. The discrepancy may be partly attributed to the fact that in these step-change methods, the material has not reached a steady state at the time observations are taken [20]. It may also be shown that the analysis used to derive m from such a test is not rigorous. It is assumed that the slope of the logarithmic curve at a point is m, but if m is varying with strain-rate then so is σ_0 and the true slope be readily shown to be

$$\frac{d(\ln \sigma_e)}{d(\ln \dot{\varepsilon}_e)} = \dot{\varepsilon}_e \left(\frac{dm}{d\dot{\varepsilon}} \ln \dot{\varepsilon}_e + \frac{1}{\sigma_0} \frac{d\sigma_0}{d\dot{\varepsilon}_e} \right) + m. \tag{5}$$

Other constitutive equations could be used especially where the logarithmic $\sigma_e/\dot{\varepsilon}_e$ diagram is not linear in the strain-rate range of interest. An appropriate equation for the situation in which the strain-rate sensitivity decreases with increasing strain-rate is

$$\sigma_e = \sigma_0 \sinh^{-1}(\dot{\varepsilon}_e/\dot{\varepsilon}_0) \tag{6}$$

and this has been used in the analysis of creep [21]. This relation, originally suggested by Prandtl, is certainly preferable to a modified form of equation (4) in which σ_0 and m are taken to be functions of the strain-rate.

Hardening

In a constant crosshead speed tensile test, the strain-rate diminishes as the testpiece extends. Assuming a known, constant value of m, a constant true strain-rate curve can be derived. Nicholson [2] fitted an equation,

$$\sigma = f(\dot{\varepsilon}) \cdot \dot{\varepsilon}^n \tag{7}$$

to such a curve and obtained a value of the strain hardening index, n, of 0.06 for eutectoid zinc-aluminium. Watts and Stowell [22] carried out true constant strain-rate tests on copper-aluminium eutectic and found the results to be well fitted by an equation,

$$\sigma/\sigma_0 = 1 + f(\dot{\varepsilon}) \cdot \varepsilon. \tag{8}$$

The degree of linear strain hardening was appreciable and was attributed to microstructural coarsening.

Some degree of hardening generally takes place during superplastic deformation but at the present time, this cannot be shown, incontrovertibly, to be strain dependent. Grain growth will strongly influence the properties as is evident from equations (1), (2) and (3) but this too can be either strain or time dependent. The difficulty of introducing a time function is well known [23] in creep and in superplasticity, until more data are available, a simple model as indicated by equation (4) appears justified.

Temperature
The effect of temperature is well indicated by equation (3) which is of a similar form to that applying to the creep of conventional alloys. Superplastic forming temperatures are generally in the region of one-half the absolute melting point of the alloy and frequently this differs from the temperature for hot working conventional alloys of similar chemical composition. All existing analyses of superplastic forming deal with constant temperature processes, although in practice a deliberate attempt is often made to carry the material through a transformation temperature towards the end of the forming process to obtain grain growth and consequent strengthening of the formed part [24].

Anisotropy
Anisotropy has been reported [25–27] in uniaxial tests; this is manifested by unequal transverse strains and a dependence of total elongation on the orientation of the testpiece with the direction of rolling. In general, however, the level of the stress/strain-rate curves obtained from test pieces of different orientation are very similar and in any analysis, the use of an isotropic model should not introduce serious error.

Prior history
Fields and Stewart [24] have clearly demonstrated dynamic softening and other history dependent effects in certain samples of zinc-aluminium alloys which have been subjected to a variety of thermal and mechanical treatments immediately prior to testing. Roberts [28] has also demonstrated in superplastic zinc-aluminium that dynamic softening occurs after a step increase in applied torque, even in specimens which previously had been held at the test temperature for a time which was long compared with the duration of the test.

These results are technologically important and the subject of certain patents [29] but no attempt can yet be made to allow for such effects in the analysis of processes.

Plasticity theories
The behaviour of superplastic alloys under multiaxial stress systems has not been examined extensively, but it is generally assumed that the same laws apply as those originally proposed for stationary creep by Odqvist [23] and Bailey. The generalized stress and strain-rate functions in equation (4) may

be expressed, in the usual tensor notation, as

$$\sigma_e^2 = \tfrac{3}{2}\sigma'_{ij}\cdot\sigma'_{ij}, \tag{9a}$$

$$\dot{\varepsilon}_e^2 = \tfrac{2}{3}\dot{\varepsilon}_{ij}\cdot\dot{\varepsilon}_{ij} \tag{9b}$$

and the flow rule is

$$\dot{\varepsilon}_{ij} = d\lambda\sigma'_{ij} \tag{10}$$

where σ'_{ij} is the stress deviation tensor.

Uniaxial tension and compression tests and torsion tests have been reasonably correlated on this basis for tin-lead by Ghosh [30] and for zinc-aluminium by Roberts [28] as is shown in Fig. 2.

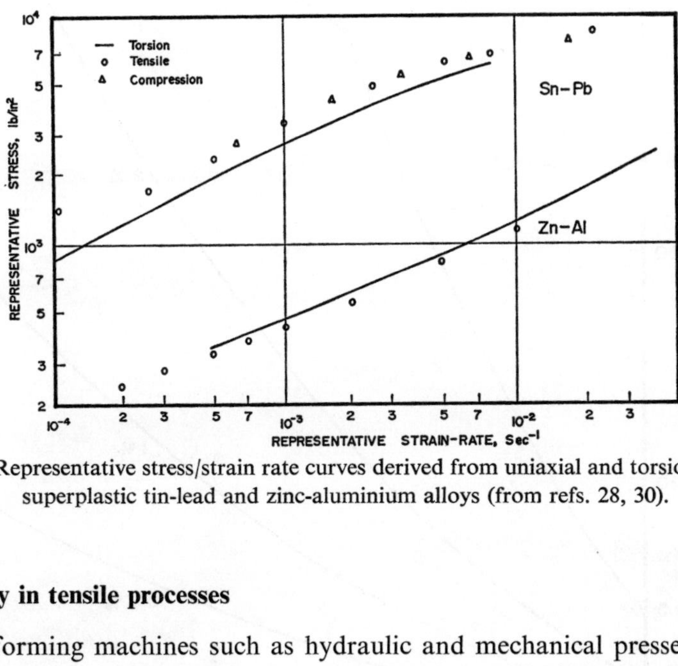

Fig. 2. Representative stress/strain rate curves derived from uniaxial and torsion tests on superplastic tin-lead and zinc-aluminium alloys (from refs. 28, 30).

Stability in tensile processes

Metal forming machines such as hydraulic and mechanical presses are essentially rate-controlled and will operate satisfactorily under both rising and falling loads. The occurrence of a maximum in the deforming force is therefore not particularly significant unless it coincides with the development of an objectionable concentration of strain in the deforming part. In many sheet metal pressing and stretch-forming operations diffuse necking does appear to start at a load maximum but frequently the neck is so widespread and gradual that the forming process may be safely carried well beyond this point.

In cold forming of sheet metal, where the material is considered to be strain-hardening but not time-dependent, the development of strain in an already diffusely necked region is well understood, largely following the work of Marciniak [31]. In superplastic forming, a similar understanding must be developed because frequently the load maximum occurs at the start of the

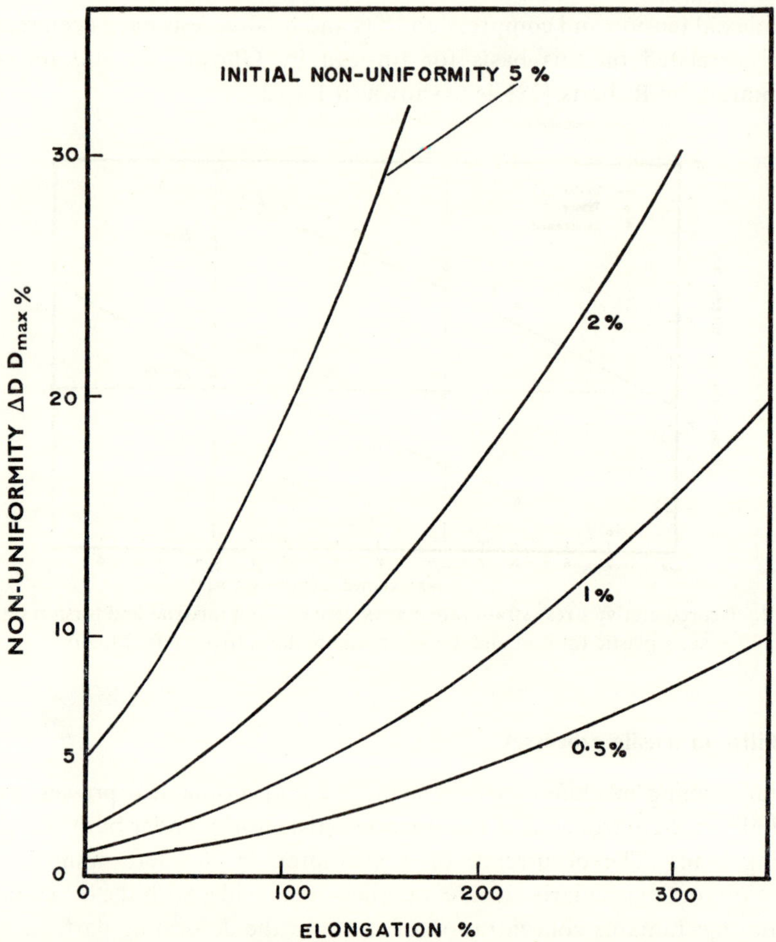

Fig. 3. Variation of non-uniformity with elongation under constant load for bars of the same material, $m = 0.5$, but having different initial non-uniformity.

process and deformation occurs entirely in a region which is, in a sense, diffusely necked. Large overall deformations can be achieved without objectionable non-uniformities in strain because the viscous behaviour of the material controls the rate of development of necking.

Plastic instability of time-dependent materials in simple tension has been studied by various authors. Nadai and Manjoine [32] examined both linear viscous and linear viscoplastic materials and concluded that necking will not occur in the linear viscous material and that the bar will be drawn out to a filament as in pulling a glass rod in a hot flame. In materials which are both strain-hardening and rate-sensitive, regions of stable deformation exist and

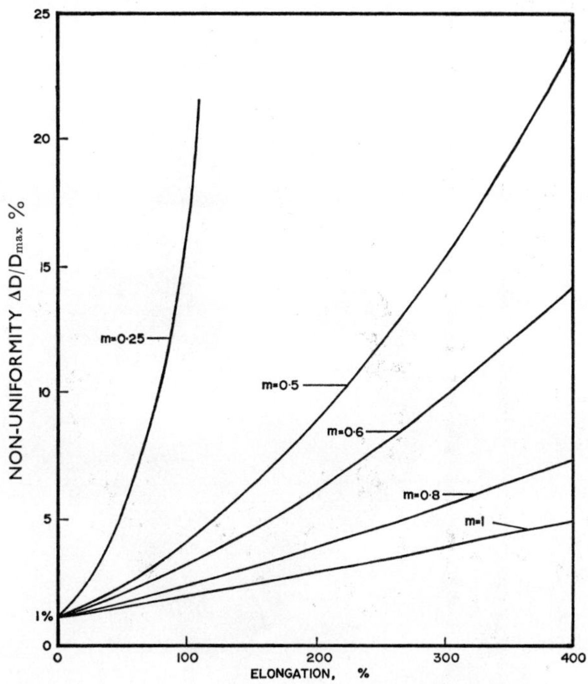

Fig. 4. Growth of non-uniformity with total elongation for testpieces having an initial non-uniformity of 1% and deformed under constant load.

the onset of instability has been studied by Campbell [33], Hart [34] and Klepaczko [35]. The geometry of developing necks has been treated by Nadai and Manjoine [32], Hoff [36] and Campbell [33] and more recently by Wray [37] who provides valuable experimental results and makes the important distinction between the development of multiple, concurrent non-uniformities and a single catastrophic neck. For a superplastic material which obeys equation (4), an initially perfect bar will retain a uniform cross-section in a tension test and in order to obtain a realistic solution it is necessary to assume some initial non-uniformity in the bar. An examination of the references given above, indicated that while controlling equations for

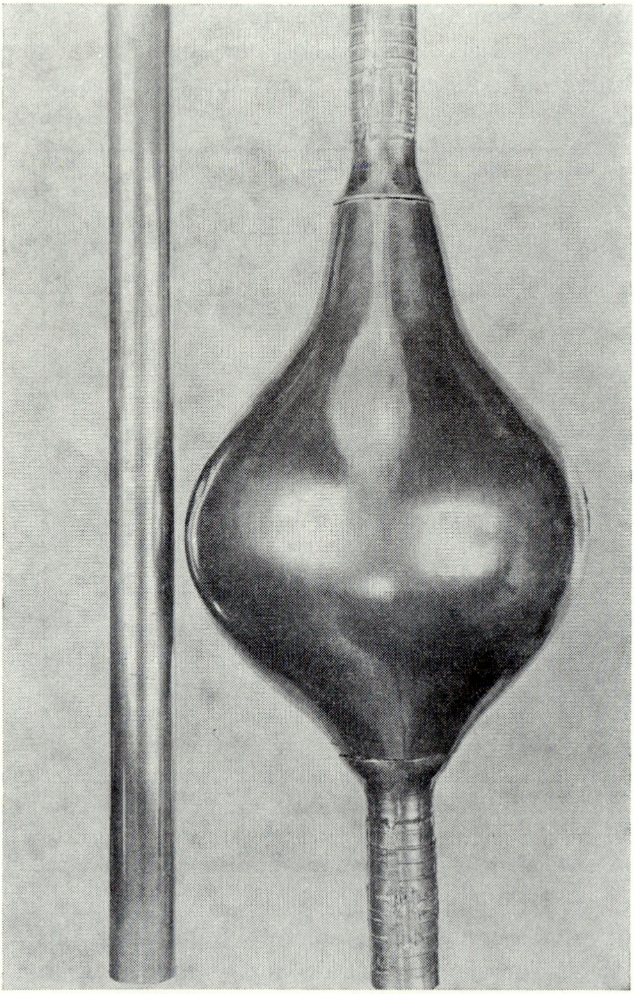

Fig. 5. Photograph of a superplastic tin-lead tube fixed at each end and expanded by internal pressure (from ref. 13).

the process are given, solutions can only be obtained using idealized conditions and particular constitutive equations.

In order to demonstrate the way in which non-uniformities might develop in superplastic forming a numerical analysis was performed on a slightly tapered bar of circular cross section. Initially the shape was that of a frustum of a cone. The material was assumed to obey equation (4) and it was found that changes in geometry were similar for both constant load and constant average strain-rate processes. (This result appears to be a consequence of the constitutive equation chosen with σ_0 and m constant and is not general.) The growth of the non-uniformity, expressed as the ratio of the diameter difference ΔD to the current maximum diameter D max., is shown in Figs. 3 and 4 for constant load tests. In Fig. 3, an initial non-uniformity of 0.5% corresponds to a taper of 0.0075 in. (0.19 mm) in a bar of 0.5 in. (12.7 mm) initial diameter. In general, the non-uniformity in a properly prepared test-piece would be less than this, so it is not surprising that large, nearly uniform elongations greater than 500% are obtained in superplastic tests. Fig. 4 indicates the strong influence of strain-rate sensitivity on uniformity. In hot tensile tests of conventional alloys, m is generally less than 0.25 and, as indicated by Fig. 4, total elongations are generally not much greater than 100–200%.

The expansion of a thin-walled tube is technologically a more interesting process than uniaxial tension and Fig. 5, from ref. [13], shows the degree of free expansion which can be achieved in a superplastic tube. Two kinds of non-uniformity can develop in this process. One type, the variation of diametral expansion along the axis is clearly evident in the photograph; the other type, variation of strain around a circumference will also occur, if the initial wall thickness is non-uniform. To illustrate this, a numerical analysis was performed for a tube having the initial cross-section shown in Fig. 6. It was found that particles on the mid-wall radius will have both radial and circumferential components of displacements. Fig. 7 shows the growth of the current wall thickness variation as given by,

$$H = \frac{h_{max} - h_{min}}{h_{max}}$$

where h denotes the wall thickness, with the diametral expansion of the tube,

$$R = \frac{r_m - r_{m0}}{r_{m0}}$$

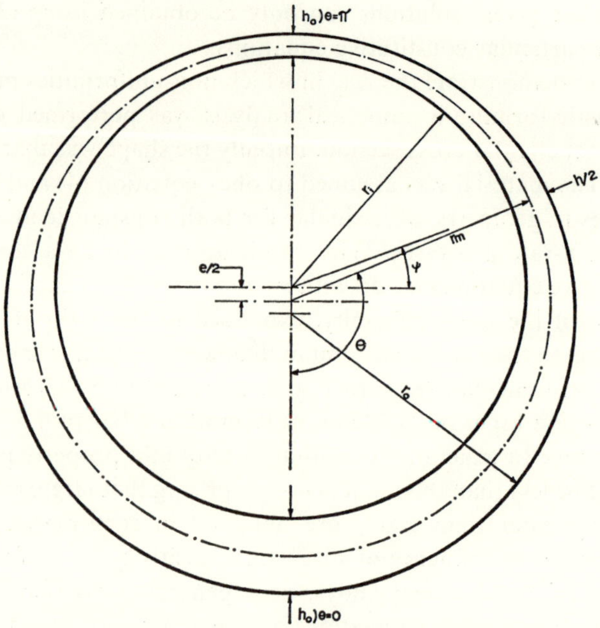

Fig. 6. Schematic section of the underformed non-uniform tube.

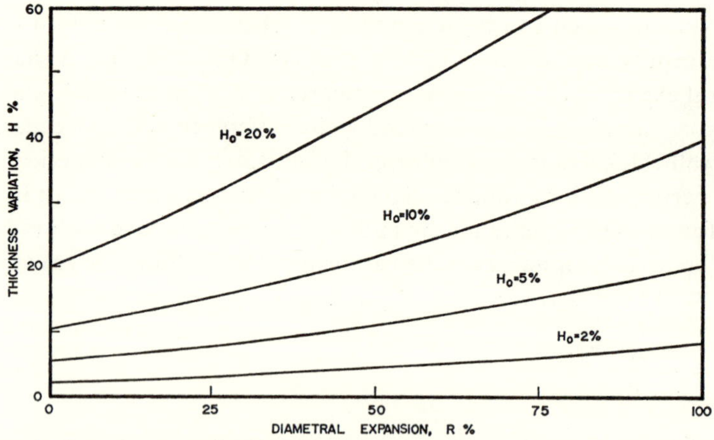

Fig. 7. Increase in non-uniformity with expansion for tubes of the same material $m = 0.5$ but different initial non-uniformity.

where r_{m0} and r_m are the initial and the current mid-wall radii of the tube respectively. Commercial tolerances on superplastic tubes would probably be about 2% and the diagram indicates that acceptable products could be formed by expanding tubes by about 100% or perhaps greater.

Conclusion

The plasticity theories and constitutive equations used in the analysis of stationary creep in structures appear to be applicable to the study of superplastic deformation with certain limitations which have been indicated above. A non-linear viscous model as given by the equation,

$$(\sigma_e/\sigma_0) = (\dot{\varepsilon}_e/\dot{\varepsilon}_0)^m$$

has been used successfully to analyse certain processes. The range of material parameters in superplasticity differs from that in creep and, in particular, the strain-rate sensitivity is unusually high and in the range 0.35–0.9. Unusual deformation processes can be performed with these materials although in tensile operations non-uniformities will develop continuously but slowly during the process.

A few analyses of superplastic forming processes have already appeared in the literature and the results of some numerical analyses of tensile processes have been presented here. Work of this nature provides useful information but it is apparent that there is scope for greater application of the methods of continuum mechanics not only to new superplastic forming processes but also to the re-examination of a number of statements to be found in the literature which, at present, depend largely on qualitative arguments.

Acknowledgements

The authors would like to thank the National Research Council of Canada for their financial support of this work.

References

[1] Pearson, C. E., The viscous properties of extruded eutectic alloys of lead-tin and bismuth-tin. *Journal of the Institute of Metals*, *39* (1928) 111–123.

[2] Nicholson, R. B., *Plasticity and superplasticity*, Inst. of Metallurgists, Review Course, Ser. 2, No. 3, Eastbourne, U. K., October 1969.

[3] Rosenhain, W., et al., Zinc alloys with aluminium and copper. *J. Institute of Metals*, *23* (1920) 261–317.
[4] Hargreaves, F., The ball hardness and the cold-working of soft metals and eutectics. *J. Institute of Metals*, *39* (1928) 301–327.
[5] Bochvar, A. A., and Z. A. Sviderskaia, Superplasticity phenomena in zinc-aluminium alloys. *Izv. Akad. Nauk, S.S.S.R., Otdel, Tekh, Nauk.*, *9* (1945) 821, (In Russian).
[6] Presnyakov, A. A., and V. V. Chernyakova, *Fiz. Metal, Metaloved*, *8* (1959) 114, (In Russian).
[7] Presnyakov, A. A., and G. V. Starekova, *Fiz. Metal, Metaloved*, *12* (1961) 873, (In Russian).
[8] Underwood, E. E., A review of superplasticity. *Journal of Metals*, *14* (1962) 914–919.
[9] Backofen, W. A., I. R. Turner, and D. H. Avery, Superplasticity in an Al-Zn alloy. *Transactions of the ASM*, *57* (1964) 980–990.
[10] Holt, D. L. and W. A. Backofen, Superplasticity in the Al-Cu eutectic alloy. *Transactions of the ASM*, *59* (1966) 755–768.
[11] Avery, D. H. and W. A. Backofen, A structural basis for superplasticity. *Transactions of the ASM*, *58* (1965) 551–562.
[12] Backofen, W. A. et al, *Superplasticity*. ASM Pre-congress Seminar: Ductility-limitations, exploitation, utilization, October 15, 1967.
[13] Al-Naib, T. Y. M., and J. L. Duncan, Superplastic metal forming. *International Journal Mech. Sci.*, *12* (1970) 463–477.
[14] Petch, N. J., The cleavage strength of polycrystals. *Journal of the iron and steel institute*, *174* (1953) 25–28.
[15] Packer, C. M., and O. D. Sherby, An interpretation of the superplasticity phenomenon in two-phase alloys. *Transactions of the ASM*, *60* (1967) 21–28.
[16] Packer, C. M., *Superplasticity in metals*. Ph. D. Thesis, Stanford University, 1968.
[17] Padmanabhan, K. A., and G. J. Davies, Numerical analysis of superplasticity data for use in metal forming applications. *J. Mech. Phys. Solids*, *18* (1970) 261–275.
[18] Drucker, D. C., *Introduction to mechanics of deformable bodies*. McGraw Hill, N.Y., (1967) 42.
[19] Dunlop, G. L., and D. M. R. Taplin, The tensile properties of a superplastic aluminium bronze. SM Report No. 83, University of Waterloo, Canada, (1971). (To be published by *Journal of Material Science*.)
[20] Nuttall, K., Letter to the editor, Strain-rate sensitivity in superplastic metals. *Int. J. Mech. Sci.*, *13* (1971) 83–85.
[21] Nadai, A., Theory of flow and fracture of solids. *Enging. Soc. Mono.*, McGraw Hill, II (1963) 521.
[22] Watts, B. M., and M. J. Stowell, The variation in flow stress and microstructure during superplastic deformation of the Al-Cu eutectic. *Journal of Material Science*, *6* (1971) 228–237.
[23] Odqvist, F. K., *Mathematical theory of creep and creep rupture*. Oxford at the Clarendon Press, (1966) 15.
[24] Fields, D. S., and T. J. Stewart, Strain effects in the superplastic deformation of 78 Zn–22 Al. *Int. J. Mech. Sci.*, *13* (1971) 63–75.
[25] Naziri, H., and R. Pearce, Anisotropic effects in superplastic Zn-0.4% Al sheet. *Journal of the Institute of Metals*, *98* (1970) 71–77.

[26] Packer, C. M., R. H. Johnson, and O. D. Sherby, Evidence for the importance of crystallographic slip during superplastic deformation of eutectic zinc-aluminium. *Transactions of the Metallurgical Society of AIME, 242* (1968) 2485–2489.

[27] Dunlop, G. L., and D. M. R. Taplin, *Textures and anisotropic flow of a superplastic aluminium bronze*. SM Report, No. 85, University of Waterloo, Canada, June (1971).

[28] Roberts, G. T., M. Sc. Diss. UMIST, Manchester University (1971).

[29] Thermoforming of metals, IBM pat. 783954, Canada.

[30] Ghosh, A., and J. L. Duncan, Torsion tests on superplastic tin-lead alloy. *Int. J. Mech. Sci., 12* (1970) 499–511.

[31] Marciniak, Z., and K. Kuczynski, Limit strains in the progress of stretch forming sheet metal. *Int. J. Mech. Sci., 9* (1967) 609–620.

[32] Nadai, A., and M. J. Manjoine, High speed tension tests at elevated temperatures. *Journal of Applied Mech., 63* (1941) A77–A91.

[33] Campell, J. D., Plastic instability in rate-dependent materials. *Journal of Mechanics and Physics of Solids, 15* (1967) 359–370.

[34] Hart, E. W., Theory of the tensile test. *Acta Metallurgica, 15* (1967) 351–355.

[35] Klepaczko, J., Generalized conditions for stability in tension tests. *Int. J. Mech. Sci., 10* (1968) 296–313.

[36] Hoff, N. J., The necking and the rupture of rods subjected to constant tensile loads. *Journal of Applied Mechanics, 20* (1953) 105–108.

[37] Wray, P. J., Tensile plastic instability at an elevated temperature and its dependence upon strain rate. *Journal of Applied Physics, 41* (1970) 3352–3347.

Yield under high hydrostatic pressure

P. K. Fung, D. J. Burns and N. C. Lind

University of Waterloo
Waterloo, Ontario, Canada

Early experimental evidence suggests that the yield of a metal is unaffected by a moderate hydrostatic pressure; that a yield locus can be represented by a cylindrical surface in stress space with generators parallel to the hydrostatic axis. Some contrary test results have been reported more recently (e.g., by Hu).

However, existing experimental methods have several shortcomings. Some of these have been eliminated in the present study. A trixial testing apparatus has been built to study the initial and subsequent yield surfaces under hydrostatic pressure up to 100 000 psi (7000 kg/cm^2).

The design provides uniform triaxial compression to the specimen in a pressurizing vessel and keeps a constant pressure during loading of the specimen in tension. A clip gage is placed on the specimen inside the pressure vessel. A Morrison seal effectively reduces the friction between the vessel and the moving parts penetrating the wall; the friction and specimen loads are measured using an external load cell.

Three series of tests were conducted to study the effect of tensile plastic deformation on the yield surface for different hydrostatic pressures. The effects of high hydrostatic pressure on the initial and subsequent yield surfaces and stress-strain relationships are presented and discussed.

Introduction

Investigations of the effects of hydrostatic pressure on the stress strain behaviour of materials have been summarized by Bridgman [1], Crossland [2] and Pugh [3]. It has been shown that, in general, the ductility of metals increases with hydrostatic pressure. There is, however, some disagreement as to the effect of hydrostatic pressure on yield. Crossland [2] and Pugh [3] concluded that hydrostatic pressure had little or no effect on the initial yield and flow of ductile metals, whereas Ratner [4] and Hu [5, 6] found that increasing the hydrostatic pressure raised the stress/strain curve. However, the methods of measuring stress used by the last two workers are open to criticism.

Hu [7] also studied the effect of hydrostatic pressure on subsequent yield surfaces for a soft brass. Specimens were prestressed in tension to 18 310, 31 380 or 39 240 psi; then subjected to a hydrostatic pressure and retested in tension. These tests showed that the subsequent yield stress of specimen, prestressed plastically in tension at atmospheric pressure, decreased as the superimposed hydrostatic pressure increased. Hu concluded that his results could be explained by the formation of a local bulge on the yield surface as shown in Fig. 1. It is difficult to accept this conclusion; it is contrary to the hypothesis of a convex yield surface [8], and the method of stress measurement is open to criticism since there was no measure of the friction in the

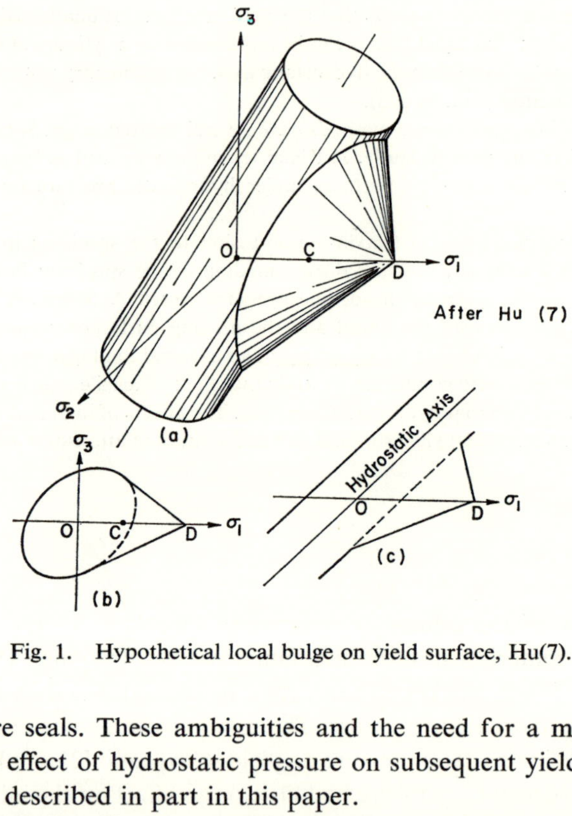

Fig. 1. Hypothetical local bulge on yield surface, Hu(7).

high pressure seals. These ambiguities and the need for a more extensive study of the effect of hydrostatic pressure on subsequent yield surfaces led to the study described in part in this paper.

A machine is described for studying initial and subsequent yield surfaces under hydrostatic pressure up to 100 000 psi (7000 kg/cm^2). The results of a study of a brass, similar to that tested by Hu [6, 7] are presented.

Apparatus

A schematic of the apparatus is shown as Fig. 2. The pressure vessel has a minimum bore diameter of $2\frac{1}{2}$ inches and an outside diameter of 9 inches. It is 24 inches long. The yoke used to mount the specimen in the pressure vessel is shown in detail in Fig. 3. Since the upper and lower rods of this yoke have the same diameter, there is no end load exerted on the specimen when

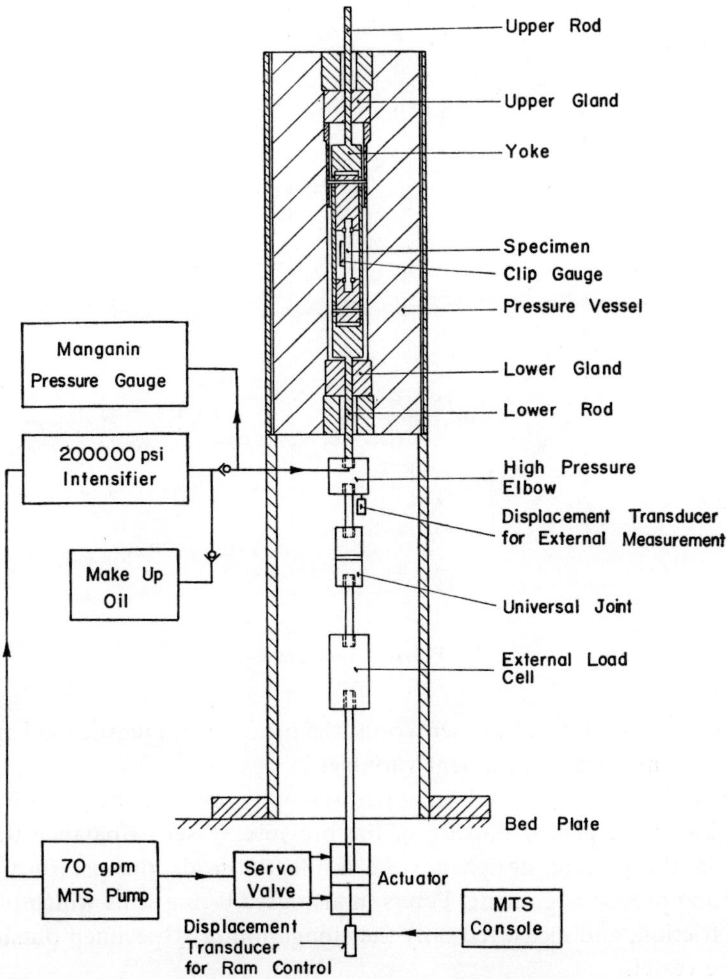

Fig. 2. Schematic of high pressure system.

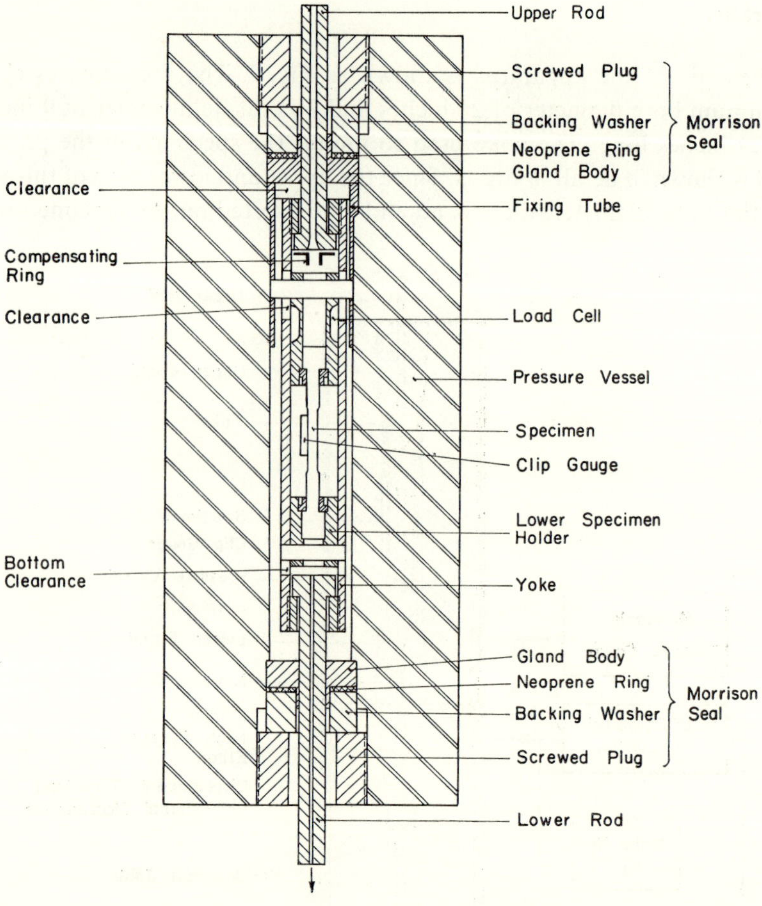

Fig. 3. Pressure vessel assembly.

the vessel is pressurized. In other words, the pressure and tensile load applied to the specimen are independent variables.

Paterson [9] has described an apparatus of a similar principle; however, he used a frame placed outside of the pressure vessel to balance the end loads; in the present design a yoke is placed inside the pressure vessel facilitating precise alignment; Paterson also used 0-ring seals which provide higher friction, and measured only the elongation of a specimen outside the pressure vessel.

The specimens used by Hu [7] had threaded ends. Morrison [10] and others have shown that threaded connections can introduce large bending stresses in a specimen. To avoid this problem, the specimens used in the present tests had plug ends, Fig. 4. The specimen is held in the upper and lower specimen holders by split collets and screwed plugs. The specimen holders are a sliding fit in the yoke.

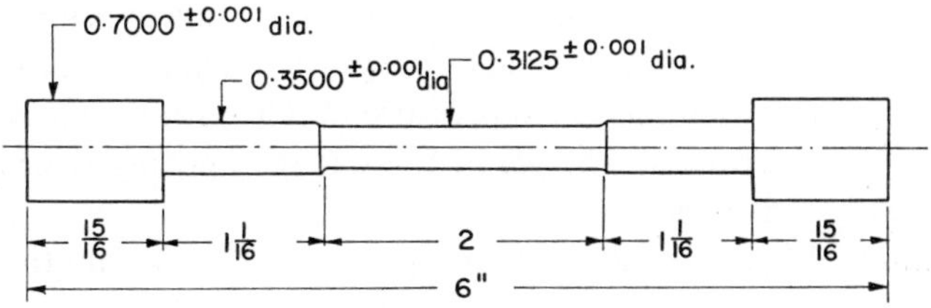

Fig. 4. Tension specimen.

The upper specimen holder can be fitted with electrical resistance strains gauges and used as a load cell. The compensating gauges for the load cell are on a small ring mounted below the upper rod as shown in Fig. 3. This load cell is redundant if the Morrison high pressure seals are functioning correctly, i.e., if friction is small and/or essentially constant during a test.

The clip gauge mounted on the specimen, Fig. 3, has strain gauges mounted on its top and bottom surfaces to make a full bridge. The pressure effect on the electric resistance strain gauges is compensated in the circuit. The linearity of the clip gauge was very good when calibrated at atmospheric pressure. The leads from the load cell and from the clip gauge are carried out of the vessel through the upper rod of the yoke. The cone seal used in this rod was designed by Harwood [11] to carry ten wires and has been tested to 100000 psi pressure.

The lower rod of the yoke is hollow to allow high pressure oil into the vessel from the 200000 lb/in^2 Belfast intensifier [12]. A Harwood Manganin cell is used to measure the pressure [11]. Shell Tellus 69 was used as the pressure medium (so-called make-up oil on Fig. 2) for tests up to 55000 lb/in^2. At higher pressures, it is necessary to use other fluids such as Esso Univis p-38 since Tellus 69 'freezes', i.e., tends to solidify. If this occurs, one no longer has a true measure of the pressure in the vessel and the high pressure seals may seize.

The high pressure seals shown in Fig. 3 were developed by Morrison, Crossland and Parry [2], [13–15] for liquid applications where the pressure is constant or slowly fluctuating, and reciprocating or rotary motion has to be transmitted into the vessel with very little leakage and low friction. The gland body is forced by the liquid pressure onto a neoprene rubber packing ring which is supported through a backing washer by a screwed plug. At high pressures, the packing ring tends to act as a perfectly plastic material and exert a hydrostatic pressure, P_e, on the cylindrical lip of the gland body. Since the annular cross-section of the packing ring is always significantly less than the area of the high pressure end of the gland body, the packing pressure, P_e, is always greater than the liquid pressure, p. Therefore, no leakage occurs around the outer surface of the gland body if the packing is properly supported by extrusion rings.

The cylindrical lip of the gland body is subjected externally to a pressure, p_e, over the packing thickness and internally to a pressure distribution ranging from atmospheric to liquid pressure, p. This combination of pressure controls the bore diameter of the gland lip. The pressure distribution acting between the gland body and rod also causes a local reduction in rod diameter. The aforementioned deflections and the initial clearance between gland body and rod dictate the leakage and friction characteristics of the Morrison seal. The seals used in the present tests had a gland bore diameter of 0.75 inches and an initial clearance of 0.004 inches.

Experimental procedure

The ASTM-B16 (SAE 72) brass had a nominal composition of 61.5% Cu, 3.25%Pb with Zn as the balance. The specimens were machined from 7/8 inch diameter bar and then heat-treated at 650°C for 5 hours. This produced a soft grade material with a grain size of about 0.024 mm.

Three types of tests were made. In the first series, the specimen was subjected to a hydrostatic pressure, p, and then loaded in tension to establish the yield curve. In the second and third series of tests, the specimens were strained beyond yield in tension at atmospheric pressure. After unloading, they were subjected to a hydrostatic pressure, p, and then loaded in tension to give the subsequent yield curve.

In each series, specimens were tested at pressures, p, of atmospheric, 20000, 40000 and 60000 lb/in^2. In series II and III, the prestresses were 21000 and 28000 lb/in^2, respectively. In all tests, the tension was applied by a 3 kip MTS servo-hydraulic actuator operating in stroke control with a ram speed of 0.02 inches/minute.

At pressure, the yoke and its upper and lower rod, act as a single member. Therefore, a measure of the friction in the Morrison seals can be obtained immediately before tensioning the specimen by starting the test with the yoke lifted up to its 'top position'. Figure 3 shows the yoke in the position it occupies while the specimen is being tensioned. The so-called 'bottom clearance' of about 0.5 inches shown in Fig. 3 is the amount the yoke can be lifted. If external load vs. time and clip gauge strain vs. time records are started from the top position one has a clear record of seal friction variation immediately before the test. A similar procedure can be followed after the tension test, i.e., unloading, lifting the yoke and then reloading to check that the friction characteristics of the seals have not changed significantly.

The strain gauges on the internal load cell were damaged at the start of the tests reported herein so the aforementioned measurements of seal friction before and after a tension test were very necessary.

Results and discussion

Typical external load vs. internal, i.e., clip gauge, displacement curves are shown in Figs. 5 and 6. The Series I tests showed that a superimposed hydrostatic pressure tends to eliminate the initial yield step in this material, c.f. Figs. 5(a) and (b). Serrated yielding was observed in the Series II and III tests irrespective of the ambient pressure, c.f. Figs. 6(a) and (b). Fortunately, this serrated yielding did not interfere seriously with the definition of subsequent yield stress in these tests.

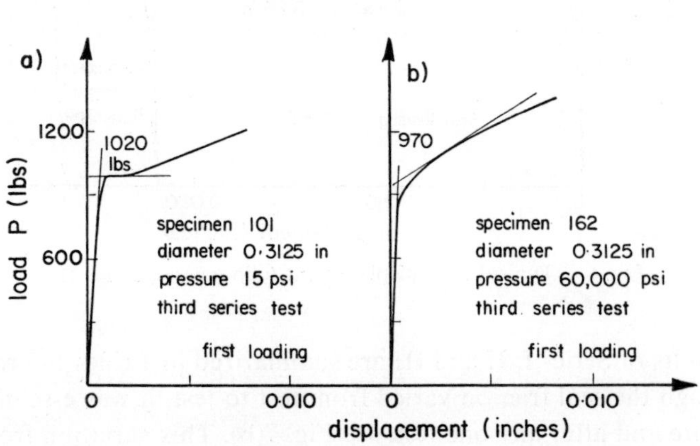

Fig. 5. Typical load-clip gauge displacement curves, series I.

Typical external load vs. time and external displacement vs. time curves are shown in Fig. 7. Fig. 7(b) shows, from right to left: – pretensioning and unloading at atmospheric pressure; reset displacement pen to zero; measurement of seal friction and subsequent yielding at 60000 lb/in^2; and finally, after unloading, lifting the yoke and resetting the displacement pen, another check on the seal friction.

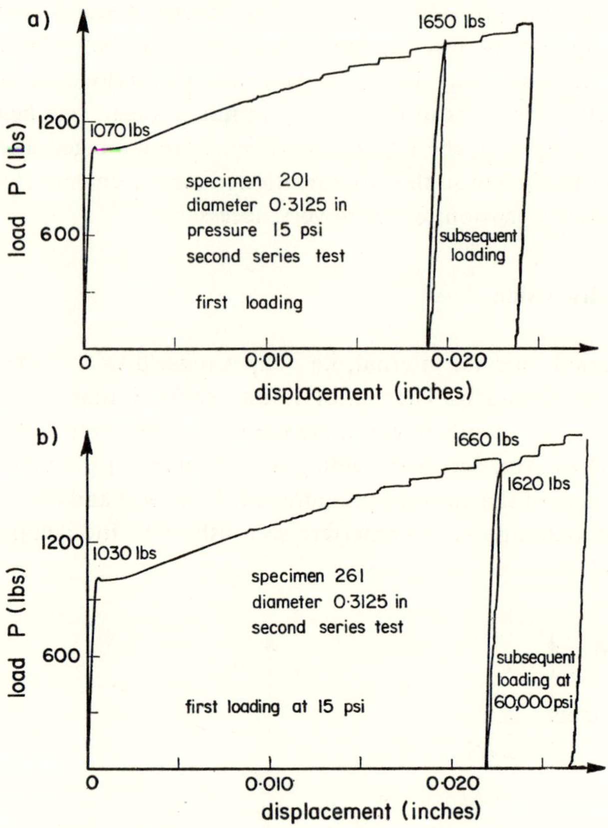

Fig. 6. Typical load-displacement (clip gauge), series II.

The results of Series I, II and III are summarized in Tables 1–3 respectively. Although the seal friction varied from test to test, it was essentially constant before and after any one test, see Fig. 7(b). This variation from test to test probably occurs because the seals have to be dismantled to change the

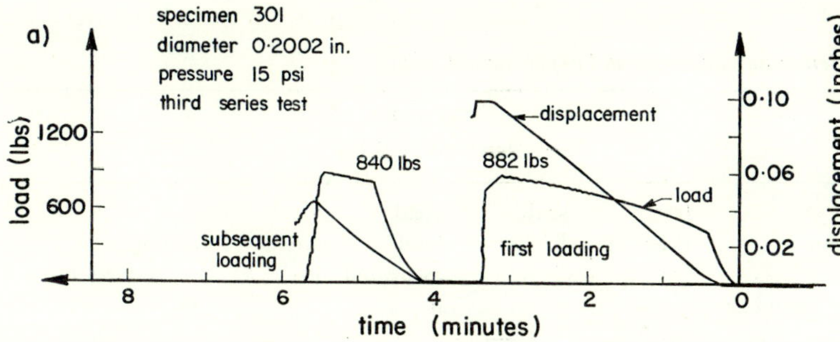

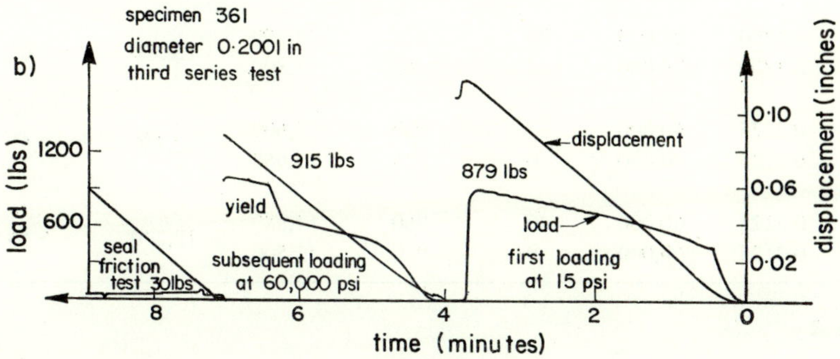

Fig. 7. Typical load-time curves and external displacement-time curves, series III.

specimen. Since the rods and gland bodies are never perfectly round and are not necessarily replaced in the same angular position, the clearance annulus and friction characteristics may vary slightly from test to test. The variation in friction characteristics with pressure could have been eliminated by varying the packing thickness and/or initial clearance to find a suitable combination [16].

The intersection of elastic and post-yield slopes has been used to define the yield loads listed in Tables 1–3. The yield stresses listed are the yield load, minus the average seal friction, divided by the original cross-sectional area of the specimen. The last column in Table 1 shows the percentage change in yield stress with respect to the yield stress at atmospheric pressure. Tables 2 and 3 show the percentage difference between subsequent yield and prestress.

Deformations of the specimen were recorded with a clip gage and a displacement differential transformer placed inside and outside of the pressure

TABLE 1

Seal friction measurement and experimental results

Series I Prestress = 0 Psi							
specimen No.	diameter	test pressure psi	seal friction lb.	yield load measured lb.	yield stress psi	f_y average yield stress psi	$1\text{-}f_y/f'$ % change
101	0.3125	15	0	1020	13,300	13,100	0
102	1.3125	15	0	990	12,900	$= f'$	
121	0.1800	20,000	90	426	13,200	13,500	+3.05
122	0.1800	20,000	114	465	13,800		
141	0.3125	40,000	24	990	12,600	12,750	−2.67
142	0.3125	40,000	30	1025	12,950		
161	0.3125	60,000	40	960	11,980	12,240	−6.56
162	0.3125	60,000	9	970	12,500		

TABLE 2

Seal friction measurement and experimental results

Series II Prestress = 21,000 Psi = f_o							
specimen No.	diameter	test pressure psi	seal friction lb.	yield load measured lb.	yield stress psi	f_y average yield stress psi	% change $100\times(1-f_y/f_o)$
201	0.3125	15	0	1650	21,500	21,500	+2.4
202	0.3125	15	0	1650	21,500		
221	0.1800	20,000	219	750	20,900	20,900	−0.5
241	0.3125	40,000	44	1716	21,700	21,100	+0.5
242	0.2000	40,000	210	855	20,500		
261	0.3125	60,000	0	1620	21,100	20,700	−1.4
262	0.3125	60,000	0	1560	20,300		

TABLE 3

Seal friction measurement and experimental results

Series III Prestress = 28,000 Psi = f_o							
specimen No.	diameter	test pressure psi	seal friction lb.	yield load measured lb.	yield stress psi	f_y average yield stress psi	% change $100 \times (1 - f_y/f_o)$
301	0.2002	15	0	860	27,400	28,500	+1.8
302	0.2000	15	0	930	29,600		
321	0.1800	20,000	162	870	27,900	27,900	−0.3
341	0.2000	40,000	60	960	28,700	28,700	+2.5
361	0.2001	60,000	30	920	28,400	28,450	+1.6
362	0.2000	60,000	30	925	28,500		

vessel respectively. The difference of yield loads determined by load vs. (inside) displacement and load vs. time curves was always less than one per cent. Tables 1–3 show high friction in the seals at 20000 psi pressure and lower friction at 60000 psi; it was therefore necessary to measure the seal friction as the load was measured outside the pressure vessel in each test. The load cell was calibrated prior to the tests; the maximum error did not exceed 0.5 per cent of the load rating. Differential pre-stress in the series II and series III tests was less than 0.7 per cent of the prestress. Of course, the present change quoted in the last columns of Tables 1–3 involves a difference between two large quantities and is somewhat sensitive to measurement error, but the change is evidently small.

A careful examination of Tables 1–3 suggests that hydrostatic pressure has little or no effect on the initial and subsequent yield behaviour of the material. This agrees with the usual assumption in the theory of plasticity.Therefore, the yield surface can be represented by a cylindrical surface in the three-dimensional stress state and as hardening proceeds, this cylindrical yield surface may simply expand or shift parallel to its hydrostatic axis or a corner may occur [17]. The hypothesis of a local bulge illustrated in Fig. 1 is rejected.

The data listed in Tables 1–3 contradicts that reported in reference [7] for a similar material. Since Hu had no measure of the friction in his high pressure seals, it is believed that his external load data is not a reliable measure of the loads applied to the specimen.

Acknowledgements

This research was supported by a grant from the National Research Council of Canada. The advice and help of U. H. Mohaupt during the design and development of the high pressure assembly is gratefully acknowledged. The brass was donated by Noranda Copper Mills Ltd.

References

[1] Bridgman, P. W., *Studies in large plastic flow and fracture*. London and New York› McGraw Hill, (1952).

[2] Crossland, B., The effect of fluid pressure on the shear properties of metals. *Proc. Inst. Mech. Engrs., London, 168* (1954) 935–946.

[3] Pugh, H. L. D. and D. Green, The effect of hydrostatic pressure on the plastic flow and fracture of metals. *Proc. Inst. Mech. Engrs.*, (1956).

[4] Ratner, S. I., Change of mechanical properties of metals under hydrostatic pressure. (in Russian) *Zhurnal Techniseskoi Physiki, 19*, No. 3, (1949) 408.

[5] Hu, L. W., Development of a triaxial stress testing machine and triaxial stress experiments. *Proc. SESA, 16*, (2), (1959) 27–37.

[6] Hu, L. W., Plastic stress-strain relations and hydrostatic stress. *Proc. 2nd Symp. on Naval Structural Mechanics*. Pergamon Press (1960) 194–201.

[7] Hu, L. W., J. Markowitz and T. A. Bartush, A triaxial-stress experiment on yield condition in plasticity. *Exp. Mech., 23* (1966) 58–64.

[8] Drucker, D. C., Plasticity. *Proc. 1st Symp. Naval Structural Mechanics*, (1960) 407–455.

[9] Paterson, M. S., Triaxial testing of materials at pressure up to 10 000 Kg/cm^2 (150 000 lb./sq. in.). *J. Instn. Engrs., 36* (1964) 23–29.

[10] Morrison, J. L. M., The yield of mild steel with particular reference to the effect of size of specimens. *Proc. Instn. Mech. Engrs., 142* (1939) 193–223.

[11] Harwood Engineering Co., Walpole Mass, private communications.

[12] Crossland, B., W. J. Skelton and W. R. D. Wilson, High pressure intensifiers: Development of a unit for pressures of 200 000 lbf/in^2 and possible improvement in design. *Proc. Instn. Mech. Engrs., 182* (1967–1968), Part 3C, High Pressure Engineering, 175–187.

[13] White, D. J., B. Crossland and J. L. M. Morrison, Effect of hydrostatic pressure on the direct-stress fatigue strength of an alloy steel. *J. Mech. Eng. Sc., 1* (1959) 39–49.

[14] Burns, D. J. and J. S. C. Parry, Effect of large hydrostatic pressure on the torsional fatigue strength of two steels. *J. Mech. Eng. Sc.*, *6* (1964) 293–305.
[15] Libertiny, G. Z., High pressure low speed push-pull fatigue machine. *Proc. Inst. Mech. Eng., London*, *182* (1967–68), No. 3C, 115–121.
[16] Pick, R. J. and D. J. Burns, Dimensional parameters influencing the behaviour of the Morrison unsupported Aun seal. Depart. Mech. Eng., University of Waterloo (1970).
[17] Mair, W. M. and H. L. D. Pugh, Effect of pre-strain on yield surfaces in copper. *J. Mech. Eng. Sc.* *6* (1964) 150–163.

A unified theory on elastic-plastic wave propagation of combined stress

T. C. T. Ting

University of Illinois at Chicago Circle, Chicago, Illinois, U.S.A.

A unified approach is presented for a class of waves propagating through an elastic-plastic medium such that the governing equations contain only one space variable but the stress state can have more than one stress component. Plane waves and cylindrical waves belong to this class as well as combined longitudinal and torsional waves in a thin-walled tube. The coefficient matrices of the governing differential equations can be divided into submatrices each having a special form which simplifies the analysis substantially. This is illustrated by the determination of wave speeds, eigenvectors, characteristic conditions, and discontinuity conditions. It is shown that the right eigenvector $\boldsymbol{d}$ defined in the paper plays a universal role in determining the discontinuity conditions across both the characteristics and the elastic plastic boundary. This vector $\boldsymbol{d}$ is also tangent to the stress path in stress space for simple wave solutions.

1. Introduction

Elastic-plastic waves under combined stress have been studied by a number of authors for various types of combined loadings using a variety of constitutive equations. Plane waves in a half-space were considered by Rakhmatulin [1], Cristescu [2], Bleich and Nelson [3], and Ting and Nan [4] for pressure-shear combinations; by Ting [5] for two-shear loadings; by Goel and Malvern [6] for pressure-two shear stresses. Wave propagation in a thin-walled tube was first studied by Clifton [7] and Fukuoka [23] for an isotropic work-hardening material and by Goel and Malvern [8] for a more general material. Cylindrical waves of combined stress were considered by Cristescu [9] and the formulation of the equations for plane as well as cylindrical waves with various combined loadings was given by Fong [10], Nan [11] and Ting [12].

Clifton's work [7] on waves due to combined stress in a thin-walled tube has stimulated a wide interest in the study of combined stress waves. By considering the torsional stress and the longitudinal stress at the same time, the governing equations yield two plastic wave speeds, the faster of which can have a value comparable to the longitudinal elastic wave speed in the tube. Thus the propagation of the plastic strain at the elastic wave speed, which has been attributed to the strain rate effect, can also be due to the combined stress effect. By considering the simple wave solution due to combined step loadings at the end of the tube, Clifton observed that an unexpected unloading occurred before a plastic loading was initiated in the tube. This unexpected result was also observed by Ting for continuous loadings [13] and for general discontinuos loadings [14].

One of the special features in the analysis of the combined stress waves is that the coefficient matrices of the governing differential equation consists of submatrices each having a special form. This feature simplifies the analysis considerably as is evident from the results obtained here. It should be noted that many other wave propagation problems such as longitudinal waves in a thin rod and spherically symmetric waves are special cases of the combined stress waves considered here.

The purpose of this paper is to present a unified approach which is applicable to all combined stress waves considered in the literature as well as others which have not yet been studied in detail. Some of the results presented in this paper have been obtained in [12], but for clarity and in order to make the paper self-contained, they are presented briefly here. The special case of a thin-walled tube considered by Clifton is used occasionally to illustrate the general results. For sake of simplicity in presentation, the material is assumed to be isotropic work-hardening. Extension to other materials can be obtained with some modifications.

2. Basic equations

The vectors $\boldsymbol{\sigma}$ and $\boldsymbol{\varepsilon}$ are used for the stress and strain respectively. If m is the number of components in both $\boldsymbol{\sigma}$ and $\boldsymbol{\varepsilon}$, $m \leqq 6$. (In the case of a thin-walled tube, there are only two non-zero stress components, hence $m = 2$). Let $\boldsymbol{v}$ be the particle velocity and n be the number of its components. For plane waves or cylindrical waves only one space variable is involved and the equations of motion can be written in matrix notation as

$$\boldsymbol{M}\boldsymbol{\sigma}_x + \boldsymbol{b}_1 = \rho \boldsymbol{v}_t \tag{1}$$

where ρ is the mass density, t denotes time, x specifies the space variable.

$\boldsymbol{M}$ is an $n \times m$ matrix whose elements are constants and $\boldsymbol{b}_1$ is a column vector whose elements are functions of $\boldsymbol{\sigma}$ and x. For plane waves $\boldsymbol{b}_1 \equiv \boldsymbol{0}$ and for cylindrical waves x measures the radial distance from the axis of symmetry.

The equations of continuity can be written as

$$\boldsymbol{\varepsilon}_t = \boldsymbol{N}\boldsymbol{v}_x + \boldsymbol{b}_2 \tag{2}$$

in which $\boldsymbol{N}$ is an $m \times n$ matrix whose elements are constants. $\boldsymbol{b}_2$ is a column vector whose elements are functions of $\boldsymbol{v}$ and x. In the case of plane waves, $\boldsymbol{b}_2 \equiv \boldsymbol{0}$.

The stress-strain relationship for rate-independent elastic-plastic materials can be written as

$$\boldsymbol{\varepsilon}_t = \boldsymbol{S}\boldsymbol{\sigma}_t \tag{3}$$

where $\boldsymbol{S}$ is an $m \times m$ square matrix. In the elastic region $\boldsymbol{S}$ is denoted by $\boldsymbol{S}^e$ whose elements depend on the elastic constants only. In the plastic region, $\boldsymbol{S}$ is denoted by $\boldsymbol{S}^p$ and for isotropic work-hardening material it can be written as, [17],

$$\boldsymbol{S}^p = \boldsymbol{S}^e + G(k)(\nabla f)(\nabla f)^T \tag{4}$$

with the yield condition

$$f(\boldsymbol{\sigma}) = k^2. \tag{5}$$

In Eq. (4), G is a function of the yield stress k, ∇f is the gradient of f with respect to the components of $\boldsymbol{\sigma}$, and T stands for the transpose.

Eliminating $\boldsymbol{\varepsilon}$, Eqs. (1), (2) and (3) can be reduced to

$$\boldsymbol{A}\boldsymbol{w}_t + \boldsymbol{B}\boldsymbol{w}_x = \boldsymbol{b} \tag{6}$$

where

$$\boldsymbol{A} = \begin{bmatrix} \rho \boldsymbol{I} & \boldsymbol{0} \\ \boldsymbol{0} & \boldsymbol{S} \end{bmatrix}, \quad \boldsymbol{w} = \begin{bmatrix} \boldsymbol{v} \\ \boldsymbol{\sigma} \end{bmatrix},$$
$$\boldsymbol{B} = \begin{bmatrix} \boldsymbol{0} & -\boldsymbol{M} \\ -\boldsymbol{N} & \boldsymbol{0} \end{bmatrix}, \quad \boldsymbol{b} = \begin{bmatrix} \boldsymbol{b}_1 \\ \boldsymbol{b}_2 \end{bmatrix} \tag{7}$$

and $\boldsymbol{I}$ is the unit matrix. Although both $\boldsymbol{A}$ and $\boldsymbol{B}$ are symmetric in all combined stress waves considered in the literatures (see [12]), the following analysis is not restricted by symmetry of $\boldsymbol{A}$ and $\boldsymbol{B}$ except section 8.

Equation (6) is a standard form for a system of first order differential equations which has been studied extensively elsewhere (see [15, 16] for example). Thus the eigenvalues, (or the characteristic wave speeds), c are determined by the roots of the determinant:

$$\|c\boldsymbol{A}-\boldsymbol{B}\| = 0 \tag{8}$$

while the characteristic condition is given by

$$\boldsymbol{l}^T\boldsymbol{B}\,\frac{\mathrm{d}\boldsymbol{w}}{\mathrm{d}t} = c\boldsymbol{l}^T\boldsymbol{b}, \qquad c \neq 0, \tag{9a}$$

or

$$\boldsymbol{l}^T\boldsymbol{A}\,\frac{\mathrm{d}\boldsymbol{w}}{\mathrm{d}t} = \boldsymbol{l}^T\boldsymbol{b}, \qquad \text{for all } c, \tag{9b}$$

along $\mathrm{d}x/\mathrm{d}t = c$, where

$$\frac{\mathrm{d}\boldsymbol{w}}{\mathrm{d}t} = c\boldsymbol{w}_x+\boldsymbol{w}_t, \tag{10}$$

$$\boldsymbol{l}^T(c\boldsymbol{A}-\boldsymbol{B}) = \boldsymbol{0}. \tag{11a}$$

Here $\mathrm{d}\boldsymbol{w}/\mathrm{d}t$ is the total derivative along the characteristic, $\boldsymbol{l}$ is the left eigenvector of the system, Eq. (9b) applies to all c but Eq. (9a) applies to $c \neq 0$ only. Associated with $\boldsymbol{l}$ is the right eigenvector $\boldsymbol{r}$ defined by

$$(c\boldsymbol{A}-\boldsymbol{B})\boldsymbol{r} = \boldsymbol{0} \tag{11b}$$

which is useful in finding simple wave solutions. $\boldsymbol{r}$ is also related to the discontinuity in $\boldsymbol{w}_x$ or $\boldsymbol{w}_t$ across the characteristics.

If $\boldsymbol{r}_i$ is the right eigenvector associated with the eigenvalue c_i and $\boldsymbol{l}_j$ is the left eigenvector associated with c_j where $c_j \neq c_i$, it can be shown that $\boldsymbol{r}_i$ and $\boldsymbol{l}_j$ are biorthogonal with respect to $\boldsymbol{B}$, i.e.,

$$\boldsymbol{l}_j^T\boldsymbol{B}\boldsymbol{r}_i = 0. \tag{12a}$$

In view of Eq. (11a) or (11b),

$$\boldsymbol{l}_j^T\boldsymbol{A}\boldsymbol{r}_i = 0. \tag{12b}$$

In this paper, use is made of the fact that $\boldsymbol{A}$ and $\boldsymbol{B}$ defined in Eq. (6), have the special forms given in Eq. (7) which simplifies the analysis of Eq. (6) substantially.

3. Eigenvalues c

With $\boldsymbol{A}$ and $\boldsymbol{B}$ given by Eq. (7)

$$
\begin{aligned}
c\boldsymbol{A}-\boldsymbol{B} &= \begin{bmatrix} \rho c\boldsymbol{I} & \boldsymbol{M} \\ \boldsymbol{N} & c\boldsymbol{S} \end{bmatrix} = \begin{bmatrix} \rho c\boldsymbol{I} & \boldsymbol{0} \\ \boldsymbol{N} & \boldsymbol{D} \end{bmatrix} \begin{bmatrix} \boldsymbol{I} & \boldsymbol{M}/\rho c \\ \boldsymbol{0} & \boldsymbol{I}/\rho c \end{bmatrix} \\
&= \begin{bmatrix} \boldsymbol{I} & \boldsymbol{0} \\ \boldsymbol{N}/\rho c & \boldsymbol{I}/\rho c \end{bmatrix} \begin{bmatrix} \rho c\boldsymbol{I} & \boldsymbol{M} \\ \boldsymbol{0} & \boldsymbol{D} \end{bmatrix}
\end{aligned} \tag{13}
$$

where

$$\boldsymbol{D}(c) = \rho c^2\boldsymbol{S}-\boldsymbol{NM}. \tag{14}$$

Hence

$$\|c\boldsymbol{A}-\boldsymbol{B}\| = (\rho c)^{-(m-n)}\|\boldsymbol{D}(c)\|. \tag{15}$$

Thus instead of finding the roots of Eq. (8), one can determine c by evaluating the roots of

$$\|\boldsymbol{D}(c)\| = 0. \tag{16}$$

Since $\|c\boldsymbol{A}-\boldsymbol{B}\|$ is a determinant of order $(m+n)$ while $\|\boldsymbol{D}\|$ is a determinant of order m, a simplification in the algebraic expansion of the determinants is feasiable. For a thin-walled tube, $n = m = 2$, for cylindrical waves due to pressure combined with two shear, $n = 3$, $m = 5$. In general $m \geqq n$, and Eq. (15) shows that $c = 0$ is a root of $\|\boldsymbol{D}\| = 0$ unless $n = m$ even though $c = 0$ may not be a root of $\|c\boldsymbol{A}-\boldsymbol{B}\| = 0$. Notice also that the roots of $\|\boldsymbol{D}\| = 0$ appear in pairs because $\boldsymbol{D}$ is a function of c^2 as defined by Eq. (14). Since $\|\boldsymbol{D}(c)\| = 0$ has $2m$ roots while $\|c\boldsymbol{A}-\boldsymbol{B}\| = 0$ has $(n+m)$ roots, $\|\boldsymbol{D}(c)\| = 0$ has $(m-n)$ more roots than $\|c\boldsymbol{A}-\boldsymbol{B}\| = 0$. These $(m-n)$ extra roots clearly have the value zero as is evident from Eq. (15).

If superscripts e and p denote elastic and plastic regions respectively

$$\boldsymbol{D}^p(c) = \boldsymbol{D}^e(c)+\rho c^2G(\nabla f)(\nabla f)^T \tag{17}$$

in view of Eq. (4).

Using this, it can be shown that (see [12])

$$\|\boldsymbol{D}^p(c)\| = \|\boldsymbol{D}^e(c)\|+\rho c^2G(\nabla f)^T(\boldsymbol{D}^e)^*(\nabla f) \tag{18}$$

where $(\boldsymbol{D}^e)^*$ is the adjoint of $\boldsymbol{D}^e$, i.e.

$$\boldsymbol{D}^e(\boldsymbol{D}^e)^{*T} = (\boldsymbol{D}^e)^{*T}\boldsymbol{D}^e = \|\boldsymbol{D}^e\|\boldsymbol{I}. \tag{19}$$

Equation (18) is a relationship useful for comparing the positions of the roots of $\|\boldsymbol{D}^p\| = 0$ and $\|\boldsymbol{D}^e\| = 0$.

In general, $\|\boldsymbol{D}^p\| = 0$ will have three distinct roots in the plastic region: the fast wave speed c_f, the slow wave speed c_s, and the intermediate wave speed c_2 which is identical to the shear wave speed in the elastic region (see [21, 24, 25]). By letting $c = c_f$ and $c = c_s$ respectively in Eq. (18)

$$\rho c_f^2 G = -\frac{\|\boldsymbol{D}^e(c_f)\|}{(\nabla f)^T(\boldsymbol{D}^e(c_f))^*(\nabla f)}, \tag{20a}$$

$$\rho c_s^2 G = -\frac{\|\boldsymbol{D}^e(c_s)\|}{(\nabla f)^T(\boldsymbol{D}^e(c_s))^*(\nabla f)} \tag{20b}$$

and hence

$$\frac{\|\boldsymbol{D}^e(c_f)\|}{\|\boldsymbol{D}^e(c_s)\|} = \frac{c_f^2}{c_s^2}\,\frac{(\nabla f)^T(\boldsymbol{D}^e(c_f))^*(\nabla f)}{(\nabla f)^T(\boldsymbol{D}^e(c_s))^*(\nabla f)}. \tag{21}$$

Equations (20a), (20b) give G in terms of c_f or c_s while Eq. (21) represents a relationship between c_f and c_s. In the particular case of a thin-walled tube,

$$\boldsymbol{D}^e(c) = \begin{bmatrix} \left(\dfrac{c^2}{c_0^2} - 1\right) & 0 \\ 0 & \left(\dfrac{c^2}{c_2^2} - 1\right) \end{bmatrix}, \qquad (\boldsymbol{D}^e(c))^* = \begin{bmatrix} \left(\dfrac{c^2}{c_2^2} - 1\right) & 0 \\ 0 & \left(\dfrac{c^2}{c_0^2} - 1\right) \end{bmatrix} \tag{22a, b}$$

where $c_0^2 = E/\rho$ and E is the Young's modulus, Eqs. (20a), (20b) and (21) reduce to Eqs. (21), (22) and (23) of [13].

4. The eigenvectors

It is not difficult to show that if $\boldsymbol{d}$ and $\boldsymbol{\delta}$ are the right and left eigenvectors of $\boldsymbol{D}$, i.e., if

$$\boldsymbol{D}\boldsymbol{d} = \boldsymbol{0}, \qquad \boldsymbol{\delta}^T\boldsymbol{D} = \boldsymbol{0}, \tag{23a, b}$$

$\boldsymbol{r}$ and $\boldsymbol{l}$ defined by Eqs. (11a), (11b) are given by

$$\boldsymbol{r} = \alpha\begin{bmatrix} -\boldsymbol{M}\boldsymbol{d} \\ \rho c\boldsymbol{d} \end{bmatrix}, \qquad \boldsymbol{l} = \alpha'\begin{bmatrix} -\boldsymbol{N}^T\boldsymbol{\delta} \\ \rho c\boldsymbol{\delta} \end{bmatrix} \tag{24a, b}$$

where α and α' are proportionality factors. The proof may be given by substituting Eqs. (24) into Eqs. (11). Although Eqs. (24) can be applied to the case when $c = 0$ is a root of $\|c\boldsymbol{A}-\boldsymbol{B}\| = 0$, the result can lead to the trivial solution $\boldsymbol{r} = \boldsymbol{l} = \boldsymbol{0}$ because $\boldsymbol{Md}$ and $\boldsymbol{N}^T\boldsymbol{\delta}$ may vanish. $\boldsymbol{r}$ and $\boldsymbol{l}$ can be obtained easily from

$$\boldsymbol{Br} = \boldsymbol{0}, \qquad \boldsymbol{l}^T\boldsymbol{B} = \boldsymbol{0}, \quad \text{if} \quad c = 0. \tag{24c}$$

It should be pointed out that c can be positive or negative in Eqs. (24a, b) which may give two solutions for each $\boldsymbol{d}$ and $\boldsymbol{\delta}$.

Using Eqs. (24a, b), the problem is reduced to finding $\boldsymbol{d}$ and $\boldsymbol{\delta}$ defined in Eqs. (23). We will show next that $\boldsymbol{d}$ and $\boldsymbol{\delta}$ satisfy Eqs. (23) if both are given by the following expressions

$$\boldsymbol{d}(c) = \beta(\boldsymbol{D}^e(c))^{*T}\nabla f, \tag{25a}$$

$$\boldsymbol{\delta}(c) = \beta'(\boldsymbol{D}^e(c))^{*}\nabla f \tag{25b}$$

where β and β' are proportionality factors. To prove this, consider first the elastic region where $\boldsymbol{D} = \boldsymbol{D}^e$ and $\|\boldsymbol{D}^e\| = 0$. With $\boldsymbol{d}$ given by Eq. (25a), it follows in view of Eq. (19) that

$$\boldsymbol{D}^e\boldsymbol{d} = \beta\boldsymbol{D}^e(\boldsymbol{D}^e)^{*T}\nabla f = \beta\|\boldsymbol{D}^e\|\nabla f = \boldsymbol{0}.$$

Hence $\boldsymbol{d}$ defined by Eq. (25a) satisfies Eq. (23a). In the plastic region $\boldsymbol{D} = \boldsymbol{D}^p$, $\|\boldsymbol{D}^p\| = 0$ and using Eqs. (17), (19)

$$\begin{aligned}\boldsymbol{D}^p\boldsymbol{d} &= \{\boldsymbol{D}^e+\rho c^2G(\nabla f)(\nabla f)^T\}\beta(\boldsymbol{D}^e)^{*T}(\nabla f)\\ &= \beta\|\boldsymbol{D}^e\|(\nabla f)+\beta(\nabla f)\{\rho c^2G(\nabla f)^T(\boldsymbol{D}^e)^{*T}(\nabla f)\}.\end{aligned}$$

Since the quantity in the last bracket is a scalar,

$$(\nabla f)^T(\boldsymbol{D}^e)^{*T}(\nabla f) = (\nabla f)^T(\boldsymbol{D}^e)^{*}(\nabla f),$$

and consequently in view of Eq. (18),

$$\boldsymbol{D}^p\boldsymbol{d} = \beta\|\boldsymbol{D}^p\|\nabla f = \boldsymbol{0}.$$

This completes the proof for Eq. (25a), and since the proof for Eq. (25b) is similar, it can be seen that Eqs. (25) apply to both the elastic region and the plastic region. It is advantageous to use Eqs. (25) instead of Eqs. (23) for the plastic region because $\boldsymbol{D}^e$ is in general simple in form and $(\boldsymbol{D}^e)^*$ can be readily obtained. (See Eqs. (22)). For the elastic region Eqs. (23) can be solved easily and it makes little difference whether one uses Eqs. (23) or (25).

For a thin-walled tube, $f = (\sigma/\theta)^2+\tau^2$, where $\theta = \sqrt{3}$ for the von Mises yield condition and $\theta = 2$ for the Tresca yield condition. Hence $(\nabla f)^T =$

$= (2\sigma/\theta^2, 2\tau)$, and substitution of Eq. (22b) into Eqs. (25) yields, with a proper choice of β and β',

$$\boldsymbol{d} = \boldsymbol{\delta} = \begin{bmatrix}\psi\\1\end{bmatrix}, \qquad \psi = \frac{((c^2/c_2^2)-1)}{((c^2/c_0^2)-1)}\,\frac{\sigma}{\theta^2\tau}$$

for $c = c_f$ and c_s, and

$$\boldsymbol{d} = \boldsymbol{\delta} = \begin{bmatrix}1\\0\end{bmatrix}, \qquad \text{for} \quad c = c_0$$

$$\boldsymbol{d} = \boldsymbol{\delta} = \begin{bmatrix}0\\1\end{bmatrix}, \qquad \text{for} \quad c = c_2$$

Whether or not every solution of Eqs. (23) permits the expressions given by Eqs. (25), i.e., the 'completeness' of the solution, will be discussed in Section 8.

The following relations may be stated as counterparts of Eqs. (12):

$$\boldsymbol{\delta}_j^T\boldsymbol{N}\boldsymbol{M}\boldsymbol{d}_i = 0 \tag{26a}$$

$$\boldsymbol{\delta}_j^T\boldsymbol{S}\boldsymbol{d}_i = 0 \tag{26b}$$

For a thin-walled tube, $\boldsymbol{N} = \boldsymbol{M} = \boldsymbol{I}$ and Eqs. (26a) reduce to the fact that $\boldsymbol{\delta}_j$ and $\boldsymbol{d}_i$ are orthogonal to each other (see [7]).

5. The characteristic conditions

By substituting Eqs. (7) and (24b) into Eqs. (9a, b), the following characteristic conditions are obtained

$$\boldsymbol{\delta}^T\boldsymbol{N}\left(\boldsymbol{M}\frac{\mathrm{d}\boldsymbol{\sigma}}{\mathrm{d}t} - \rho c\frac{\mathrm{d}\boldsymbol{v}}{\mathrm{d}t}\right) = -c\boldsymbol{\delta}^T(\boldsymbol{N}\boldsymbol{b}_1 - \rho c\boldsymbol{b}_2) \tag{27a}$$

or

$$\rho c\boldsymbol{\delta}^T\left(\boldsymbol{S}\frac{\mathrm{d}\boldsymbol{\sigma}}{\mathrm{d}t} - \boldsymbol{N}\frac{\mathrm{d}\boldsymbol{v}}{\mathrm{d}t}\right) = -\boldsymbol{\delta}^T(\boldsymbol{N}\boldsymbol{b}_1 - \rho c\boldsymbol{b}_2) \tag{27b}$$

for $c \neq 0$. If $c = 0$, Eq. (9b) applies.

For the plane waves, $\boldsymbol{b}_1 = \boldsymbol{b}_2 = \boldsymbol{0}$ and it follows from Eq. (27a) that

$$\boldsymbol{\delta}^T\boldsymbol{N}(\boldsymbol{M}\mathrm{d}\boldsymbol{\sigma} - \rho c\mathrm{d}\boldsymbol{v}) = \boldsymbol{0},\ c \neq 0. \tag{28}$$

In the case of a thin-walled tube, $\boldsymbol{N} = \boldsymbol{M} = \boldsymbol{I}$, $\boldsymbol{d} = \boldsymbol{\delta}$ and Eq. (28) reduces to

$$\boldsymbol{d}^T(\mathrm{d}\boldsymbol{\sigma} - \rho c \mathrm{d}\boldsymbol{v}) = \boldsymbol{0}$$

which means that $(\mathrm{d}\boldsymbol{\sigma} - \rho c \mathrm{d}\boldsymbol{v})$ is orthogonal to $\boldsymbol{d}$. This is interesting because $(\mathrm{d}\boldsymbol{\sigma} - \rho c \mathrm{d}\boldsymbol{v})$, though in vector form, also appears in the characteristic condition $(\mathrm{d}\sigma - \rho c \mathrm{d}v) = 0$ which is valid for longitudinal waves in a thin rod.

6. Simple waves

A solution of Eq. (6) with $\boldsymbol{b} = \boldsymbol{0}$ is called a 'simple wave solution' if $\boldsymbol{w}$ is constant along the characteristics. Since the slope of the characteristic is a function of $\boldsymbol{\sigma}$ and hence of $\boldsymbol{w}$, the characteristics are straight lines in the $\{x-t\}$ plane. For a simple wave solution one may regard $\boldsymbol{w}$ as a function of the characteristic wave speed c and according to the theory of simple waves [15, 16], $\mathrm{d}\boldsymbol{w}/\mathrm{d}c$ is proportional to the right eigenvector $\boldsymbol{r}$. Using Eqs. (7) and (24a, b)

$$\frac{\mathrm{d}\boldsymbol{\sigma}}{\mathrm{d}c} = \boldsymbol{d}(c), \tag{29}$$

$$\frac{\mathrm{d}\boldsymbol{v}}{\mathrm{d}c} = -\boldsymbol{M}\boldsymbol{d}(c)/\rho c, \tag{30}$$

$\boldsymbol{d}$ is given by Eq. (25a) in which β is the proportionality factor now expressed as a function of c. $\beta(c)$ can be determined by using

$$1 = (\nabla c)^T \frac{\mathrm{d}\boldsymbol{\sigma}}{\mathrm{d}c} = (\nabla c)^T\boldsymbol{d} = \beta(\nabla f)^T\boldsymbol{D}^{e*}(\nabla c). \tag{31}$$

It follows from the stress-strain relationship, given in Eq. (3), that

$$\frac{\mathrm{d}\boldsymbol{\varepsilon}}{\mathrm{d}c} = \boldsymbol{S}\frac{\mathrm{d}\boldsymbol{\sigma}}{\mathrm{d}c} = \boldsymbol{S}\boldsymbol{d}.$$

Post-multiplying Eq. (14) by $\boldsymbol{d}$ and eliminating $\boldsymbol{S}\boldsymbol{d}$ from the above equation results in

$$\frac{\mathrm{d}\boldsymbol{\varepsilon}}{\mathrm{d}c} = \frac{1}{\rho c^2}\boldsymbol{N}\boldsymbol{M}\boldsymbol{d}. \tag{32}$$

Equations (29), (30) and (32) furnish a system of ordinary differential equations for the simple wave solutions. If the initial and boundary conditions are given in terms of $\boldsymbol{\sigma}$ only, Eq. (29) can be integrated independent of Eqs. (30) and (32). If we regard c in Eq. (29) as a parameter, integration of Eq. (29) yields a stress path in stress space for simple wave solutions and $\boldsymbol{d}$ is a vector tangent to the stress path.

7. The elastic-plastic boundary

The total derivative of $\boldsymbol{w}$ along a curve Γ in the $x-t$ plane is (see Eq. (10))

$$\frac{\mathrm{d}\boldsymbol{w}}{\mathrm{d}t} = \lambda \boldsymbol{w}_x + \boldsymbol{w}_t \tag{33}$$

where $\lambda = \mathrm{d}x/\mathrm{d}t$ is the slope of the curve Γ. Elimination of $\boldsymbol{w}_x$ between Eqs. (6) and (33) yields

$$(\lambda \boldsymbol{A} - \boldsymbol{B})\boldsymbol{w}_t = \lambda \boldsymbol{b} - \boldsymbol{B}\frac{\mathrm{d}\boldsymbol{w}}{\mathrm{d}t}. \tag{34}$$

With $(\lambda \boldsymbol{A} - \boldsymbol{B})$ given by Eq. (13) and noticing that

$$\begin{bmatrix} \boldsymbol{I} & \boldsymbol{0} \\ \boldsymbol{N}/\rho\lambda & \boldsymbol{I}/\rho\lambda \end{bmatrix}^{-1} = \begin{bmatrix} \boldsymbol{I} & \boldsymbol{0} \\ -\boldsymbol{N} & \rho\lambda \boldsymbol{I} \end{bmatrix}, \tag{35}$$

Eq. (34) can be written as

$$\begin{bmatrix} \rho\lambda \boldsymbol{I} & \boldsymbol{M} \\ \boldsymbol{0} & \boldsymbol{D} \end{bmatrix} \begin{bmatrix} \boldsymbol{v}_t \\ \boldsymbol{\sigma}_t \end{bmatrix} = \begin{bmatrix} \boldsymbol{I} & \boldsymbol{0} \\ -\boldsymbol{N} & \rho\lambda \boldsymbol{I} \end{bmatrix} \left(\lambda \boldsymbol{b} - \boldsymbol{B}\frac{\mathrm{d}\boldsymbol{w}}{\mathrm{d}t} \right). \tag{36}$$

If Γ is not a characteristic, the matrix on the left can be inverted and

$$\begin{bmatrix} \boldsymbol{v}_t \\ \boldsymbol{\sigma}_t \end{bmatrix} = \begin{bmatrix} \boldsymbol{I}/\rho\lambda & -\boldsymbol{M}\boldsymbol{D}^{-1}/\rho\lambda \\ \boldsymbol{0} & \boldsymbol{D}^{-1} \end{bmatrix} \begin{bmatrix} \boldsymbol{I} & \boldsymbol{0} \\ -\boldsymbol{N} & \rho\lambda \boldsymbol{I} \end{bmatrix} \left(\lambda \boldsymbol{b} - \boldsymbol{B}\frac{\mathrm{d}\boldsymbol{w}}{\mathrm{d}t} \right) \tag{37}$$

which gives $\boldsymbol{v}_t$ and $\boldsymbol{\sigma}_t$ in terms of $\mathrm{d}\boldsymbol{w}/\mathrm{d}t$ along Γ.

If Γ is the boundary between an elastic region and a plastic region, (Eq. 36) applies to both sides of Γ provided the superscripts e and p are introduced to distinguish between elastic and plastic regions. Since $\boldsymbol{\sigma}$ and $\boldsymbol{v}$ are continuous across an elastic-plastic boundary, the right hand side of Eq. (36) remains the same and

$$\begin{bmatrix} \rho\lambda \boldsymbol{I} & \boldsymbol{M} \\ \boldsymbol{0} & \boldsymbol{D}^e \end{bmatrix} \begin{bmatrix} \boldsymbol{v}_t^e \\ \boldsymbol{\sigma}_t^e \end{bmatrix} = \begin{bmatrix} \rho\lambda \boldsymbol{I} & \boldsymbol{M} \\ \boldsymbol{0} & \boldsymbol{D}^p \end{bmatrix} \begin{bmatrix} \boldsymbol{v}_t^p \\ \boldsymbol{\sigma}_t^p \end{bmatrix}, \tag{38}$$

or

$$\rho\lambda(\boldsymbol{v}_t^e - \boldsymbol{v}_t^p) + \boldsymbol{M}(\boldsymbol{\sigma}_t^e - \boldsymbol{\sigma}_t^p) = \boldsymbol{0}, \tag{39a}$$

$$\boldsymbol{D}^e\boldsymbol{\sigma}_t^e = \boldsymbol{D}^p\boldsymbol{\sigma}_t^p. \tag{39b}$$

The last equation can be rewritten, using Eqs. (17) and (5), as

$$\boldsymbol{D}^e(\boldsymbol{\sigma}_t^e - \boldsymbol{\sigma}_t^p) = 2\rho\lambda^2 Gkk_t^p(\nabla f), \tag{40}$$

or, using Eq. (19) as

$$\boldsymbol{\sigma}_t^e - \boldsymbol{\sigma}_t^p = 2\rho\lambda^2 Gkk_t^p(\boldsymbol{D}^e)^{*T}(\nabla f)/\|\boldsymbol{D}^e\|. \tag{41}$$

If

$$\beta = 2\rho\lambda^2 Gkk_t^p/\|\boldsymbol{D}^e\| \tag{42}$$

it follows that in view of Eq. (25a)

$$\boldsymbol{\sigma}_t^e - \boldsymbol{\sigma}_t^p = \boldsymbol{d}(\lambda) \tag{43}$$

and Eq. (39a) may be expressed as

$$\boldsymbol{v}_t^e - \boldsymbol{v}_t^p = -\boldsymbol{M}\boldsymbol{d}(\lambda)/\rho\lambda. \tag{44}$$

Equations (43) and (44) furnish the discontinuity in the time derivative of $\boldsymbol{\sigma}$ and $\boldsymbol{v}$ across an elastic-plastic boundary.

The results obtained so far, lead to the universal relationship

$$[\boldsymbol{\sigma}_t] = \boldsymbol{d}(\lambda), [\boldsymbol{v}_t] = -\boldsymbol{M}\boldsymbol{d}(\lambda)/\rho\lambda, [\boldsymbol{\varepsilon}_t] = \boldsymbol{N}\boldsymbol{M}\boldsymbol{d}(\lambda)/\rho\lambda^2 \tag{45}$$

where $[\boldsymbol{\sigma}_t]$ denotes the discontinuity in $\boldsymbol{\sigma}_t$ across a curve Γ of slope λ. If Γ is an elastic-plastic boundary, λ can be the loading wave speed c_l or the unloading wave speed c_u. If Γ is the characteristic, λ can be c_f, c_2, or c_s in the plastic region, or c_1, c_2 in the elastic region where c_1 is the dilatational wave speed in the elastic region. For a thin-walled tube c_1 is replaced by c_0 and

$$[\boldsymbol{\sigma}_t] = \beta \begin{bmatrix} \left(\dfrac{\lambda^2}{c_2^2} - 1\right)\dfrac{2\sigma}{\theta^2} \\ \left(\dfrac{\lambda^2}{c_0^2} - 1\right)2\tau \end{bmatrix} = -\rho\lambda[\boldsymbol{v}_t] = \rho\lambda^2[\boldsymbol{\varepsilon}_t]. \tag{46}$$

Eq. (46) applies to $\lambda = c_f$, c_s, c_u, c_l, c_0, and c_2. It should be mentioned that proportionality factor β is fixed and is given by Eq. (42) when Γ is an elastic-plastic boundary.

In the next section both will be discussed the completeness of the solution $\boldsymbol{d}$ given by Eq. (25a) and the solution $\boldsymbol{\delta}$ indicated by Eq. (25b). In view of the similarity between $\boldsymbol{d}$ and $\boldsymbol{\delta}$, it is sufficient to discuss the completeness of $\boldsymbol{d}$ only.

8. Completeness of the solution *d*

The solution of Eq. (23a) in the form of Eq. (25a) is obviously not complete if $\|\boldsymbol{D}\| = 0$ has a multiple root. For instance, if $\boldsymbol{D}$ has the dimension of $m = 5$, $\|\boldsymbol{D}\| = 0$ will have 5 roots for c^2. All of these five roots cannot be distinct since there exist at the most three wave speeds in the plastic region and two wave speeds in the elastic region. Thus $\boldsymbol{d}$ given by Eq. (25a) cannot furnish five distinct $\boldsymbol{d}$.

In the elastic region, $\boldsymbol{D}^e$ has in general a simple form and all $\boldsymbol{d}$ associated with a multiple root can be obtained by solving Eq. (23a) directly. It is in the plastic region that Eq. (25a) becomes most useful. If $c = c_f$ and c_s are simple roots of $\|\boldsymbol{D}^p\| = 0$, $\boldsymbol{d}$ given by Eq. (25a) is the only solution of Eq. (23a) for $c = c_f$ and c_s. It will be shown that this is the case if $\boldsymbol{N} = \boldsymbol{M}^T$ which is true for all waves due to combined stress considered in the literatures (see [12]). More precisely, if c_1 and c_2 are the non-zero roots of $\|\boldsymbol{D}^e\| = 0$ with the multiplicity of a and b respectively, i.e., if

$$\|\boldsymbol{D}^e(c)\| = \gamma^{2(m-a-b)}(c^2 - c_1^2)^a(c^2 - c_2^2)^b \tag{47}$$

where γ is independent of c, it may be shown that

$$\|\boldsymbol{D}^p(c)\| = \gamma' c^{2(m-a-b)}(c^2 - c_1^2)^{a-1}(c^2 - c_2^2)^{b-1}(c^2 - c_f^2)(c^2 - c_s^2) \tag{48}$$

provided $\boldsymbol{N} = \boldsymbol{M}^T$.

To prove Eq. (48), first notice that $\boldsymbol{S}^e$ is symmetric and positive definite for elastic materials. Using the assumption that $\boldsymbol{N} = \boldsymbol{M}^T$,

$$\boldsymbol{D}^e(c) = \rho c^2 \boldsymbol{S}^e - \boldsymbol{M}^T \boldsymbol{M} \tag{49}$$

which is also symmetric. Let $\boldsymbol{d}_i$ be the eigenvector associated with c_i where $\|\boldsymbol{D}^e(c_i)\| = 0$, so that

$$\rho c_i^2 \boldsymbol{S}^e \boldsymbol{d}_i = \boldsymbol{M}^T \boldsymbol{M} \boldsymbol{d}_i. \tag{50}$$

Notice that c_i is either equal to 0, c_1 or c_2. Defining the matrix

$$\boldsymbol{R} = \{\boldsymbol{d}_1, \boldsymbol{d}_2, \ldots, \boldsymbol{d}_m\}, \tag{51}$$

since both $\boldsymbol{S}^e$ and $(\boldsymbol{M}^T\boldsymbol{M})$ are symmetric and $\boldsymbol{S}^e$ is positive definite it can be shown that (see [22])

$$\boldsymbol{R}^T\boldsymbol{S}^e\boldsymbol{R} = \boldsymbol{I} \tag{52}$$

using a proper normalization on $\boldsymbol{d}_i$. If $\boldsymbol{C}$ is the diagonal matrix with elements c_i^2, Eq. (50) can be written as

$$\rho\boldsymbol{S}^e\boldsymbol{R}\boldsymbol{C} = \boldsymbol{M}^T\boldsymbol{M}\boldsymbol{R} \tag{53}$$

or, using Eq. (52) as

$$\rho\boldsymbol{C} = \boldsymbol{R}^T\boldsymbol{M}^T\boldsymbol{M}\boldsymbol{R}. \tag{54}$$

It follows from Eq. (49) that in view of Eqs. (52) and (54)

$$\boldsymbol{R}^T\boldsymbol{D}^e(c)\boldsymbol{R} = \boldsymbol{R}^T(\rho c^2\boldsymbol{S}^e-\boldsymbol{M}^T\boldsymbol{M})\boldsymbol{R} = \rho(c^2\boldsymbol{I}-\boldsymbol{C})$$

Since $\|\boldsymbol{S}^e\|$ is finite, $\|\boldsymbol{R}\| \neq 0$ by Eq. (52) and $\boldsymbol{R}^{-1}$ exists such that

$$\boldsymbol{R}^{-1}(\boldsymbol{D}^e(c))^{-1}(\boldsymbol{R}^T)^{-1} = \rho^{-1}(c^2\boldsymbol{I}-\boldsymbol{C})^{-1}$$

or

$$(\boldsymbol{D}^e(c))^{-1} = \rho^{-1}\boldsymbol{R}(c^2\boldsymbol{I}-\boldsymbol{C})^{-1}\boldsymbol{R}^T.$$

But $(\boldsymbol{D}^e)^{-1} = (\boldsymbol{D}^e)^*/\|\boldsymbol{D}^e\|$ in view of Eq. (19) and

$$(\boldsymbol{D}^e(c))^* = \rho^{-1}\|\boldsymbol{D}^e\|\boldsymbol{R}(c^2\boldsymbol{I}-\boldsymbol{C})^{-1}\boldsymbol{R}^T. \tag{55}$$

Substituting Eq. (55) into Eq. (18) yields

$$\|\boldsymbol{D}^p(c)\| = \|\boldsymbol{D}^e(c)\| + \boldsymbol{\xi}^T\{c^2(c^2\boldsymbol{I}-\boldsymbol{C})^{-1}\|\boldsymbol{D}^e(c)\|\}\boldsymbol{\xi} \tag{56}$$

where

$$\boldsymbol{\xi} = \sqrt{G}\boldsymbol{R}^T\nabla f. \tag{57}$$

Since $(c^2\boldsymbol{I}-\boldsymbol{C})$ is a diagonal matrix whose elements are either c^2, $c^2-c_1^2$ or $c^2-c_2^2$, the $c^2(c^2\boldsymbol{I}-\boldsymbol{C})^{-1}$ is a diagonal matrix whose elements are either 1, $c^2(c^2-c_1^2)^{-1}$ or $c^2(c^2-c_2^2)^{-1}$. In view of Eq. (47), the right hand side of Eq. (56) has the following common factor

$$c^{2(m-a-b)}(c^2-c_1^2)^{a-1}(c^2-c_2^2)^{b-1}. \tag{58}$$

This completes the proof of Eq. (48).

Now, since $c = c_1$ cannot be a root of $\|\boldsymbol{D}^p\| = 0$, the factor $(c^2-c_1^2)$ in Eq. (48) should be absent. Thus we must have

$$a = 1 \tag{59}$$

in Eq. (48) and also in Eqs. (47) and (58).

It should be mentioned that $c = c_2$ can be a root of both $\|\boldsymbol{D}^p\| = 0$ and $\|\boldsymbol{D}^e\| = 0$ when $b \geqq 2$. The vector $\boldsymbol{d}$ associated with c_2 has different expressions depending on whether c_2 is considered as an elastic wave speed or as a plastic wave speed. (see [6] for an example). From Eq. (55), with the use of Eqs. (47) and (59), the elements of $(\boldsymbol{D}^e(c))^*$ have the following common factor:

$$c^{2(m-b-2)}(c^2-c_2^2)^{b-1}. \tag{60}$$

Thus $(\boldsymbol{D}^e(c_2))^* = \boldsymbol{0}$ if $b \geqq 2$ and $\boldsymbol{d}(c)$ as given by Eq. (25a) yields the trivial solution $\boldsymbol{d}(c_2) = \boldsymbol{0}$. Therefore when c_2 is a multiple root of $\|\boldsymbol{D}^e\| = 0$, $\boldsymbol{d}(c_2)$ should be obtained by solving Eq. (23a).

Finally, $c = 0$ is a root of $\|\boldsymbol{D}^e\| = 0$ if $\|\boldsymbol{NM}\| = 0$. In all problems considered in the literatures, $\boldsymbol{NM}$ has a singularity of order $m-n$. Hence $c = 0$ is a root of $\|\boldsymbol{D}^e\| = 0$ of multiplicity $2(m-n)$. In view of Eqs. (47) and (59), we have the relation

$$n = 1+b. \tag{61}$$

9. Concluding remarks

With minor modifications (see [12]), the analysis represented is not restricted to isotropic work-hardening materials and may be applied to combined kinematic and isotropic work hardening materials considered by Goel and Malvern [8].

The restriction on the speed of an unloading and loading wave was studied in [18, 19] for longitudinal waves in a thin rod and in [20] for combined stress waves in a thin-walled tube. A unified treatment on combined stress waves was given in [12] with results agreeing with the general three-dimensional case considered by Mandel [21].

It is hoped that the present paper, in connection with [12], provides a unified treatment which will be useful in analyzing a more complicated combined stress waves in rate-independent elastic-plastic materials.

Acknowledgment

The work presented here was supported by the U.S. Army Research Office-Durham under DA-ARO-D-31-124-72-G87

References

[1] Rakhmatulin, kh. A., On the propagation of elastic-plastic waves owing to combined loadings. *Appl. Math. Mech. (PMM)*, *22* (1958) 1079–1088.

[2] Cristescu, N., On the propagation of elastic-plastic waves for combined stresses. *Appl. Math. Mech. (PMM)*, *23* (1959) 1605–1612.

[3] Bleich, H. H. and I. Nelson, Plane waves in an elastic-plastic half space due to combined surface pressure and shear. *J. Appl. Mech.*, *33* (1966) 149–158.

[4] Ting, T. C. T. and Ning Nan, Plane waves due to combined compressive and shear stresses in a half-space. *J. Appl. Mech.*, *36* (1969) 189–197.

[5] Ting, T. C. T., Interaction of shock waves due to combined two shear loadings. *Int. J. Solids Structures*, *5* (1969) 415–435.

[6] Goel, R. P. and L. E. Malvern, Elastic-plastic plane waves with combined compressive and two shear stresses in a half space. *J. Appl. Mech.*, *38* (1971) 895–898.

[7] Clifton, R. J., An analysis of combined longitudinal and torsional plastic waves in a thin-walled tube. Fifth U.S. National Congress of *Appl. Mech., Proceedings. ASME, N.Y.*, (1966) 465–480.

[8] Goel, R. P. and L. E. Malvern, Biaxial plastic simple waves with combined kinematic and isotropic work hardening. *J. Appl. Mech.*, *37* (1970) 1100–1106.

[9] Cristescu, N., Some observations about the case of plane axial-symmetric deformations in the dynamic problem of plasticity (Prandtl-Reuss theory). *Com. Acad. R. P. Romine* (1), *6* (1956) 19–28.

[10] Fong, Jeffrey F., *Elastic-plastic wave in a half-space of a linearly work-hardening material for coupled shear loadings*. Tech. Rept. No. 161, Div. Eng. Mech., Stanford University, Stanford, Calif., 1966.

[11] Nan, Ning, *Elastic-plastic waves for combined stress*. Tech. Rept. No. 184, Dept. Appl. Mech., Stanford University, Stanford, Calif., 1968.

[12] Ting, T. C. T., Elastic-plastic boundaries in the propagation of plane and cylindrical waves of combined stress. *Q. Appl. Math.*, *47* (1970) 441–449.

[13] Ting, T. C. T., On the initial slope of elastic-plastic boundaries in combined longitudinal and torsional wave propagation. *J. Appl. Mech.*, *36* (1969) 203–211.

[14] Ting, T. C. T., The initiation of combined stress waves in a thin-walled tube due to impact loadings. *Int. J. Solids Structures*, *8* (1972) 269–293.

[15] Courant, R. and Hilbert, D., *Methods of mathematical physics. II*, Interscience Publ., N.Y., 1962.

[16] Jeffrey, A. and T. Taniutti, Nonlinear wave propagation with applications to physics and magnetohydrodynamics. Academic Press, N. Y., 1964.

[17] Hill, R., *The mathematical theory of plasticity*. Clarendon Press, Oxford, 1950.

[18] Lee, E. H., A boundary value problem in the theory of plastic wave propagation. *Q. Appl. Math.*, *10* (1953) 335–346.

[19] Clifton, R. J. and T. C. T. Ting, The elastic-plastic boundary in one-dimensional wave propagation. *J. Appl. Mech.*, *35* (1968) 812–814.

[20] Clifton, R. J., Elastic-plastic boundaries in combined longitudinal and torsional plastic wave propagation. *J. Appl. Mech.*, *35* (1968) 782–786.

[21] Mandel, Jean, Ondes plastiques dans un milieu indefini a trois dimensions. *J. Mecanique* (1), *1* (1962) 3–30.

[22] Hilderbrand, F. B., *Methods of applied mathematics*. Prentice-Hall, N. Y., 74–79.
[23] Fukuoka, Hidekazu, Infinitesimal plane waves in elastic-plastic tubes under combined tension-torsion loads, *Proceedings of the Sixteenth Japan National Congress for Applied Mechanics* (1966) 109–113.
[24] Hill, R., Acceleration waves in solids, *J. Mech. Physics Solids, 10* (1962) 1–16.
[25] Craggs, J. W., Plastic Waves, in *Progress in Solid Mechanics, 2* (1961) 141–197.

A note on the strength of the combined tension-torsion waves in elastic-plastic tubes

Hidekazu Fukuoka

Osaka University, Toyonaka, Osaka, Japan

The singular surfaces and characteristic equations for waves propagating in thin-walled tubes made of rate independent elastic-plastic material under combined tension-torsion are studied by means of Hadamard's discontinuity relations.

The variations in strength of the acceleration waves as they propagate in tubes loaded statically by combined stress are determined using the second order compatibility conditions developed by Thomas.

1. Introduction

Elastic-plastic waves in an infinite medium of Prandtl-Reuss material have been investigated by Thomas [1] and Mandel [2] by considering waves as propagating singular surfaces. Speeds of propagation of fast and slow waves in a thin-walled tube have been determined by Clifton [3] and Fukuoka [4]. Clifton [3][5] and Vitiello and Clifton [6] have solved several boundary value problems for waves in a thin-walled tube. Ting [7] has given an analytical solution for the initial slope of elastic-plastic boundaries in combined longitudinal and torsional wave propagation. Some experimental results of the combined wave propagation in a thin-walled tube were reported by Lipkin and Clifton [8], and Fukuoka and Masui [9].

In this paper, plane acceleration waves in a thin-walled tube made of rate independent elastic-plastic material and loaded by combined tension-torsion are studied by considering the waves as propagating singular surfaces. Growth and decay of the acceleration waves are investigated for combined loading.

2. Governing equations

It is assumed that the stress state is two dimensional and that the deformation is so small that the difference between material and spatial coordinates is negligible. Radial inertia is also neglected. The x, y and z axes are orientated in the axial, circumferential and radial directions of the tube respectively and $\sigma_1 = \sigma_x, \sigma_2 = \tau_{xy}, v_1 = v_x, v_2 = v_y$. Latin indices have the range 1, 2 throughout the paper and summation is implied by repeated indices. The equations of motion are

$$\rho \partial v_i/\partial t = \partial \sigma_i/\partial x \tag{1}$$

where ρ is the mass density of the medium. The constitutive equations of Prandtl-Reuss materials for this case of combined stress are given in [4] by

$$\partial \sigma_i/\partial t = a_{ik} \partial v_k/\partial x. \tag{2}$$

In the elastic state the coefficients a_{ik} are the constants

$$a_{11} = 2(1+\nu)\mu, \qquad a_{12} = a_{21} = 0, \qquad a_{22} = \mu \tag{3}$$

where ν is Poisson's ratio and μ is the modulus of rigidity. In the plastic state these coefficients are given by

$$\begin{aligned} a_{11} &= \frac{2(1+\nu)\mu(H'+3\mu \sin^2\theta)}{3\mu+H'-(1-2\nu)\mu\cos^2\theta}, \\ a_{12} = a_{21} &= -\frac{2\sqrt{3}(1+\nu)\mu^2 \cos\theta \sin\theta}{3\mu+H'-(1-2\nu)\mu\cos^2\theta}, \\ a_{22} &= \frac{\mu H' + 2(1+\nu)\mu^2\cos^2\theta}{3\mu+H'-(1-2\nu)\mu\cos^2\theta} \end{aligned} \tag{4}$$

where H' is the slope of the equivalent stress-plastic strain curve and θ is the stress ratio defined by

$$\tan\theta = \sqrt{3}\sigma_2/\sigma_1. \tag{5}$$

The equivalent stress σ_{eq} is given by

$$\sigma_{eq}^2 = \sigma_1^2 + 3\sigma_2^2. \tag{6}$$

3. Compatibility conditions for discontinuities and characteristic equations

Consider a singular surface Σ across which the velocities and stresses are continuous while their first derivatives may be discontinuous. Let G denote the speed of propagation of the surface Σ, in the direction normal to itself. Plane waves whose normals coincide with the tube axis are considered in the present case. Then, compatibility conditions of the first order are written as follows:

$$[\partial v_i/\partial x] = \eta_i, \qquad [\partial v_i/\partial t] = -G\eta_i \tag{7}$$

for velocities and

$$[\partial \sigma_i/\partial x] = \xi_i, \quad [\partial \sigma_i/\partial t] = -G\xi_i \tag{8}$$

for stresses. The square bracket denotes a jump in the quantity enclosed, for example if

$$[f] = f_2 - f_1, \tag{9}$$

f_1 and f_2 indicate the values of f on the front and on the rear of the surface Σ, respectively. In (7) and (8) η_i and ξ_i are functions defined over the surface Σ.

Compatibility conditions of the second order given by Thomas [1] for the plane acceleration waves which propagate with constant speed reduce to

$$\begin{aligned} &[\partial^2 v_i/\partial x^2] = \bar{\eta}_i, \\ &[\partial^2 v_i/\partial x \partial t] = -G\bar{\eta}_i + (\delta \eta_i/\delta t), \\ &[\partial^2 v_i/\partial t^2] = G^2 \bar{\eta}_i - 2G(\delta \eta_i/\delta t) \end{aligned} \tag{10}$$

for velocities and

$$\begin{aligned} &[\partial^2 \sigma_i/\partial x^2] = \bar{\xi}_i, \\ &[\partial^2 \sigma_i/\partial x \partial t] = -G\bar{\xi}_i + (\delta \xi_i/\delta t), \\ &[\partial^2 \sigma_i/\partial t^2] = G^2 \bar{\xi}_i - 2G(\delta \xi_i/\delta t) \end{aligned} \tag{11}$$

for stresses, where $\bar{\eta}_i$ and $\bar{\xi}_i$ are functions defined over the surface Σ and $\delta/\delta t$ denotes differentiation following the moving surface Σ.

Evaluating the equations of motion (1) on the rear and on the front of the surface Σ, subtracting the corresponding members of the two equations thus obtained, and applying compatibility conditions (7), (8) yields

$$\xi_i = -\rho G \eta_i \tag{12}$$

in which both the density ρ and its derivative are assumed to be continuous across the surface Σ. Similarly, evaluating the constitutive equations (2) on both sides of Σ and subtracting the corresponding members of the two equations it follows that

$$[\partial \sigma_i/\partial t] = [a_{ik}](\partial v_k/\partial x)_1 + (a_{ik})_2[\partial v_k/\partial x]. \tag{13}$$

Applying the compatibility conditions to equations (13)

$$-G\xi_i = [a_{ik}](\partial v_k/\partial x)_1 + (a_{ik})_2\eta_k. \tag{14}$$

On account of equations (12), equations (14) may be rewritten as

$$\rho G^2\eta_i = [a_{ik}](\partial v_k/\partial x)_1 + (a_{ik})_2\eta_k$$

or equivalently

$$\{\rho G^2\delta_{ik} - (a_{ik})_2\}\eta_k = [a_{ik}](\partial v_k/\partial x)_1 \tag{15}$$

where δ_{ik} is the Kronecker delta. When the wave propagates into a region of constant state, the characteristic equation results in the form

$$\{\rho G^2\delta_{ik} - (a_{ik})_2\}\eta_k = 0. \tag{16}$$

4. Strength of wave

Differentiating equations (1) and (2) with respect to time, evaluating the derivatives on the both sides of Σ and subtracting the corresponding members of the two equations, one obtains

$$\rho[\partial^2 v_i/\partial t^2] = [\partial^2\sigma_i/\partial x\partial t] \tag{17}$$

and

$$\begin{aligned}[\partial^2\sigma_i/\partial t^2] = {} & [\partial a_{ik}/\partial t][\partial v_k/\partial x] + [\partial a_{ik}/\partial t](\partial v_k/\partial x)_1 + \\ & + [\partial v_k/\partial x](\partial a_{ik}/\partial t)_1 + [a_{ik}][\partial^2 v_k/\partial x\partial t] + \\ & + [a_{ik}](\partial^2 v_k/\partial x\partial t)_1 + [\partial^2 v_k/\partial x\partial t](a_{ik})_1.\end{aligned} \tag{18}$$

It is now assumed that the medium in front of the surface Σ is the region of constant state under static combined loading. Then, equations (18) reduce to

$$[\partial^2\sigma_i/\partial t^2] = [\partial a_{ik}/\partial t][\partial v_k/\partial x] + [\partial^2 v_k/\partial x\partial t](a_{ik})_2. \tag{19}$$

But equations (17) and (19) can be written in the form

$$\rho G^2\bar{\eta}_i - 2\rho G(\delta\eta_i/\delta t) = -G\bar{\xi}_i + (\delta\xi_i/\delta t) \tag{20}$$

and

$$G^2\bar{\xi}_i - 2G(\delta\xi_i/\delta t) = [\partial a_{ik}/\partial t]\eta_k - G\bar{\eta}_k(a_{ik})_2 + (\delta\eta_k/\delta t)(a_{ik})_2 \tag{21}$$

if use is made of the compatibility conditions (10) and (11). It follows from equations (12), (20) and (21) that

$$2\rho G^2(\delta\eta_i/\delta t)-[\partial a_{ik}/\partial t]\eta_k = \{\rho G^2\delta_{ik}-(a_{ik})_2\}\{G\bar{\eta}_k-(\delta\eta_k/\delta t)\}. \qquad (22)$$

If equations (22) are multiplied by η_i and summed over the repeated index i by making use of the homogeneous equations (16) for the case where the side 1 is the region of constant state, one finally obtains

$$2\rho G^2\eta_i(\delta\eta_i/\delta t)-[\partial a_{ik}/\partial t]\eta_i\eta_k = 0. \qquad (23)$$

Let us define the strength of the wave by

$$\eta^2 = \eta_1^2+3\eta_2^2 \qquad (24)$$

and the angle ϕ by

$$\tan\phi = \sqrt{3}\eta_2/\eta_1. \qquad (25)$$

When both sides of the surface Σ are in elastic states,

$$\partial a_{ik}/\partial t = 0,$$

and then it follows that

$$\eta = \text{constant}.$$

Otherwise, values of $[\partial a_{ik}/\partial t]$ in the equation (23) are obtained from equations (4) as

$$\left[\frac{\partial a_{11}}{\partial t}\right] = \frac{4(1+\nu)^2\mu^2}{\{3\mu+H'-(1-2\nu)\mu\cos^2\theta\}^2}\left[\!\left[\cos^2\theta\left[\frac{\partial H'}{\partial t}\right]+\right.\right.$$
$$\left.\left.+(3\mu+H')\sin 2\theta\left[\frac{\partial\theta}{\partial t}\right]\right]\!\right],$$
$$\left[\frac{\partial a_{12}}{\partial t}\right] = \frac{2\sqrt{3}(1+\nu)\mu^2}{\{3\mu+H'-(1-2\nu)\mu\cos^2\theta\}^2}\left[\!\left[\sin\theta\cos\theta\left[\frac{\partial H'}{\partial t}\right]+\right.\right.$$
$$\left.\left.+\{(1-2\nu)\mu\cos^2\theta-(3\mu+H')\cos 2\theta\}\left[\frac{\partial\theta}{\partial t}\right]\right]\!\right], \qquad (26)$$
$$\left[\frac{\partial a_{22}}{\partial t}\right] = \frac{3\mu^2}{\{3\mu+H'-(1-2\nu)\mu\cos^2\theta\}^2}\left[\!\left[\sin^2\theta\left[\frac{\partial H'}{\partial t}\right]-\right.\right.$$
$$\left.\left.-\{2(1+\nu)\mu+H'\}\sin 2\theta\left[\frac{\partial\theta}{\partial t}\right]\right]\!\right].$$

$[\partial H'/\partial t]$ on the right hand side of the above expression is given by

$$[\partial H'/\partial t] = (H''/H')[\dot{\sigma}_{eq}]. \tag{27}$$

Differentiating equation (6) with respect to time and making use of both the compatibility conditions (8) and equations (12) leads to

$$[\dot{\sigma}_{eq}] = \rho G^2(\eta_1 \cos\theta + \sqrt{3}\eta_2 \sin\theta). \tag{28}$$

It follows from equations (27), (28) and (25) that

$$[\partial H'/\partial t] = \rho G^2(H''/H')\eta \cos(\phi-\theta). \tag{29}$$

$[\partial\theta/\partial t]$ is obtainable by differentiating equation (5) with respect to time and making use of both the compatibility conditions and equation (25). Thus

$$[\partial\theta/\partial t] = (\rho G^2/\sigma_{eq})\eta \sin(\phi-\theta). \tag{30}$$

Substituting equations (25), (26), (29) and (30) into equation (23), the equation for the strength of the wave may be expressed by

$$(\delta\eta/\delta t) - K(\theta)\eta^2 = 0 \tag{31}$$

where

$$K(\theta) = \frac{\sin\theta\sin\phi + 2(1+\nu)\cos\theta\cos\phi}{\{3+h'-(1-2\nu)\cos^2\theta\}^2} \times \times \left[\frac{h''\{\sin\theta\sin\phi + 2(1+\nu)\cos\theta\cos\phi}{2h'} \cdot \cos(\phi-\theta) + + \frac{2(1+\nu)(3+h')\sin\theta\cos\phi - \{2(1+\nu)+h'\}\cos\theta\sin\phi}{s} \cdot \sin(\phi-\theta)\right] \tag{32}$$

in which $h' = H'/\mu$, $h'' = H''/\mu$ and $s = \sigma_{eq}/\mu$. ϕ may be related to θ using equations (25), (16), (4) and (5). Examples of ϕ, θ relationships are given by the curves in Fig. 1. Equation (31) may be integrated at once for the present case in which the medium in front of the surface Σ is the region of constant state under static combined loading. Thus

$$1/\eta = (1/\eta_0) - K(\theta)t \tag{33}$$

or

$$1/\eta = (1/\eta_0) - \{K(\theta)/G\}x \tag{34}$$

where η_0 is the initial value of η for $x = 0$.

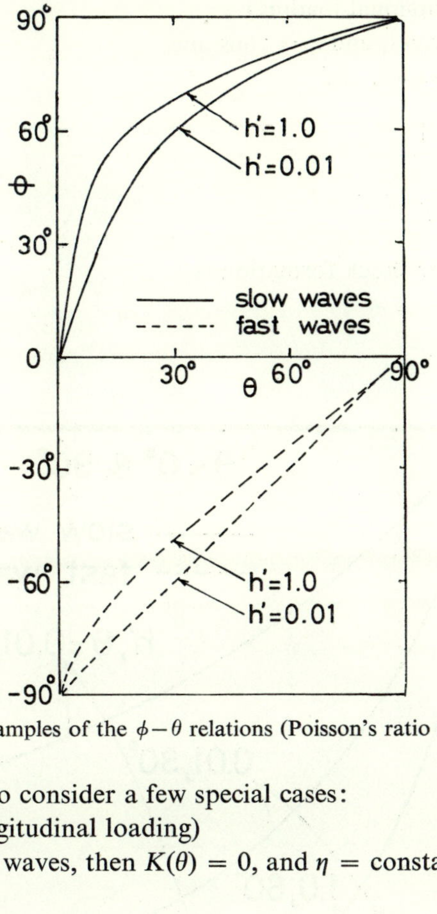

Fig. 1. Examples of the $\phi-\theta$ relations (Poisson's ratio $\nu = 0.33$).

It is of interest to consider a few special cases:

1. $\theta = 0$ (pure longitudinal loading)

$\phi = -90°$ for fast waves, then $K(\theta) = 0$, and η = constant as in the elastic case.

As $\phi = 0$ for slow waves,

$$\eta = \frac{1}{\dfrac{1}{\eta_0} - \dfrac{h''}{2h'}\left\{\dfrac{2(1+\nu)}{2(1+\nu)+h'}\right\}^2 t}.$$

When $h'' < 0$, η decays to zero in infinite time. On the other hand, when $h'' > 0$, η will become infinite in finite time

$$t_c = \frac{2h'}{\eta_0 h''}\left\{\frac{2(1+\nu)+h'}{2(1+\nu)}\right\}^2,$$

and the acceleration wave will be changed to a shock wave.

2. $\theta = 90°$ (pure torsional loading)
$\phi = 0$ for fast waves, and η = constant.
$\phi = 90°$ for slow waves, and

$$\eta = \frac{1}{\dfrac{1}{\eta_0} - \dfrac{h''}{2h'(3+h')^2}t}.$$

The critical time for shock formation is

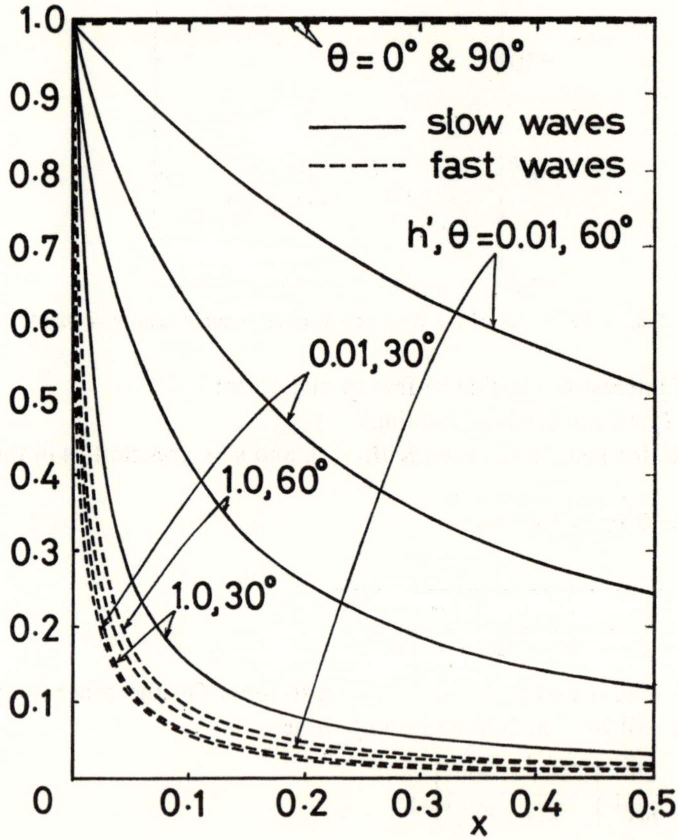

Fig. 2. Decay of the strength of the combined wave for linear work-hardening material.

$$t_c = \frac{2h'(3+h')^2}{\eta_0 h''}.$$

3. $h'' = 0$ (linear work-hardening).

Examples of variation of the strength η as the wave propagates in a tube are shown in Fig. 2 for some values of the hardening index h' and the stress ratio θ. Values of Poisson's ratio ν and the nondimensional stresses are fixed as 0.33 and 0.0038 respectively. Initial values η_0 are taken as unity. The solid curves indicate values for slow waves and the dotted curves for fast waves. It can be seen that, in general, fast waves decay very quickly.

Acknowledgement

I would like to acknowledge to Dr. J. Bejda of Warsaw for his helpful comments on this note.

References

[1] Thomas, T. Y., *Plastic flow and fracture in solids*. Academic Press, New York, 1961.
[2] Mandel, J., Ondes plastiques dans un milieu indéfini à trois dimensions. *Journal de Mécanique*, *1* (1962) 3–30.
[3] Clifton, R. J., *An analysis of combined longitudinal and torsional plastic waves in a thin-walled tube*. Proceedings of the 5th U.S. National Congress of Applied Mechanics, ASME, June 1966, 465–480.
[4] Fukuoka, H., *Infinitesimal plane waves in elastic-plastic tubes under combined tension-torsion loads*. Proceedings of the 16th Japan National Congress for Applied Mechanics, (Tokyo, 1966), 109–113.
[5] Clifton, R. J., Elastic-plastic boundaries in combined longitudinal and torsional plastic wave propagation. *Journal of Applied Mechanics*, *35* (1968) 782–786.
[6] Vitiello, D. L. and R. J. Clifton, *The numerical solution of a problem in the propagation of plastic waves of combined stress*. Brown University, Division of Applied Mathematics Report, July, 1967.
[7] Ting, T. C. T., On the initial slope of elastic-plastic boundaries in combined longitudinal and torsional wave propagation. *Journal of Applied Mechanics*, *36* (1969) 203–211.
[8] Lipkin, J. and R. J. Clifton, *An experimental study of combined longitudinal and torsional plastic waves in a thin-walled tube*. 12th International Congress of Applied Mechanics, Stanford University, August 1968.
[9] Fukuoka, H. and K. Masui, *Experiment on incremental impact loading of plastically prestressed aluminum*. Proceedings of the 13th Japan Congress of Materials Research, March 1970, 106–109.

Uniqueness and stability at finite deformation of rigid-plastic thick-walled cylinders under hydrostatic pressure

Hans Strifors and Bertil Storåkers

The Royal Institute of Technology, Stockholm, Sweden

The rate-problem for finite deformations of a strain-hardening plastic solid under hydrostatic surface pressure is rigorously formulated. Basic relationships are employed in a search for necessary and sufficient conditions in case of bifurcating solutions for closed-end pressurized cylinders undergoing deformation which is uniform in the circumferential direction. It is found that buckling may take place both under internal and external pressure. Some explicit numerical results are given for cylinders manufactured from materials with different strain-hardening characteristics. The tangent modulus theory based on the Kirchhoff hypotheses, is shown to provide correct results for the external buckling load if the geometry of a shell approaches that of a thin shell.

Introduction

The problem of stress and strain in a rigid-plastic solid exposed to complex loading is in general non-linear and may be analysed only by solving a series of rate problems, since the instantaneous, constitutive continuum properties depend on the loading history. If finite straining occurs, further non-linearities appear for static and geometric reasons. The analytical difficulties may be overcome by the application of numerical methods e.g. the finite element technique, which has reached a high degree of refinement (cf e.g. [1]). The important question remains however, if at a generic stage of deformation, the equilibrium configuration of a structure is stable in the classical sense and if the structural response to prescribed changes of boundary loads and velocities is unique.

Solutions to problems of uniqueness of deformation and of stability are few and at different levels for rigid-plastic structures, undergoing finite deformation. A basic continuum theory was given in a series of papers by

Hill [2–6] in 1956–1957 but in spite of the merits of these uniqueness and stability criteria they have been rarely employed in the general treatment of practical problems. Some special shell problems have recently been dealt with, [7, 8], but a wide class of problems is still open for exploration, which from a practical view-point ought to be attractive e.g. due to the increasing interest for minimum weight design.

The title problem has recently been treated by Storåkers [8] through Hill's criteria. These give sufficient bounds for uniqueness and stability and it turns out that uniqueness implies stability but that the inverse is not necessarily true. The present study is undertaken with the intention of obtaining more precise limits for uniqueness of deformation through a complete solution of the rate-problem, with geometric effects taken fully into account, for a thick-walled, closed-end cylinder manufactured from a rigid-plastic, strain-hardening material and subjected to internal and external hydrostatic pressure.

A three-dimensional approach is adopted which provides exact results for thick-walled cylinders and makes it possible to investigate the applicability of the Kirchhoff assumptions, used for thin shells, to the present problem. The results are conservative as the occurrence of a bifurcation may not necessarily indicate that the load-carrying capacity is exhausted, i.e. a structure collapses.

The present shell problem has been considered, from the stability point of view, by Onat [9] for a von Mises solid at incipient yield. General treatments of bifurcation problems in the case of finite plastic strain have mostly been concerned with homogeneous steady stress states generated in bars subjected to uniform uniaxial tension or compression (cf. e.g. [10, 11]).

The rate-problem for a rigid-plastic strain-hardening solid in a state of finite strain

At a generic instant the internal stress distribution, the deformation state of the solid developed under hydrostatic pressure loading (not necessarily uniform), and the mechanical material properties, are supposed to be known. The interest is focused on a complete solution for any possible incipient response of the solid to a change of the pressure load.

Equilibrium equations are formulated in terms of the first Piola-Kirchhoff (nominal) stress tensor defined by

$$s^{\alpha}{}_{i} = \frac{\rho_R}{\rho}\,\sigma^j_i\,\frac{\partial X^\alpha}{\partial x^j} \tag{1}$$

where σ^i_j are the mixed components of the Cauchy (true) stress and X^α is the label in a reference configuration (material frame) of a generic particle occupying the place x^j in the current configuration. ρ_R is the mass density in the reference configuration and ρ is the current density. The stress rates are then related through

$$\dot{s}^{\alpha}{}_{i} = \frac{\rho_R}{\rho}\,[\dot{\sigma}^j_i + \sigma^j_i v^k{}_{,k} - \sigma^k_i v^j{}_{,k}]\,\frac{\partial X^\alpha}{\partial x^j} \tag{2}$$

where v^i are particle velocities and a dot and a comma denote material rate of change and covariant differentiation, respectively. Conditions for instantaneous local translational and rotational balance throughout the continuum may be expressed by (with a semicolon denoting total covariant differentiation)

$$s^{\alpha i}{}_{;\alpha} = 0 \tag{3}$$

and

$$s^{\alpha i}\,\frac{\partial x^j}{\partial X^\alpha} = s^{\beta j}\,\frac{\partial x^i}{\partial X^\beta}. \tag{4}$$

For continued balance

$$\dot{s}^{\alpha i}{}_{;\alpha} = 0 \tag{5}$$

and

$$\dot{s}^{\alpha i}\,\frac{\partial x^j}{\partial X^\alpha} + \frac{\rho_R}{\rho}\,\sigma^{ki} v^j{}_{,k} = \dot{s}^{\beta j}\,\frac{\partial x^i}{\partial X^\beta} + \frac{\rho_R}{\rho}\,\sigma^{kj} v^i{}_{,k}. \tag{6}$$

The boundary condition is given as

$$\sigma^i{}_j n_i = -p n_j, \tag{7}$$

in which n_i describes the unit surface normal and p the hydrostatic surface pressure.

Introducing (1) into (7) and taking the material time derivative yields for continued deformation

$$\frac{\rho}{\rho_R}\,\dot{s}^\alpha_i\,\frac{\partial x^j}{\partial X^\alpha}\,n_j = -\dot{p}n_i + pv^j{}_{,i}n_j - pv^k{}_{,k}n_i. \tag{8}$$

By assuming incompressibility and letting the material and spatial frames coincide at the considered instant, i.e. setting

$$\frac{\partial x^i}{\partial X^\alpha} = \delta^i_\alpha$$

eqs. (2), (6) and (8) reduce to

$$\dot{s}^j{}_i = \dot{\sigma}^j{}_i - \sigma^k{}_i v^j{}_{,k} \tag{9}$$

$$\dot{s}^{ij} + \sigma^{kj} v^i{}_{,k} = \dot{s}^{ji} + \sigma^{ki} v^j{}_{,k} \tag{10}$$

and

$$\dot{s}^j{}_i n_j = -\dot{p} n_i + p v^j{}_{,i} n_j, \tag{11}$$

respectively.

For a given steady stress state, the equilibrium equations (5), (10) and the boundary condition (11) furnish the basic equations for adjacent equilibria, provided that the constitutively admissible velocity field is known. The solid is assumed to have a coinciding flow potential and yield locus $f(\sigma_{ij}, \kappa)$, being an isotropic function of Cauchy stress σ_{ij} and $\varkappa$ a scalar parameter representing a functional of the deformation history. Introducing a scalar measure h for the current rate of strain-hardening the constitutive properties may be summarized as

$$hd_{ij} = \begin{cases} \dfrac{(\partial f/\partial \sigma_{kl})\tilde{\sigma}_{kl}}{(\partial f/\partial \sigma^{mn})(\partial f/\partial \sigma_{mn})} \dfrac{\partial f}{\partial \sigma^{ij}}, & f = 0 \text{ and } \dfrac{\partial f}{\partial \sigma_{kl}} \tilde{\sigma}_{kl} > 0 \\ 0 & f < 0 \text{ or } f = 0 \text{ and } \dfrac{\partial f}{\partial \sigma_{kl}} \tilde{\sigma}_{kl} \leqq 0 \end{cases} \tag{12}$$

where

$$d_{ij} = \tfrac{1}{2}(v_{i,j} + v_{j,i}) \tag{13}$$

is the stretching tensor and $\tilde{\sigma}_{kl}$ is the co-rotational (Jaumann) rate of change of Cauchy stress. Due to the assumed isotropy properties of the plastic potential, $\partial f/\partial \sigma_{ij}$ and σ_{ij} will have the same principal vectors which implies that

$$\frac{\partial f}{\partial \sigma_{kl}} \tilde{\sigma}_{kl} = \frac{\partial f}{\partial \sigma_{ij}} \dot{\sigma}_{ij}. \tag{14}$$

Eqs. (5) and (10) to (14) form the set necessary for a complete solution of the present rate-problem.

The steady state problem for a pressurized cylinder

It is well-known from numerous theoretical investigations that for a sufficiently long, closed-end, isotropic cylinder the assumption of material incompressibility leads to a very simple solution for the steady mode of deformation consisting of an expansion under plane axial deformation and circumferentially uniform pure shear. Under the assumption that a pure shear deformation gives rise to a stress deviator state of pure shear (which does not necessarily follow from the isotropy assumption), the steady state of stress is that of pure shear with a superposed hydrostatic stress component except at the cylinder ends, and in a cylindrical coordinate system (r, φ, z) the physical components of the principal stresses are $\sigma_{rr}(r)$, $\sigma_{\varphi\varphi}(r)$, and $\sigma_{zz}(r)$, where σ_{zz} is the hydrostatic component.

Hence, local circumferential and axial equilibrium is satisfied and, irrespective of its general form, the yield condition reduces to

$$\sigma_{\varphi\varphi} - \sigma_{rr} = 2k \tag{15}$$

where k denotes the yield stress in pure shear.

Substituting (15) into the condition for local radial equilibrium yields

$$r\frac{\partial \sigma_{rr}}{\partial r} - 2k = 0. \tag{16}$$

Under internal pressure p_i and external pressure p_e the boundary conditions $\sigma_{rr}(a) = -p_i$ and $\sigma_{rr}(b) = -p_e$ determine the current magnitudes of internal and external radii described by a and b respectively at a generic instant in the integrated form of (16), i.e.

$$p_i - p_e = \int_a^b \frac{2k}{r}\,\mathrm{d}r. \tag{17}$$

Since for incompressible materials this pressure difference only is of significance p_e is put to zero and the internal pressure p_i denoted by p may be positive or negative. Hence in view of this simplification

$$\sigma_{rr} = -p + \int_a^r \frac{2k(\rho)}{\rho}\,\mathrm{d}\rho \tag{18}$$

but it should be remembered that k, the yield shear stress, changes sign with p in eqs. (15) to (18).

It can be shown that in view of (15), (16), and the boundary conditions for

σ_{rr} that if

$$\sigma_{zz} = \tfrac{1}{2}(\sigma_{rr}+\sigma_{\varphi\varphi}) \tag{19}$$

the condition for axial equilibrium at the cylinder ends can be satisfied in an integral form.

Thus for a cylinder exposed to internal pressure p with original radii a_0 and b_0, the true stress and deformation states are fully determined by eqs.(15) to (19) if the incompressibility condition is satisfied.

A bifurcating solution

The relationship describing the rate-problem will now be employed in order to investigate under which conditions a bifurcation of the circumferentially symmetric solution may occur.

Axially non-homogeneous modes are not considered as it has been shown in [8] that for sufficiently long cylinders only planar velocity modes may appear during quasi-static deformation. Using the above cylindrical coordinate system the conditions for local, radial and circumferential equilibrium follow from (5) in the form

$$\frac{\partial \dot{s}_{rr}}{\partial r} + \frac{1}{r}\frac{\partial \dot{s}_{\varphi r}}{\partial \varphi} + \frac{\dot{s}_{rr}-\dot{s}_{\varphi\varphi}}{r} = 0 \tag{20}$$

and

$$\frac{\partial \dot{s}_{r\varphi}}{\partial r} + \frac{1}{r}\frac{\partial \dot{s}_{\varphi\varphi}}{\partial \varphi} + \frac{\dot{s}_{r\varphi}+\dot{s}_{\varphi r}}{r} = 0. \tag{21}$$

Axial rotational balance requires with the present stress distribution according to (10), that

$$\dot{s}_{r\varphi}+\sigma_{\varphi\varphi}\left(\frac{1}{r}\frac{\partial v_r}{\partial \varphi} - \frac{v_\varphi}{r}\right) = \dot{s}_{\varphi r}+\sigma_{rr}\frac{\partial v_\varphi}{\partial r} \tag{22}$$

where v_r and v_φ are physical velocity components. The boundary conditions are in view of (11),

$$\dot{s}_{rr}(a) = -\dot{p}+p\frac{\partial v_r}{\partial r}, \quad \dot{s}_{r\varphi}(a) = p\left(\frac{1}{r}\frac{\partial v_r}{\partial \varphi} - \frac{v_\varphi}{r}\right) \tag{23}$$

and

$$\dot{s}_{rr}(b) = 0, \quad \dot{s}_{r\varphi}(b) = 0 \tag{24}$$

Since $d_{ii} = 0$ and in view of the adopted constitutive equation, deformation proceeds instantaneously in a mode of pure shear with principal deformation increments in radial and circumferential direction such that $d_{r\varphi} = 0$. Thus,

$$\frac{\partial v_r}{\partial r} + \frac{1}{r}\frac{\partial v_\varphi}{\partial \varphi} + \frac{v_r}{r} = 0,$$
$$\frac{1}{r}\frac{\partial v_r}{\partial \varphi} + \frac{\partial v_\varphi}{\partial r} - \frac{v_\varphi}{r} = 0. \tag{25}$$

Eqs. (12) to (14) yield the magnitude of the radial deformation increment

$$h\frac{\partial v_r}{\partial r} = \tfrac{1}{2}(\dot{\sigma}_{rr} - \dot{\sigma}_{\varphi\varphi}) \tag{26}$$

where in accordance with (9),

$$\dot{\sigma}_{rr} = \dot{s}_{rr} + \sigma_{rr}\frac{\partial v_r}{\partial r},$$
$$\dot{\sigma}_{\varphi\varphi} = \dot{s}_{\varphi\varphi} + \sigma_{\varphi\varphi}\left(\frac{1}{r}\frac{\partial v_\varphi}{\partial \varphi} + \frac{v_r}{r}\right). \tag{27}$$

Considering (25) and (22), the equilibrium equations (20) and (21) can be identically satisfied by introducing a scalar stress function $\Phi(r, \varphi)$ such that

$$\dot{s}_{rr} = \frac{1}{r}\frac{\partial \Phi}{\partial r} + \frac{1}{r^2}\frac{\partial^2 \Phi}{\partial \varphi^2} + \frac{1}{r^2}\int_c^r \rho(\sigma_{rr} + \sigma_{\varphi\varphi})\frac{\partial^2 v_\varphi}{\partial \rho \partial \varphi}\,\mathrm{d}\rho,$$
$$\dot{s}_{r\varphi} = -\frac{\partial}{\partial r}\left(\frac{1}{r}\frac{\partial \Phi}{\partial \varphi}\right) + \frac{1}{r^2}\int_c^r \rho(\sigma_{rr} + \sigma_{\varphi\varphi})\frac{\partial v_\varphi}{\partial \rho}\,\mathrm{d}\rho, \tag{28}$$
$$\dot{s}_{\varphi\varphi} = \frac{\partial^2 \Phi}{\partial r^2}.$$

Using this result in (26) and considering (25), (27)

$$\frac{\partial^2 \Phi}{\partial r^2} - \frac{1}{r}\frac{\partial \Phi}{\partial r} - \frac{1}{r^2}\frac{\partial^2 \Phi}{\partial \varphi^2} = \frac{1}{r^2}\int_c^r \rho(\sigma_{rr} + \sigma_{\varphi\varphi})\frac{\partial^2 v_\varphi}{\partial \rho \partial \varphi}\,\mathrm{d}\rho +$$
$$+(\sigma_{rr} + \sigma_{\varphi\varphi} - 2h)\frac{\partial v_r}{\partial r}. \tag{29}$$

A Fourier series solution to the velocity field equations (25) was given earlier by Skrzypek and Życzkowski [12] in the form

$$v_r = \sum_{n=0}^{\infty} u_n \cos(n\varphi),$$
$$v_\varphi = \sum_{n=1}^{\infty} v_n \sin(n\varphi) \tag{30}$$

where

$$u_0 = \frac{a_0}{r},$$
$$u_1 = a_1\left(\log\frac{r}{r_1} - 1\right),$$
$$u_n = -\frac{a_n}{n}\left\{\cos\left[(n^2-1)^{\frac{1}{2}}\log\frac{r}{r_n}\right] + (n^2-1)^{\frac{1}{2}}\sin\left[(n^2-1)^{\frac{1}{2}}\log\frac{r}{r_n}\right]\right\} \quad \text{for} \quad n>1, \tag{31}$$
$$v_1 = -a_1\log\frac{r}{r_1},$$
$$v_n = a_n\cos\left[(n^2-1)^{\frac{1}{2}}\log\frac{r}{r_n}\right] \quad \text{for} \quad n>1,$$

and a_0, a_1, a_n are amplitudes of constitutively admissible velocity modes and r_1, r_n arbitrary constants.

For sake of simplicity, the right-hand side of (29) may be denoted by

$$\sum_{n=0}^{\infty} f_n(r)\cos(n\varphi)$$

where the functions $f_n(r)$ are given by (31). A solution for Φ may be sought in series form

$$\Phi = \sum_{n=0}^{\infty} \Phi_n(r)\cos(n\varphi)$$

and substituting this into (29) yields a solution obtainable by standard methods

$$\Phi = \frac{1}{2}\int_{c_0}^{r}\rho\left(\frac{r^2}{\rho^2}-1\right)f_0(\rho)\mathrm{d}\rho + r\int_{c_1}^{r}\log\left(\frac{r}{\rho}\right)f_1(\rho)\mathrm{d}\rho\cos\varphi +$$
$$+\sum_{n=2}^{\infty} r(n^2-1)^{-\frac{1}{2}}\int_{c_n}^{r}\sin\left[(n^2-1)^{\frac{1}{2}}\log\frac{r}{\rho}\right]f_n(\rho)\mathrm{d}\rho\cos(n\varphi) \tag{32}$$

where c_0, c_1 and c_n are arbitrary constants.

Substitution of (32) into the expressions for $\dot{s}_{rr}$ and $\dot{s}_{r\varphi}$ given in (28), inserting the proper expressions for f_0, f_1, f_n and using the appropriate velocity components gives

$$\dot{s}_{rr} = a_0 \int_{c_0}^{r} \frac{2h-\sigma_{rr}-\sigma_{\varphi\varphi}}{\rho^3}\, \mathrm{d}\rho - \frac{a_1}{r}\left\{\int_{c_1}^{r}\left[\frac{1}{\rho^2}\int_{c}^{\rho}(\sigma_{rr}+\sigma_{\varphi\varphi})\mathrm{d}r' + \right.\right.$$

$$\left.\left. + \frac{2h-\sigma_{rr}-\sigma_{\varphi\varphi}}{\rho}\right]\mathrm{d}\rho + \frac{1}{r}\int_{c}^{r}(\sigma_{rr}+\sigma_{\varphi\varphi})\mathrm{d}\rho\right\}\cos\varphi -$$

$$- \sum_{n=2}^{\infty} \frac{a_n(n^2-1)^{\frac{1}{2}}}{r}\left[\int_{c_n}^{r}\left(\left\{\cos\left[(n^2-1)^{\frac{1}{2}}\log\frac{r}{\rho}\right] - \right.\right.\right.$$

$$\left. -(n^2-1)^{\frac{1}{2}}\sin\left[(n^2-1)^{\frac{1}{2}}\log\frac{r}{\rho}\right]\right\}$$

$$\left\{\frac{n}{\rho^2}\int_{c}^{\rho}(\sigma_{rr}+\sigma_{\varphi\varphi})\sin\left[(n^2-1)^{\frac{1}{2}}\log\frac{r'}{r_n}\right]\mathrm{d}r'\right\} +$$

$$+ \frac{2h-\sigma_{rr}-\sigma_{\varphi\varphi}}{n\rho}\left\{\sin\left[(n^2-1)^{\frac{1}{2}}\log\frac{\rho}{r_n}\right] - \right.$$

$$\left.\left. - (n^2-1)^{\frac{1}{2}}\cos\left[(n^2-1)^{\frac{1}{2}}\log\frac{\rho}{r_n}\right]\right\}\right)\mathrm{d}\rho +$$

$$\left. + \frac{n}{r}\int_{c}^{r}(\sigma_{rr}+\sigma_{\varphi\varphi})\sin\left[(n^2-1)^{\frac{1}{2}}\log\frac{r_n}{\rho}\right]\mathrm{d}\rho\right]\cos(n\varphi) \tag{33}$$

and

$$\dot{s}_{r\varphi} = -\frac{a_1}{r}\left\{\int_{c_1}^{r}\left[\frac{1}{\rho^2}\int_{c}^{\rho}(\sigma_{rr}+\sigma_{\varphi\varphi})\mathrm{d}r' + \frac{2h-\sigma_{rr}-\sigma_{\varphi\varphi}}{\rho}\right]\mathrm{d}\rho + \right.$$

$$\left. + \frac{1}{r}\int_{c_1}^{r}(\sigma_{rr}+\sigma_{\varphi\varphi})\mathrm{d}\rho\right\}\sin\varphi -$$

$$- \sum_{n=2}^{\infty}\frac{a_n(n^2-1)^{\frac{1}{2}}}{r}\left[\int_{c_n}^{r}\cos\left[(n^2-1)^{\frac{1}{2}}\log\frac{r}{\rho}\right]\right.$$

$$\left(\frac{n^2}{\rho^2}\int_{c}^{\rho}(\sigma_{rr}+\sigma_{\varphi\varphi})\sin\left[(n^2-1)^{\frac{1}{2}}\log\frac{r'}{r_n}\right]\mathrm{d}r' + \right.$$

$$+ \frac{2h-\sigma_{rr}-\sigma_{\varphi\varphi}}{\rho}\left\{\sin\,(n^2-1)^{\frac{1}{2}}\log\frac{\rho}{r_n}\right] -$$

$$-(n^2-1)^{\frac{1}{2}}\cos\left[(n^2-1)^{\frac{1}{2}}\log\frac{\rho}{r_n}\right]\Bigg\}\Bigg)\mathrm{d}\rho+$$

$$+\frac{1}{r}\int_c^r(\sigma_{rr}+\sigma_{\varphi\varphi})\sin\left[(n^2-1)^{\frac{1}{2}}\log\frac{\rho}{r_n}\right]\mathrm{d}\rho\Bigg]\sin(n\varphi). \tag{34}$$

Partial, radial integrations of (33), (34) and some lengthy but ordinary manipulations yields, in view of (15) and (16),

$$\dot{s}_{rr} = a_0\left(\Big|_{c_0}^{r}\frac{\sigma_{rr}}{\rho^2}+2\int_{c_0}^{r}\frac{h-2k}{\rho^3}\,\mathrm{d}\rho\right)-\frac{2a_1}{r}\left[\frac{1}{c_1}\int_c^{c_1}(\sigma_{rr}+k)\mathrm{d}\rho+\right.$$

$$\left.+\int_{c_1}^{r}\frac{h}{\rho}\,\mathrm{d}\rho\right]\cos\varphi-\sum_{n=2}^{\infty}\frac{a_n(n^2-1)^{\frac{1}{2}}}{r}$$

$$\left\{\frac{2n}{c_n}\cos\left[(n^2-1)^{\frac{1}{2}}\log\frac{r}{c_n}\right]\int_c^{c_n}(\sigma_{rr}+k)\sin\right.$$

$$\left[(n^2-1)^{\frac{1}{2}}\log\frac{\rho}{r_n}\right]\mathrm{d}\rho-$$

$$-\frac{1}{n}\left[\Big|_{c_n}^{r}\sigma_{rr}\left\{(n^2-1)^{\frac{1}{2}}\cos\left[(n^2-1)^{\frac{1}{2}}\log\frac{rr_n}{\rho^2}\right]+\right.\right.$$

$$\left.+\sin\left[(n^2-1)^{\frac{1}{2}}\log\frac{rr_n}{\rho^2}\right]\right\}+$$

$$+\int_{c_n}^{r}\left(\frac{h-2k}{\rho}\left\{2(n^2-1)^{\frac{1}{2}}\cos\left[(n^2-1)^{\frac{1}{2}}\log\frac{rr_n}{\rho^2}\right]-\right.\right.$$

$$\left.-(n^2-2)\sin\left[(n^2-1)^{\frac{1}{2}}\log\frac{rr_n}{\rho^2}\right]\right\}-$$

$$\left.\left.-n^2\frac{h}{\rho}\sin\left[(n^2-1)^{\frac{1}{2}}\log\frac{r}{r_n}\right]\right)\mathrm{d}\rho\right]\right\}\cos(n\varphi) \tag{35}$$

and

$$\dot{s}_{r\varphi} = -\frac{2a_1}{r}\left[\frac{1}{c_1}\int_c^{c_1}(\sigma_{rr}+k)\mathrm{d}\rho+\int_{c_1}^{r}\frac{h}{\rho}\,\mathrm{d}\rho\right]\sin\varphi-$$

$$-\sum_{n=2}^{\infty}\frac{a_n(n^2-1)^{\frac{1}{2}}}{r}\left(\frac{2}{c_n}\left\{\cos\left[(n^2-1)^{\frac{1}{2}}\log\frac{r}{c_n}\right]+(n^2-1)^{\frac{1}{2}}\times\right.\right.$$

$$\times \sin\left[(n^2-1)^{\frac{1}{2}} \log \frac{r}{c_n}\right]\Big\}\int_c^{c_n} (\sigma_{rr}+k) \sin\left[(n^2-1)^{\frac{1}{2}} \log \frac{\rho}{r_n}\right] \mathrm{d}\rho -$$

$$-\Bigg|_{c_n}^{r} \sigma_{rr} \sin\left[(n^2-1)^{\frac{1}{2}} \log \frac{rr_n}{\rho^2}\right] - \int_{c_n}^{r} \left(\frac{h-2k}{\rho}\Big\{\sin\left[(n^2-1)^{\frac{1}{2}}\right.\right.$$

$$\left.\log \frac{rr_n}{\rho^2}\right] + (n^2-1)^{\frac{1}{2}} \cos\left[(n^2-1)^{\frac{1}{2}} \log \frac{rr_n}{\rho^2}\right]\Big\} +$$

$$+\frac{h}{\rho} \sin\left[(n^2-1)^{\frac{1}{2}} \log \frac{r}{r_n}\right] - (n^2-1)^{\frac{1}{2}} \cos\left[(n^2-1)^{\frac{1}{2}}\right.$$

$$\left.\left.\log \frac{r}{r_n}\right]\Big\}\right) \mathrm{d}\rho \Big) \sin(n\varphi). \tag{36}$$

Introduction of admissible velocity fields into the boundary conditions (26) yields

$$\dot{s}_{rr}(a) = -\dot{p} - a_0 \frac{p}{a^2} + a_1 \frac{p}{a} \cos\varphi + \sum_{n=2}^{\infty} p \frac{a_n(n^2-1)^{\frac{1}{2}}}{na}$$

$$\Big\{\sin\left[(n^2-1)^{\frac{1}{2}} \log \frac{a}{r_n}\right] -$$

$$-(n^2-1)^{\frac{1}{2}} \cos\left[(n^2-1)^{\frac{1}{2}} \log \frac{a}{r_n}\right]\Big\} \cos(n\varphi) \tag{37}$$

$$\dot{s}_{r\varphi}(a) = a_1 \frac{p}{a} \sin\varphi + \sum_{n=2}^{\infty} p \frac{a_n(n^2-1)^{\frac{1}{2}}}{a} \sin\left[(n^2-1)^{\frac{1}{2}} \log \frac{a}{r_n}\right] \sin(n\varphi). \tag{38}$$

Correlation of the general stress-rate solution (35), (36) and the boundary conditions (24), (37), (38) requires for a continuation of the steady state mode that $c_0 = b$ and

$$a_0 = \frac{\dot{p}}{2}\left(\int_a^b \frac{h-2k}{r^3} \mathrm{d}r\right)^{-1}. \tag{39}$$

The condition for a non-trivial solution to the velocity components with amplitude a_1 reduces, in view of (17), to

$$\int_a^b \frac{h-k}{r} \mathrm{d}r = 0 \tag{40}$$

irrespective of the magnitude of the constant c. For all $n > 1$, necessary and sufficient conditions for the occurrence of any non-zero amplitude a_n follow as

$$\frac{2n^2}{c_n} \cos\left[(n^2-1)^{\frac{1}{2}} \log \frac{b}{c_n}\right]\int_c^{c_n} (\sigma_{rr}+k) \sin\left[(n^2-1)^{\frac{1}{2}} \log \frac{\rho}{r_n}\right] \mathrm{d}\rho +$$

$$+\sigma_{rr}(c_n)\left\{(n^2-1)^{\frac{1}{2}} \cos\left[(n^2-1)^{\frac{1}{2}} \log \frac{br_n}{c_n^2}\right] +\right.$$

$$\left.+ \sin\left[(n^2-1)^{\frac{1}{2}} \log \frac{br_n}{c_n^2}\right]\right\} - \int_{c_n}^{b} \left(\frac{h-2k}{\rho} \left\{2(n^2-1)^{\frac{1}{2}}\right.\right.$$

$$\left.\cos\left[(n^2-1)^{\frac{1}{2}} \log \frac{br_n}{\rho^2}\right] - (n^2-2) \sin\left[(n^2-1)^{\frac{1}{2}} \log \frac{br_n}{\rho^2}\right]\right\} +$$

$$\left. + n^2 \frac{h}{\rho} \sin\left[(n^2-1)^{\frac{1}{2}} \log \frac{b}{r_n}\right]\right) \mathrm{d}\rho = 0 \tag{41}$$

$$\frac{2}{c_n}\left\{\cos\left[(n^2-1)^{\frac{1}{2}} \log \frac{b}{c_n}\right] + (n^2-1)^{\frac{1}{2}}\right.$$

$$\left.\sin\left[(n^2-1)^{\frac{1}{2}} \log \frac{b}{c_n}\right]\right\}\int_c^{c_n} (\sigma_{rr}+k) \sin\left[(n^2-1)^{\frac{1}{2}} \log \frac{\rho}{r_n}\right] \mathrm{d}\rho + \sigma_{rr}(c_n)$$

$$\sin\left[(n^2-1)^{\frac{1}{2}} \log \frac{br_n}{c_n^2}\right] - \int_{c_n}^{b} \left(\frac{h-2k}{\rho} \left\{\sin\left[(n^2-1)^{\frac{1}{2}}\right.\right.\right.$$

$$\left.\left.\log \frac{br_n}{\rho^2}\right] + (n^2-1)^{\frac{1}{2}} \cos\left[(n^2-1)^{\frac{1}{2}} \log \frac{br_n}{\rho^2}\right]\right\} + \frac{h}{\rho}\left\{\sin\left[(n^2-1)^{\frac{1}{2}}\right.\right.$$

$$\left.\left.\log \frac{b}{r_n}\right] - (n^2-1)^{\frac{1}{2}} \cos\left[(n^2-1)^{\frac{1}{2}} \log \frac{b}{r_n}\right]\right\}\right) \mathrm{d}\rho = 0 \tag{42}$$

and two similar equations with b replaced by a.

A straightforward, though tedious, elimination of c and c_n from these four equations yields

$$\int_a^b \left(\frac{h}{r} \cos\left[(n^2-1)^{\frac{1}{2}} \log \frac{b}{r_n}\right] + \frac{h-2k}{n^2 r}\left\{2(n^2-1)^{\frac{1}{2}} \sin\left[(n^2-1)^{\frac{1}{2}}\right.\right.\right.$$

$$\left.\left.\log \frac{br_n}{r^2}\right] + (n^2-2) \cos\left[(n^2-1)^{\frac{1}{2}} \log \frac{br_n}{r^2}\right]\right\}\right) \mathrm{d}r = 0 \tag{43}$$

and

$$\int_a^b \left(\frac{h}{r} \sin\left[(n^2-1)^{\frac{1}{2}} \log \frac{b}{r_n} \right] - \frac{h-2k}{n^2 r} \left\{ 2(n^2-1)^{\frac{1}{2}} \cos\left[(n^2-1)^{\frac{1}{2}} \log \frac{br_n}{r^2} \right] - (n^2-2) \sin\left[(n^2-1)^{\frac{1}{2}} \log \frac{br_n}{r^2} \right] \right\} \right) \mathrm{d}r = 0. \tag{44}$$

This result is valid irrespective of the magnitude of $\sigma_{rr}(c_n)$, which is to be expected since the present analysis is independent of a superposed hydrostatic stress component.

Finally, elimination of r_n from (43) and (44) results in

$$\int_a^b \frac{h}{r}\,\mathrm{d}r - \left(\left\{ \int_a^b \frac{h-2k}{r} \sin\left[2(n^2-1)^{\frac{1}{2}} \log \frac{r}{b} \right] \mathrm{d}r \right\}^2 + \left\{ \int_a^b \frac{h-2k}{r} \cos\left[2(n^2-1)^{\frac{1}{2}} \log \frac{r}{b} \right] \mathrm{d}r \right\}^2 \right)^{\frac{1}{2}} = 0. \tag{45}$$

Discussion

Eq. (39) provides the amplitude for a continuous, steady state mode under a pressure increase at a generic instant, while eqs. (40) and (45) are necessary and sufficient requirements for bifurcating eigenmodes to occur.

It may be concluded from (39) that when

$$\int_a^b \frac{h-2k}{r^3}\,\mathrm{d}r = 0 \tag{46}$$

the cylinder can not resist a pressure increase and consequently stability may be lost in a mode resembling the perfect cylinder shape. It is interesting to note here that (46), (40) and (45) are identical to conditions derived in [8] which are based on Hill's criteria even though they are expressed slightly differently ((45) being insensitive to any phase angle appearing in the transcendental functions). Thus, for the present example, Hill's approach generates uniqueness criteria also of necessity.

It is of immediate interest to investigate if a pressure maximum may be preceded by a quasi-static bifurcation indicated by (40) or (45). For the construction of bifurcating velocity fields and associated stress-rate fields, the constitutive equation (12) must be taken into account ensuring that the rate of dissipation is non-negative throughout the cylinder. It is clear from (12) and

(26) that for the stress distribution under consideration this may be satisfied by determining the amplitudes of critical, peturbing velocity fields in (30) and (31) in such a way that, for $a \leqq r \leqq b$, $\partial v_r/\partial r \geqq 0$ for compression and $\partial v_r/\partial r \leqq 0$ for expansion of the cylinder. When external pressure predominates, the sign of k should be reversed everywhere indicating that (40) and (46) can not be satisfied, because both h and k are positive for a stable material. In this case the interest may be focused on (45).

In order to evaluate the integrals in (45), a more precise definition must be given of the strain-hardening parameter h. For the present, degenerate, steady strain history of pure shear, h may be adopted as being solely a function of some shear strain measure γ. If γ is specified in such a way that its material rate of change is compatible with the shear component of the stretching tensor it follows from (12) that in the present case

$$h = \frac{\mathrm{d}k}{\mathrm{d}\gamma}. \tag{47}$$

During the steady state deformation history presently considered

$$\dot{\gamma} = d_{rr} = -d_{\varphi\varphi} = -\frac{\dot{r}}{r} \tag{48}$$

and consequently at a generic instant

$$\gamma = \log \frac{r_0}{r} \tag{49}$$

applies for a particle currently at r and initially at r_0.

Introducing the incompressibility condition $r^2 - a^2 = r_0^2 - a_0^2$, (49) yields

$$\gamma = \tfrac{1}{2} \log \frac{r^2 - a^2 + a_0^2}{r^2}. \tag{50}$$

Thus on account of (47) and (50) the left-hand side of (45) may be computed provided the yield shear stress dependence on strain is known.

Figs. 1 and 2, obtained by numerical quadrature of (45), show critical strains and external pressures for a frequently used material model with a strain-hardening law given by

$$k = k_0 \gamma^m \tag{51}$$

where k_0 and m are material constants. In the investigated range of wall

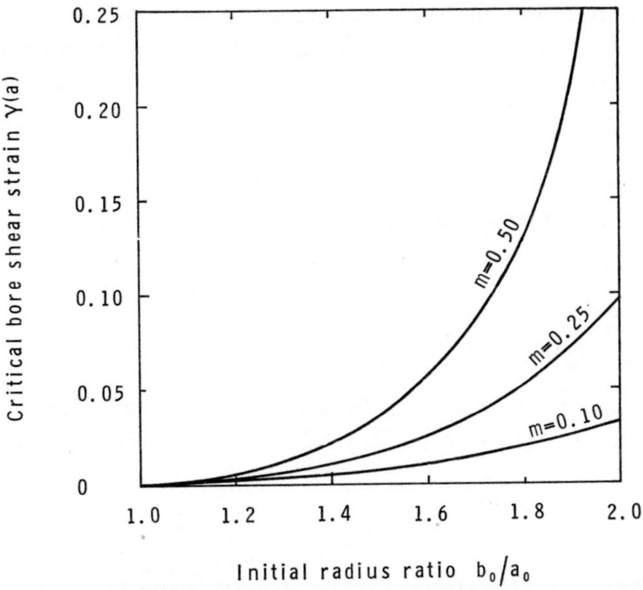

Fig. 1. Theoretical critical bore shear strain versus initial radius ratio for cylinders of some different material strain-hardening indices.

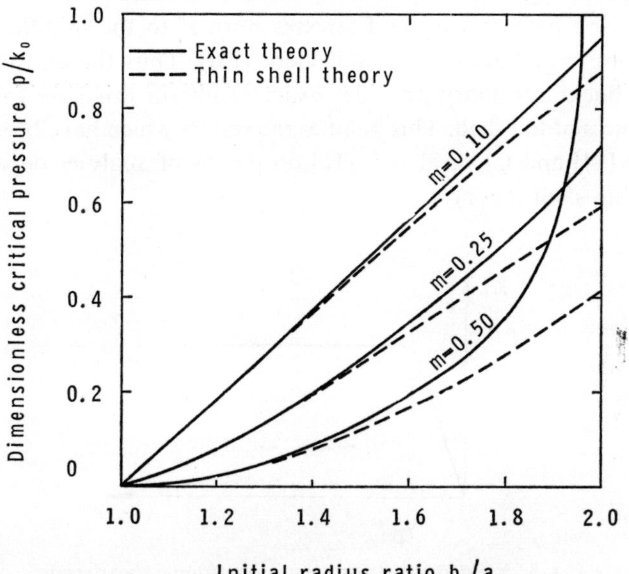

Fig. 2. Theoretical dimensionless critical hydrostatic pressure versus initial radius ratio for cylinders of some different material strain-hardening indices.

ratios the critical deformation mode proved to be the one with wave number 2 in (30). It can be seen that the critical magnitudes of relevant parameters are quite sensitive to the strain-hardening exponent m of the material and it may be concluded from the case $m = 0.5$, that above a certain radius ratio, a slight increase in this ratio increases the buckling pressure very significantly.

For the case of a thin-walled shell a series expansion of the integrands in (45) followed by integration yields (when only first order terms containing h and k are maintained),

$$h = \frac{3k}{n^2-1}\left(\frac{b+a}{b-a}\right)^2. \tag{52}$$

If the radial stress is neglected, the hoop stress $\sigma_{\varphi\varphi}$ equals $2k$ according to (15) and introducing this into (52) yields in view of (47) and by setting $n=2$

$$\sigma_{\varphi\varphi} = 2\,\frac{\mathrm{d}k}{\mathrm{d}\gamma}\left(\frac{b-a}{b+a}\right)^2. \tag{53}$$

Replacing $\mathrm{d}k/\mathrm{d}\gamma$ by the shear modulus for a linearly elastic solid it may be concluded that (53) is compatible with the result for a thin-walled incompressibly elastic shell obtained on an analysis based on Kirchhoff's assumptions stating that plane sections remain plane and normal to the middle surface during deformation and stresses normal to the middle surface are negligible, (cf. e.g. Timoshenko and Gere [13]). Thus the classical, tangent modulus, buckling theory provides exact results in this case for vanishing wall thickness of the shell. This justifies the results which have been presented by Dubey [14] and Chakrabarty [15] on means of analyses departing from classical thin-shell theory.

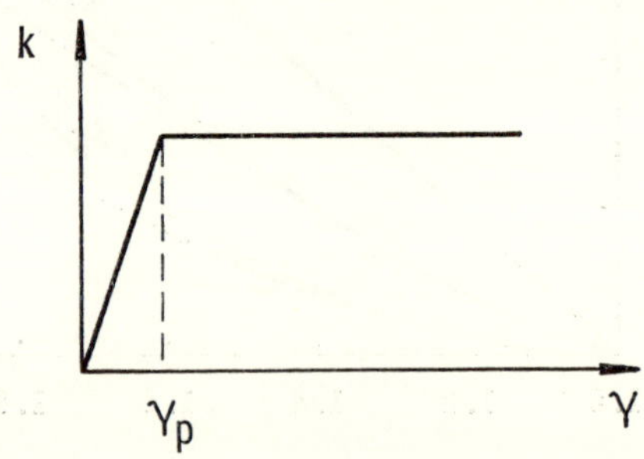

Fig. 3. Yield shear stress versus logarithmic shear strain.

On account of (17), (47), (51), (52) and the incompressibility condition, approximate critical pressure values have been computed as shown in Fig. 2. Comparison with the exact result indicate that the employed version of the tangent modulus theory gives conservative and, rather accurate estimates of buckling pressure for radius ratios here investigated.

For the case of predominating internal pressure, a possibility of loss of stability exists in the form of a circumferentially uniform deformation mode. A condition for this to occur may be obtained in a simpler form by integration of (46) using the strain-hardening and strain measures introduced in (47) and (50) respectively. This results in

$$k[\gamma(a)] \exp[-2\gamma(a)] + \sum_{i=1}^{n} \Delta k[\gamma(r_i)] \exp[-2\gamma(r_i)] - \\ -k[\gamma(b)] \exp[-2\gamma(b)] = 0 \qquad (54)$$

where $k = k(\gamma)$ is assumed to be a piecewise continuous function of radius r having some n discrete discontinuities $\Delta k(r_i)$ at different radii r_i. This restricts the solution to piecewise homogeneous material properties, as might be the case e.g. for composite shells.

For a cylinder made of locally stable and homogeneously hardening material it is readily proved [8] that (40) is always violated whenever (46) is during steady state deformation if h is a monotonically decreasing function of γ. Consequently the attention is again drawn to (45) for bifurcating solutions. An investigation of the weakness of (45) in comparison with (46) was earlier made by Storåkers [8] and some results may be summarized shortly.

For cylinders of parabolically strain-hardening material, the condition shown in (46) was found to be critical using a trial and error procedure at least for radius ratios b/a extending from 1 to 20 in the strain-hardening exponent range $0.01 \leqq m \leqq 0.50$.

However, for a cylinder material which strain-hardens in accordance with Fig. 3, it was observed that (45) was critical for a velocity field with wave number 4 in (30) when $\gamma_p < 0.065$ assuming a wall-ratio of 2. This reveals a possibility of the occurence of superposed transcendental deformation modes during the steady expansion process. Although the assumed strain-hardening characteristics are not very realistic as far as commercially used materials are concerned, this formal result should be worthy of consideration for investigators of stability of internally pressurized cylinders, since it is demonstrated that the assumption of unique deformation up to loss of stability may be non-conservative.

In the present example the situation is qualitatively similar to that in a partially yielding cylinder having an ideally plastic core contained in an elastic envelope. In a recent stability analysis Chakrabarty and Alexander [16] have drawn attention to the significance of material elasticity in thick-walled cylinders. The present investigation indicates that this may be essential from the uniqueness point of view, too. Due to the fundamentally different material laws of elasticity and plasticity no quantitative conclusions my be drawn at present but a fully three-dimensional elastic-plastic analysis of the present problem seems to be highly desirable.

This raises the question of validity of the adopted rigid-plastic material model to predict bifurcation and instability phenomena in cylinders made of real metals. When part of the cylinder is stressed below the yield-point, consideration of the elastic zone is necessary. But for a fully plastic shell it appears to be reasonable to neglect elastic deformation, as elastic and plastic moduli are of different orders of magnitude for common metals. It must be kept in mind, that the conventional flow laws of plasticity impose severe restrictions on the admissible set of particle velocity fields. Thus deviations from idealized assumptions regarding e.g. material behaviour and structural geometry may seriously affect the critical values of relevant parameters as demonstrated by Dubey [14]. For the present problem the differential equations governing the velocity field are hyperbolic and permit a rather free choice of velocity fields. In case of unsustained uniaxial stress, hyperbolic compatibility equations generate a much smaller class of solutions which accounts for the fact that e.g. necking of a rod under unconstrained uniaxial tension [6] and buckling of a spherical shell under external pressure [7] may not be explained on the rigid-plastic theory.

In order to evaluate explicitly the uniqueness conditions derived above, the cylinder material properties must be known as to the relationship between yield shear stress and shear strain under finite deformation. It has been suggested (cf. e.g. [17]) that such relationships may be obtained from ordinary torsion tests of thin-walled tubes. However, in such tests the finite deformation history of a material element is not strictly the same as in the present problem since in a torsion test the principal directions of the stretching tensor rotate in a material frame.

In order to employ the information obtained from such a test for the present problem, it is necessary to relate the strain-hardening parameter h

to some norm of the deformation history. The most simple way of obtaining the desired information without any further assumptions seems to be to subject a thin-walled closed-end cylinder to internal hydrostatic pressure.

References

[1] Hibbitt, H. D., P. V. Marcal and J. R. Rice, A finite element formulation for problems of large strain and large displacement. *Int. J. Solids Structures*, *6* (1970) 1103.

[2–5] Hill, R., On the problem of uniqueness in the theory of a rigid-plastic solid I–IV. *J. Mech. Phys. Solids 4* (1956) 247, *5* (1956) 1,*5* (1957) 153 and 302.

[6] Hill, R., Stability of rigid-plastic solids. *J. Mech. Phys. Solids*, *6* (1957) 1.

[7] Miles, J. P., Bifurcation in rigid-plastic materials under spherically symmetric loading conditions. *J. Mech. Phys. Solids*, *17* (1969) 303.

[8] Storåkers, B., Bifurcation and instability modes in thick-walled rigid-plastic cylinders under pressure. *J. Mech. Phys. Solids*, *19* (1971) 339.

[9] Onat, E. T., *The effects of non-homogeneity caused by strain-hardening on the small deformations of a rigid-plastic solid.* Proc. IUTAM Symposium, Warsaw, Pergamon Press, (1959) 171.

[10] Cowper, G. R. and E. T. Onat, The initiation of necking and buckling in plane plastic flow. *Proc. 4th US Nat. Congr. Appl. Mech., ASME*, (1962) 1023.

[11] Ariaratnam, S. T. and R. N. Dubey, Some cases of bifurcation in elastic-plastic solids in plane strain. *Quart. Appl. Math., XXVII*, (1969) 349.

[12] Skrzypek, J. and M. Życzkowski, Non-symmetrical plastic collapse modes of a symmetrically loaded circular cylinder. *Acta Mechanica*, *7* (1969) 213.

[13] Timoshenko, S. P. and J. M. Gere, *Theory of elastic stability*. Mc Graw-Hill, (1961) 293.

[14] Dubey, R. N., Instabilities in thin elastic-plastic tubes. *Int. J. Solids Structures*, *5* (1969) 699.

[15] Chakrabarty, J., On uniqueness and stability in rigid-plastic solids. *Int. J. Mech. Sci.*, *11* (1969) 723.

[16] Chakrabarty, J. and J. M. Alexander, Plastic instability of thick-walled tubes with closed ends. *Int. J. Mech. Sci.*, *11* (1969) 175.

[17] Crossland, B. and J. A. Bones, The ultimate strength of thickwalled cylinders subjected to internal pressure. *Engineering*, *179* (1955) 80.

Theory of probability and limit analysis of structures under multi-parameter loading

Giuliano Augusti

Università di Firenze, Florence, Italy

and

Alessandro Baratta

Università di Napoli, Naples, Italy

Assuming that the fundamental hypotheses of limit analysis hold, it is proved that the probability of collapse of a structure with stochastic local yield condition, subjected to a given set of loads, can be bounded from above and below, respectively by calculating the probability of finding a *statically admissible* stress field, and a mechanism for which the load system is *kinematically sufficient*. An example is presented in order to illustrate the possibility of practical application of the procedure. Its use for approximate reliability analysis of structures is discussed in the final Section.

Essential nomenclature

D_p	domain defined by $P_c[W] \leqq P$.
$D_p^{(OC)}$	smallest convex domain containing D_p
$D_p^{(IC)}$	largest convex domain contained in D_p
i, j, l	indices
k	a number, $0 \leqq k < \infty$
P	a constant, $0 \leqq P \leqq 1$
$P_c[W]$	probability of collapse under the loads $[W]$
$P_\gamma[W]$, $P_\psi[W]$	probabilities defined by eqs. (6) and (10)
$p_w[W]$	joint probability density of applied loads
Q	(generalized) stress vector
R	reliability
$\dot{q}$	strain rate vector
$\dot{u}$	velocity vector
V_i	unit base vector in load space

$\boldsymbol{W}$	load vector
W_i	component of load vector
$\boldsymbol{x}$	position vector
Γ_p, Ψ_p	domains defined by $P_\gamma[\boldsymbol{W}] \leqq P$ and $P_\psi[\boldsymbol{W}] \leqq P$ resp.
$\varphi[\boldsymbol{Q}]$	yield function
$[\boldsymbol{x}]$	terminal point of vector $\boldsymbol{x}$
$\{C\}$, $\{D\}$ etc.	event C, event D, etc.
$(\tilde{\ })$	random quantity

Scope

Modern methods of calculation and automatic computation, enable the engineer nowadays to solve almost any structural problem, provided enough information is given for the behaviour of the materials, on the condition and history of loading, and so on.

All these quantities are essentially *random*. Therefore any further progress in structural analysis and design may prove to be illusory because errors due to uncertainties of data are in many cases already larger than the errors due to computations. Whence the increasing interest in the development of the probabilistic approach to structural problems i.e. of methods by which the random variability of the data can be introduced and rationally taken into account.

Essential for this approach is the knowledge of the *probability of crisis* of the structure under given loads. If the loading condition is known, and its intensity is measured by a parameter k only, the probability of crisis is given, for any value of k, by a *probability distribution function*

$$P(k) = \text{Prob}\,\{\tilde{s} \leqq k\}. \tag{1}$$

In (1), $\tilde{s}$ is the *random variable* which measures the strength of the structure with respect to the loading condition under consideration.

If the load intensity parameter is also a random variable $\tilde{k}$ and its *probability density function* $p_w(k)$ is known, the overall probability of collapse of the structure is given by

$$P_{\text{coll}} = \int_0^\infty p_w(k)P(k)\mathrm{d}k \tag{2}$$

provided that k may assume positive values only. The *reliability* (or *probability of survival*) R is given by

$$R = 1 - P_{coll} = 1 - \int_0^\infty p_w(k)P(k)\mathrm{d}k. \tag{3}$$

When the structure under consideration behaves in accordance with the basic hypotheses of classical limit analysis, except possibly for some randomness in the local yield conditions, the *crisis* whose probability is defined in (1) may be identified with *plastic collapse*.

Under this assumption the authors have recently proposed a simple but rigorous procedure by which the strength distribution (1) and the reliability (3) can be bounded from above *and* below [1][2]. In some instances very close bounds can be obtained [3].

The main purpose of the present paper consists in generalization of the results of [1] and [2] to loads that depend on more than one independent parameters, which may be deterministic or random.

Firstly, the theorems of probabilistic limit analysis, already formulated in [1] [2], are fully demonstrated for multi-parameter loading. Then, some suggestions for the development of practically applicable numerical procedures are presented together with a numerical example. In the concluding section the actual use of the results in reliability analysis and probabilistic design is discussed, however without entering into the complex problems of the choice of a reliability level, which involve socio-economical questions well outside the realm of structural engineering and applied mechanics.

Basic theorems and general results

This paper deals with structures that behave in accordance with the theory of limit analysis, subject to loads depending on n independent parameters W_i ($i = 1, 2, \ldots n$). In an n-dimensional Euclidean load space $\mathscr{W}_n$ with unit base vectors $\boldsymbol{V}_i$, the loads acting on the structure are represented by the vector

$$\boldsymbol{W} = W_1\boldsymbol{V}_1 + W_2\boldsymbol{V}_2 + \ldots + W_n\boldsymbol{V}_n. \tag{4}$$

For a structure and loading thus defined, the classical *statical* (or *equilibrium*) and *kinematic* theorems of limit analysis must be reformulated in probabilistic terms if the local strengths of the structure are *stochastic*,[1]) as shown in the following.

[1]) The actual structure can then be thought of as a *sample* structure taken out of a *population* of nominally identical structures.

a. *Equilibrium theorem.*

For a given load vector $\boldsymbol{W}$, a *statically admissible* stress field $Q(\boldsymbol{x})$ is, by a well known definition, a stress field in equilibrium with the load $\boldsymbol{W}$ and not violating the yield condition

$$\varphi[Q(\boldsymbol{x})] \leqq 0 \qquad \forall \boldsymbol{x} : [\boldsymbol{x}] \in B \tag{5}$$

where B is the volume occupied by the structure, and $\varphi[Q]$ represents the yield function, which may be random.[2])

Imagine now that only a finite number of stress fields in equilibrium with the given load $\boldsymbol{W}$ are investigated for the validity of ineq. (5). Either of two events can occur, namely

event $\{C\}$: (at least) one statically admissible stress field is found among the investigated fields;

event $\{D\}$: none of the investigated fields is statically admissible.

Let the probability of the latter event be denoted by

$$P_{\psi}[\boldsymbol{W}] = \text{Prob}\,\{D\} \tag{6}$$

and $P_c[\boldsymbol{W}]$ indicate the *probability of collapse* of the structure when subjected to the load $\boldsymbol{W}$. Consequently $1-P_c[\boldsymbol{W}]$ is the corresponding *probability of survival.*

According to the equilibrium theorem of limit analysis, the structure does not collapse if $\{C\}$ occurs, but nothing can be said in case of the event $\{D\}$.

Since $\{C\}$ implies *survival* of the structure under the load $\boldsymbol{W}$,

$$\begin{aligned} &\text{Prob}\,\{C\} \leqq 1-P_c[\boldsymbol{W}] \\ &\text{Prob}\,\{D\} = 1-\text{Prob}\,\{C\} \geqq 1-(1-P_c[\boldsymbol{W}]) \end{aligned} \tag{7}$$

and finally the *equilibrium theorem of probabilistic limit analysis* is obtained in the form

$$P_{\psi}[\boldsymbol{W}] \geqq P_c[\boldsymbol{W}]. \tag{8}$$

b. *Kinematic theorem.*

A load vector $\boldsymbol{W}$ is said to be *kinematically sufficient* if there exists a strain rate field $\dot{\boldsymbol{q}}(\boldsymbol{x})$, compatible with the internal and external constraints imposed

[2]) If the yield function is well determined (*deterministic*), a given stress field either is or is not statically admissible. If the yield function is *random*, there is a certain probability, comprised between 0 and 1, that the field is statically admissible, and a certain probability that it is not. Random variations of structural geometry are not considered explicitly in this paper, but would have an analogous effect.

on the structure and such that the power of the load exceeds the plastic power dissipated in the body of the structure, i.e.

$$\boldsymbol{W}\cdot\dot{\boldsymbol{u}} > \mathscr{L}(\dot{\boldsymbol{q}}) = \int_B \mathscr{L}^{(1)}(\dot{\boldsymbol{q}}, \boldsymbol{x})\mathrm{d}B \tag{9}$$

where $\dot{\boldsymbol{u}}$ is the *velocity vector* associated with the load $\boldsymbol{W}$ and the field $\dot{\boldsymbol{q}}(\boldsymbol{x})$, and $\mathscr{L}^{(1)}(\dot{\boldsymbol{q}}, \boldsymbol{x})$ is the dissipated plastic power per unit volume.

Assume now that for a given load $\boldsymbol{W}$, a number of compatible *mechanisms* are investigated, denote by $\{F\}$ the event that ineq. (9) holds for at least one of these mechanisms, and indicate the probability of $\{F\}$ by [3])

$$P_\gamma[\boldsymbol{W}] = \text{Prob}\,\{F\}. \tag{10}$$

According to the kinematic theorem of limit analysis, the event $\{F\}$ implies collapse of the structure, whence the *kinematic probabilistic theorem* [4])

$$P_\gamma[\boldsymbol{W}] \leqq P_c[\boldsymbol{W}]. \tag{11}$$

c. *Representation of the functions* $P_c[\boldsymbol{W}]$, $P_\psi[\boldsymbol{W}]$, $P_\gamma[\boldsymbol{W}]$.

In the applications, the functions $P_c[\boldsymbol{W}]$, $P_\psi[\boldsymbol{W}]$ and $P_\gamma[\boldsymbol{W}]$ will most conveniently be represented by their *level surfaces*. For any given constant P: $(0 \leqq P \leqq 1)$, these surfaces are the boundaries of the three domains in load space defined by:

$$\begin{aligned} &D_p\colon (\{[\boldsymbol{W}]\in D_p\} \Leftrightarrow \{P_c[\boldsymbol{W}] \leqq P\})\\ &\Psi_p\colon (\{[\boldsymbol{W}]\in \Psi_p\} \Leftrightarrow \{P_\psi[\boldsymbol{W}] \leqq P\})\\ &\Gamma_p\colon (\{[\boldsymbol{W}]\in \Gamma_p\} \Leftrightarrow \{P_\gamma[\boldsymbol{W}] \leqq P\}). \end{aligned} \tag{12}$$

It can be seen that for any P the domains D_p, Ψ_p and Γ_p comprise the origin of the load space $[\boldsymbol{W}] = 0$, and that by definition

$$\{P_1 \leqq P_2\} \Leftrightarrow \{D_{p_1} \subseteq D_{p_2};\, \Psi_{p_1} \subseteq \Psi_{p_2};\, \Gamma_{p_1} \subseteq \Gamma_{p_2}\}. \tag{13}$$

Ineqs. (8) and (11) then lead to

$$\{\Psi_p \subseteq D_p \subseteq \Gamma_p\} \qquad \forall P\colon (0 \leqq P \leqq 1). \tag{14}$$

[3]) As in footnote[2]), this probability is either 0 or 1 if the yield function, and consequently the dissipated plastic power, are *deterministic*; but it can assume any value if they are *random*.

[4]) Note that inequalities analogous to (8) and (11) hold for the probability of *incremental collapse*, in which case the load parameters W_i, rather than being the components of a set of contemporary loads, denote loads that can act at successive times in arbitrary cycles.

In the deterministic limit analysis, the probabilities $P_c[W]$, $P_\psi[W]$ and $P_\gamma[W]$ are discontinuous functions, that assume only the values 0 and 1 throughout the load space $\mathscr{W}_n$.[5]) Denoting by D^*, Ψ^* and Γ^* the domains bounded by the surfaces of discontinuity of $P_c[W]$, $P_\psi[W]$ and $P_\gamma[W]$ respectively, the definitions (12) become

$$\left.\begin{aligned} D_p &= D^* \\ \Psi_p &= \Psi^* \\ \Gamma_p &= \Gamma^* \end{aligned}\right\} \forall P\colon (0 \leqq P < 1) \qquad (15)$$

$$D_p = \Psi_p = \Gamma_p = \mathscr{W}_n \quad \text{when} \quad P = 1.$$

Then, (14) yields the well known relationship

$$\Psi^* \subseteq D^* \subseteq \Gamma^*. \qquad (16)$$

In view of applications, it appears important to note that,

$$D^{(OC)}_{0.5p} \subseteq D_p \subseteq D^{(IC)}_{2p} \qquad (17)$$

where $D^{(OC)}_{0.5p}$ and $D^{(IC)}_{2p}$ are respectively the smallest *convex* domain containing $D_{0.5p}$ and the largest convex domain contained in D_{2p}. This result, which will be referred to as the *semiconvexity* of D_p, is demonstrated in the Appendix.

In the deterministic limit analysis, formulae (15) and (17) yield

$$D^{*(OC)} = D^* = D^{*(IC)} \qquad (18)$$

i.e. the well known *convexity* of the yield surfaces in load or stress spaces.

Calculation of the probability bounding functions

In order to illustrate the practical applicability of the theorems demonstrated in the previous Section, two simple procedures for obtaining either the functions $P_\gamma[W]$ and $P_\psi[W]$ or the domains Γ_p and Ψ_p will now be presented.

a. *Kinematical procedure.*

The simplest procedure to find domains Γ_p, defined by (12), is to consider a certain number of collapse mechanisms $\dot{q}_j(x)$, independently from each other.

[5]) Cf. footnotes [2]) and [3]).

For any fixed mechanism $\dot{\boldsymbol{q}}_j(\boldsymbol{x})$, let

$$P_{\gamma j}[\boldsymbol{W}] = \text{Prob}\,\{\boldsymbol{W}\cdot\dot{\boldsymbol{u}}_j > \mathscr{L}_j(\dot{\boldsymbol{q}})\} \tag{19}$$

be the probability that the load $\boldsymbol{W}$ is kinematically sufficient for that mechanism. The dissipated plastic power $\mathscr{L}_j(\dot{\boldsymbol{q}})$ is a *random variable*, whose probability distribution

$$P_j(k) = \text{Prob}\,\{\mathscr{L}_j(\dot{\boldsymbol{q}}) \leqq k\} \qquad \forall k\colon (0 \leqq k < \infty) \tag{20}$$

depends on the stochastic variability of the structure, essentially the local yield condition [6]).

Since for any distribution law

$$\frac{\mathrm{d}P_j(k)}{\mathrm{d}k} \geqq 0 \qquad \forall k\colon (0 \leqq k < \infty) \tag{21}$$

the definitions (19) and (20) yield immediately

$$\{\boldsymbol{W}\cdot\dot{\boldsymbol{u}}_j \leqq k\} \Leftrightarrow \{P_{\gamma j}[\boldsymbol{W}] \leqq P_j(k)\}. \tag{22}$$

Denoting by $P_j^{-1}(P)$ the inverse function of $P_j(k)$, (22) implies that the probability of the load $\boldsymbol{W}$ to be kinematically sufficient for the mechanism $\dot{\boldsymbol{q}}_j(\boldsymbol{x})$, does not exceed P in the semispace defined by

$$\boldsymbol{W}\cdot\dot{\boldsymbol{u}}_j \leqq P_j^{-1}(P) \qquad \forall P\colon (0 \leqq P \leqq 1). \tag{23}$$

The semi-space (23) therefore is the Γ_p domain (12) relative to the mechanism $\dot{\boldsymbol{q}}_j(\boldsymbol{x})$ and the probability P.

By varying the mechanism $\dot{\boldsymbol{q}}_j(\boldsymbol{x})$, the semispace (23) changes because, in general, both the velocity vector $\dot{\boldsymbol{u}}_j$ and the function $P_j^{-1}(P)$ vary. Since, according to (14), the domain D_p must be contained in each of the semispaces (23), it will be contained in their *intersection*. Note in particular that the intersection of the semispaces relative to any two mechanisms $\dot{\boldsymbol{q}}_j(\boldsymbol{x})$ and $-\dot{\boldsymbol{q}}_j(\boldsymbol{x})$, is the *strip*

$$-P_{-j}^{-1}(P) \leqq \boldsymbol{W}\cdot\dot{\boldsymbol{u}}_j \leqq P_j^{-1}(P) \tag{24}$$

bounded by two parallel hyperplanes. $P_j^{-1}(P)$ and $P_{-j}^{-1}(P)$ can be equal in the particular case of *symmetric* strengths, e.g. equal positive and negative limit moments.

[6]) In all definitions such as (20), and wherever else appropriate throughout this paper, it is tacitly supposed that non-dimensional quantities are introduced.

It is thus demonstrated that a convex outer bound to D_p can be easily determined by considering, independently from each other, a number of collapse mechanisms and *intersecting* the corresponding domains (23), provided that for every mechanism the distribution law of the dissipated plastic power (20) can be calculated. Much research, well beyond the limited scope of this paper, is still needed before this calculation will be possible in the general case: the example presented later takes advantage of usual simplifying assumptions.

If all the mechanisms were jointly considered, the probability $P_\gamma[\boldsymbol{W}]$ defined in (10) would be introduced: since

$$P_c[\boldsymbol{W}] \geqq P_\gamma[\boldsymbol{W}] \geqq P_{\gamma j}[\boldsymbol{W}] \qquad \forall j \tag{25}$$

a better outer approximating domain would be obtained. However, the probabilities $P_{\gamma j}[\boldsymbol{W}]$ do not refer to statistically independent events, and therefore the overall lower bound probability $P_\gamma[\boldsymbol{W}]$ can only be calculated in a rather cumbersome way by taking account of the *correlations* between the individual mechanisms (see, e.g. [4]).

b. Statical or 'equilibrium' procedure.

While the kinematic procedures, just discussed, have already been dealt with, in various forms, by some other authors, the present writers believe to have been the first proponents of the equilibrium approach in the probabilistic limit analysis [1][2].

The basis for this approach is the static theorem (8), or the equivalent relation $D_p \supseteq \Psi_p$ (14), which may be more evident for multi-parameter loading. The actual implementation of the static approach requires consideration of a number of stress fields

$$\boldsymbol{Q}_j^{\boldsymbol{W}} = \boldsymbol{Q}_j^{\boldsymbol{W}}(x) \tag{26}$$

each defined throughout the structure and in equilibrium with the load $\boldsymbol{W}$. The probability $P_{\psi j}[\boldsymbol{W}]$ that the j-th field $\boldsymbol{Q}_j^{\boldsymbol{W}}$ is *not* statically admissible is given by

$$P_{\psi j}[\boldsymbol{W}] = 1 - \text{Prob}\,\{\varphi[\boldsymbol{Q}_j^{\boldsymbol{W}}(x_l)] \leqq 0 \qquad \forall\, l \in (1, \ldots, N)\} \tag{27}$$

where N is the number of points in the structure, where it seems to be appropriate to check the yield condition (5).

There are several simple ways of constructing equilibrated stress fields $\boldsymbol{Q}_j^{\boldsymbol{W}}$. For example, let

$$\boldsymbol{Q}_{ij} = \boldsymbol{Q}_{ij}(x) \qquad (i = 1, 2, \ldots n) \tag{28}$$

be a set of n stress fields, each in equilibrium with the i-th unit load $V_i = 1$, but otherwise arbitrary. For any load $\boldsymbol{W}$, remembering (4), any set (28) defines an equilibrated field

$$\boldsymbol{Q}_j^{\boldsymbol{W}} = \sum_{i=1}^{n} W_i \boldsymbol{Q}_{ij}. \tag{29}$$

If m sets like (28) are chosen, eq. (29) defines m sets $\boldsymbol{Q}_j^{\boldsymbol{W}}$ for each $\boldsymbol{W}$.

If a structure is simple, and the number n of the independent load parameters is small, it may be more convenient to obtain each equilibrated field $\boldsymbol{Q}_j^{\boldsymbol{W}}$ from a stress field associable with a collapse mechanism. This procedure is actually followed in the numerical example below.[7])

Given a definition of $\boldsymbol{Q}_j^{\boldsymbol{W}}$, let Ψ_{pj} be, in the load space $\mathscr{W}_n$, the domain defined by

$$\{[\boldsymbol{W}] \in \Psi_{pj}\} \Leftrightarrow \{P_{\psi j}[\boldsymbol{W}] \leqq P\} \tag{30}$$

where P is a constant. According to (14), D_p contains every Ψ_{pj} and their union

$$D_p \supseteq \bigcup_{j=1}^{m} \Psi_{pj}. \tag{31}$$

An inner bound to D_p is thus obtained. A larger, i.e. better approximating domain can be found from the probability $P_\psi[\boldsymbol{W}]$, as defined in (6). Actual calculation of $P_\psi[\boldsymbol{W}]$ involves a number of terms that increases rapidly with m.

c. *A numerical example.*[8])

A first example of application of the described procedures has been developed on the simple portal frame shown in Fig. 1, which was already studied, under a one-parameter loading condition, in [1] and [2].

The two independent load parameters are now the magnitudes of the vertical force, W_1 (positive downwards), and of the horizontal force, W_2 (positive rightwards). In the calculations, it is assumed that the possible plastic deformations are limited to rotations concentrated in the 11 indicated

[7]) A current research shows that the approximation can be further improved by letting the (random) limit condition $\varphi[\boldsymbol{Q}(\boldsymbol{x})] = 0$ (5) be automatically satisfied in an appropriate number of points of the structure [3].

[8]) The calculations have been performed at the Computing Center of the Facoltà di Scienze – Università di Napoli, directed by Prof. Vinciguerra.

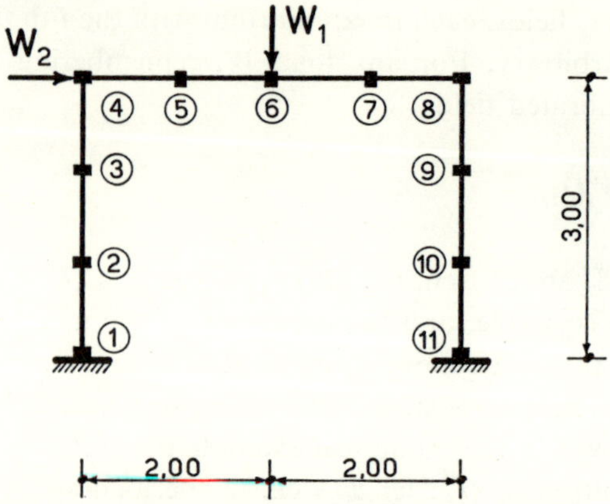

Fig. 1. Example frame (two-parameter loading)

cross-sections (Fig. 1). Under the further hypothesis of equal positive and negative limit moments M_{0i} in each section, the yield condition (5) becomes:

$$|M_i| - M_{0i} \leqq 0 \qquad (i = 1, 2, \ldots, 11). \tag{32}$$

Each M_{0i} is, in the present context, a random variable (random strength), *described* by a *probability distribution* function[9])

$$P_{0i}(k) = \text{Prob}\{\tilde{M}_{0i} \leqq k\}; \qquad (0 \leqq k < \infty) \tag{33a}$$

so that

$$\text{Prob}\{|M_i| - \tilde{M}_{0i} \leqq 0\} = \text{Prob}\{\tilde{M}_{0i} \geqq |M_i|\} = 1 - P_{0i}(|M_i|). \tag{33b}$$

In the example, $P_{0i}(k)$ is taken to be the same for all 11 sections, namely the *Gaussian* (*normal*) distribution with mean $\overline{M}_{0i} = \overline{M}_0 = 10$ and variance $\text{Var}(M) = (0.10\,\overline{M}_0)^2$. It is also assumed that the 11 limit moments $\tilde{M}_{0i}$ are statistically independent.

Because of the previous assumptions, the structure behaves symmetrically with respect to the sign of both loads, so that

$$P_c(W_1, W_2) = P_c(-W_1, W_2) = P_c(-W_1, -W_2) = P_c(W_1, -W_2). \tag{34}$$

The following calculations are therefore limited to the first quadrant in the load space ($W_1 \geqq 0, W_2 \geqq 0$).

The *kinematic procedure* is developed with reference to the three collapse mechanisms shown in Fig. 2a, which are the significant mechanisms of a

[9]) Cf. footnote [6]).

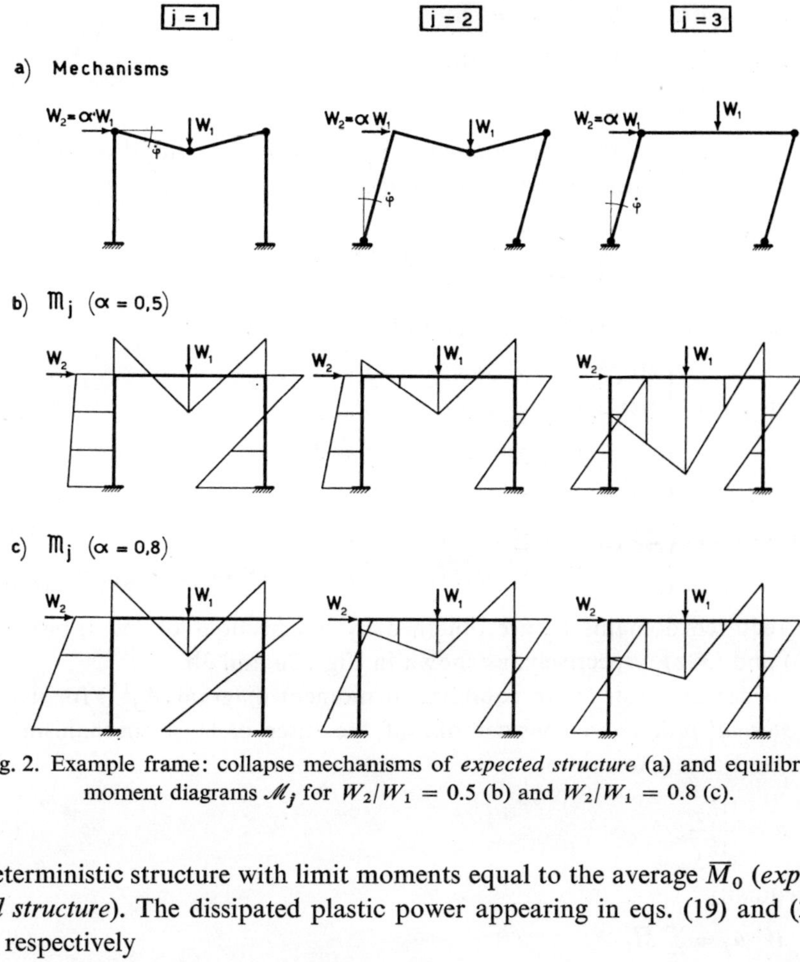

Fig. 2. Example frame: collapse mechanisms of *expected structure* (a) and equilibrated moment diagrams $\mathscr{M}_j$ for $W_2/W_1 = 0.5$ (b) and $W_2/W_1 = 0.8$ (c).

deterministic structure with limit moments equal to the average $\overline{M}_0$ (*expected structure*). The dissipated plastic power appearing in eqs. (19) and (20), is respectively

$$
\begin{aligned}
j = 1: &\quad \mathscr{L}_1 = \tilde{M}_{04} + 2\tilde{M}_{06} + \tilde{M}_{08} \\
j = 2: &\quad \mathscr{L}_2 = \tilde{M}_{01} + 2\tilde{M}_{06} + 2\tilde{M}_{08} + \tilde{M}_{011} \qquad (35) \\
j = 3: &\quad \mathscr{L}_3 = \tilde{M}_{01} + \tilde{M}_{04} + \tilde{M}_{08} + \tilde{M}_{011}
\end{aligned}
$$

if $\dot{\varphi} = 1$ (Fig. 2a)[9])

Since the $\tilde{M}_{0i}$ have been assumed to be normally distributed, the $\mathscr{L}_j$ are also normally distributed, with means and variances given by[10])

[10]) Remember that, given two *normally distributed* random variables $\tilde{x}$ and $\tilde{y}$ with mean values $\bar{x}$, $\bar{y}$ and correlation coefficient ρ_{xy},

$$\overline{\mathscr{L}}_1 = 40; \qquad \mathrm{Var}(\mathscr{L}_1) = 6;$$
$$\overline{\mathscr{L}}_2 = 60; \qquad \mathrm{Var}(\mathscr{L}_2) = 10; \tag{36}$$
$$\overline{\mathscr{L}}_3 = 40; \qquad \mathrm{Var}(\mathscr{L}_3) = 4.$$

For example, to obtain the domains Γ_p for $P = 0.1$ and $P = 0.01$ take first from the normal distribution tables

$$\frac{k-\overline{\mathscr{L}}_j}{\sqrt{\mathrm{Var}(\mathscr{L}_j)}} = \begin{Bmatrix} -1.28 \\ -2.33 \end{Bmatrix} \text{ when } P_j(k) = P = \begin{Bmatrix} 0.1 \\ 0.01 \end{Bmatrix} \tag{37}$$

Eq. (23) provides then for each considered mechanism

$$P = 0.1 \quad \begin{cases} 2W_1 \leqq 36.85 & (j = 1) \\ 2W_1 + 3W_2 \leqq 55.95 & (j = 2) \\ 3W_2 \leqq 37.44 & (j = 3) \end{cases} \tag{38a}$$

$$P = 0.01 \quad \begin{cases} 2W_1 \leqq 34.30 & (j = 1) \\ 2W_1 + 3W_2 \leqq 52.64 & (j = 2) \\ 3W_2 \leqq 35.34 & (j = 3) \end{cases} \tag{38b}$$

The required domains Γ_p are obtained by intersections of the semispaces (38a) and (38b) respectively, as shown in Figs. 3a and 3b.

In order to construct an equilibrated moment diagram $\mathscr{M}_j$[11]) for use in the *statical procedure*, consider one of the three collapse mechanisms in Fig. 2a. For a given ratio

$$\alpha = \frac{W_2}{W_1} \tag{39}$$

the equality

$$\boldsymbol{W} \cdot \dot{\boldsymbol{u}}_j = \sum_l \overline{M}_{0l} \dot{\varphi}_l \tag{40a}$$

$$\overline{(x+y)} = \bar{x} + \bar{y} \tag{a}$$

$$\mathrm{Var}(\tilde{x}+\tilde{y}) = \mathrm{Var}(\tilde{x}) + \mathrm{Var}(\tilde{y}) + 2\rho_{xy}\sqrt{\mathrm{Var}(\tilde{x})\,\mathrm{Var}(\tilde{y})}. \tag{b}$$

Therefore, if $\tilde{x}$ and $\tilde{y}$ are statistically independent ($\rho_{xy} = 0$)

$$\mathrm{Var}(\tilde{x}+\tilde{y}) = \mathrm{Var}(\tilde{x}) + \mathrm{Var}(\tilde{y}) \tag{c}$$

while, if $\tilde{y} = \tilde{x}$ (i.e. $\rho_{xy} = 1$)

$$\mathrm{Var}(2\tilde{x}) = 4\,\mathrm{Var}(\tilde{x}). \tag{d}$$

[11]) Of such *diagrams*, only the values M_{jl} in the 11 pre-fixed cross-sections are actually significant.

a)

b)

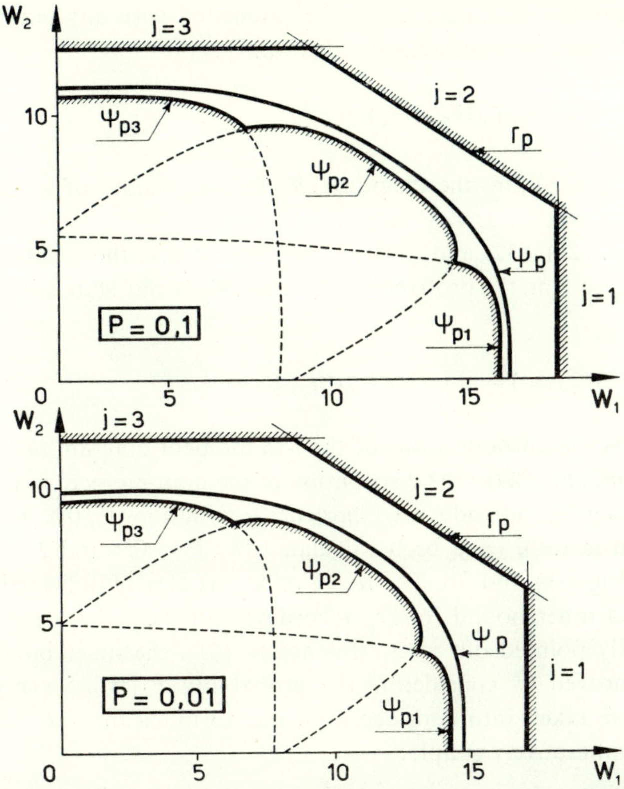

Fig. 3. Example frame: outer and inner approximate domains for probabilities of collapse P = 0.1 and P = 0.01.

defines the kinematically sufficient loads

$$W_1 = W_{1j}^{(\alpha)} \qquad W_2 = \alpha W_{1j}^{(\alpha)}. \tag{40b}$$

Following the usual procedure of deterministic limit analysis, it is easy to construct the moment diagram $\bar{\mathcal{M}}_j^{(\alpha)}$, say, which is in equilibrium with the loads (40) and attains the value $\pm\bar{M}_0$ at the active plastic hinges. This diagram is unique for mechanisms $j = 2$ and $j = 3$. An infinity of diagrams that satify the above condition exist for $j = 1$ (which is a partial collapse mechanism): the diagram satisfying the further condition

$$\sum_{i=1}^{11} \bar{M}_{1i}^2 = \min \tag{41}$$

has been chosen in this example.

Three equilibrated diagrams can be associated with any point $\{W_1, W_2\}$ in the load space by homotetical reduction

$$\mathcal{M}_j(W_1, W_2) = \mathcal{M}_j(W_1, \alpha W_1) = \bar{\mathcal{M}}_j^{(\alpha)} \frac{W_1}{W_{1j}^{(\alpha)}}. \tag{42}$$

By way of illustration, the diagrams $\mathcal{M}_j$ for two values of α are shown in Figs. 2b, c to within a *scale* factor.

From eqs. (27), (32) and (33b), remembering that the 11 $\tilde{M}_{0i}$ are statistically independent, the probability $P_{\psi j}$ that $\mathcal{M}_j$ is not statically admissible, is given by

$$P_{\psi j}(W_1, W_2) = 1 - \prod_{i=1}^{11} [1 - P_{0i}(M_{ji})] \tag{43}$$

where M_{ji} is the absolute value of the j-th moment diagram (42) in the i-th cross-section, and $P_{0i}(k)$ the distribution of the limit moment in that section.

Introducing the already used *normal* distribution $P_{0i}(k)$, the domains Ψ_{pj} defined in (30), have been calculated for $P = 0.1$ and $P = 0.01$, and plotted in Figs. 3a and 3b. The *union* of Ψ_{pj} is clearly indicated and gives the required inner bound for D_p according to (31).

As already pointed out with reference to (31), the inner bound domain can be improved by considering the probability $P_\psi[W]$. Since only three diagrams are taken into consideration, the formula for the total probability is comparatively simple:

$$1 - P_\psi = (1 - P_{\psi 1}) + (1 - P_{\psi 2}) + (1 - P_{\psi 3}) - P_{1,2}^a - P_{2,3}^a - P_{3,1}^a + P_{1,2,3}^a \tag{44}$$

where $P_{r,s}^a$ is the probability that both $\mathcal{M}_r$ and $\mathcal{M}_s$ are statically admissible, and $P_{1,2,3}^a$ the probability that all three diagrams (42) are statically admissible. It is easy to show that

$$\begin{aligned} P_{r,s}^a &= \prod_{i=1}^{11} [1 - P_{0i}(\max\{M_{ri}, M_{si}\})] \\ P_{1,2,3}^a &= \prod_{i=1}^{11} [1 - P_{0i}(\max\{M_{1i}, M_{2i}, M_{3i}\})] \end{aligned} \tag{45}$$

where, as in (43), the absolute values of M_{ji} are introduced.

The domain Ψ_p, defined by eqs. (12) and (44), is also plotted in Figs. 3a, 3b. The improvement over the *union* of individual domains Ψ_{pj} is not very significant.

It is also important to note that the validity of eqs. (43) and (45), on which the evaluation of Ψ_{pj} and Ψ_p has been based, depend on the assumed statistical independence of the $\tilde{M}_{0i}$, but not on the choice of specific distribution laws $P_{0i}(k)$. On the contrary, the assumption of normal distribution of $\tilde{M}_{0j}$ is essential for calculating the Γ_p-domains without introducing manyfold integrals.

Practical application of the results. Evaluation of the 'reliability'.

In the present paper, it has been demonstrated that two theorems, corresponding to the fundamental theorems of limit analysis, allow the fully rigorous determination of two bounds, $P_\gamma[\boldsymbol{W}]$ and $P_\psi[\boldsymbol{W}]$, to the probability of collapse $P_c[\boldsymbol{W}]$ under a given load $\boldsymbol{W}$, of a ductile structure sufficiently *described* from the stochastic point of view. $P_\gamma[\boldsymbol{W}]$ has been used by other investigators, without recognizing that it is inherently an *unsafe* bound, because it yields a probability of collapse smaller than the actual one, as shown by ineq. (11).

In a practical semi-probabilistic problem, it may be required that the probability of collapse do not exceed a given value P. The domain of the load vectors $\boldsymbol{W}$ satisfying this condition is the domain D_p defined by (12). A *safe* approximation of D_p is its inner bound domain Ψ_p, obtained via the static procedure. The *unsafe* approximation obtained by the kinematic method does not require great computational difficulties, only if the collapse mechanism is known with a very large degree of confidence, i.e. if the probability of collapse of any other mechanism is negligible. Even in the simple example of Figs. 1–3 one mechanism does predominate for most loading conditions, but at least two mechanisms must be taken into consideration when the ratio W_2/W_1 is in the ranges 0.30–0.35 and 1.35–1.40 approximately (Fig. 3). Moreover, it is worth remarking that, because of random variations of strength within the structure, other mechanisms than those of the *expected structure* may become significant [3]: however, in the few published examples of probabilistic limit analysis not much attention has so far been paid to this possibility.

In conclusion, the writers believe that many more structural examples must be investigated before the relative merits of the *kinematic* and the *static* approaches can be assessed in general.

So far, probabilities of collapse under *given* loads have been considered. If the loads are described by a random vector $\tilde{\boldsymbol{W}}$ with *probability density*

$$p_w(W) = \lim_{\substack{\Delta W_1 \to 0 \\ \cdots \\ \Delta W_n \to 0}} \frac{\text{Prob}\{W_1 \leqq \tilde{W}_1 \leqq W_1 + \Delta W_1; \ldots W_n \leqq \tilde{W}_n \leqq W_n + \Delta W_n\}}{\Delta W_1 \cdot \Delta W_2 \ldots \Delta W_n} \tag{46}$$

the *probability of survival* or *reliability* of the structure is obtained from a generalization of eq. (3), i.e.

$$R = 1 - \int_{-\infty}^{+\infty} \cdots \int_{-\infty}^{+\infty} P_c[W] p_w[W] \mathrm{d}W_1 \ldots \mathrm{d}W_n. \tag{47}$$

Upper and lower bounds to the reliability R are given by

$$\begin{aligned} R_\psi &= 1 - \int_{-\infty}^{+\infty} \cdots \int_{-\infty}^{+\infty} P_\psi[W] p_w[W] \mathrm{d}W_1 \ldots \mathrm{d}W_n \\ R_\gamma &= 1 - \int_{-\infty}^{+\infty} \cdots \int_{-\infty}^{+\infty} P_\gamma[W] p_w[W] \mathrm{d}W_1 \ldots \mathrm{d}W_n. \end{aligned} \tag{48}$$

which, because of (8) and (11), satisfy the inequalities

$$R_\psi \leqq R \leqq R_\gamma. \tag{49}$$

Acknowledgement

Research partially supported by the Italian National Research Council (C.N.R.).

Appendix

Proof of the semi-convexity of D_p.

Let W_1 and W_2 be two load vectors, and W_3 their linear combination

$$W_3 = \lambda W_1 + (1-\lambda) W_2 \quad \text{with} \quad 0 \leqq \lambda \leqq 1. \tag{A.1}$$

For any assumed compatible *mechanism* (*deformation field*) $\dot{q}(x)$,

$$\begin{aligned} &\{W_1 \cdot \dot{u} \leqq \mathscr{L}(\dot{q}) \quad \text{and} \quad W_2 \cdot \dot{u} \leqq \mathscr{L}(\dot{q})\} \Rightarrow \\ &\quad \Rightarrow \{W_3 \cdot \dot{u} = \lambda W_1 \cdot \dot{u} + (1-\lambda) W_2 \cdot \dot{u} \leqq \lambda \mathscr{L}(\dot{q}) + (1-\lambda)\mathscr{L}(\dot{q}) = \mathscr{L}(\dot{q})\} \end{aligned} \tag{A.2}$$

and viceversa

$$\{W_3 \cdot \dot{u} > \mathscr{L}(\dot{q})\} \Rightarrow \{W_1 \cdot \dot{u} > \mathscr{L}(\dot{q}) \quad \text{or} \quad W_2 \cdot \dot{u} > \mathscr{L}(\dot{q})\}. \tag{A.3}$$

In other words, if there exists a *mechanism* $\dot{q}(x)$ such that ineq. (9) holds for the load W_3, the same inequality applies to W_1 or W_2: i.e., collapse under W_3 implies collapse under *either* W_1 *or* W_2. Whence

$$P_c[W_3] \leqq P_c[W_1] + P_c[W_2]. \tag{A.4}$$

If now

$$[W_1] \in D_p; \qquad [W_2] \in D_p \tag{A.5}$$

where D_p is the domain defined by (12), eqs. (A.1) and (A.4) yield respectively

$$[W_3] \in D_p^{(OC)} \qquad [W_3] \in D_{2p}. \tag{A.6}$$

By an appropriate choice of W_1, W_2 and λ, W_3 can represent any arbitrary point of $D_p^{(OC)}$, and therefore

$$D_p^{(OC)} \subseteq D_{2p}. \tag{A.7}$$

An obvious corollary is that

$$D_{2p}^{(IC)} \supseteq D_p^{(OC)} \supseteq D_p \tag{A.8}$$

and therefore (17) is fully proved.

It is also possible to prove that D_p is a starlike domain with respect to the origin $[W] \equiv \mathbf{0}$, where, as already noted,

$$P_c[\mathbf{0}] = 0. \tag{A.9}$$

In fact, take any load W such that

$$[W] \in D_p \tag{A.10}$$

then, (A.4) implies that any point of the segment from $\mathbf{0}$ to $[W]$ also belongs to D_p, q.d.e.

References

[1] Augusti, G. and A. Baratta, Analisi limite di strutture con variazioni aleatorie di resistenza. *Proceedings of the 1st Italian National Congress of Theoretical and Applied Mechanics*, (Udine 1971) AIMETA, 1971.

[2] Augusti, G. and A. Baratta, Limit analysis of structures with stochastic strength variations. *Journal of Structural Mechanics*, *1*, No. 1, 1972.
[3] Baratta, A., An improvement in the static method for limit analysis of structures with stochastic strength variations. *Journal of Structural Mechanics* (in the press).
[4] Stevenson, J. and F. Moses, Reliability analysis of framed structures. *Journal of the Structural Division*, ASCE, *96*, No. ST 11, 1970.

Traction-displacement boundary conditions for plastic fracture by hole growth

V. Nagpal and F. A. McClintock
Massachusetts Institute of Technology

C. A. Berg
University of Pittsburgh

M. Subudhi
Brooklyn Polytechnic Institute, U.S.A.

From a plane strain model with cylindrical holes in a non-hardening matrix, criteria are developed for the localization of plastic flow in terms of the shape, size, and spacing of the holes. The onset of this localization determines the ductility of unnotched specimens or parts. The subsequent traction-displacement relation across the sheet provides a boundary condition for the calculation of crack growth. Such a relation is derived in the form of a constitutive equation giving the traction in terms of the porosity and the direction of relative displacement across the sheet, and a process equation telling how the porosity changes with increments in relative displacement. Once localization has occurred, the work of separation per unit crack surface turns out to be very nearly the yield strength in shear times the half-spacing of the holes for high triaxiality and perhaps $^1/_4$ that value when the plane of localization is at 90° to a free surface. In both cases the direction of relative motion across the band, within ±45° of the normal direction, has remarkably little effect.

1. Introduction

Metallurgical evidence for ductile fracture by the growth and coalescence of holes has been known for a number of years, starting with the work of Tipper (1949), Plateau et al. (1958), and Puttick (1959). More recently, the scanning electron microscope has provided convincing evidence that the process occurs even in steels up to a tensile strength of 250 000 psi (Brook, 1970) and that the deformation of the surfaces of the holes is continuous enough so that insight into their growth can be obtained with the aid of continuum, isotropic, homogeneous plasticity. This is particularly true in copper where the results of Puttick (1969) and Backofen et al. (1954) have been simulated in Plasticene by Rhines, as reported by McClintock (1968a).

The mechanics of the problem is less well understood. The growth of individual, circularly cylindrical holes has been studied by McClintock (1968a). The initial growth of spherical holes has been reported by Rice and Tracey (1969). Arrays of holes have more recently been studied by Needleman (1972). The ductility estimated from these analyses on the assumption of macroscopically homogeneous strain is too high by a factor of up to 5 (for example, see McClintock, 1968b). Apparently the reason for such an overestimate is that the analyses have not included the localization of flow on the incipient fracture surface. Evidence for the importance of this localization is given by the results of Bluhm and Morrissey (1965), who found that in low carbon steel, while the growth of holes was evident soon after the point of maximum load (the tensile strength), the sudden decrease of load associated with the final reduction of area did not introduce any new metallurgical mode of behavior. This strongly suggests that the critical event in fracture by hole growth is a localization of plastic flow. Hill (1956, 1957) determined sufficient conditions for uniqueness and stability of incompressible plastic flow, but specifically excluded considerations of velocity discontinuities on the grounds that they are tantamount to fracture. Hill later (1961) considered the development of sheets of velocity discontinuities in generalized plastic deformation (including dilational). In his work, which is an extension of Hadamard's (1903) classical examination of discontinuities in continuous bodies, Hill noted that velocity discontinuities may occur only across non-deforming surfaces (those in which strain increments vanish). However, Hill did not point out the special relevance of plane dilational flow to fracture. Berg (1970, 1971) also discussed the necessary and sufficient conditions for localization in dilational plastic flow. (The sufficiency of the conditions follows on physical grounds from consideration of any material inhomogeneity in the deforming field.) The criterion is:

Plastic flow localizes at a point when there exists

a) a non-deforming surface through that point, and
b) some field of relative motion across a thin layer parallel to the surface that does not require further deformation in the material on either side of the layer.

This criterion is the analogue in three dimensions of the criterion for necking in a uniaxial tensile specimen where the surrounding bar becomes rigid, or for necking over a region of the order of the sheet thickness in strips under plane stress, where the direction of the line of necking is inextensible. Note that this criterion requires plane strain, at least locally, so that for a direction z that lies in the non-deforming surface and is normal to the relative displacement across the layer, $\varepsilon_z = 0$ and $u, v \neq f(z)$. The direction in the z-plane parallel to the relative motion must also be inextensible. In a nearly incompressible, non-hardening material this inextensibility is provided by either of the slip lines or by a rigid region. As regards continued rigidity of the surroundings, strain hardening may prevent band formation by increasing the effective shear strength of the layer while hole growth decreases it. As a first approximation we shall seek the hole size at which localization can occur in the absence of such effects. Furthermore, the rate of decrease of traction with flow in the localized band will give insight into the possible effects of strain hardening. Finally, we seek the traction-displacement relation across the band, which serves as a boundary condition for the calculations of stress, strain and displacement in the adjacent body, and thus provides a coherent theory of fracture at the scale of a homogeneous, dilating continuum.

In summary, the object of a study of the traction-displacement relations across a band of localized flow is to determine the traction in order to know whether such localized flow can occur without deformation in the surrounding material. The traction will be a function of the directions and magnitudes of the major and minor axes $2\boldsymbol{p}_i$ of the elongated, roughly ellipsoidal holes, the bases $2\boldsymbol{L}_i$ of the lattice on which the centers of the holes are arranged, and the direction of relative motion:

$$\boldsymbol{t} = \boldsymbol{t}(\boldsymbol{p}_i, \boldsymbol{L}_i, u/|u|). \tag{1}$$

To determine how the traction changes with relative motion requires not only eq. 1, but also what might be called a process equation telling how the parameters $\boldsymbol{p}_i$, which describe the porosity, change with an increment of relative motion:

$$d\boldsymbol{p}_i/d|u| = d\boldsymbol{p}_i/d|u|(\boldsymbol{p}_i, \boldsymbol{L}_i, \boldsymbol{u}/|u|). \tag{2}$$

The fundamental complexity of eqs. 1 and 2 makes it desirable to find acceptable approximations. We shall investigate the possibility that one need consider only circularly cylindrical holes, for which the only parameters are the radius p and the spacing $2L$. Furthermore, it may be sufficient to determine the initial traction, the initial rate of decrease of traction with displacement, and a rough estimate of the total work W done in separating the surfaces. We hope, then, to obtain useful approximate expressions of the form

$$t_{s,n} = t_{s,n}(p/L, u_s/u_n), \tag{3a}$$

$$\frac{d(t/2k)}{d|u|} = \frac{d(t/2k)}{d|u|}(p/L, u_s/u_n), \tag{3b}$$

$$W/2kL = (W/2kL)[p_0/L, u_s/u_n]. \tag{3c}$$

2. Example of the localization of flow

Consider the localization of flow in a specimen of non-hardening material with a blunted crack tip, as shown in Fig. 1. Such blunting is well known, although the details of the shape remain controversial (Brook, 1971; Elliott

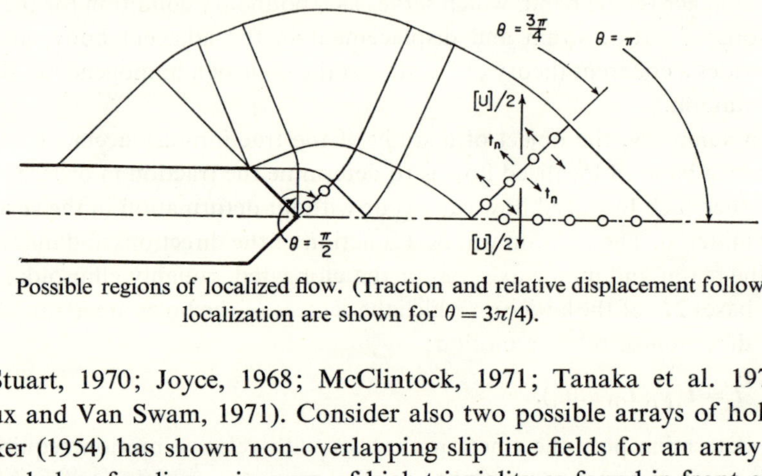

Fig. 1. Possible regions of localized flow. (Traction and relative displacement following localization are shown for $\theta = 3\pi/4$).

and Stuart, 1970; Joyce, 1968; McClintock, 1971; Tanaka et al. 1970; Pelloux and Van Swam, 1971). Consider also two possible arrays of holes. Drucker (1954) has shown non-overlapping slip line fields for an array of circular holes of radius p; in areas of high triaxiality as found in front of a crack the logarithmic spiral regions can overlap, giving the configurations shown in Fig. 2. Note also that these slip line fields are suitable for relative

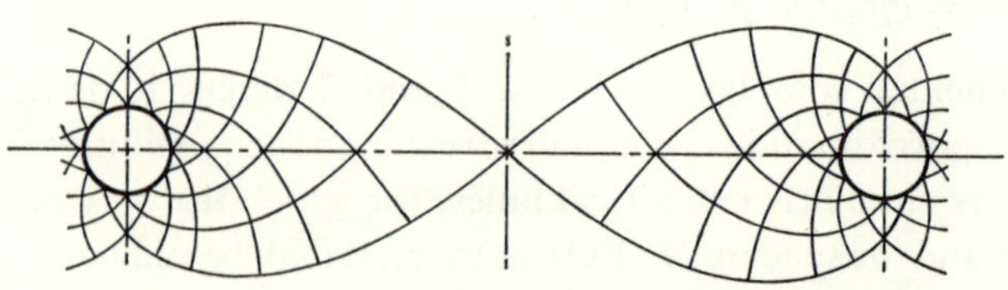

Fig. 2. Overlapping flow fields around circular holes in a band.

motion at 45° to the row of holes as well as normal to it, so they are appropriate for all three arrays of holes shown in Fig. 1. These three cases can be thought of as special examples of a general case of localized flow across a band lying at an angle θ to a free surface. When the relative motion across the band is within 45° of the normal direction, the spiral flow field of Fig. 2 still applies, and the traction across the plane of localization is purely normal. The allowable value of this traction without causing general yielding around corners at the crack tip has been given by Hill (1954). It can be worked out from the slip line field of Fig. 1. At the same time, assume that the density of holes reduces the average shear stress k required for homogeneous plastic deformation throughout the body by the factor $1-p^2/L^2$, corresponding roughly to the volume fraction of holes:

$$t_n = 2k(1-p^2/L^2)(1+\theta-\pi/2),\ t_s = 0. \tag{4}$$

The next question is how close together the cylindrical holes must be to reduce the mean traction carried by the band to the value given by eq. 4. For circularly cylindrical holes the surrounding slip line field consists of logarithmic spirals, which in general must overlap in order to reach the maximum triaxiality found in front of a sharp crack ($\theta = \pi$ in eq. 4). As shown in Fig. 3a, let Φ be the angle at which a slip line breaks through to the surface of the hole and let Φ_m be the angle for the outermost slip line. The traction where the slip line meets the plane of localization is purely normal and is found from the equilibrium equations along the slip line subtending the angle Φ:

$$\sigma_{nn} = 2k(1+\Phi). \tag{5}$$

The radius from the center of the hole to the point in question is found from the equation for the logarithmic spiral:

$$\Phi = \ln(r/p). \tag{6}$$

The mean traction on the plane of localization is now found from eqs. 5 and 6 by integration:

$$t_n = -\frac{1}{L}\int_0^L \sigma_{nn} dr = -2k \ln p/L. \tag{7}$$

Introducing eq. 4 gives the porosity required for any direction θ of the band of localized flow:

$$t_n/2k = -\ln(p/L) = (1+\theta-\pi/2)(1-p^2/L^2). \tag{8}$$

Table 1. Band angle and normal traction for localization of plastic flow in porous media.

Relative porosity p/L	Angle of band from free surface θ	Maximum stress $(\sigma_{nn})_{\max}/2k = 1-\ln(p/L)$	Mean traction $(t_{nn})_{av}/2k = -\ln(p/L)$
0.071	185°	3.64	2.64
0.1	166°	3.3	2.3
0.14	145°	2.92	1.92
0.2	129°	2.61	1.61
0.283	111°	2.26	1.26
0.4	95°	1.915	0.915
0.566	80.7°	1.57	0.57
0.8	68°	1.223	0.223

The results of eqs. 4 through 8 are presented in Table 1. Directly ahead of the crack only a $\frac{1}{2}\%$ change in density is required. Note also that the local stress at the tip of the logarithmic spiral field, σ_{nn}, exceeds that normally thought of as being the maximum available with plastic constraint. To be sure, the flow field calculated here provides only an upper bound to the limit load. It appears plausible, however, that the presumably rigid region could be proven to be rigid, using the method of Bishop (1953) to develop an equilibrium stress distribution which does not exceed yield. If so, this flow field would be exact and unique (Bishop, et al., 1956).

Once localization occurs, the next step is to determine the traction-displacement relation across the band of localized flow. Under these circumstances circular holes will not remain circular, so we next consider more general combinations of shapes of holes and directions of relative motion.

3. Slip line fields for bands of localized flow

For circular holes at various angles of motion, the slip line fields are shown in Fig. 3. In the limiting case of pure shear the slip line field becomes that found by Green (1954). The slip line fields for a row of slits perpendicular to the plane of localization are shown in Fig. 4 for the two limiting cases of pure tension and pure shear. The field for pure shear was given by Green (1954), although a re-calculation by Carson (1970) gave a field with fan angles of 8°, radii of 0.025 times the slit length, and a shear band zone of 0.740 times the slit length. The field for tension (relative motions at an angle

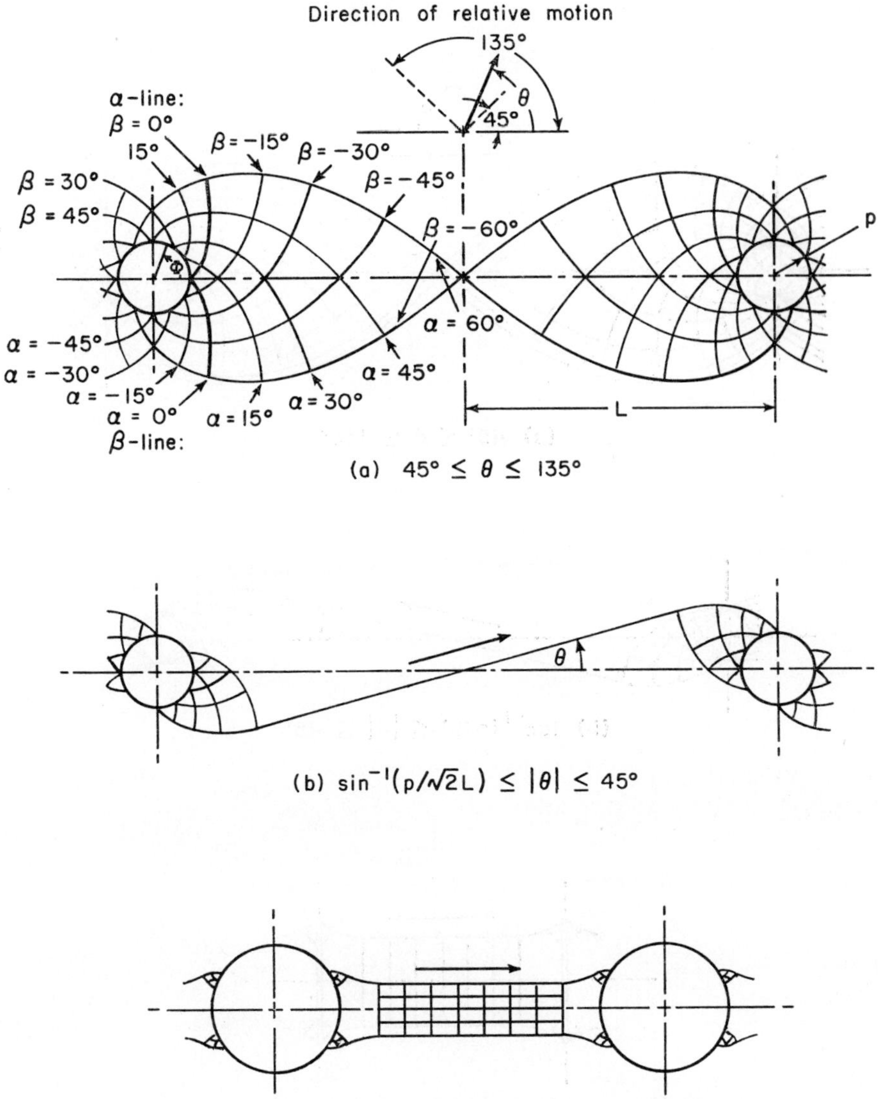

Fig. 3. Slip line fields for circular holes.

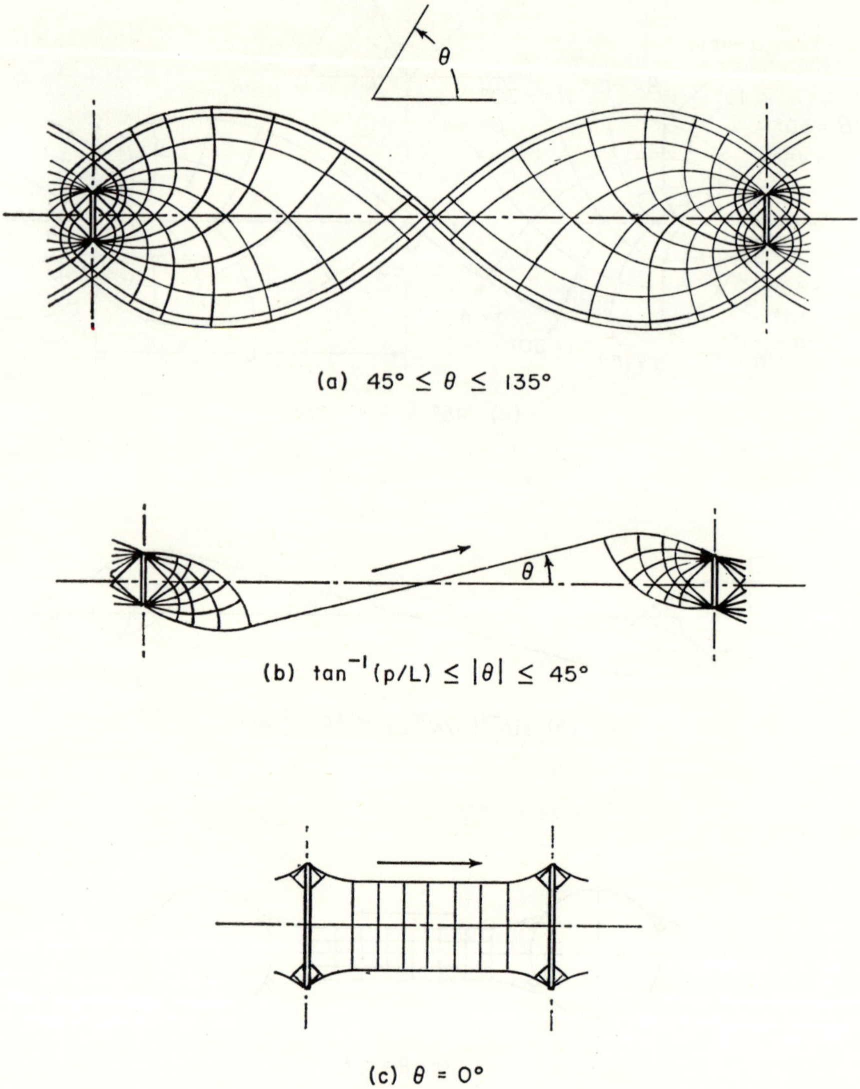

Fig. 4. Slip line fields for vertical slits.

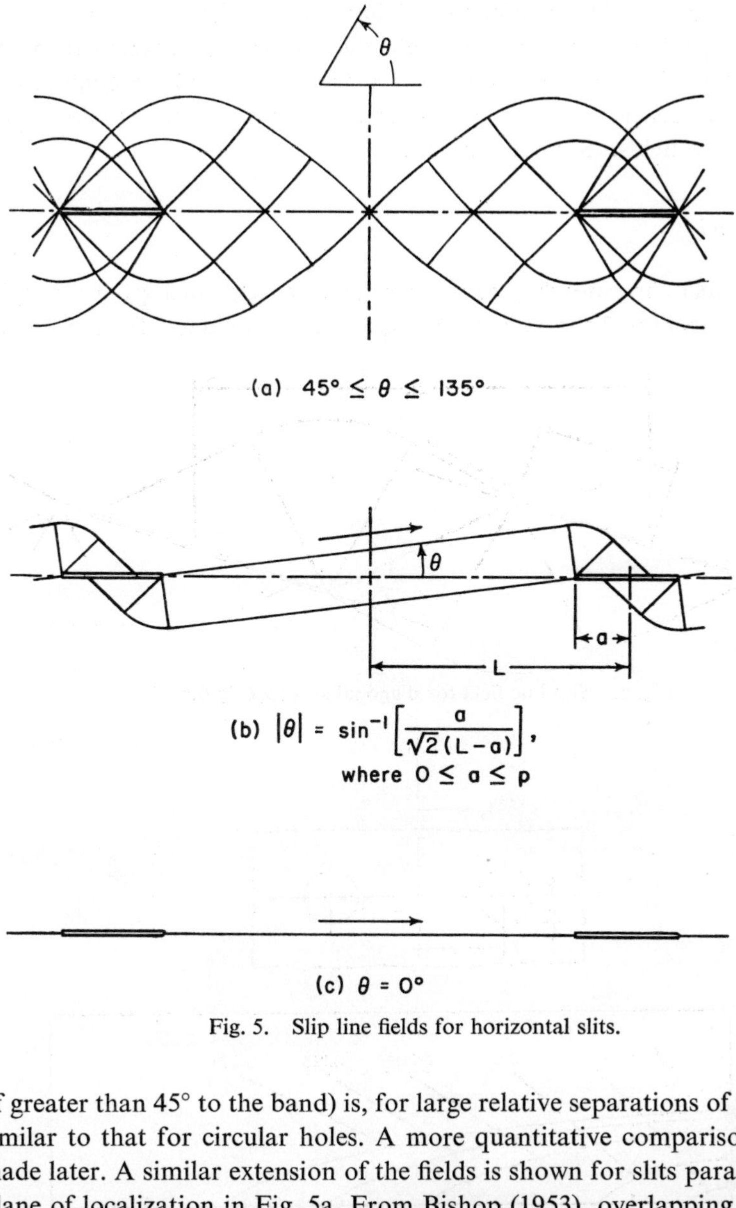

Fig. 5. Slip line fields for horizontal slits.

of greater than 45° to the band) is, for large relative separations of the holes, similar to that for circular holes. A more quantitative comparison will be made later. A similar extension of the fields is shown for slits parallel to the plane of localization in Fig. 5a. From Bishop (1953), overlapping fields are not required for a porosity greater than 0.116. The field for pure shear, Fig. 5c, is again due to Green. Fig. 6 gives the character of the slip line field for a class of deformations involving oblique slits under tensile loads that give opening at each point within the slit. Two solutions can be obtained, depending on whether it is the α or β lines that loop continuously around the slits in an S-shaped pattern. Lower porosities would require different flow fields. (Subudhi (1970)).

4. Yield loci

The yield loci corresponding to the various flow fields of Figs. 3 through 6 are plotted in Figs. 7 through 10. As expected, the yield loci for the slits are

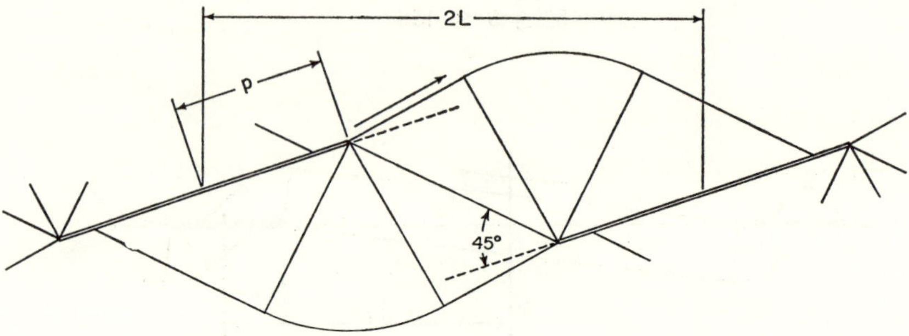

Fig. 6. Slip line field for diagonal slits ($p/L \geqq 0.634$).

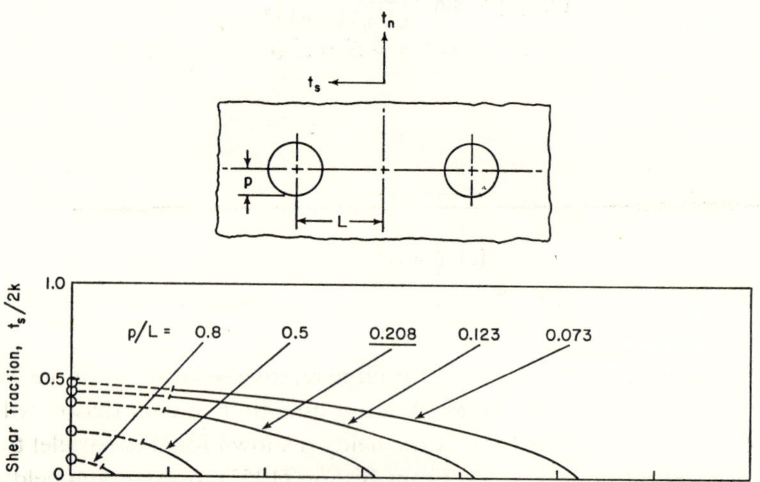

Fig. 7. Yield loci for circular holes.

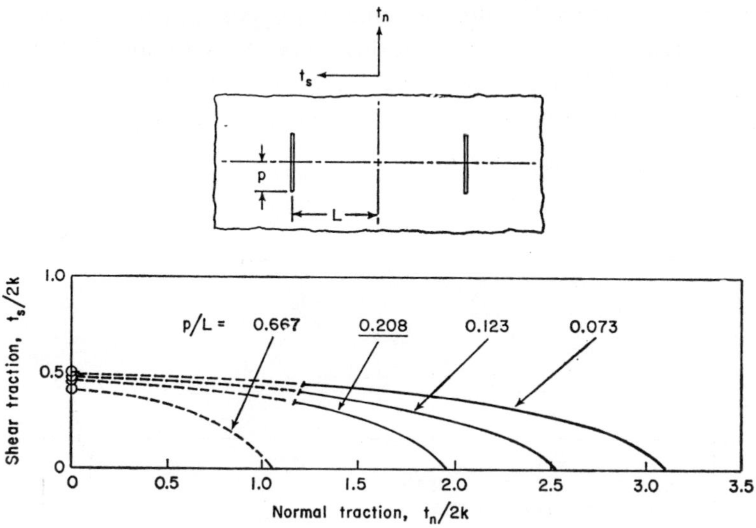

Fig. 8. Yield loci for vertical slits.

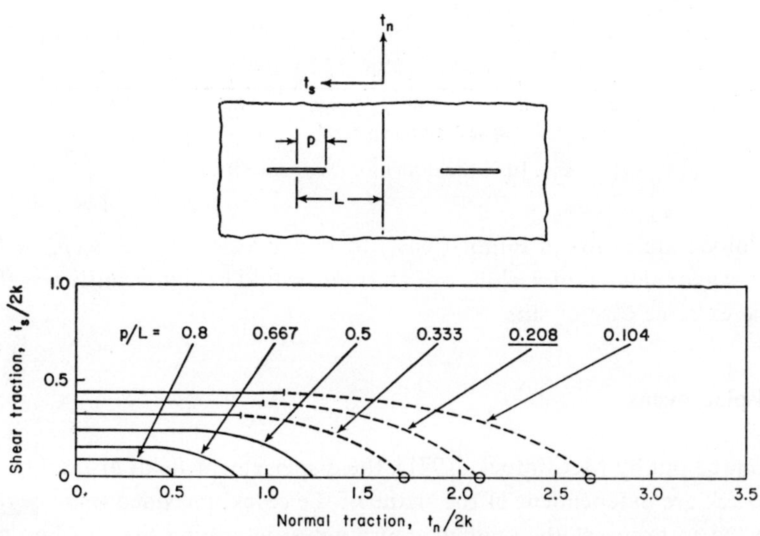

Fig. 9. Yield loci for horizontal slits.

higher than those for holes with the same major axis p_m. Roughly speaking, the mean traction carried by arrays of slits is greater by k than that for holes, although for closely spaced holes, $p_m/L \rightarrow 1$, the tractions for both the circular holes and the parallel slits approach zero, but at differing rates. The effects of

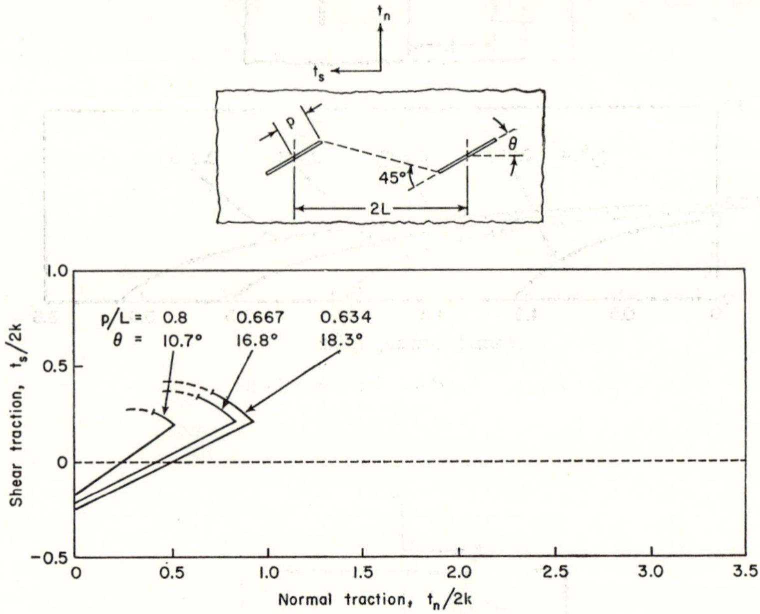

Fig. 10. Yield loci for diagonal slits.

more moderate ratios of minor (least) to major axes, such as $p_l/p_m = 0.5$, or of hexagonal holes of various orientations, would be significantly less than for the extreme case of slits.

5. Displacements

As pointed out by McClintock (1971), the displacement fields around the tip of a crack are independent of the shape of the crack, provided that they are expressed in terms of the angular coordinates describing the slip line field rather than in terms of the linear coordinates. It is easiest to present this relation by plotting the α and β coordinates of the slip line field (expressed in terms of angles, e.g. Hill, 1950) as functions of the two components of

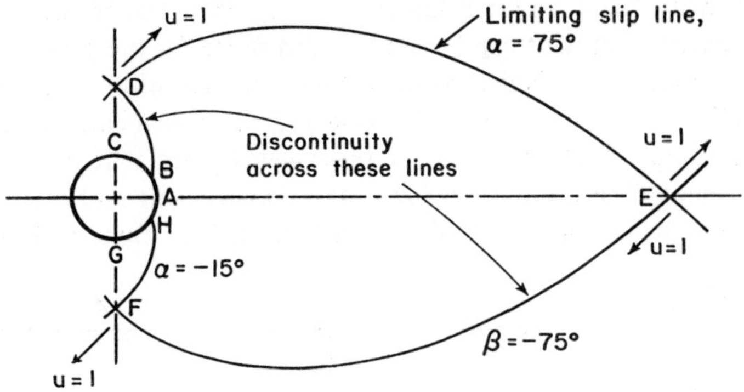

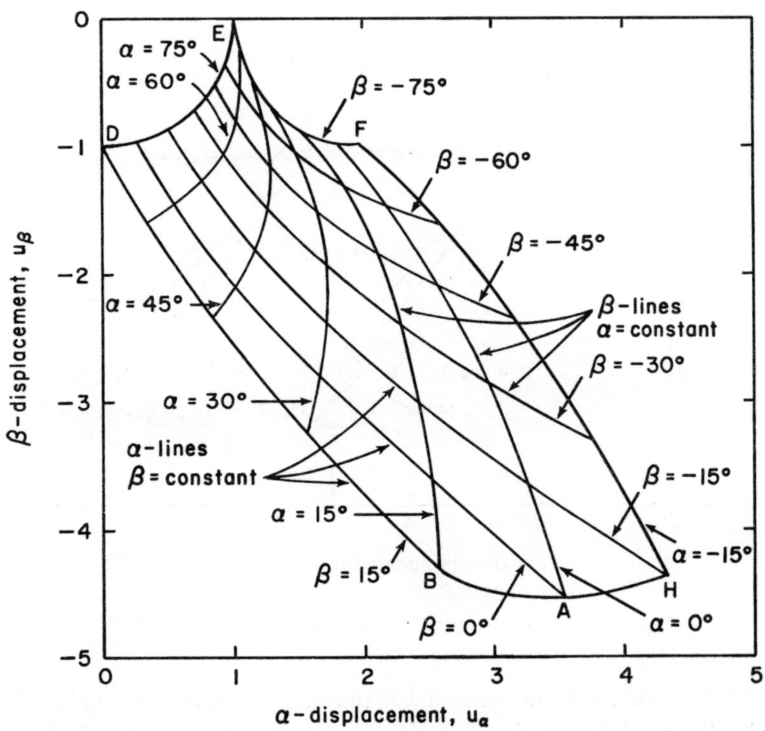

Fig. 11a. Hodograph for slip line field of Figure 3(a) with limiting slip lines of 75°.

displacement. Such a plot is called a hodograph. The results for a single displacement increment can be used to find those for the general case if the single component of displacement is chosen in a 45° direction. That is, symmetry allows obtaining the results for a 135° displacement, and then superimposing appropriate mixtures of these fields gives that for an arbitrary displacement. The hodograph for a 45° displacement and a slip line field subtending an arc of 150° is shown in Fig. 11. The numerical calculations

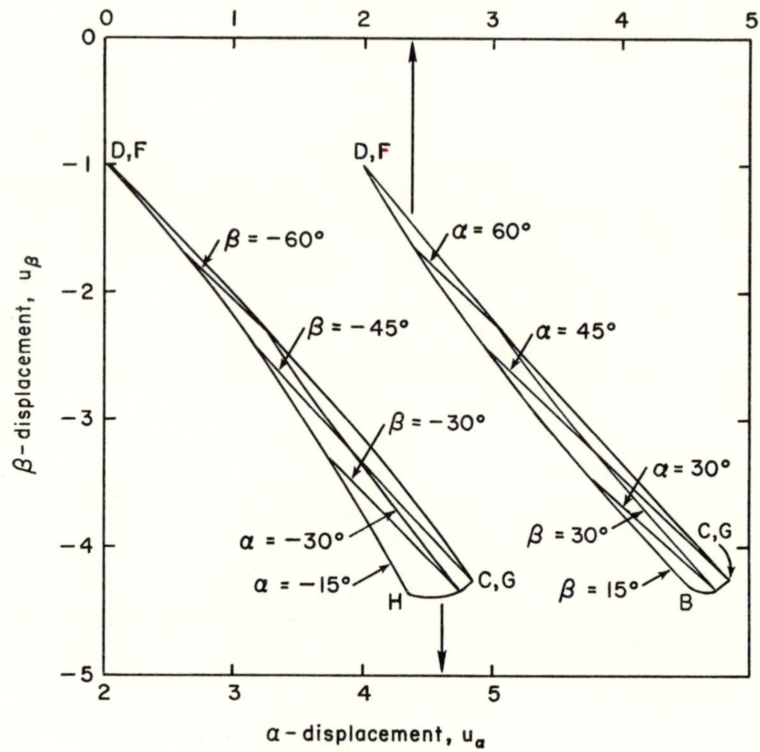

Fig. 11b. Hodographs for overlapping flow fields.

were carried out by hand using a 15° mesh. The results are within 2% of those for the special case of a 0° notched specimen under normal displacement developed by Wang (1954) and plotted as a hodograph by McClintock (1971).

In order to illustrate the slip line fields and the displacements associated with the growing holes, consider the hexagonal hole shown in Fig. 12. The choice of a polygon, as opposed to an ellipse, is suggested by the generation of corners by the instability associated with alternating sliding off, as observed by Joyce (1968) and discussed by McClintock (1971). The polygonal shape has the advantage that the relative displacements are concentrated

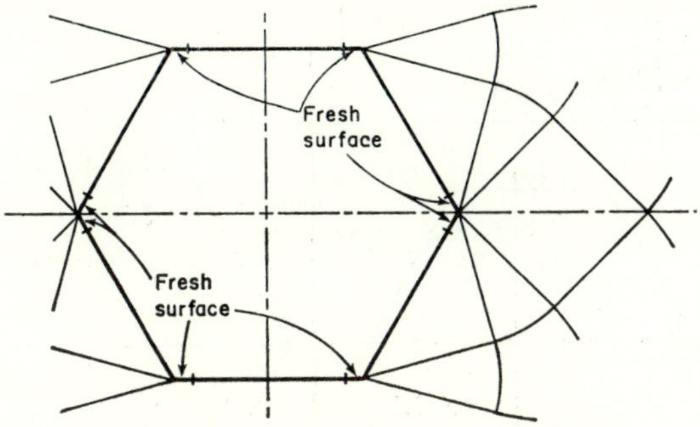

Fig. 12. Flow field for a hexagonal hole.

at the corners and hence the change in shape of the hole can be approximated by a discrete number of calculations. The choice of a hexagon, as opposed to a square, is general enough to exhibit changes in shape. It is assumed here that the relative displacement across a fan at any one of the corners produces a corresponding growth in length of the side of the hexagon opposite the fan. The effects of strain hardening and the asymmetry of the fan tend to cause it to generate a new surface not quite in the direction of the side of the hexagon, but for simplicity, the holes will be assumed to grow as exact hexagons. Thus the growth of the hole can be found by considering the rigid-body motions of the sides, and extrapolating the sides to find the new corners.

The resulting shape changes are shown in Fig. 13 for normal and 45° displacement increments and for two relatively extreme porosity ratios. Note that with the lower porosity the growth of the holes tends to be in the normal direction, whereas for the higher porosity the growth shifts to the transverse direction. This change in shape has also been observed by Needleman (1972). The shape change also indicates that the approximation of the hole by a circle or regular hexagon will be a reasonably close approximation over a relatively wide range. The decrease in traction associated with this hole growth is plotted in Fig. 14 and can be closely approximated by the empirical

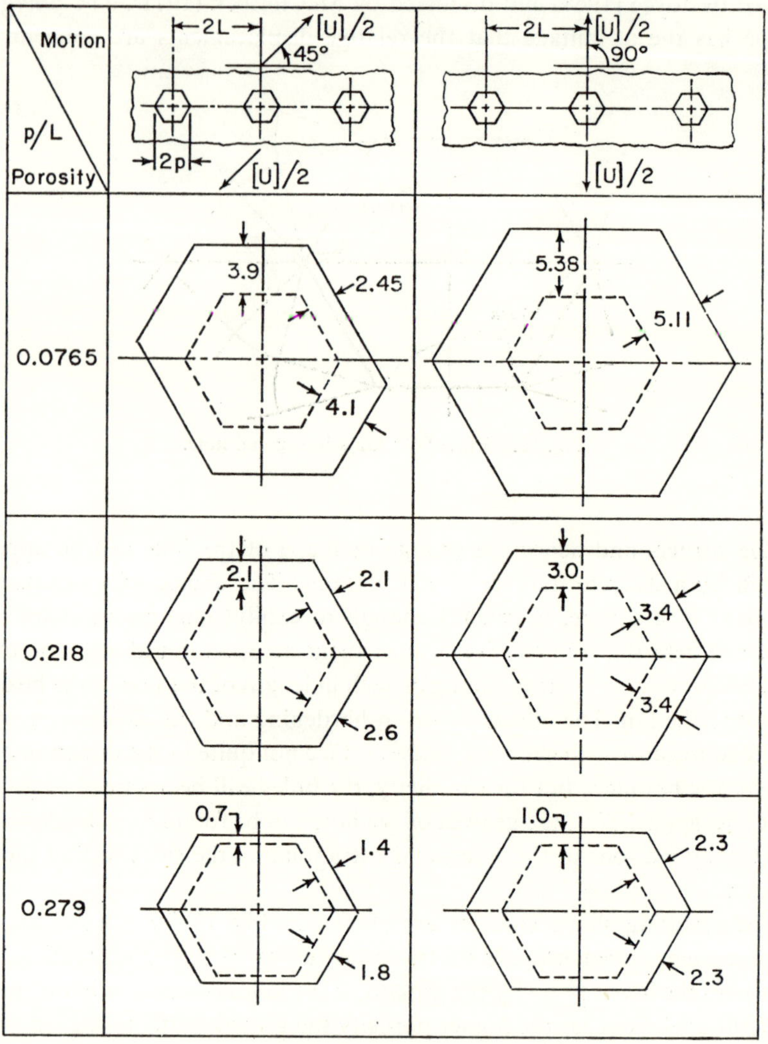

Fig. 13. Change in shape of hexagonal holes with imposed unit displacement ($[U]/2=1$).

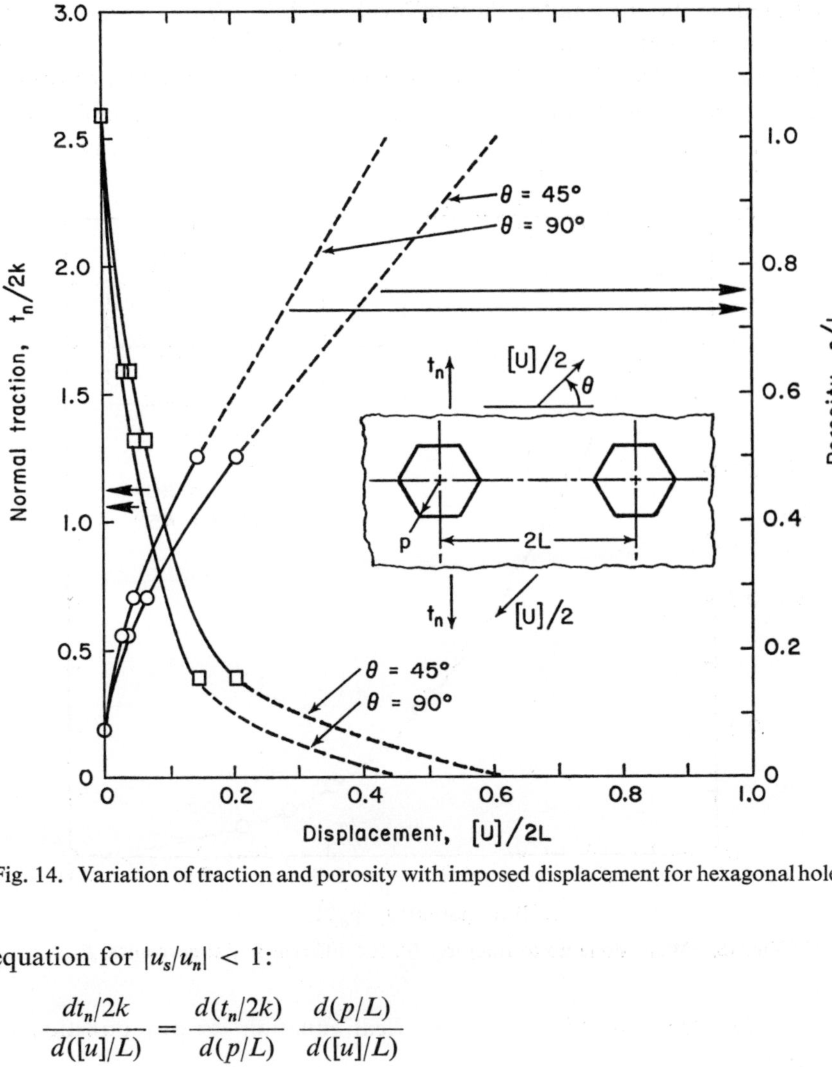

Fig. 14. Variation of traction and porosity with imposed displacement for hexagonal holes.

equation for $|u_s/u_n| < 1$:

$$\frac{dt_n/2k}{d([u]/L)} = \frac{d(t_n/2k)}{d(p/L)}\,\frac{d(p/L)}{d([u]/L)}$$

$$\approx\left(-10-19\,\frac{p}{L}\right)\left(8-23\,\frac{p}{L}+22\left(\frac{p}{L}\right)^2\right),\quad \text{for}\quad \frac{p}{L}<0.5, \quad (9)$$

$$t_s/2k = 0.$$

The rate of decrease of traction with displacement is so great that strain hardening is unlikely to prevent localization, once equality of eqs. 4 and 8 is satisfied, so that flow in the sheet of localized flow will not require flow in the surrounding material.

To get the total work, traction-displacement relation was numerically integrated for two directions of motion across hexagonal holes, assuming after each step circular holes with $p = p_m$. The results are shown in Fig. 15 and

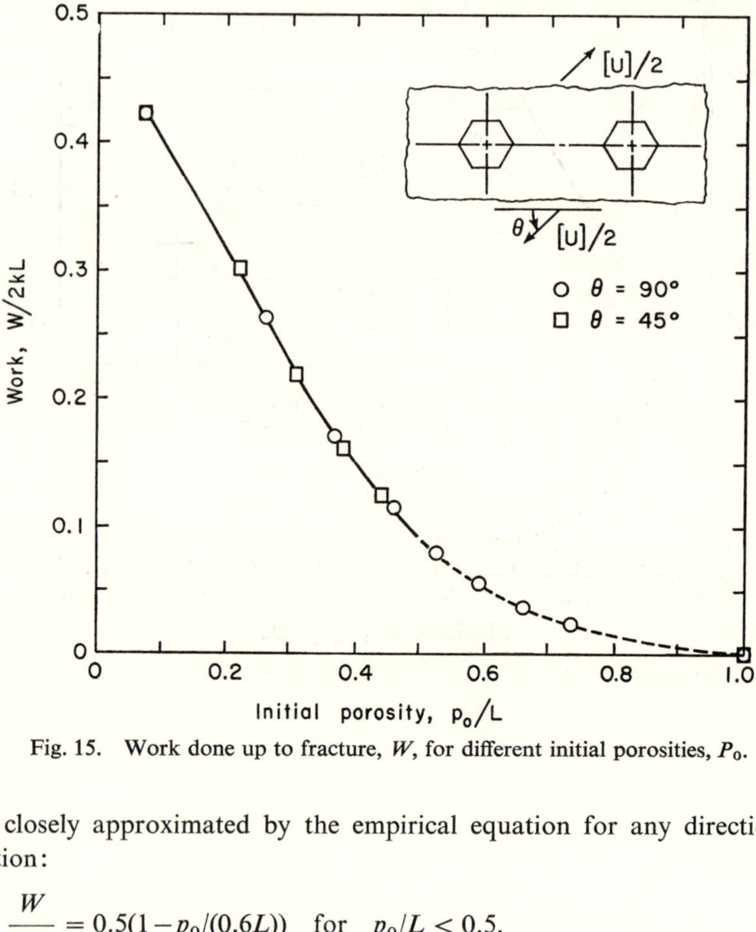

Fig. 15. Work done up to fracture, W, for different initial porosities, P_0.

are closely approximated by the empirical equation for any direction of motion:

$$\frac{W}{2kL} = 0.5(1-p_0/(0.6L)) \quad \text{for} \quad p_0/L < 0.5. \tag{10}$$

Note that for high triaxiality ($p_0|L \ll 1$), the work done in separating the sheet of localized flow becomes independent of initial triaxiality:

$$W = kL. \tag{11}$$

The work to propagate a crack gives the stress intensity factor for crack propagation, *assuming* elastic surroundings of modulus E:

$$k_{Ic} = \sigma_p\sqrt{c} = \sqrt{\frac{EW}{\pi(1-\nu^2)}} = \sqrt{\frac{Ekl}{\pi(1-\nu^2)}}. \tag{12}$$

For typical values of inclusion spacing $2L$ and shear strength k this gives values decidedly lower than typical values of the observed critical stress intensity factor k_{Ic}. It therefore appears that either the surroundings are plastic, or the stress to initiate is greater than that for propagation.

6. Discussion and conclusions

Approximate equations have been developed for fracture by the development of a sheet of localized flow. Crack *initiation* (by forming the sheet) seems to depend on the force balance between eqs. 4 and 7 rather than on the balance between strain hardening and hole growth. Once formed, the sheet of localized flow provides boundary conditions on the surrounding body. The solution of the elastic-plastic behavior of the body subject to these boundary conditions as well as to the tractions or displacements over other parts of the body will provide a quantitative theory of crack growth by hole coalescence.

The traction-displacement relations for the plane of localized flow can be thought of in two parts which may be called *constitutive* and *process* equations. The *constitutive* equation, eq. 7, gives the traction in terms of the direction of displacement increment and internal parameters which are here taken to be simply the ratio of the mean radius p of the growing holes to their spacing, L. A *process* relation (Fig. 13) tells how the porosity p/L changes with the relative displacement across the plane of localized flow. If solution of the behavior of the body subject to these boundary conditions indicates that the body itself remains elastic, the fracture toughness can be calculated by equating the work done on the sheet of localized flow, eq. 7, to the strain energy release rate as in eq. 12. This simple case does not appear likely, for the magnitude of the critical stress intensity factor is usually much greater than calculated from eq. 12. If it is initiation that is limiting, not only must a stress and strain be attained for localization, but also these must be attained over a region of some size comparable to the spacing between the holes. For some materials and conditions the strain gradients may be so steep that the concept of a plane of localization does not occur at all, and the crack growth must be analyzed by considering the interaction of holes with a crack tip one at a time, as partly treated by McClintock (1969). More work will be reported shortly.

Acknowledgement

The support of the National Science Foundation, through Grant GK–1875–X2, is deeply appreciated.

References

[1] Backofen, W. A., A. J. Shaler and B. B. Hundy, Mechanical anisotropy in copper. *Trans. Am. Soc. Met. 46* (1954) 655–675.

[2] Berg, C. A., Plastic dilatation and void interaction. *Inelastic Behavior of Solids*, M. F. Kanninen et al. (eds.), McGraw-Hill, New York, (1970) 171–210.

[3] Berg, C. A., *On plane dilatational plasticity* (1972). In press.

[4] Bishop, J. F. W., On the complete solution to problems of deformation of a plastic rigid material. *J. Mech. Phys. Solids 2* (1970) 43–53.

[5] Bishop, J. F. W., A. P. Green and R. Hill, A note on the deformable region in a rigid plastic body. *J. Mech. Phys. Solids 4* (1956) 256–258.

[6] Bluhm, J. I. and R. J. Morrissey, Fracture in a tensile specimen. *Proceedings of First International Conf. on Fracture, Senadi*, Japanese Society for Strength and Fracture of Materials *3* (1966) 1739–1780.

[7] Broek, D., *A study on ductile fracture*. Ph. D. Thesis, Technische Hogeschool Delft, Nederlandse Boekdruk Industrie N.V., 's-Hertogenbosch, Netherlands. (1971)

[8] Brook, R., Private communication. See also Brook's micrographs in Berg (1970).

[9] Carson, J. W., *A study of plane strain ductile fracture*. Ph. D. Thesis, Dept. of Mechanical Engineering, MIT, Cambridge, Mass. (1970).

[10] Drucker, D. C., On obtaining plane strain or plane stress conditions in plasticity. *Proc. 2nd U.S. Nat'1. Cong. of Applied Mechanics* (1954) 485–488.

[11] Elliott, D. and H. Stuart, Fractographic features associated with general yielding in low strength steels. *Proc. 3rd Conf. on Scanning Electron Microscopy*, Illinois Institute of Technology, Chicago, Ill. (1970). (See also personal communication to R. M. N. Pelloux, confirming the convexity of their surfaces).

[12] Green, A. P., The plastic yielding of metal junctions due to combined shear and pressure. *J. Mech. Phys. Solids 2* (1954) 197–211.

[13] Hadamard, J., *Leçons sur la propagation des ondes et les equations des l'hydrodynamique*. Paris: Hermann (1903).

[14] Hill, R., *The mathematical theory of plasticity*. Clarendon Press, Oxford, England (1950).

[15] Hill, R., On the limits set by plastic yielding to the intensity of singularities of stress. *J. Mech. Phys. Solids 2* (1954) 278–285.

[16] Hill, R., On the problem of uniqueness in the theory of a rigid-plastic solid – I. *J. Mech. Phys. Solids 4* (1956) 247–255.

[17] Hill, R., Stability of rigid-plastic solids. *J. Mech. Phys. Solids 6* (1957) 1–8.

[18] Hill, R., Discontinuity relations in mechanics of solids. *Progress in Solid Mechanics 2* (1961). I. N. Sneddon and R. Hill, eds., North Holland, Amsterdam 247–276.

[19] Joyce, J. A., *Tensile plastic deformation of notch roots.* M. S. Thesis, Dept. of Mechanical Engineering, MIT, Cambridge, Mass. (1968).

[20] McClintock, F. A., A criterion for ductile fracture by the growth of holes. *J. Appl. Mech. 35* (1968a) 363–371.

[21] McClintock, F. A., On the mechanics of fracture from inclusions. *Ductility*, Am. Soc. Metals, Metals Park, Ohio (1968b) 255–277.

[22] McClintock, F. A., Crack growth in fully plastic grooved tensile specimens. *Physics of strength and plasticity*, Orowan Anniversary Volume, A. S. Argon, ed., MIT Press (1969) 307–326.

[23] McClintock, F. A., Plasticity Aspects of Fracture. *Fracture, 3*, H. Liebowitz, ed., Academic Press, New York (1971) 24–225.

[24] Needleman, A., Void growth in an elastic-plastic medium. *J. Appl. Mech.* (1972) (in press).

[25] Pelloux, R. M. N. and L. van Swam, personal communication (1971).

[26] Plateau, J., G. Henry and C. Crussard, Complements a l'interpretation des images fournies par la microfractographies. *Metaux, Corrosion Industries 33* (1958) 141–162.

[27] Puttick, K. E., Ductile fracture in metals. *Philosophical Magazine, Series 8, Vol. 4* (1959) 964–969.

[28] Rice, J. R. and D. M. Tracey, On the ductile enlargement of voids in triaxial stress fields. *J. Mech. Phys. Solids 17* (1969) 201–217.

[29] Subudhi, M., *Plasticity of anisotropic porous metals.* M. S. Thesis, Mechanical Engineering Dept. MIT, Cambridge, Mass. (1970).

[30] Tanaka, J. P., C. A. Pampillo and J. R. Jr. Low, Fractographic analysis of the low energy fracture of an aluminium alloy. *Review of developments in plane strain fracture toughness testing*, ASTM STP 463, American Society for Testing and Materials, W. F. Brown, ed., Philadelphia (1970) 191–215.

[31] Tipper, C. F., The fracture of metals. *Metallurgia 39* (1949) 133–137.

[32] Wang, A. J., Plastic flow in a deeply notched bar with semicircular root. *Quart. Appl. Math. 11* (1954) 427–438.

Relations de comportement des milieux élastiques-plastiques et élastiques-viscoplastiques. Notion de repère directeur

Jean Mandel

Ecole Polytechnique, Ecole des Mines, Paris, France

Le but de cette étude est d'obtenir des équations générales constituant un cadre physiquement correct et mathématiquement cohérent pour décrire le comportement des corps élastiques-viscoplastiques, cadre renfermant en outre comme cas spécial la Plasticité classique indépendante du temps.

La difficulté d'exploiter une représentation fonctionnelle de l'histoire conduit à introduire des variables cachées, dont il faut définir l'évolution par des équations différentielles à ajouter à celles qui décrivent le comportement rhéologique. Nous montrons que ceci implique l'existence d'un repère privilégié (pour la description du comportement) que nous appelons repère directeur.

Nous étudions la décomposition de la transformation thermoélastique-viscoplastique en introduisant une configuration intermédiaire relâchée. Nous posons les relations de comportement pour la transformation viscoplastique (finie), puis pour la transformation complète. La nouveauté la plus frappante de ces relations est qu'elles introduisent la vitesse de rotation par rapport au repère directeur et pas seulement la vitesse de déformation.

Les corps élastiques-viscoplastiques

Ce sont les corps pour lesquels on peut définir à chaque instant dans l'espace des contraintes un domaine élastique borné tel que si la contrainte reste intérieure à ce domaine les déformations sont réversibles, quelle que soit leur vitesse.

Lorsque la contrainte sort de ce domaine, des déformations permanentes (ou plastiques) se produisent. A ce moment la contrainte dépend du trajet de déformation et en général de la vitesse à laquelle il est décrit (corps viscoplastiques). Le cas spécial d'une contrainte indépendante de la vitesse de description du trajet de déformation (plasticité classique) peut cependant constituer un schéma mathématique utile.

Les difficultés d'une représentation fonctionnelle de l'histoire des déformations conduisent à postuler que l'état thermodynamique actuel peut être défini par la donnée des variables observables (déformations, température) et des valeurs actuelles d'un certain nombre de variables cachées (ou internes) que nous appellerons variables d'écrouissage. Notamment le seuil de plasticité actuel dépend des variables d'écrouissage.

Les valeurs de ces variables changent au cours des déformations plastiques. Mais on constate que le seuil actuel de plasticité varie même en l'absence de déformations permanentes. Le corps étant déchargé, le seuil s'abaisse au cours du temps (ce phénomène de radoucissement s'explique dans un monocristal par la disparition progressive de dislocations actives par suite de l'agitation thermique). En conséquence on admettra que les variables d'écrouissage peuvent évoluer même en l'absence de déformations permanentes.

Cinématique de la transformation élastique-viscoplastique

Soit (0) la configuration occupée par un élément macroscopique de matière à l'instant 0, sous contrainte nulle et à la température T_0, (*a*) la configuration qu'il occupe à l'instant actuel *t* sous une contrainte $\boldsymbol{\sigma}$ et à la température *T*. On introduit une configuration intermédiaire (κ) dite configuration relâchée (ou neutre) actuelle, en supposant qu'à l'instant *t* on décharge rapidement l'élément et on le ramène à sa température initiale. La configuration *x* n'est définie qu'à une rotation près.

Soit $\boldsymbol{F}$ le gradient de la transformation complète $0 \rightarrow a$, $\boldsymbol{P}$ le gradient de la transformation dite plastique $0 \rightarrow \kappa$, $\boldsymbol{E}$ le gradient de la transformation élastique $\kappa \rightarrow a$. On a:

$$\boldsymbol{F} = \boldsymbol{EP}, \tag{1}$$

$$\text{grad}\, \boldsymbol{v} = \dot{\boldsymbol{F}}\boldsymbol{F}^{-1} = \dot{\boldsymbol{E}}\boldsymbol{E}^{-1} + \boldsymbol{E}\dot{\boldsymbol{P}}\boldsymbol{P}^{-1}\boldsymbol{E}^{-1}, \tag{2}$$

$$\mathscr{D} = \{\text{grad}\, \boldsymbol{v}\} = \{\dot{\boldsymbol{E}}\boldsymbol{E}^{-1}\} + \{\boldsymbol{E}\dot{\boldsymbol{P}}\boldsymbol{P}^{-1}\boldsymbol{E}^{-1}\}, \tag{3}$$

$\boldsymbol{v}$ désigne la vitesse, grad $\boldsymbol{v}$ son gradient par rapport aux coordonnées actuelles *x*, $\mathscr{D}$ la vitesse de déformation (rapportée à la configuration actuelle), { } la partie symétrique d'un tenseur.

Chacun des termes du 2ème membre de 2 ou 3 dépend de la configuration relâchée choisie, plus exactement de la façon dont elle varie entre *t* et $t+dt$.

En effet, si $\boldsymbol{E}' = \boldsymbol{E}\boldsymbol{V}$, ($\boldsymbol{V}$ orthogonal); on a:

$$\dot{\boldsymbol{E}}'\boldsymbol{E}'^{-1} = \dot{\boldsymbol{E}}\boldsymbol{E}^{-1} + \boldsymbol{E}\dot{\boldsymbol{V}}\boldsymbol{V}^T\boldsymbol{E}^{-1}. \tag{4}$$

C'est seulement pour un choix bien déterminé du mouvement de la configuration κ que l'expression «vitesse de déformation élastique» pour le terme $\{\dot{\boldsymbol{E}}\boldsymbol{E}^{-1}\}$ au second membre de (3) est physiquement correcte. Une étude thermodynamique ultérieure permettra de préciser ce point.

Variables d'écrouissage et repère directeur

On suppose que l'état thermodynamique de l'élément de matière est défini par les variables suivantes:
Δ = tenseur de déformation de Green dans le passage de (κ) à (a)[1])

$$1+2\Delta = \boldsymbol{E}^T\boldsymbol{E} = \boldsymbol{S}^2, \tag{5}$$

T = température (absolue),
$\boldsymbol{\alpha}_j$ = variables d'écrouissage, variables cachées scalaires ou tensorielles.
On peut aussi prendre comme variables: T, $\boldsymbol{\alpha}_j$ et:
$\boldsymbol{\pi}$ = tenseur des contraintes de Kirchhoff par rapport à la configuration κ, lié au tenseur eulérien σ des contraintes par:

$$\boldsymbol{\pi} = (\text{dét}\,\boldsymbol{E})\boldsymbol{E}^{-1}\boldsymbol{\sigma}\boldsymbol{E}^{T-1}. \tag{6}$$

L'hypothèse des variables d'écrouissage *implique* l'existence d'une suite privilégiée de configurations relâchées, ou encore d'un (au moins) repère privilégié 0 pour les configurations relâchées, distinct du repère du mouvement, et que nous appellerons *repère directeur* (parce que sa position précise l'orientation de l'élément de matière).

En effet, prenons un autre repère S dont D se déduit par la transformation orthogonale

$$\boldsymbol{x} = \boldsymbol{\beta}(t)\hat{\boldsymbol{x}} \tag{7}$$

(l'accent ^ est réservé aux grandeurs relatives à D). Les tenseurs deviennent:

$$\Delta = \boldsymbol{\beta}\hat{\Delta}\boldsymbol{\beta}^T \rightarrow \hat{\Delta} = \boldsymbol{\beta}^T\Delta\boldsymbol{\beta}.$$

Donc l'état thermodynamique est maintenant défini par Δ. T, $\boldsymbol{\alpha}_j$ *et* $\boldsymbol{\beta}(t)$.

[1]) Il n'est pas indiqué de définir l'état par T, α_j et le tenseur de déformation de Green dans le passage de l'état initial à l'état actuel (ou ce qui revient au meme $\boldsymbol{F}^T\boldsymbol{F}$). En effet, s'il n'y a aucun écrouissage, l'état est le même dans toutes les configurations relâchées successives, quoique P varie.

Soit f une fonction scalaire de l'état thermodynamique. Si l'on prend des configurations relâchées telles que le trièdre D conserve une orientation fixe:

$$f = \hat{f}(\hat{\boldsymbol{\Delta}}, T, \hat{\boldsymbol{\alpha}}_j) \tag{8}$$

est une fonction de forme invariante des variables thermodynamiques.

Avec d'autres configurations relâchées:

$$f = \hat{f}(\boldsymbol{\beta}^T \boldsymbol{\Delta} \boldsymbol{\beta}, T, \boldsymbol{\beta}^T \boldsymbol{\alpha}_j \boldsymbol{\beta}) = f(\boldsymbol{\Delta}, T, \boldsymbol{\alpha}_j, \boldsymbol{\beta}(t)) \tag{9}$$

n'est plus une fonction de forme invariante des variables thermodynamiques (à moins qu'elle ne soit isotrope).

Dans un monocristal les variables d'écrouissage sont les variables qui définissent la distribution des dislocations et plus généralement des défauts du réseau atomique. Le repère directeur est un repère attaché au réseau (pris dans l'état relâché).

Dans un polycristal on peut prendre comme trièdre directeur un trièdre attaché au réseau de l'un quelconque des cristaux constitutifs (supposé ramené à l'état neutre par déformation pure). On peut aussi prendre comme trièdre directeur un trièdre dont la vitesse de rotation est la moyenne des vitesses des réseaux des différents cristaux (chacun de ceux-ci étant séparément ramené à l'état neutre par déformation pure). Les variables d'écrouissage sont: les variables internes des N monocristaux, les N contraintes résiduelles s'exerçant entre les cristaux (variables tensorielles – non indépendantes: conditions d'équilibre), enfin les variables nécessaires pour définir la géométrie actuelle de l'assemblage, soit les $N-1$ rotations relatives des réseaux (ramenés à l'état neutre par déformation pure) et les N transformations $\boldsymbol{P}$ des cristaux par rapport à leurs réseaux (non indépendantes: conditions de compatibilité, oú interviennent aussi les variables précédentes).

Définition thermodynamique de la vitesse de déformation élastique

Soit u l'énergie interne, s l'entropie, $\varphi = u - Ts$ l'énergie libre, par unité de masse. En utilisant des configurations relâchées telles que le trièdre directeur conserve une orientation fixe, u, s, φ sont des fonctions de forme invariante de $\boldsymbol{\Delta}$, T, $\boldsymbol{\alpha}_k$. Donc, par exemple:

$$\dot{\varphi} = \operatorname{tr}\left(\frac{\partial \varphi}{\partial \boldsymbol{\Delta}}\, \dot{\boldsymbol{\Delta}}\right) + \frac{\partial \varphi}{\partial T}\, \dot{T} + \operatorname{tr}\left(\frac{\partial \varphi}{\partial \boldsymbol{\alpha}_k}\, \dot{\boldsymbol{\alpha}}_k^T\right) \tag{10}$$

$\partial\varphi/\partial\boldsymbol{\Delta}$ désigne le tenseur de coordonnées $\partial\varphi/\partial\Delta_{ij}$ et $\operatorname{tr}(\partial\varphi/\partial\boldsymbol{\Delta})\dot{\boldsymbol{\Delta}})$ la trace du produit de ce tenseur par le tenseur $\dot{\boldsymbol{\Delta}}$.

Utilisons l'inégalité de la puissance dissipée (puissance transformée en chaleur).[1])

$$\phi = \operatorname{tr}\left(\frac{\sigma\mathscr{D}}{\rho}\right) - (\dot{\varphi} + s\dot{T}) \geqq 0. \tag{11}$$

Le premier terme du second membre (puissance des contraintes par unité de masse) s'écrit encore (en utilisant (6), (3) et (5):

$$\mathscr{P} = \operatorname{tr}\left(\frac{\boldsymbol{\pi}}{\rho_\kappa}\, \dot{\boldsymbol{\Delta}}\right) + \operatorname{tr}\left(\frac{\boldsymbol{\pi}}{\rho_\kappa}\, \boldsymbol{S}^2 \dot{\boldsymbol{P}} \boldsymbol{P}^{-1}\right). \tag{12}$$

La déformation $\boldsymbol{\Delta}$ a été définie (5) en imaginant une décharge élastique rapide de la configuration (a) à la configuration (κ). Dans ce cas on déduit de l'inégalité (11), par un raisonnement classique:

$$\frac{\boldsymbol{\pi}}{\rho_\kappa} = \frac{\partial \varphi}{\partial \boldsymbol{\Delta}}, \qquad s = -\frac{\partial \varphi}{\partial T}. \tag{13}$$

Le ler terme de la puissance des contraintes[2]):

$$\mathscr{P}_1 = \operatorname{tr}\left(\frac{\boldsymbol{\pi}\dot{\boldsymbol{\Delta}}}{\rho_\kappa}\right) = \operatorname{tr}\left(\frac{\boldsymbol{\sigma}\dot{\boldsymbol{E}}\boldsymbol{E}^{-1}}{\rho}\right)$$

est donc égal, d'après (10), au taux d'accroissement de l'énergie libre à T et $\boldsymbol{\alpha}_k$ constants (pour les $\boldsymbol{\alpha}_k$, s'ils sont tensoriels, constants par rapport au repère directeur). Il est naturel de définir la *vitesse de déformation élastique* par cette condition concernant la puissance des contraintes. Donc *en adoptant des configurations relâchées telles que le trièdre directeur conserve une orientation fixe,* $\dot{\boldsymbol{\Delta}}$ peut être appelé vitesse de déformation élastique en variables de Lagrange. En variables d'Euler cette vitesse devient:

$$\mathscr{D}_{el} = \{\dot{\boldsymbol{E}}\boldsymbol{E}^{-1}\}.$$

La relation (4) montre qu'avec un autre choix des configurations relâchées

[1]) Elle se déduit de l'inégalité de Clausius-Duhem, expression générale du 2ème principe de la thermodynamique, en prenant grad $T = 0$.

[2]) Quant au 2ème terme, il est, d'après (11), égal à $\phi + \operatorname{tr}(\partial\varphi/\partial\boldsymbol{\alpha}_k)\dot{\boldsymbol{\alpha}}_k^T$). Donc la puissance dépensée dans la déformation plastique est la somme de la puissance dissipée ϕ et du terme $\operatorname{tr}(\partial\varphi/\partial\boldsymbol{\alpha}_k)\dot{\boldsymbol{\alpha}}_k^T$) qui représente le taux de variation de l'énergie libre 'bloquée' (correspondant aux dislocations et aux contraintes résiduelles).

$\{\dot{\boldsymbol{E}}\boldsymbol{E}^{-1}\}$ ne peut pas, au sens que nous venons de préciser, être appelé vitesse de déformation élastique.[1])

Dérivées par rapport à *t* dans le mouvement par rapport au trièdre directeur

Dans ce qui précède nous avons été amenés à choisir des configurations relâchées κ_0 telles que le trièdre directeur D reste d'orientation fixe. En fait ce choix, commode, n'est pas indispensable. Ce qui importe c'est de prendre les dérivées par rapport à t des grandeurs tensorielles attachées à la configuration κ, non dans le mouvement par rapport aux axes fixes S, mais dans le mouvement par rapport au trièdre D. Du même coup, se trouve résolue la question, restée obscure en plasticité classique, de la dérivée à introduire dans les relations de comportement: dérivée de Jaumann, dérivée de Truesdell, ...? Réponse: il faut prendre la dérivée dans le mouvement par rapport au trièdre directeur D.

Si D se déduit des axes S par la transformation orthogonale 7, sa vitesse de rotation par rapport aux axes S a pour matrice dans les axes S:

$$\omega_D = \dot{\boldsymbol{\beta}}\boldsymbol{\beta}^{-1} \tag{14}$$

et les dérivées par rapport à t, mais dans le mouvement de par rapport à D, de grandeurs tensorielles attachées à κ et repérées dans S sont:

$$\text{pour un vecteur } v\text{:} \quad \frac{D\boldsymbol{v}}{Dt} = \dot{\boldsymbol{v}} - \boldsymbol{\omega}_D\boldsymbol{v}, \tag{15}$$

$$\text{pour un tenseur du 2ème ordre } \boldsymbol{u}\text{:} \quad \frac{D\boldsymbol{u}}{Dt} = \dot{\boldsymbol{u}} - \boldsymbol{\omega}_D\boldsymbol{u} + \boldsymbol{u}\boldsymbol{\omega}_D, \tag{16}$$

$\dot{\boldsymbol{v}}$, $\dot{\boldsymbol{u}}$ sont les dérivées dans le mouvement de κ par rapport aux axes S. Ces dérivées dépendent des configurations κ choisies pour un repère S donné ou, ce qui revient au même, du repère choisi (i.e. de l'observateur) pour une suite de x donnée. Au contraire les dérivée D/Dt sont *objectives.*

La dérivée d'une fonction scalaire, qui maintenant dépend de $\boldsymbol{\Delta}$, T, $\boldsymbol{\alpha}_j$ et $\boldsymbol{\beta}(t)$, mais vérifie la relation (9), est donnée par:

$$\dot{f} = \operatorname{tr}\left(\frac{\partial f}{\partial \boldsymbol{\Delta}} \; \frac{D\boldsymbol{\Delta}}{Dt}\right) + \frac{\partial f}{\partial T}\dot{T} + \operatorname{tr}\left(\frac{\partial f}{\partial \boldsymbol{\alpha}_j} \; \frac{D\boldsymbol{\alpha}_j^T}{Dt}\right) \tag{17}$$

[1]) à moins qu'il n'y ait isotropie dans les configurations relâchées. Dans ce cas $\boldsymbol{\sigma}$ est une fonction isotrope de $\boldsymbol{E}\boldsymbol{E}^T$, donc $\boldsymbol{E}^{-1}\boldsymbol{\sigma}\boldsymbol{E}$ est symétrique, d'oú $\operatorname{tr}(\boldsymbol{E}^{-1}\boldsymbol{\sigma}\boldsymbol{E}\dot{\boldsymbol{V}}\boldsymbol{V}^T) = 0$.

et toutes les relations contenant des vitesses se transposent immédiatement.

La vitesse de déformation s'écrit au lieu de (3):

$$\mathscr{D} = \left\{\frac{DE}{Dt}\,E^{-1}\right\} + \left\{E\,\frac{DP}{Dt}\,P^{-1}E^{-1}\right\} \tag{18}$$

et *les 2 termes du 2ème membre sont maintenant indépendants du choix de la configuration* κ.[1]) Le premier est la vitesse de déformation élastique.

Frontière élastique et frontière de plasticité instantanée

Soit π le tenseur des contraintes de Kirchhoff relatif à une configuration relâchée κ, pour laquelle le trièdre directeur se déduit des axes fixes S par la transformation orthogonale (7).

D'après la définition du corps élastique-viscoplastique, il existe à chaque instant, dans l'espace π, T, un domaine:

$$G_\kappa(\pi, T, \alpha_j, \beta) < 0 \tag{19}$$

éventuellement réduit à une ligne, tel que, si le point π, T se trouve dans ce domaine, aucune déformation plastique n'est possible, et un domaine plus étendu:

$$f_\kappa(\pi, T, \alpha_j, \beta) < 0, \qquad (G_\kappa < 0 \Rightarrow f_\kappa < 0) \tag{20}$$

éventuellement l'espace π, T tout entier, tel que, si le point π, T se trouve dans ce domaine, aucune déformation plastique *instantanée* n'est possible.

L'hypersurface $G_\kappa = 0$ est la frontière élastique; l'hypersurface $f_\kappa = 0$ sera appelée frontière de plasticité instantanée.[2])

Si l'on utilise des configurations κ telles que le trièdre directeur conserve une orientation fixe, G, f sont des fonctions de forme invariante de $\hat{\pi}$, T, $\hat{\alpha}_j$. Avec d'autres configurations relâchées, G et f sont des fonctions qui vérifient la relation (9) (où Δ est remplacé par π). La notion de trièdre directeur a ainsi permis de clarifier la question du critère d'écoulement d'un corps aniso-

[1]) Dans les changements de la configuration κ, E et P, qui ne se rattachent à κ que par un de leurs indices (le premier pour P, le deuxième pour E) ne se transforment pas comme des tenseurs. P et E^T se transforment comme des vecteurs.

[2]) Les métaux n'ont pas de déformations plastiques instantanées; mais leurs déformations plastiques peuvent devenir très rapides et l'introduction d'une frontière de plasticité instantanée peut être un schéma mathématique commode. C'est le schéma de la plasticité classique.

trope, question classique, mais qui restait un peu obscure dans le cas où l'on ne négligeait pas les déformations élastiques.

On va maintenant étudier la transformation viscoplastique $\boldsymbol{P}$, puis sa composition avec la transformation élastique.

La transformation viscoplastique. Cas où les déformations plastiques instantanées sont impossibles ($f < 0$).

La vitesse $\dot{\boldsymbol{P}}\boldsymbol{P}^{-1}$ de la transformation plastique, rapportée à la configuration κ actuelle, ne dépend que des valeurs actuelles des variables thermodynamiques et de leurs vitesses. Dans le cas envisagé ici elle ne peut pas dépendre de la vitesse de contrainte. Nous admettrons donc qu'elle dépend seulement de la contrainte, de la température et des variables d'écrouissage, par une relation de forme invariante dans un repère directeur. i.e. en adoptant des configurations relâchées telles que le trièdre directeur reste d'orientation fixe:

$$\dot{\boldsymbol{P}}\boldsymbol{P}^{-1} = \hat{A}(\boldsymbol{\pi}, T, \boldsymbol{\alpha}_j)H(\hat{G}), \tag{21}$$

H désigne la fonction d'Heaviside ($H = 0$ si $\hat{G} < 0$, $H = 1$ si $\hat{G} > 0$). $\hat{A}$ s'annule en même temps que $\hat{G}$. Dans la suite, pour alléger l'écriture, on omettra le facteur $H(\hat{G})$.

Il faut définir la variation en fonction de t des variables d'écrouissage α_j. Nous admettrons qu'elles sont déterminées par des équations différentielles:

$$\dot{\boldsymbol{\alpha}}_k = \hat{\boldsymbol{h}}_k(\boldsymbol{\pi}, T, \boldsymbol{\alpha}_j) \tag{22}$$

de forme invariante dans un repère directeur.[1] [2]

Puisqu'on n'exclut pas la possibilité de variations des $\boldsymbol{\alpha}$ dans le domaine élastique, $\hat{\boldsymbol{h}}_k$ ne s'annule pas nécessairement en même temps que $\hat{G}$.

Il faut noter que, par (21), c'est la vitesse de transformation toute entière (9 composantes) par rapport au trièdre directeur[3], pas seulement la vitesse de déformation (6 composantes), qui est déterminée. L'exemple du monocristal le montre clairement. Si la déformation plastique d'un monocristal

[1]) L'idée de relations analogues à (22), pour définir l'évolution des variables d'écrouissage, paraît avoir été formulée pour la lère fois par J. Zarka en 1967, cf [4].

[2]) Dans le cas de déformations plastiques avec variation de volume, on peut faire intervenir la masse volumique ρ_κ dans la configuration relâchée comme une variable d'écrouissage. L'équation (22) correspondante est l'équation de continuité:

$$\dot{\rho}_\kappa = -\rho_\kappa \operatorname{tr}(\dot{\boldsymbol{P}}\boldsymbol{P}^{-1}) = -\rho_\kappa \operatorname{tr}\hat{\boldsymbol{A}}.$$

[3]) mais, bien entendu, la rotation du trièdre directeur reste indéterminée.

résulte d'un glissement d'amplitude γ, suivant un plan atomique de normale $\boldsymbol{n}$ et dans une direction atomique de vecteur unitaire $\boldsymbol{m}$, la vitesse en un point de coordonnées ξ_j dans un trièdre attaché au réseau atomique est donnée par:

$$v_i = \dot{\gamma} m_i n_j \xi_j$$

d'où:

$$\dot{\boldsymbol{P}}\boldsymbol{P}^{-1} = \text{grad}\, \boldsymbol{v} = \dot{\gamma}\boldsymbol{m} \otimes \boldsymbol{n}$$

($\dot{\gamma}$ fonction de $\boldsymbol{\pi}$, T, $\boldsymbol{\alpha}_j$). Par rapport au réseau atomique, non seulement la vitesse de déformation, mais aussi la vitesse de rotation dans la transformation plastique sont donc déterminées.

Si l'on utilise des configurations relâchées κ quelconques, se déduisant des précédentes par une transformation orthogonale $\boldsymbol{\beta}(t)$, les relations précédentes deviennent:

$$\frac{D\boldsymbol{P}}{Dt}\boldsymbol{P}^{-1} = \boldsymbol{A}(\pi, T, \boldsymbol{\alpha}_j, \boldsymbol{\beta}) \tag{23}$$

$$\frac{D\boldsymbol{\alpha}_k}{Dt} = \boldsymbol{h}_k(\pi, T, \boldsymbol{\alpha}_j, \boldsymbol{\beta}) \tag{24}$$

les fonctions tensorielles $u(\boldsymbol{A}, \boldsymbol{h}_k$ ou $G)$ se déduisant des fonctions correspondantes $\hat{u}$ relatives au trièdre directeur par la relation:

$$\boldsymbol{u}(\pi, T, \boldsymbol{\alpha}_j, \boldsymbol{\beta}) = \boldsymbol{\beta}\hat{\boldsymbol{u}}(\boldsymbol{\beta}^T\boldsymbol{\pi}\boldsymbol{\beta}, T, \boldsymbol{\beta}^T\boldsymbol{\alpha}_j\boldsymbol{\beta})\boldsymbol{\beta}^T \tag{25}$$

et les dérivées D/Dt étant définies par:

$$\frac{D\boldsymbol{P}}{Dt} = \dot{\boldsymbol{P}} - \omega_D \boldsymbol{P} \qquad \text{(cf. note page 7)}, \tag{26}$$

$$\frac{D\boldsymbol{\alpha}}{Dt} = \dot{\boldsymbol{\alpha}} - \omega_D\boldsymbol{\alpha} + \boldsymbol{\alpha}\omega_D \qquad \text{où} \quad \omega_D = \dot{\boldsymbol{\beta}}\boldsymbol{\beta}^T. \tag{27}$$

Si l'élément de matière est isotrope dans un état initial (neutre), les fonctions $\hat{G}$, $\hat{\boldsymbol{A}}$, $\hat{\boldsymbol{h}}$ ne changent pas lorsqu'on lui impose préalablement, dans cet état initial, une transformation orthogonale quelconque. Ce sont donc des fonctions isotropes de $\boldsymbol{\pi}$ et des $\boldsymbol{\alpha}_j$. Alors d'après (25), G, $\boldsymbol{A}$, $\boldsymbol{h}$ (repère quelconque) coincident avec $\hat{G}$, $\hat{\boldsymbol{A}}$, $\hat{\boldsymbol{h}}$, donc sont indépendants de l'orientation $\boldsymbol{\beta}$.

En particulier si tous les paramètres d'écrouissage sont *scalaires*, G, $\boldsymbol{A}$, $\boldsymbol{h}$ sont des fonctions isotropes de $\boldsymbol{\pi}$, ce qui signifie que l'isotropie subsiste dans toutes les configurations relâchées. L'*écrouissage est isotrope*. Dans ce cas $\boldsymbol{A}$ est un tenseur symétrique et (21) montre que la rotation de l'élément par rapport au trièdre directeur est nulle. Autrement dit le repère directeur est ici corotationnel à l'élément (pris bien entendu dans les configurations relâchées). On notera que, si l'écrouissage est isotrope, il n'y a pas d'effet Bauschinger. Pour rendre compte de cet effet on doit donc introduire des variables d'écrouissage tensorielles.

Cas où il y a des déformations plastiques non retardées

De telles déformations se produisent lorsque le point de charge $\boldsymbol{\pi}$, T atteint la frontière de plasticité instantanée $f = 0$ avec une vitesse dirigée vers les f positifs.

Soit λ la vitesse de f lorsqu'il n'y a pas de déformations instantanées:

$$\lambda = \operatorname{tr}\left(\frac{\partial f}{\partial \boldsymbol{\pi}} \frac{D\boldsymbol{\pi}}{Dt}\right) + \frac{\partial f}{\partial T}\dot{T} + \operatorname{tr}\left(\frac{\partial f}{\partial \boldsymbol{\alpha}_k}\boldsymbol{h}_k^T\right). \tag{28}$$

Si $\lambda > 0$, le point $\boldsymbol{\pi}$, T reste sur la frontière précédente et l'entraîne (il ne peut pas la franchir puisqu'elle correspond au cas limite de vitesses de déformation infinies). Si $\lambda < 0$, il revient à l'intérieur du domaine $f < 0$.

Les termes correspondant aux déformations instantanées dans les expressions de $(D\boldsymbol{P}/Dt)\boldsymbol{P}^{-1}$ et $D\boldsymbol{\alpha}_k/Dt$ s'annulent donc en même temps que λ. On peut donc poser:

$$\frac{D\boldsymbol{P}}{Dt}\boldsymbol{P}^{-1} = \boldsymbol{A}(\boldsymbol{\pi}, T, \boldsymbol{\alpha}_j, \boldsymbol{\beta}) + \lambda \boldsymbol{B}(\boldsymbol{\pi}, T, \boldsymbol{\alpha}_j, \boldsymbol{\beta})\ ^{1}), \tag{29}$$

$$\frac{D\boldsymbol{\alpha}_k}{Dt} = \boldsymbol{h}_k(\boldsymbol{\pi}, T, \boldsymbol{\alpha}_j, \boldsymbol{\beta}) + \lambda \boldsymbol{l}_k(\boldsymbol{\pi}, T, \boldsymbol{\alpha}_j, \boldsymbol{\beta}) \tag{30}$$

si

$$f(\boldsymbol{\pi}, T, \boldsymbol{\alpha}_j, \boldsymbol{\beta}) = 0 \qquad \text{et} \qquad \lambda \geqq 0$$

tandis que (23) et (24) s'appliquent si $\lambda < 0$ ou $f < 0$.

[1]) En ce qui concerne la forme des tenseurs $\boldsymbol{A}$ et $\boldsymbol{B}$, disons seulement ici que, pour les métaux, en partant des lois des glissements intractristallins, on établit (cf. [3]) l'existence du potentiel plastique classique f et celle d'un potentiel viscoplastique $\Omega(\boldsymbol{\pi}, T, \alpha_j, \beta)$ (J. R. Rice, 1970), tous 2 fonctions convexes de $\boldsymbol{\pi}$ et telles que:

$$\{\boldsymbol{A}\} = \frac{\partial \Omega}{\partial \boldsymbol{\pi}}, \qquad \{\boldsymbol{B}\} = \mu \frac{\partial f}{\partial \boldsymbol{\pi}}, \quad \mu > 0.$$

La dépendance en $\boldsymbol{\beta}$ de f, $\boldsymbol{B}$, $\boldsymbol{l}_k$ est toujours exprimée par la relation (25).

Les fonctions $\boldsymbol{l}_k$ sont liées entre elles par une relation qui résulte du fait que si $\lambda > 0$ f reste nul, d'où:

$$\operatorname{tr}\left(\frac{\partial f}{\partial \boldsymbol{\pi}}\frac{D\boldsymbol{\pi}}{Dt}\right)+\frac{\partial f}{\partial T}\dot{T}+\operatorname{tr}\left[\frac{\partial f}{\partial \boldsymbol{\alpha}_k}(\boldsymbol{h}_k^T+\lambda \boldsymbol{l}_k^T)\right]=\lambda\left[1+\operatorname{tr}\left(\frac{\partial f}{\partial \boldsymbol{\alpha}_k}\boldsymbol{l}_k^T\right)\right]=0$$

soit

$$1+\operatorname{tr}\left(\frac{\partial f}{\partial \boldsymbol{\alpha}_k}\boldsymbol{l}_k^T\right)=0. \tag{31}$$

La surface $f = 0$ peut présenter des points pyramidaux ou coniques correspondant à l'entrée en jeu simultanée de plusieurs mécanismes non retardés. En un point où se coupent n nappes de la surface on devra dans (29) et (30) faire intervenir n multiplicateurs λ. On aura alors n conditions (31).

Avec $\boldsymbol{A} = 0$ et $\boldsymbol{h}_k = 0$ on obtient les règles d'écoulement de la plasticité classique précisées (influence de $\boldsymbol{\beta}$ et $\boldsymbol{\omega}_D$) et complètées (partie antisymétrique de (29)).

Composition des transformations élastique et plastique

(κ) étant une configuration relâchée (actuelle) arbitraire, (a) la configuration contrainte actuelle, la transformation élastique $\boldsymbol{E}$, qui fait passer de (κ) à (a), peut être décomposée sous les 2 formes:

$$\boldsymbol{E} = \boldsymbol{RS} = \boldsymbol{S}_1\boldsymbol{R} \qquad \boldsymbol{R} \text{ orthogonal (propre)},$$
$$\boldsymbol{S},\ \boldsymbol{S}_1 \text{ symétriques.}$$

La 2ème introduit la configuration relâchée particulière (κ_1), telle que l'on passe de (κ_1) à (a) par une déformation pure $\boldsymbol{S}_1$. On passe de (κ) à (κ_1) par la transformation $\boldsymbol{R}$. Pour (κ) le trièdre directeur se déduit des axes fixes par la transformation $\boldsymbol{\beta}$, pour (κ_1) il s'en déduit donc par la transformation $\boldsymbol{\beta}_1 = \boldsymbol{R\beta}$ et par conséquent, d'après (14), sa vitesse de rotation par rapport aux axes fixes a pour matrice, dans ces axes:

$$\boldsymbol{\omega}_D^1 = \dot{\boldsymbol{\beta}}_1\boldsymbol{\beta}_1^{-1} = \boldsymbol{R\omega}_D\boldsymbol{R}^T+\dot{\boldsymbol{R}}\boldsymbol{R}^T. \tag{32}$$

En utilisant maintenant la décomposition $\boldsymbol{E} = \boldsymbol{RS}$, la formule (2) devient:

$$\operatorname{grad} \boldsymbol{v} = \dot{\boldsymbol{R}}\boldsymbol{R}^T+\boldsymbol{R}(\dot{\boldsymbol{S}}\boldsymbol{S}^{-1}+\boldsymbol{S}\dot{\boldsymbol{P}}\boldsymbol{P}^{-1}\boldsymbol{S}^{-1})\boldsymbol{R}^T.$$

En introduisant les dérivées de $\boldsymbol{S}$, $\boldsymbol{P}$ dans le mouvement par rapport au trièdre directeur:

$$\frac{D\boldsymbol{P}}{Dt} = \dot{\boldsymbol{P}} - \boldsymbol{\omega}_D \boldsymbol{P},$$

$$\frac{D\boldsymbol{S}}{Dt} = \dot{\boldsymbol{S}} - \boldsymbol{\omega}_D \boldsymbol{S} + \boldsymbol{S}\boldsymbol{\omega}_D$$

on obtient:

$$\operatorname{grad} \boldsymbol{v} = \boldsymbol{\omega}_D^1 + \boldsymbol{R} \frac{D\boldsymbol{S}}{Dt} \boldsymbol{S}^{-1}\boldsymbol{R}^T + \boldsymbol{R}\boldsymbol{S} \frac{D\boldsymbol{P}}{Dt} \boldsymbol{P}^{-1}\boldsymbol{S}^{-1}\boldsymbol{R}^T \tag{33}$$

où $\boldsymbol{\omega}_D^1$ est donné par (32).

Les 3 termes du 2ème membre de (33) sont *indépendants de la configuration relâchée choisie* κ.

Dans (33) on remplacera $(D\boldsymbol{P}/Dt)\boldsymbol{P}^{-1}$ par l'expression (29) (ou (23) selon les cas) et $\boldsymbol{S} = (1+2\boldsymbol{\Delta})^{\frac{1}{2}}$ par l'expression obtenue en inversant (13), soit:

$$\boldsymbol{S} = \mathscr{S}(\boldsymbol{\pi}, T, \boldsymbol{\alpha}_j, \boldsymbol{\beta}), \tag{34}$$

la fonction $\mathscr{S}$ vérifiant la relation (25). Lorsque les déformations thermoélastiques sont infiniment petites on peut remplacer $\boldsymbol{S}$ par 1 dans le 3ème terme de (33), par $\mathbf{1}+\boldsymbol{\Delta}$ dans le 2ème, d'où:

$$\operatorname{grad} \boldsymbol{v} = \boldsymbol{\omega}_D^1 + \boldsymbol{R} \frac{D\boldsymbol{\Delta}}{Dt} \boldsymbol{R}^T + \boldsymbol{R}(\boldsymbol{A}+\lambda\boldsymbol{B})\boldsymbol{R}^T \tag{35}$$

où

$$\boldsymbol{\Delta} = \boldsymbol{\Lambda}(\boldsymbol{\alpha}_j, T_0, \boldsymbol{\beta})\boldsymbol{\pi} + \boldsymbol{k}(\boldsymbol{\alpha}_j, T_0, \boldsymbol{\beta})(T-T_0),$$

$$\frac{D\boldsymbol{\Delta}}{Dt} = \boldsymbol{\Lambda} \frac{D\boldsymbol{\pi}}{Dt} + \boldsymbol{k}\dot{T} + \frac{\partial \boldsymbol{\Lambda}}{\partial \boldsymbol{\alpha}_j} \frac{D\boldsymbol{\alpha}_j^T}{Dt} \boldsymbol{\pi} + \frac{\partial \boldsymbol{k}}{\partial \boldsymbol{\alpha}_j} \frac{D\boldsymbol{\alpha}_j^T}{Dt} (T-T_0).$$

On notera la présence des termes en $D\boldsymbol{\alpha}_j/Dt$ lorsque la complaisance thermoélastique $\boldsymbol{\Lambda}$, $\boldsymbol{k}$ dépend de l'écrouissage.

Pour la configuration κ on peut notamment faire 2 choix:

a) la configuration κ_1: dans ce cas $\boldsymbol{R} = \mathbf{1}$, mais les expressions de $\boldsymbol{A}$, $\boldsymbol{B}$, $\boldsymbol{S}$, $\boldsymbol{\omega}_D$ contiennent la vriable $\boldsymbol{\beta}$ (sauf dans le cas d'isotropie dans un état initial),

b) la configuration κ_0 telle que le trièdre D reste parallèle aux axes (ce choix se recommande lorsque le corps est anisotrope). Dans ce cas $\boldsymbol{\beta} = 1$. Les dérivées D/Dt coïncident avec les dérivées ordinaires. $\omega_D^1 = \dot{\boldsymbol{R}}\boldsymbol{R}^T$. $\boldsymbol{R}$ est variable.

Cette rotation $\boldsymbol{R}$ est à priori indéterminée. Mais on ne peut pas l'éliminer en prenant tout simplement la partie symétrique des 2 membres de (33). L'équation obtenue, contenant encore $\boldsymbol{R}$, serait insuffisante pour déterminer le comportement. Autrement dit, et ceci bouleverse les habitudes acquises des exemples classiques, la relation de comportement d'un milieu élastique-viscoplastique, n'est pas une relation entre la vitesse de déformation, les variables $\boldsymbol{\pi}$, T, $\boldsymbol{\alpha}_j$ et leurs dérivées temporelles. Elle introduit également la rotation $\boldsymbol{R}$ qui fait passer de κ_0 (configuration à laquelle sont rapportés $\boldsymbol{\pi}$ et $\boldsymbol{\alpha}_j$) à κ_1, et la différence entre la vitesse de rotation de l'élément $\boldsymbol{\omega} = \frac{1}{2}(\text{grad}\, \boldsymbol{v} - \text{grad}^T \boldsymbol{v})$ et celle $\boldsymbol{\omega}_D^1 = \dot{\boldsymbol{R}}\boldsymbol{R}^T$ du trièdre directeur pris dans la configuration κ_1.[1])

Toutefois lorsque les déformations élastiques *et les déformations plastiques* sont infiniment petites (déformations plastiques contenues), on peut prendre $\boldsymbol{R} = \mathbf{1} + \boldsymbol{\varepsilon}$, $\boldsymbol{\varepsilon}$ infiniment petit. L'équation (35) se scinde alors en:

$$\mathcal{D} = \dot{\Delta} + \{\boldsymbol{A} + \lambda \boldsymbol{B}\} \qquad \{\ \}\ \text{partie symétrique,}$$

$$\boldsymbol{\omega} = \dot{\boldsymbol{R}} + \}\boldsymbol{A} + \lambda \boldsymbol{B}\{ \qquad \}\ \{\ \text{partie antisymétrique}$$

et seule en fait est utile la lère de ces équations, dont la forme est la forme classique des relations de Prandtl-Reuss (la 2ème servant uniquement à déterminer $\boldsymbol{R}$).

Cas d'isotropie dans un état initial

$\boldsymbol{A}$, $\boldsymbol{B}$, $\boldsymbol{S}$ sont des fonctions isotropes de $\boldsymbol{\pi}$ et des $\boldsymbol{\alpha}_j$, donc aussi:

$$\underline{A} = \boldsymbol{S}\boldsymbol{A}\boldsymbol{S}^{-1}, \qquad \underline{B} = \boldsymbol{S}\boldsymbol{B}\boldsymbol{S}^{-1}$$

et par suite:

$$\boldsymbol{R}\underline{A}(\boldsymbol{\pi}, T, \boldsymbol{\alpha}_j)\boldsymbol{R}^T = \underline{A}(\boldsymbol{\pi}_1, T, \boldsymbol{\alpha}_j^{(1)})$$

$\boldsymbol{\pi}_1$, $\boldsymbol{\alpha}_j^{(1)}$ étant les matrices relatives à la configuration κ_1 ($\boldsymbol{\pi}_1 = \boldsymbol{R}\boldsymbol{\pi}\boldsymbol{R}^T$). La formule (33) s'écrit alors:

$$\text{grad}\, \boldsymbol{v} = \boldsymbol{\omega}_D^1 + \frac{D\boldsymbol{S}_1}{Dt}\boldsymbol{S}_1^{-1} + \underline{A}(\boldsymbol{\pi}_1, T, \boldsymbol{\alpha}_j^{(1)}) + \lambda \underline{B}(\boldsymbol{\pi}_1, T, \boldsymbol{\alpha}_j^{(1)}) \qquad (36)$$

[1]) donc elle satisfait bien au principe d'objectivité.

avec:

$$\boldsymbol{S}_1 = \hat{\mathscr{S}}(\boldsymbol{\pi}_1, T, \boldsymbol{\alpha}_j^{(1)})$$

$$\frac{D\boldsymbol{S}_1}{Dt} = \dot{\boldsymbol{S}}_1 - \boldsymbol{\omega}_D^1 \boldsymbol{S}_1 + \boldsymbol{S}_1 \boldsymbol{\omega}_D^1$$

(36) s'obtenait directement de (33) en prenant pour κ la configuration κ_1. $\boldsymbol{\omega}_D^1$ est à priori indéterminé, mais on ne l'élimine pas en prenant les parties symétriques des 2 membres de (36), puisqu'il figure dans DS^1/Dt.

Si les déformations thermoélastiques sont infiniment petites, on peut confondre $\boldsymbol{\pi}_1$ et $\boldsymbol{\sigma}$ (contrainte eulérienne), $\underline{\boldsymbol{A}}$ et $\boldsymbol{A}$, $\underline{\boldsymbol{B}}$ et $\boldsymbol{B}$, $(DS/Dt)S_1^{-1}$ et DS_1/Dt. Dans ce cas en prenant les parties antisymétriques de (36) on obtient:

$$\boldsymbol{\omega} = \boldsymbol{\omega}_D^1 + \}\boldsymbol{A} + \lambda \boldsymbol{B}\{.$$

Si l'écrouissage est isotrope (paramètres α_j scalaires), $\boldsymbol{A}$ et $\boldsymbol{B}$ sont symétriques, donc $\boldsymbol{\omega}_D^1 = \boldsymbol{\omega}$ (repère directeur corotationnel). La dérivée D/Dt est alors la dérivée classique de Jaumann.

Conclusion

On a obtenu une forme générale pour les relations de comportement des milieux élastiques-viscoplastiques (sans exclure le cas classique de plasticité non visqueuse). Les variables d'écrouissage y jouent un rôle important. Mais en outre ces relations font intervenir la rotation d'un trièdre directeur (ce qui leur confère uh aspect insolite). La notion de repère directeur s'est révélée essentielle dans toute la théorie. Elle a permis de clarifier certains points qui restaient obscurs en plasticité classique.

Références

[1] Lee, E. H., Elastic-plastic deformation at finite strains. *J. App. Mech. 36* (1969) 361.

[2] Mandel, J., Sur la décomposition d'une transformation élastoplastique. *C. R. Acad. Sc. Paris*, t. *272* (1971) 276–279.
Sur les relations de comportement d'un milieu élastique-viscoplastique. *C. R. Acad. Sc. Paris*, t. *272* (1971) 1596–1598, t. *273* (1971) 44–46.

[3] Mandel, J., *Plasticité classique et viscoplasticité*. Cours au C.I.S.M., Udine (Italie), 1971.

[4] Zarka, J., *Sur la viscoplasticité des métaux*. Thèse (1968). Mémorial des Sciences et Techniques de l'Armement, 2ème fasc. 1970, p. 223–292.

Finite-strain theory of inelastic behavior of crystalline solids

J. Kratochvíl

Institute of Solid State Physics, Prague, Czechoslovakia

Elastic, viscous and plastic behavior of solids is described within the thermodynamic framework of internal state variables suggested by Coleman and Gurtin and independently by Valanis. The theory includes changes in both, geometry and physical properties induced by crystal defects and phase-transformation which may occur during thermomechanical treatment of materials. Attention is paid to the role of rotations in elastic-inelastic solids subjected to general deformation.

Introduction

An attempt is made to develop an internally consistent theory of inelastic behavior for crystalline materials in accordance with both microscopic behavior and macroscopic experiments. The theory is motivated by the following physical picture.

Mechanical and thermal treatments produce rearrangements of atoms in crystalline materials and corresponding variations in physical properties. The rearrangements may be divided into three categories which are schematically illustrated in Fig. 1.

First consider small displacements of the atoms from the equilibrium positions (Fig. 1a). The atoms return to the original positions when the cause of the rearrangement is removed. This arrangement will be called *elastic* since materials of this type are macroscopically identified as elastic.

In Fig. 1b the form of a characteristic atomic rearrangement during plastic deformation is indicated. Such arrangements are usually rate sensitive and will be called *visco-plastic*. Note that during a visco-plastic rearrangement the atoms are just moved into equivalent positions, i.e. surroundings of the atoms are identical before and after such movements. Therefore, variations in physical properties are independent of such rearrangements.

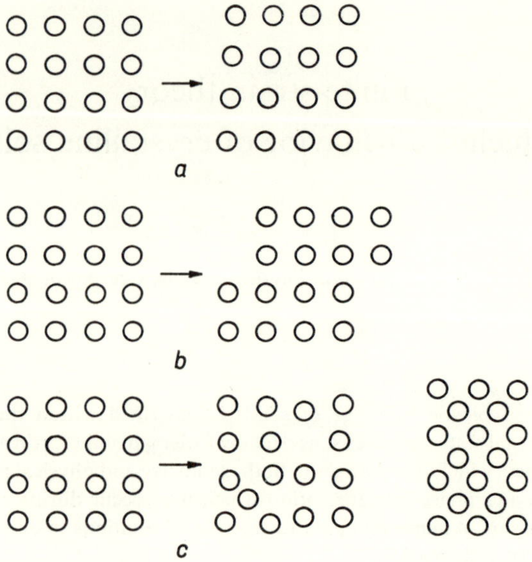

Fig. 1

The most drastic rearrangements accompany the generation and elimination of crystal defects and phase-transformations. These types of rearrangements, indicated schematically in Fig. 1c, will be called *structural* rearrangements. Crystal defects in the form of dislocations, twin boundaries etc. fall into the category of structural rearrangements. They have a decisive influence on variations in physical properties [1], whereas the elastic rearrangements cause only mild changes of properties of a material and the visco-plastic rearrangements do not influence physical properties at all.

An important result of solid state physics is that geometrical changes of the body and changes in physical properties induced by different types of rearrangements are governed by different physical laws. This fact should be reflected in a phenomenological theory of general deformation of crystalline solids in which all the three categories of atomic rearrangements may occur.

Motivated by solid state physics the concepts of motion, elastic motion, and quasi-elastic motion will be introduced in the next section. Roughly speaking the *elastic motion* describes macroscopic, geometrical changes of the body induced by elastic rearrangements only. The *quasi-elastic motion* includes both, elastic and structural rearrangements. The *motion* represents total macroscopic geometrical change of the body in which all the rearrangements may take place.

The concept of motion which will be introduced here is identical with that employed in continuum physics. However, since in general elastic and structural rearrangements can not take place independent of the visco-plastic rearrangements because of loss of continuity, elastic and quasi-elastic motions will be formulated using local concepts.

In the next section, elastic, visco-plastic, and structural deformation gradients will be introduced and related to the motions. Using the microscopic concept these gradients will be interpreted as quantities which describe the geometrical changes of the body induced by elastic, visco-plastic, and structural rearrangements respectively.

Current values of elastic, visco-plastic and structural deformation gradients together with current values of temperature and temperature gradients will enter constitutive equations suggested in the next section. But these quantities are not sufficient to determine the response of the material. The variations in properties of crystalline solids are sensitive to the details of the structural rearrangements. In the present framework these details will be described by n parameters α_i, $i = 1, \ldots, n$ which will be called *structural parameters*. We may interpret α_i as characteristic parameters of the material that appear in theories of work-hardening (see e.g. Kröner [2]). Only scalar quantities α_i which are indifferent with respect to the material frame will be considered here. The possibility to include α_i of other tensorial ranks has been discussed in [3] where α_i are called work-hardening parameters.

The present development is an extension of both the physical model described in [4] and the theory suggested in [3] where the macroscopic geometrical changes caused by the structural rearrangements have been neglected. As in [3] we utilize the framework of thermodynamics using internal state variables formulated by Coleman and Gurtin [5], and Valanis [6]. The internal state variables in the present theory are chosen to be the visco-plastic deformation gradient, the structural deformation gradient and the structural parameters.

Basic assumptions

Concept of elastic and quasi-elastic motions. Consider a body B with particles X. Functions $\chi(X, t)$ and $\theta(X, t)$ of X and time t will be interpreted as a *motion* of B and a time-dependent *temperature distribution* over B respectively.

Each material point X will be identified with its position $\boldsymbol{X}$ in a fixed reference configuration $\varkappa$ at a reference time t_0, $\boldsymbol{X} = \varkappa(X) \equiv \chi(X, t_0)$.

Assume that for any pair $\chi(X, \tau)$, $\theta(X, \tau)$, $\tau \in (-\infty, t\rangle$, and any point X in B there exist two continuous sequences of local configurations in the neighbourhood of X labeled $\chi'_t(X, \tau)$ and $\hat{\chi}_t(X, \tau)$ with the following properties: 1. For each pair $\chi(X, \tau)$, $\theta(X, \tau)$, $\tau \in (-\infty, t\rangle$, the continuous sequences $\chi'_t(X, \tau)$ and $\hat{\chi}_t(X, \tau)$, $\tau \in (-\infty, t\rangle$, are unique. 2. There is mutual one to one correspondence among $\chi(X, \tau)$, $\chi'_t(X, \tau)$, and $\hat{\chi}_t(X, \tau)$ for all τ, $\tau \in (-\infty, t\rangle$ and all X in B. 3. At time t

$$\chi(X, t) = \chi'_t(X, t) = \hat{\chi}_t(X, t). \tag{2.1}$$

The property (2.1) requires that χ'_t and χ_t have global character at $\tau = t$; this need not to be true at any other time $\tau < t$.

The continuous sequence $\chi'_t(X, \tau)$, $\tau \in (-\infty, t\rangle$ will be called a local *quasi-elastic motion* at X related to the pair $\chi(X, \tau)$, $\theta(X, \tau)$, $\tau(-\infty, t\rangle$. Similarly the continuous sequence $\hat{\chi}_t(X, \tau)$, $\tau \in (-\infty, t\rangle$ will be called a local *elastic motion* at X related to the pair $\chi(X, \tau)$, $\theta(X, \tau)$, $\tau \in (-\infty, t\rangle$. If the reference time t_0 satisfies the relation $t_0 \leqq t$, the local configurations $\varkappa'_t(X) \equiv \hat{\chi}'_t(X, t_0)$ and $\hat{\varkappa}_t \equiv \hat{\chi}_t(X, t_0)$ may be introduced as a moving local quasi-elastic reference configuration at X and a moving local elastic reference configuration at X respectively.

The deformation gradients[1]) $\boldsymbol{F}$, $\boldsymbol{E}$, $\boldsymbol{M}$, and $\boldsymbol{P}$ which describe the mappings $\boldsymbol{F}(X, t)\colon \varkappa(X) \to \chi(X, t)$, $\boldsymbol{E}(X, t)\colon \hat{\varkappa}_t(X) \to \chi(X, t)$, $\boldsymbol{M}(X, t)\colon \varkappa'_t(X) \to \hat{\varkappa}_t(X)$, and $\boldsymbol{P}(X, t)\colon x(X) \to x'_t(X)$ respectively (see Fig. 2) satisfy the relation

$$\boldsymbol{F}(X, t) = \boldsymbol{E}(X, t)\boldsymbol{M}(X, t)\boldsymbol{P}(X, t). \tag{2.2}$$

In accordance with the physical concepts mentioned in the introduction we shall interpret $\boldsymbol{E}$, $\boldsymbol{M}$, and $\boldsymbol{P}$ as the *elastic*, *structural*, and *visco-plastic deformation gradients*, respectively. We shall assume that $\boldsymbol{F}$, $\boldsymbol{E}$, $\boldsymbol{M}$, and $\boldsymbol{P}$ are nonsingular.

The relative deformation gradients $\boldsymbol{F}_{(t)}(X, \tau)$ and $\boldsymbol{E}_{(t)}(X, \tau)$, and the deformation gradients $\boldsymbol{M}_t$ and $\boldsymbol{P}_t$ which describe the local mappings $\boldsymbol{M}_t(X, \tau)$: $\hat{\chi}_t(X, \tau) \to \chi'_t(X, \tau)$, and $\boldsymbol{P}_t(X, \tau)\colon \chi'_t(X, \tau) \to \chi(X, \tau)$ are related by

$$\boldsymbol{F}_{(t)}(X, \tau) = \boldsymbol{P}_t(X, \tau)\boldsymbol{M}_t(X, \tau)\boldsymbol{E}_{(t)}(X, \tau). \tag{2.3}$$

[1]) To use the term 'gradient' for $\boldsymbol{E}$, $\boldsymbol{M}$, and $\boldsymbol{P}$ is not quite appropriate as the mappings $\boldsymbol{E}$, $\boldsymbol{M}$, $\boldsymbol{P}$ need not to be integrable.

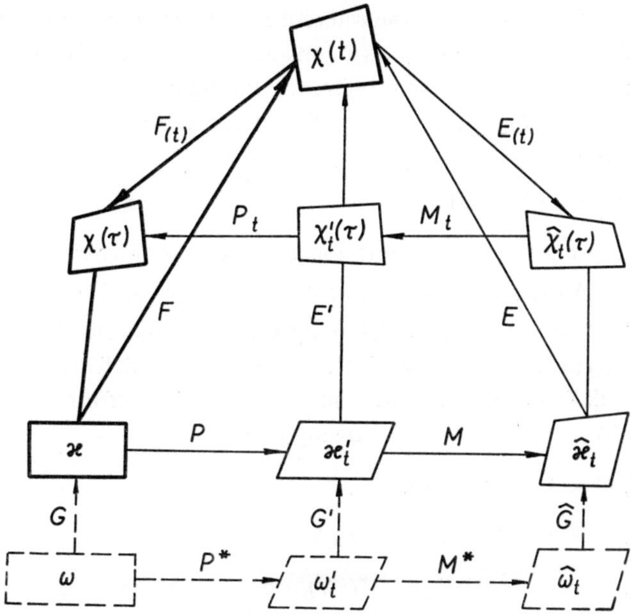

Fig. 2

Introducing the velocity gradients $\boldsymbol{L}$, $\boldsymbol{L}_E$, $\boldsymbol{L}_M$, and $\boldsymbol{L}_P$ in the form (dots on top of symbols indicate material time derivatives; dependence on X will not always be indicated)

$$\boldsymbol{L}(t) = \frac{\partial}{\partial\tau}\, \boldsymbol{F}_{(t)}(\tau)\bigg|_{\tau=t} = \dot{\boldsymbol{F}}(t)\boldsymbol{F}^{-1}(t), \tag{2.4}$$

$$\boldsymbol{L}_E(t) = \frac{\partial}{\partial\tau}\, \boldsymbol{E}_{(t)}(\tau)\bigg|_{\tau=t} = \dot{\boldsymbol{E}}(t)\boldsymbol{E}^{-1}(t), \tag{2.5}$$

$$\boldsymbol{L}_M(t) = \frac{\partial}{\partial\tau}\, \boldsymbol{M}_t(\tau)\bigg|_{\tau=t} = \boldsymbol{E}(t)\dot{\boldsymbol{M}}(t)\boldsymbol{M}^{-1}(t)\boldsymbol{E}^{-1}(t), \tag{2.6}$$

$$\boldsymbol{L}_P(t) = \frac{\partial}{\partial\tau}\, \boldsymbol{P}_t(\tau)\bigg|_{\tau=t} = \boldsymbol{E}(t)\boldsymbol{M}(t)\dot{\boldsymbol{P}}(t)\boldsymbol{P}^{-1}(t)\boldsymbol{M}^{-1}(t)\boldsymbol{E}^{-1}(t). \tag{2.7}$$

Differentiation of the relation (2.3) with respect to τ and setting $\tau = t$ yield the decomposition rule for velocity gradient $\boldsymbol{L}$

$$\boldsymbol{L}(t) = \boldsymbol{L}_E(t) + \boldsymbol{L}_M(t) + \boldsymbol{L}_P(t). \tag{2.8}$$

Further, consider a change of the reference configurations (see Fig. 2) which will be useful in material symmetry considerations. If $\boldsymbol{G}(X)$ corresponds to a mapping between two fixed reference configurations $\boldsymbol{\varkappa}(X)$ and $\boldsymbol{\omega}(X)$, $\boldsymbol{G}(X)$: $\boldsymbol{\omega}(X) \rightarrow \boldsymbol{\varkappa}(X)$, we have for the deformation gradient $\boldsymbol{F}^*(X, t)$: $\boldsymbol{\omega}(X) \rightarrow \boldsymbol{\chi}(X, t)$

$$\boldsymbol{F}^*(X, t) = \boldsymbol{F}(X, t)\boldsymbol{G}(X). \tag{2.9}$$

Let $\boldsymbol{\omega}_t'(X)$ and $\hat{\boldsymbol{\omega}}_t(X)$ be a moving local reference configurations such that $\boldsymbol{G}'(X)$ and $\hat{\boldsymbol{G}}(X)$ describe independent of time the local mappings $\boldsymbol{G}'(X)$: $\boldsymbol{\omega}_t'(X) \rightarrow \boldsymbol{\varkappa}_t'(X)$ and $\hat{\boldsymbol{G}}(X)$: $\hat{\boldsymbol{\omega}}_t(X) \rightarrow \hat{\boldsymbol{\varkappa}}_t(X)$. Then the elastic deformation gradient $\boldsymbol{E}^*(X, t)$: $\hat{\boldsymbol{\omega}}_t(X) \rightarrow \boldsymbol{\chi}(X, t)$, the structural deformation gradient $\boldsymbol{M}^*(X, t)$: $\boldsymbol{\omega}_t'(X) \rightarrow \hat{\boldsymbol{\omega}}_t(X)$, and the visco-plastic deformation gradient $\boldsymbol{P}^*(X, t)$: $\boldsymbol{\omega}(X) \rightarrow \boldsymbol{\omega}_t'(X)$ yield following relations

$$\boldsymbol{E}^*(X, t) = \boldsymbol{E}(X, t)\hat{\boldsymbol{G}}(X), \tag{2.10}$$

$$\boldsymbol{M}^*(X, t) = \hat{\boldsymbol{G}}^{-1}(X)\boldsymbol{M}(X, t)\boldsymbol{G}'(X), \tag{2.11}$$

$$\boldsymbol{P}^*(X, t) = \boldsymbol{G}'^{-1}(X)\boldsymbol{P}(X, t)\boldsymbol{G}(X). \tag{2.12}$$

Balance laws and constitutive equations. A thermodynamic process for a body B is described by the following functions of X and t: The spatial position $\boldsymbol{x} = \boldsymbol{\chi}(X, t)$ in the motion $\boldsymbol{\chi}$, the symmetric Cauchy stress tensor $\boldsymbol{T}$, the specific body force per unit mass $\boldsymbol{b}$, the specific internal energy per unit mass u, the heat flux vector $\boldsymbol{h}$, the heat supply per unit mass and unit time r, the specific entropy per unit mass η, the absolute temperature $\theta > 0$, the elastic deformation gradient $\boldsymbol{E}$, the structural deformation gradient $\boldsymbol{M}$, the visco-plastic deformation gradient $\boldsymbol{P}$, and the structural parameter vector $\boldsymbol{\alpha}$, $\boldsymbol{\alpha} = (\alpha_1, \ldots, \alpha_n)$.

A set of the functions $\boldsymbol{x}$, $\boldsymbol{T}$, $\boldsymbol{b}$, u, $\boldsymbol{h}$, r, η, θ, $\boldsymbol{E}$, $\boldsymbol{M}$, $\boldsymbol{P}$, and $\boldsymbol{\alpha}$ defined for all X in B and all t is called a thermodynamic process in B if it satisfies the relationship (2.2), and if it is compatible with the laws of balance. These laws, expressed in differential forms, are:

The balance of linear momentum

$$\rho\ddot{\boldsymbol{x}} - \operatorname{div} \boldsymbol{T} = \rho\boldsymbol{b}, \tag{2.13}$$

and the balance of energy

$$\rho\dot{u} - \operatorname{tr}(\boldsymbol{T}\boldsymbol{L}^T) + \operatorname{div} \boldsymbol{h} = \rho r. \tag{2.14}$$

In (2.13) and (2.14), ρ is the mass density in the configuration $\boldsymbol{\chi}(X, t)$.

It is postulated here that a knowledge of the current values of the deformation gradients $\boldsymbol{E}$, $\boldsymbol{M}$, and $\boldsymbol{P}$, the temperature θ, the spatial gradient of temperature $\boldsymbol{g} = \operatorname{grad} \theta$, and the vector $\boldsymbol{\alpha}$ suffice to determine the state of the body B. That is, the functions $\hat{\psi}$, $\hat{\eta}$, $\hat{\boldsymbol{T}}$, $\hat{\boldsymbol{h}}$, $\hat{\boldsymbol{L}}_M$, $\hat{\boldsymbol{L}}_P$, $\boldsymbol{f}$ characterize the response of the material at the point X at time t as follows

$$\psi = \hat{\psi}(\boldsymbol{E}, \boldsymbol{M}, \boldsymbol{P}, \theta, \boldsymbol{g}, \boldsymbol{\alpha}), \tag{2.15}$$

$$\eta = \hat{\eta}(\boldsymbol{E}, \boldsymbol{M}, \boldsymbol{P}, \theta, \boldsymbol{g}, \boldsymbol{\alpha}), \tag{2.16}$$

$$\boldsymbol{T} = \hat{\boldsymbol{T}}(\boldsymbol{E}, \boldsymbol{M}, \boldsymbol{P}, \theta, \boldsymbol{g}, \boldsymbol{\alpha}), \tag{2.17}$$

$$\boldsymbol{h} = \hat{\boldsymbol{h}}(\boldsymbol{E}, \boldsymbol{M}, \boldsymbol{P}, \theta, \boldsymbol{g}, \boldsymbol{\alpha}), \tag{2.18}$$

$$\boldsymbol{L}_M = \hat{\boldsymbol{L}}_M(\boldsymbol{E}, \boldsymbol{M}, \boldsymbol{P}, \theta, \boldsymbol{g}, \boldsymbol{\alpha}), \tag{2.19}$$

$$\boldsymbol{L}_P = \hat{\boldsymbol{L}}_P(\boldsymbol{E}, \boldsymbol{M}, \boldsymbol{P}, \theta, \boldsymbol{g}, \boldsymbol{\alpha}), \tag{2.20}$$

$$\dot{\boldsymbol{\alpha}} = \boldsymbol{f}(\boldsymbol{E}, \boldsymbol{M}, \boldsymbol{P}, \theta, \boldsymbol{g}, \boldsymbol{\alpha}). \tag{2.21}$$

In (2.15) ψ designates specific free energy per unit mass, i.e. $\psi = \psi(X, t)$ or $\psi = u - \theta\eta$. The remaining functions that specify a thermodynamic process and do not appear in (2.15)–(2.21), i.e. $\boldsymbol{x}$, $\boldsymbol{b}$, and r, are determined by (2.2), and the equations (2.13) and (2.14).

A thermodynamic process in B is called admissible if it is compatible with the constitutive equations (2.15)–(2.21).

Restricted forms of constitutive equations

The principle of material frame indifference, material symmetry, and the second law of thermodynamics impose restrictions on the response functions $\hat{\psi}$, $\hat{\eta}$, $\hat{\boldsymbol{T}}$, $\hat{\boldsymbol{h}}$, $\hat{\boldsymbol{L}}_M$, $\hat{\boldsymbol{L}}_P$, $\boldsymbol{f}$ of (2.15)–(2.21). We now seek the restricted forms of constitutive relationships in accordance with these limitations.

Principle of material frame indifference. The principle of material frame indifference states that an admissible process must remain admissible after a change of frame. Let a change of frame be characterized by a time dependent orthogonal tensor $\boldsymbol{Q}$. The scalars ψ, η, θ, α_i are unaffected by a change of frame, but $\boldsymbol{F}$, $\boldsymbol{g}$, $\boldsymbol{h}$, $\boldsymbol{T}$ transform as follows

$$\boldsymbol{F} \to \boldsymbol{QF}, \boldsymbol{g} \to \boldsymbol{Qg}, \boldsymbol{h} \to \boldsymbol{Qh}, \tag{3.1}$$

$$\boldsymbol{T} \to \boldsymbol{QTQ}^T. \tag{3.2}$$

The transformation properties of $\boldsymbol{M}$, $\boldsymbol{P}$, $\boldsymbol{L}_M$, $\boldsymbol{L}_P$ may be derived if the transformation properties of the elastic deformation gradient $\boldsymbol{E}$ and the quasi-elastic deformation gradient $\boldsymbol{E}'(\boldsymbol{E}' \equiv \boldsymbol{EM}$, see Fig. 2) are known. We assume that $\boldsymbol{E}$ and $\boldsymbol{E}'$ transform under a change of frame similarly as the deformation gradient $\boldsymbol{F}$, i.e.

$$\boldsymbol{E} \to \boldsymbol{QE}, \qquad \boldsymbol{E}' \to \boldsymbol{QE}'. \tag{3.3}$$

Then the definition of $\boldsymbol{E}'$ and the relationships (2.2), $(3.1)_1$, and (3.3) yield

$$\boldsymbol{M} \to \boldsymbol{M}, \qquad \boldsymbol{P} \to \boldsymbol{P}. \tag{3.4}$$

For the velocity gradients $\boldsymbol{L}_M$ and $\boldsymbol{L}_P$ we get from (2.6), (2.7), (3.3) and (3.4)

$$\boldsymbol{L}_M \to \boldsymbol{QL}_M\boldsymbol{Q}^T, \qquad \boldsymbol{L}_P \to \boldsymbol{QL}_P\boldsymbol{Q}^T. \tag{3.5}$$

It follows from (3.1) to (3.5) that the response functions (2.15) to (2.21) are compatible with the principle of material frame indifference if and only if they obey for each orthogonal tensor $\boldsymbol{Q}$, the following identities in $\boldsymbol{E}, \boldsymbol{M}, \boldsymbol{P}, \theta, \boldsymbol{g}, \boldsymbol{\alpha}$

$$\hat{\psi}(\boldsymbol{E}, \boldsymbol{M}, \boldsymbol{P}, \theta, \boldsymbol{g}, \boldsymbol{\alpha}) = \hat{\psi}(\boldsymbol{QE}, \boldsymbol{M}, \boldsymbol{P}, \theta, \boldsymbol{Qg}, \boldsymbol{\alpha}), \tag{3.6}$$

$$\hat{\eta}(\boldsymbol{E}, \boldsymbol{M}, \boldsymbol{P}, \theta, \boldsymbol{g}, \boldsymbol{\alpha}) = \hat{\eta}(\boldsymbol{QE}, \boldsymbol{M}, \boldsymbol{P}, \theta, \boldsymbol{Qg}, \boldsymbol{\alpha}), \tag{3.7}$$

$$\boldsymbol{Q}\hat{\boldsymbol{T}}(\boldsymbol{E}, \boldsymbol{M}, \boldsymbol{P}, \theta, \boldsymbol{g}, \boldsymbol{\alpha})\boldsymbol{Q}^T = \hat{\boldsymbol{T}}(\boldsymbol{QE}, \boldsymbol{M}, \boldsymbol{P}, \theta, \boldsymbol{Qg}, \boldsymbol{\alpha}), \tag{3.8}$$

$$\boldsymbol{Q}\hat{\boldsymbol{h}}(\boldsymbol{E}, \boldsymbol{M}, \boldsymbol{P}, \theta, \boldsymbol{g}, \boldsymbol{\alpha}) = \hat{\boldsymbol{h}}(\boldsymbol{QE}, \boldsymbol{M}, \boldsymbol{P}, \theta, \boldsymbol{Qg}, \boldsymbol{\alpha}), \tag{3.9}$$

$$\boldsymbol{Q}\hat{\boldsymbol{L}}_M(\boldsymbol{E}, \boldsymbol{M}, \boldsymbol{P}, \theta, \boldsymbol{g}, \boldsymbol{\alpha})\boldsymbol{Q}^T = \hat{\boldsymbol{L}}_M(\boldsymbol{QE}, \boldsymbol{M}, \boldsymbol{P}, \theta, \boldsymbol{Qg}, \boldsymbol{\alpha}), \tag{3.10}$$

$$\boldsymbol{Q}\hat{\boldsymbol{L}}_P(\boldsymbol{E}, \boldsymbol{M}, \boldsymbol{P}, \theta, \boldsymbol{g}, \boldsymbol{\alpha})\boldsymbol{Q}^T = \hat{\boldsymbol{L}}_P(\boldsymbol{QE}, \boldsymbol{M}, \boldsymbol{P}, \theta, \boldsymbol{Qg}, \boldsymbol{\alpha}), \tag{3.11}$$

$$f(\boldsymbol{E}, \boldsymbol{M}, \boldsymbol{P}, \theta, \boldsymbol{g}, \boldsymbol{\alpha}) = f(\boldsymbol{QE}, \boldsymbol{M}, \boldsymbol{P}, \theta, \boldsymbol{Qg}, \boldsymbol{\alpha}). \tag{3.12}$$

The identities (3.6)–(3.12) can be used to derive the reduced forms for the constitutive equations (2.15) to (2.21). One set of such reduced constitutive equations is

$$\psi = \tilde{\psi}(\boldsymbol{C}_E, \boldsymbol{M}, \boldsymbol{P}, \theta, \bar{\boldsymbol{g}}, \boldsymbol{\alpha}), \tag{3.13}$$

$$\eta = \tilde{\eta}(\boldsymbol{C}_E, \boldsymbol{M}, \boldsymbol{P}, \theta, \bar{\boldsymbol{g}}, \boldsymbol{\alpha}), \tag{3.14}$$

$$\boldsymbol{T} = \boldsymbol{R}_E\tilde{\boldsymbol{T}}(\boldsymbol{C}_E, \boldsymbol{M}, \boldsymbol{P}, \theta, \bar{\boldsymbol{g}}, \boldsymbol{\alpha})\boldsymbol{R}_E^T, \tag{3.15}$$

$$\boldsymbol{h} = \boldsymbol{R}_E\tilde{\boldsymbol{h}}(\boldsymbol{C}_E, \boldsymbol{M}, \boldsymbol{P}, \theta, \bar{\boldsymbol{g}}, \boldsymbol{\alpha}), \tag{3.16}$$

$$\boldsymbol{L}_M = \boldsymbol{R}_E\tilde{\boldsymbol{L}}_M(\boldsymbol{C}_E, \boldsymbol{M}, \boldsymbol{P}, \theta, \bar{\boldsymbol{g}}, \boldsymbol{\alpha})\boldsymbol{R}_E^T, \tag{3.17}$$

$$L_P = R_E \tilde{L}_P(C_E, M, P, \theta, \bar{g}, \alpha) R_E^T, \tag{3.18}$$

$$\dot{\alpha} = \tilde{f}(C_E, M, P, \theta, \bar{g}, \alpha), \tag{3.19}$$

where $C_E = E^T E$ is the right Cauchy-Green elastic strain tensor, R_E is the rotational part of E, and $\bar{g} = E^T g$.

Using (2.6) and (2.7) we may derive alternative forms of the constitutive equations (3.17) and (3.18)

$$\bar{L}_M \equiv \dot{M}M^{-1} = U_E^{-1}\tilde{L}_M U_E \equiv \tilde{\tilde{L}}_M(C_E, M, P, \theta, \bar{g}, \alpha), \tag{3.20}$$

$$\bar{L}_P \equiv \dot{P}P^{-1} = M^{-1}U_E^{-1}\tilde{L}_P U_E M \equiv \tilde{\tilde{L}}_P(C_E, M, P, \theta, \bar{g}, \alpha) \tag{3.21}$$

where $U_E^2 = C_E$.

It is interesting to note the difference in the role of E, M and P in the response functions (3.13) to (3.19) (the similar situation arises if we consider (3.20) and (3.21) instead of (3.17) and (3.18)). The set $(C_E, M, P, \theta, \bar{g}, \alpha)$ may be replaced by the equivalent set $(C_E, C_M, C_P, R_M, R_P, \theta, \bar{g}, \alpha)$ where C_M and C_P are right Cauchy-Green tensors corresponding to M and P respectively. Unlike the elastic rotation R_E, the rotational parts of M and P denoted by R_M and R_P appear in the set. The reason is that the rotations R_M and R_P control the orientation of distinguished directions in the material[1]), e.g. axes of anisotropy (see the example of a tensile test of a monocrystal described in the previous paper [3]). This is expressed also by (3.17) and (3.18) since not only the structural and visco-plastic stretching tensors D_M and D_P but also the structural and visco-plastic spin tensors W_M and W_P are governed by (3.17) and (3.18), ($L_M = D_M + W_M$, $L_P = D_P + W_P$).

Material symmetry. Using arguments on microscopic level we have indicated in the introduction that the visco-plastic deformation does not alter the response of elastic-inelastic materials. Here we translate this qualitative physical idea into a mathematical statement.

Consider the set S of all ordered triplets of the changes of the reference configurations $\varkappa$, $\varkappa_t'$, and $\hat{\varkappa}_t$ at X that leave the response functions of the material unaltered. If we assume that the structural parameters α_i are quantities unaffected by the choice of the reference configurations, then the set S will be identified with the set of all ordered triplets of tensors $(H, H', \hat{H})$ for which the identities

[1]) Physical meaning of visco-plastic rotations has been discussed in [4].

$$\hat{\psi}(\boldsymbol{E}, \boldsymbol{M}, \boldsymbol{P}, \theta, \boldsymbol{g}, \boldsymbol{\alpha}) = \hat{\psi}(\boldsymbol{E}\hat{\boldsymbol{H}}, \hat{\boldsymbol{H}}^{-1}\boldsymbol{M}\boldsymbol{H}', \boldsymbol{H}'^{-1}\boldsymbol{P}\boldsymbol{H}, \theta, \boldsymbol{g}, \boldsymbol{\alpha}), \tag{3.22}$$

$$\hat{\eta}(\boldsymbol{E}, \boldsymbol{M}, \boldsymbol{P}, \theta, \boldsymbol{g}, \boldsymbol{\alpha}) = \hat{\eta}(\boldsymbol{E}\hat{\boldsymbol{H}}, \hat{\boldsymbol{H}}^{-1}\boldsymbol{M}\boldsymbol{H}', \boldsymbol{H}'^{-1}\boldsymbol{P}\boldsymbol{H}, \theta, \boldsymbol{g}, \boldsymbol{\alpha}), \tag{3.23}$$

$$\hat{\boldsymbol{T}}(\boldsymbol{E}, \boldsymbol{M}, \boldsymbol{P}, \theta, \boldsymbol{g}, \boldsymbol{\alpha}) = \hat{\boldsymbol{T}}(\boldsymbol{E}\hat{\boldsymbol{H}}, \hat{\boldsymbol{H}}^{-1}\boldsymbol{M}\boldsymbol{H}', \boldsymbol{H}'^{-1}\boldsymbol{P}\boldsymbol{H}, \theta, \boldsymbol{g}, \boldsymbol{\alpha}), \tag{3.24}$$

$$\hat{\boldsymbol{h}}(\boldsymbol{E}, \boldsymbol{M}, \boldsymbol{P}, \theta, \boldsymbol{g}, \boldsymbol{\alpha}) = \hat{\boldsymbol{h}}(\boldsymbol{E}\hat{\boldsymbol{H}}, \hat{\boldsymbol{H}}^{-1}\boldsymbol{M}\boldsymbol{H}', \boldsymbol{H}'^{-1}\boldsymbol{P}\boldsymbol{H}, \theta, \boldsymbol{g}, \boldsymbol{\alpha}), \tag{3.25}$$

$$\hat{\boldsymbol{L}}_M(\boldsymbol{E}, \boldsymbol{M}, \boldsymbol{P}, \theta, \boldsymbol{g}, \boldsymbol{\alpha}) = \hat{\boldsymbol{L}}_M(\boldsymbol{E}\hat{\boldsymbol{H}}, \hat{\boldsymbol{H}}^{-1}\boldsymbol{M}\boldsymbol{H}', \boldsymbol{H}'^{-1}\boldsymbol{P}\boldsymbol{H}, \theta, \boldsymbol{g}, \boldsymbol{\alpha}), \tag{3.26}$$

$$\hat{\boldsymbol{L}}_P(\boldsymbol{E}, \boldsymbol{M}, \boldsymbol{P}, \theta, \boldsymbol{g}, \boldsymbol{\alpha}) = \hat{\boldsymbol{L}}_P(\boldsymbol{E}\hat{\boldsymbol{H}}, \hat{\boldsymbol{H}}^{-1}\boldsymbol{M}\boldsymbol{H}', \boldsymbol{H}'^{-1}\boldsymbol{P}\boldsymbol{H}, \theta, \boldsymbol{g}, \boldsymbol{\alpha}), \tag{3.27}$$

$$\boldsymbol{f}(\boldsymbol{E}, \boldsymbol{M}, \boldsymbol{P}, \theta, \boldsymbol{g}, \boldsymbol{\alpha}) = \boldsymbol{f}(\boldsymbol{E}\hat{\boldsymbol{H}}, \hat{\boldsymbol{H}}^{-1}\boldsymbol{M}\boldsymbol{H}', \boldsymbol{H}'^{-1}\boldsymbol{P}\boldsymbol{H}, \theta, \boldsymbol{g}, \boldsymbol{\alpha}) \tag{3.28}$$

hold for all $\boldsymbol{E}$, $\boldsymbol{M}$, $\boldsymbol{P}$, θ, $\boldsymbol{g}$, $\boldsymbol{\alpha}$. In (3.22) to (3.28) the transformation properties of $\boldsymbol{E}$, $\boldsymbol{M}$, and $\boldsymbol{P}$ has been employed under a change of the reference configuration represented by (2.10) to (2.12).

It is easy to check that the set S forms a group with the group operation $(\boldsymbol{H}_1, \boldsymbol{H}'_1, \hat{\boldsymbol{H}}_1)(\boldsymbol{H}_2, \boldsymbol{H}'_2, \hat{\boldsymbol{H}}_2) = (\boldsymbol{H}_1\boldsymbol{H}_2, \boldsymbol{H}'_1\boldsymbol{H}'_2, \hat{\boldsymbol{H}}_1\hat{\boldsymbol{H}}_2)$, the inverse elements $(\boldsymbol{H}^{-1}, \boldsymbol{H}'^{-1}, \hat{\boldsymbol{H}}^{-1})$ and the unit $(\boldsymbol{1}, \boldsymbol{1}, \boldsymbol{1})$.

The material defined by the assumptions introduced in the previous section and the assumptions (3.3) will be called an *elastic-inelastic solid*, if for any triplet of the configurations $\boldsymbol{\varkappa}$, $\boldsymbol{\varkappa}'_t$, and $\hat{\boldsymbol{\varkappa}}_t$ the triplet of the tensors $(\boldsymbol{P}^{-1}, \boldsymbol{1}, \boldsymbol{1})$, $\boldsymbol{P}\colon \boldsymbol{\varkappa} \to \boldsymbol{\varkappa}'_t$ belongs to the group S.

According to (3.22) to (3.28) the set $(\boldsymbol{E}, \boldsymbol{M}, \boldsymbol{P}, \theta, \boldsymbol{g}, \boldsymbol{\alpha})$ in the response functions given in (2.15) to (2.21) for an elastic-inelastic solid may be replaced by the set $(\boldsymbol{E}, \boldsymbol{M}, \boldsymbol{P}\boldsymbol{P}^{-1}, \theta, \boldsymbol{g}, \boldsymbol{\alpha})$, or an equivalent set $(\boldsymbol{E}, \boldsymbol{M}, \theta, \boldsymbol{g}, \boldsymbol{\alpha})$. Thus for elastic-inelastic solids the response functions are independent of the visco-plastic deformation gradient $\boldsymbol{P}$.

Discussing material symmetry we are often interested in those members of the group S which are triplets of orthogonal tensors. Consider the case $\boldsymbol{H} = \boldsymbol{H}' = \hat{\boldsymbol{H}} = \boldsymbol{Q}$ where $\boldsymbol{Q}$ is an orthogonal transformation. Then all three reference configurations are changed the same way and $\boldsymbol{M}$ and $\boldsymbol{P}$ transform as second order tensors $\boldsymbol{M} \to \boldsymbol{Q}^T\boldsymbol{M}\boldsymbol{Q}$, $\boldsymbol{P} \to \boldsymbol{Q}^T\boldsymbol{P}\boldsymbol{Q}$. A triplet $(\boldsymbol{Q}, \boldsymbol{Q}, \boldsymbol{Q})$ belongs to the group S of an elastic-inelastic solid if and only if

$$\hat{\psi}(\boldsymbol{E}, \boldsymbol{M}, \theta, \boldsymbol{g}, \boldsymbol{\alpha}) = \hat{\psi}(\boldsymbol{Q}\boldsymbol{E}\boldsymbol{Q}^T, \boldsymbol{Q}\boldsymbol{M}\boldsymbol{Q}^T, \theta, \boldsymbol{Q}\boldsymbol{g}, \boldsymbol{\alpha}), \tag{3.29}$$

$$\hat{\eta}(\boldsymbol{E}, \boldsymbol{M}, \theta, \boldsymbol{g}, \boldsymbol{\alpha}) = \hat{\eta}(\boldsymbol{Q}\boldsymbol{E}\boldsymbol{Q}^T, \boldsymbol{Q}\boldsymbol{M}\boldsymbol{Q}^T, \theta, \boldsymbol{Q}\boldsymbol{g}, \boldsymbol{\alpha}), \tag{3.30}$$

$$\boldsymbol{Q}\hat{\boldsymbol{T}}(\boldsymbol{E}, \boldsymbol{M}, \theta, \boldsymbol{g}, \boldsymbol{\alpha})\boldsymbol{Q}^T = \hat{\boldsymbol{T}}(\boldsymbol{Q}\boldsymbol{E}\boldsymbol{Q}^T, \boldsymbol{Q}\boldsymbol{M}\boldsymbol{Q}^T, \theta, \boldsymbol{Q}\boldsymbol{g}, \boldsymbol{\alpha}), \tag{3.31}$$

$$\boldsymbol{Q}\hat{\boldsymbol{h}}(\boldsymbol{E}, \boldsymbol{M}, \theta, \boldsymbol{g}, \boldsymbol{\alpha}) = \hat{\boldsymbol{h}}(\boldsymbol{Q}\boldsymbol{E}\boldsymbol{Q}^T, \boldsymbol{Q}\boldsymbol{M}\boldsymbol{Q}^T, \theta, \boldsymbol{Q}\boldsymbol{g}, \boldsymbol{\alpha}), \tag{3.32}$$

$$Q\hat{L}_M(E, M, \theta, g, \alpha)Q^T = \hat{L}_M(QEQ^T, QMQ^T, \theta, Qg, \alpha), \tag{3.33}$$

$$Q\hat{L}_P(E, M, \theta, g, \alpha)Q^T = \hat{L}_P(QEQ^T, QMQ^T, \theta, Qg, \alpha), \tag{3.34}$$

$$f(E, M, \theta, g, \alpha) = f(QEQ^T, QMQ^T, \theta, Qg, \alpha), \tag{3.35}$$

(3.29) to (3.35) may be proven by combining (3.6) to (3.12) and (3.22) to (3.28) and realizing that the response functions of elastic-inelastic solids are independent of $\boldsymbol{P}$.

If there is a triplet of the reference configurations ($\varkappa$, $\varkappa'$, $\hat{\varkappa}$) for which the group S contains all triplets ($\boldsymbol{Q}$, $\boldsymbol{Q}$, $\boldsymbol{Q}$) where $\boldsymbol{Q}$ is from the full orthogonal group, then the material is called *isotropic* and is said to be in an undistorted state. For an isotropic elastic-inelastic solid in an undistorted state the following constitutive equations exist in view of (3.13) to (3.19) and (3.29) to (3.35)

$$\psi = \tilde{\psi}(B_E, \overline{M}, \theta, g', \alpha), \tag{3.36}$$

$$\eta = \tilde{\eta}(B_E, \overline{M}, \theta, g', \alpha), \tag{3.37}$$

$$T = \tilde{T}(B_E, \overline{M}, \theta, g', \alpha), \tag{3.38}$$

$$h = \tilde{h}(B_E, \overline{M}, \theta, g', \alpha), \tag{3.39}$$

$$L_M = \tilde{L}_M(B_E, \overline{M}, \theta, g', \alpha) \tag{3.40}$$

$$L_P = \tilde{L}_P(B_E, \overline{M}, \theta, g', \alpha), \tag{3.41}$$

$$\dot{\alpha} = \tilde{f}(B_E, \overline{M}, \theta, g', \alpha) \tag{3.42}$$

where $B_E = EE^T$, $\overline{M} = R_E M R_E^T$, and $g' = V_E g$, ($V_E^2 = B_E$).

Second law of thermodynamics. Within the present framework the second law is given a precise meaning by the following postulate of positive entropy production.

Every admissible thermodynamic process in B must obey the inequality (in the form of Clausius-Duhem inequality)

$$-\rho(\dot{\psi} + \eta\dot{\theta}) + \operatorname{tr}(TL^T) - \theta^{-1}h \cdot g \geqq 0 \tag{3.43}$$

at each time t and for all material points X in B.

Using the Coleman-Noll procedure for finding the limitations which the postulate places on constitutive assumptions, we find the following restrictions on the response functions $\tilde{\psi}$, $\tilde{\eta}$, $\tilde{T}$, $\tilde{h}$, $\tilde{L}_M$, $\tilde{L}_P$, and $\tilde{f}$ in (3.13)–(3.19) for an elastic-inelastic solid (the proofs are similar to those given in [5, 7]; $U_E^2 = C_E$):

(i) The response functions $\tilde{\psi}$, $\tilde{T}$, and $\tilde{\eta}$ are independent of $\bar{g}$.

(ii) $\tilde{\psi}$ determines $\tilde{\eta}$ through the entropy relation $\tilde{\eta} = -\partial_\theta \tilde{\psi}$, i.e.

$$\eta = -\partial_\theta \tilde{\psi}. \tag{2.44}$$

(iii) $\tilde{\psi}$ determines $\tilde{T}$ through the stress relation $\tilde{T} = 2\rho U_E \partial_{C_E} \tilde{\psi} U_E^T$, i.e.

$$T = 2\rho E \partial_{C_E} \tilde{\psi} E^T. \tag{3.45}$$

(iv) $\tilde{\psi}$, $\tilde{L}_M$, $\tilde{L}_P$, $\tilde{f}$ and $\tilde{h}$ obey the general dissipation inequality

$$2\,\mathrm{tr}\,[U_E \partial_{C_E} \tilde{\psi} U_E (\tilde{L}_M^T + \tilde{L}_P^T)] - \mathrm{tr}\,(M \partial_M \tilde{\psi}^T U_E^{-1} \tilde{L}_M U_E) -$$
$$-\partial_\alpha \tilde{\psi} \cdot \tilde{f} - (\rho\theta)^{-1} \tilde{h} \cdot U_E^{-1} \bar{g} \geqq 0. \tag{3.46}$$

(v) The visco-plastic deformation gradient P is a unimodular tensor, i.e.

$$\det P = 1 \quad \text{therefore,} \quad \mathrm{tr}\, L_P = 0. \tag{3.47}$$

The consequence (v) is interesting. The usual assumption in the theory of plasticity that plastic deformation does not change the volume, i.e. $\det P = 1$, follows in the present theory as a consequence of the second law of thermodynamics and the material symmetry.

Discussion

For practical applications of the proposed theory we must find a solution of the system of the equations (2.2) (or (2.8)), (2.13), (2.14) and (3.13) to (3.19) (or a more restricted system e.g. (3.36) to (3.42)), i.e. $42+n$ scalar relations in the most general case. When the external body force β and the heat supply r are given as functions of place and time there are a total of $42+n$ scalar functions of both time and position to be found: the temperature θ, the entropy η, the internal energy u (or the free energy ψ), 3 components of the motion x, 6 components of the stress T, 3 components of heat flux h, 9 components of the elastic deformation gradient E, 9 components of the structural deformation gradient M, 9 components of the visco-plastic deformation gradient P, and n structural parameters $\alpha^{(i)}$, $i = 1, \ldots, n$.

A boundary-value problem is the problem of finding solutions of the system (2.2) (or (2.8)), (2.13), (2.14), and (3.13) to (3.19) that satisfy certain initial and boundary conditions.

If the initial time is taken as the reference time t_0, the motion $x = \chi(X)$ is described relative to the reference configuration $\varkappa$, and according to the definition of the moving local reference configurations $\varkappa_t'$ and $\hat{\varkappa}_t$, we have initially $\boldsymbol{E}(\boldsymbol{X}, t_0) = \boldsymbol{M}(\boldsymbol{X}, t_0) = \boldsymbol{P}(\boldsymbol{X}, t_0) = \mathbf{1}$. Further an initial temperature distribution $\theta(X, t_0)$ and initial structural parameter fields $\alpha^{(i)}(X, t_0)$, $i = 1, \dots n$, must be specified.

In finite-strain theory it is not as easy to decide what boundary conditions are resonable as in infinitesimal theory. For example, if we attempt to prescribe the stress vector $\boldsymbol{t} = \boldsymbol{t}(\boldsymbol{x}, t)$ acting on the deformed boundary, ($\boldsymbol{t} = \boldsymbol{Tn}$, where $\boldsymbol{n}$ is the outward unit vector normal to the boundary in the configuration $\boldsymbol{\chi}$), and the temperature $\theta(\boldsymbol{x}, t)$ at the deformed boundary we cannot generally do so in an unambiguous manner since the places $\boldsymbol{x}$ to be occupied by the boundary particles $\boldsymbol{X}$ are unknown. In order to obtain a definite boundary condition, we have to prescribe functions $\boldsymbol{t} = \boldsymbol{t}_C(\boldsymbol{x}, t)$ and $\theta = \theta_C(\boldsymbol{x}, t)$, $\boldsymbol{x} \in C$, for every conceivable configuration C of the boundary of the body.

Such a boundary condition may be prescribed as an advance strategy indicating how to load and heat the body no matter how it is deformed. The simplest way is to prescribe both surface tractions, measured per unit area in the reference configuration and temperature distributions at the boundary, and keep these prescribed tractions and temperature fixed irrespective of the configurations of the boundary in the deformed states.

Another example is a boundary condition of place. It corresponds to constraining the boundary ∂B of the body B to a given shape and to prescribe the temperature (or other temperature condition) on it. The shape and the surface temperature may be described by functions $\boldsymbol{\chi}_0(\boldsymbol{X}, t)$ and $\theta_0(\boldsymbol{X}, t)$, $\boldsymbol{X} \in \partial B$, and the condition may be written as $\boldsymbol{\chi}(\boldsymbol{X}, t) = \boldsymbol{\chi}_0(\boldsymbol{X}, t)$, and $\theta(\boldsymbol{x}, t) = \theta(\boldsymbol{\chi}_0(\boldsymbol{X}, t), t) = \theta_0(\boldsymbol{X}, t), \boldsymbol{X} \in \partial B$.

Two difficulties are encountered in attempting to use the proposed theory for solutions of boundary-value problems. Firstly, the microscopic theories of visco-plastic properties and phase-transformations have not been developed to the point where they provide sufficient information about the constitutive relations for a particular material. Usually, informations can only be derived under special conditions, e.g. uniaxial tensile tests, and are not available for a variety of conditions that are interesting from the practical point of view.

However, an analysis of results of microscopic theories that provides a limited basis for developing the constitutive equations for a specific material is possible in special cases. An example is studied in [8] in which the theory of elastic-inelastic solids, formulated for small strain, has been used to solve the problem of torsion of a solid circular elastic-visco-plastic titanium shaft as well as the accompanied heat conduction problem. The constitutive relationships for titanium have been deduced from experiments and the theory of work-hardening.

Secondly, in solving a boundary-value problem mathematical difficulties arise. Methods of solutions and methods of approximations developed for finite-strain elasticity and thermoelasticity may be helpfull. It is interesting to note that e.g. the methods to obtain exact solutions for special non-linear problems of isotropic imcompressible elastic bodies (see [9], for simple materials [10]) can be modified to suit the present model. For an isotropic incompressible elastic-inelastic material with no heat conduction and no temperature changes, the exact solutions can be reduced to solving a system of ordinary differential equations in case of five families of motions (namely 1. bending, stretching and shearing of a rectangular block; 2. straightening, stretching and shearing of a sector of a hollow cylinder; 3. inflation, bending torsion, extension and shearing of an annular wedge; 4. inflation or eversion of a sector of a spherical shell; 5. inflation, bending, extension and azimuthal shearing of an annular wedge). The detailed analysis of these special solutions will be not studied here; this topic will be described in a subsequent paper.

Summary

1. An attempt is made to develop a phenomenological finite strain theory of inelastic behavior of materials which incorporates microscopic theories governing visco-plastic properties and phase-transformation in crystalline solids.

2. Geometrical changes of the body caused by elastic deformations, visco-plastic deformations and structural changes are considered. The corresponding deformation gradients, i.e. elastic deformation gradient $\boldsymbol{E}$, the visco-plastic deformation gradient $\boldsymbol{P}$, and the structural deformation gradient $\boldsymbol{M}$ are related to the deformation gradient $\boldsymbol{F}$ by (2.2). The velocity gradient $\boldsymbol{L}_E$, $\boldsymbol{L}_P$, $\boldsymbol{L}_M$, and $\boldsymbol{L}$ satisfy the simple decomposition rule (2.8).

3. The restrictions on the response functions in the constitutive relations imposed by the principle of material frame indifference, material symmetry and the second law of thermodynamics are considered. The explicit forms of the restricted constitutive equations are derived for isotropic elastic-inelastic solids (equations (3.36) to (3.42)).

4. In the proposed theory, not only the visco-plastic and structural

stretching tensors $\boldsymbol{D}_P$ and $\boldsymbol{D}_M$ but also the visco-plastic and structural spin tensors $\boldsymbol{W}_P$ and $\boldsymbol{W}_M$ are governed by the constitutive equations. Both elastic and inelastic rotations play an important role in the theory.

5. The assumption usually made in plasticity that plastic deformation does not change the volume of the body follows in the present theory from the and second law of thermodynamics and the material symmetry.

6. Two main difficulties are encountered in attemping to use the theory for solutions of boundary value problems: firstly, it is involved to establish realistic forms of constitutive equations for a given material from experimental results and microscopic theories; secondly, it is complex to solve, at least approximately, the system of the equations (2.2) (or (2.8)), (2.13), (2.14), and (3.13) to (3.19) (or in the special case of isotropic elastic-inelastic solids (3.36) to (4.32)) for given initial and boundary conditions.

References

[1] Christian, J. W., *The theory of transformations in metals and alloys.* Pergamon Press, Oxford, 1965.

[2] Kröner, E., Dislocations: A new concept in the continuum theory of plasticity. *J. Mathematics and Physics 42* (1963) 27–37; Kröner, E., *Initial studies of a plasticity theory based upon statistical mechanics.* Inelastic Behavior of Solids, edited by M. F. Kanninen, W. F. Adler, A. R. Rosenfield and R. I. Jaffee, Mc Graw Hill Book Company, New York, (1970).

[3] Kratochvíl, J., On a finite strain theory of elastic-inelastic materials. *Acta Mechanica* (to be published).

[4] Kratochvíl, J., Finite-strain theory of crystalline elastic-inelastic materials. *J. Appl. Physics 42* (1971) 1104–1108.

[5] Coleman, B. D. and M. E. Gurtin, Thermodynamics with internal state variables. *J. Chemical Physics 47* (1967) 597–613.

[6] Valanis, K. C., *Unified theory of thermomechanical behavior of viscoelastic materials.* Mechanical Behavior of Materials under Dynamic Loads, edited by Lindholm, U.S., Springer-Verlag, Berlin-Heidelberg-New York, 1968.

[7] Gurtin, M. E. and W. O. Williams, On the inclusion of the complete symmetry group in the unimodular group. *Arch. Rational Mech. Anal., 23* (1966) 163–172.

[8] Kratochvíl, J. and R. J. De Angelis, Torsion of a titanium elastic-visco-plastic shaft. *J. Appl. Physics, 42* (1971) 1091–1097.

[9] Truesdell, C. and W. Noll, *The non-linear theories of mechanics.* Handbuch der Physik III/3, edited by Flügge, S., Springer-Verlag, Berlin-Heidelberg-New York, 1965.

[10] Carroll, M. M., Controllable deformations of incompressible simple materials. *Int. J. Engng. Sci. 5* (1967) 515–525.

A shakedown matrix theory allowing for workhardening and second-order geometric effects

Giulio Maier

Polytechnic of Milan, Italy

A matrix description of elastoplastic systems is adopted by referring to compatible finite element models and piecewiselinearized yield surfaces and hardening. Quasi-static structural responses to variable-repeated loads and dislocations are studied. This study covers a broad class of hardening rules and situations in which geometric changes affect the equilibrium equations, but their effects may be expressed by bilinear terms in the preexisting stresses and additional displacements. The set of statements established are intended as basic for an application-oriented theory of the above defined level of generality. They include further extensions of the static (Melan's) and kinematic (Koiter's) shakedown theorems, and provide various techniques for bounding from above the values at shakedown of local quantities, such as displacements, stresses and plastic strains.

Introduction

The recent revival of interest in the behavior of elastoplastic structures under variable-repeated loads, is perhaps mainly motivated both by a better awareness of the seriously unsafe inaccuracy of limit analysis in many of such situations (see e.g. [1]), and by the broadening of the range of numerically solvable problems due to matrix methods and computers. In this paper discrete structural models are used first for achieving a convenient compact representation without loss of generality. This discretization is combined with that of the element plastic behavior, whose description in terms of translating-interacting yield planes seems flexible enough to approximate with any accuracy most of the hardening rules proposed so far, [2].

On this basis, both the evolution of the yield surfaces and, to a limited extent, the influence of deformations on equilibrium, are easily incorporated in the theory developed. The aforementioned piecewise linearization of the former and the use of linearized equilibrium equations ('second order' effects) for the latter, are believed to be hypotheses which, though restrictive, do not deprive the results achieved of technical interest. As well expected, the theoretical safety factors s (against failure by unlimited yielding) turns out to be infinite in many cases, due to hardening or stabilizing geometric effects. Then, a realistic assessment of safety, must be founded (as it should often also for s finite and in the classical domain) on the determination of 'post-shakedown' values, or at least orders of magnitude, of local quantities (primarily some meaningful displacements and some plastic strains in critical locations). These values, however, depend on the loading history which is, as a rule, unknown except for the variation intervals. Therefore recourse to upper bounds is important and often mandatory. Some procedures of generating upper bounds are pointed out here, but their practical value and relative merits have still to be evaluated by computational experience. This and further developments are being pursued elsewhere. Relationships with previous work are indicated in the text, mostly when special cases are derived from the new results.

Discrete models and basic relations

The discretization adopted here for structures and continua is characterized by continuous, piecewise linear displacement fields, governed by the n-vector of the displacements of 'free nodes', on which all external forces are assumed to act. Other nodes are fixed by constraints. Constant-strain finite element models for three- and two-dimensional problems (tetrahedral and triangular elements, respectively), trusses and lumped-compliance models for frames, are the main examples, [3][4]. Consider a situation $\bar{\Sigma}$, under external actions $\bar{\boldsymbol{F}}$, $\bar{\boldsymbol{D}}$, with elastic stresses $\bar{\boldsymbol{Q}}$ and strains $\bar{\boldsymbol{q}} = \bar{\boldsymbol{e}}$ (elastic by a convenient but inessential hypothesis). The response $\boldsymbol{u}_t$, $\boldsymbol{q}_t$, $\boldsymbol{Q}_t$, to any further loading $\boldsymbol{F}_t$, $\boldsymbol{D}_t$ in the neighborhood of $\bar{\Sigma}$ (i.e. accounting for geometric effects of second order only), is assumed to be governed by the linearized equations of compatibility and equilibrium:

$$\boldsymbol{q}_t = \boldsymbol{C}\boldsymbol{u}_t, \tag{1}$$

$$\tilde{\boldsymbol{C}}\boldsymbol{Q}_t + \boldsymbol{G}(\bar{\boldsymbol{Q}})\boldsymbol{u}_t = \boldsymbol{F}_t. \tag{2}$$

Matrix $\boldsymbol{C}$ depends only on the geometry of $\bar{\Sigma}$; the (symmetric) geometric stiffness matrix $\boldsymbol{G}$ depends also, linearly, on $\bar{\boldsymbol{Q}}$ as emphasized in (2), but not on $\boldsymbol{Q}_t$ and $\boldsymbol{u}_t$.

The elastoplastic relations between generalized, intrinsic ('natural'[1]) stresses $\boldsymbol{Q}_t$ and strains $\boldsymbol{q}_t$, if a piecewise linearization of yield surfaces and hardening is assumed [2], can be described for all elements simultaneously, as follows (see Nomenclature):

$$\boldsymbol{q}_t = \boldsymbol{e}_t + \boldsymbol{p}_t + \boldsymbol{D}_t, \tag{3}$$

$$\boldsymbol{e}_t = \boldsymbol{E}^{-1}\boldsymbol{Q}_t, \tag{4}$$

$$\boldsymbol{p}_t = \boldsymbol{N}\boldsymbol{\lambda}_t, \tag{5}$$

$$\boldsymbol{\varphi}_t = \tilde{\boldsymbol{N}}\boldsymbol{Q}_t - \boldsymbol{H}\boldsymbol{\lambda}_t - \boldsymbol{K}, \tag{6}$$

$$\boldsymbol{\varphi}_t \leqq \boldsymbol{0}, \tag{7}$$

$$\dot{\boldsymbol{\varphi}}_t' \leqq \boldsymbol{0}, \tag{8}$$

$$\dot{\boldsymbol{\lambda}}_t \geqq \boldsymbol{0} \quad \text{(hence } \boldsymbol{\lambda}_t \geqq \boldsymbol{0}\text{)}, \tag{9}$$

$$\tilde{\boldsymbol{\varphi}}_t \dot{\boldsymbol{\lambda}}_t = 0, \tag{10}$$

$$\tilde{\dot{\boldsymbol{\varphi}}}_t \dot{\boldsymbol{\lambda}}_t = 0. \tag{11}$$

Dislocations $\boldsymbol{D}_t$ in (3) cover thermal and other imposed strains and can also incorporate, through a simple artifice, possible imposed boundary displacements. In (4) $\boldsymbol{E}$ is assumed symmetric, positive definite; (5) reflects the normality property of plastic strains; in (6) the vector $\boldsymbol{K}$ incorporates the preexisting stresses $\bar{\boldsymbol{Q}}(\boldsymbol{K} = -\tilde{\boldsymbol{N}}\bar{\boldsymbol{Q}} + \bar{\boldsymbol{K}} > \boldsymbol{0}$, $\bar{\boldsymbol{K}}$ being the original plastic

[1]) 'Natural' generalized strains $\boldsymbol{q}$ and stresses $\boldsymbol{Q}$ [4] are intrinsic quantities, i.e. unaffected by rigid body motions and equilibrated, respectively. In order to illustrate their meaning, let us consider the i-th tetrahedrical finite element in a three-dimensional discrete model. Let V^i denote its volume, $\boldsymbol{\sigma}^i$ and $\boldsymbol{\varepsilon}^i$ the vectors of the 6 independent 'true' stress and strain components (constant over V^i, $\boldsymbol{\varepsilon}^i$ being defined so that $\tilde{\boldsymbol{\sigma}}^i\dot{\boldsymbol{\varepsilon}}^i V^i$ represents the internal work rate performed in the element).

The components of $\boldsymbol{q}^i$ can be assumed as elongations of the tetrahedron edges, and consequently those of $\boldsymbol{Q}^i$ as forces acting in pairs at the vertices along the edges. The relations between local and generalized variables are:

$$\boldsymbol{q}^i = \boldsymbol{T}^i\boldsymbol{\varepsilon}^i, \qquad \boldsymbol{\sigma}^i = \frac{1}{V^i}\,\tilde{\boldsymbol{T}}^i\boldsymbol{Q}^i$$

where $\boldsymbol{T}^i$ is a nonsingular matrix easily evaluated on the basis of the vertices coordinates. These relations specify how material properties in the $\boldsymbol{\sigma}^i$ and $\boldsymbol{\varepsilon}^i$ spaces (superposed) are transformed into element properties in the $\boldsymbol{Q}^i$ and $\boldsymbol{q}^i$ spaces (superposed) of the element, and viceversa.

constants). Matrix $\boldsymbol{H}$ defines the hardening rule, i.e. how the yield planes in each element translate and interact by yielding (e.g.: $\boldsymbol{H} = \boldsymbol{0}$ corresponds to nonhardening, perfect plasticity; $\boldsymbol{H} = h\tilde{\boldsymbol{N}}\boldsymbol{N}$, with $h > 0$, to Prager's rule of kinematic hardening [2]). The prime in (8) reminds that only the 'potentially active' yield modes at instant t are referred to in the inequality, i.e. those whose plastic potentials in (7) are zero. Eqs. (10) and (11) hold also componentwise as a consequence of the sign-constraints, and express the 'consistency' conditions for plastic flow.

In the framework of the above assumptions and matrix description, the data on loads are the limits within which each component of $\boldsymbol{F}_t$, $\boldsymbol{D}_t$ varies (slowly) in time, according to an unknown law. Shakedown means that:

$$\lim_{t \to \infty} \boldsymbol{\lambda}_t \equiv \boldsymbol{\lambda}_s \quad \text{is bounded.} \tag{12}$$

Preliminary remarks

a. Purely elastic displacements and stresses can be expressed, through Eqs. (1) (2) (3), as:

$$\boldsymbol{u}_t^e = \boldsymbol{S}^{-1}\boldsymbol{F}_t + \boldsymbol{S}^{-1}\tilde{\boldsymbol{C}}\boldsymbol{E}\boldsymbol{D}_t, \quad \text{where} \quad \boldsymbol{S} = \tilde{\boldsymbol{C}}\boldsymbol{E}\boldsymbol{C} + \boldsymbol{G} \tag{13}$$

$$\boldsymbol{Q}_t^e = \boldsymbol{E}\boldsymbol{C}\boldsymbol{S}^{-1}\boldsymbol{F}_t + \boldsymbol{Z}\boldsymbol{D}_t, \quad \text{where} \quad \boldsymbol{Z} = \boldsymbol{E}\boldsymbol{C}\boldsymbol{S}^{-1}\tilde{\boldsymbol{C}}\boldsymbol{E} - \boldsymbol{E}. \tag{14}$$

It is assumed that det $\boldsymbol{S} \neq 0$, i.e. that equilibrium is not neutral in the starting situation $\bar{\Sigma}$. If $\boldsymbol{D}_t$ is replaced by $\boldsymbol{p}_t$, the second terms on the r.h.s. of (13) and (14) supply displacements and stresses due to plastic strains:

$$\boldsymbol{u}_t^p = \boldsymbol{S}^{-1}\tilde{\boldsymbol{C}}\boldsymbol{E}\boldsymbol{N}\boldsymbol{\lambda}_t \tag{13a}$$

$$\boldsymbol{Q}_t^p = \boldsymbol{Z}\boldsymbol{N}\boldsymbol{\lambda}_t. \tag{14a}$$

b. It can be readily verified by means of (1) and (2) that the equation

$$\tilde{\boldsymbol{F}}\boldsymbol{u}^* = \tilde{\boldsymbol{Q}}\boldsymbol{q}^* + \tilde{\boldsymbol{u}}\boldsymbol{G}\boldsymbol{u}^* \tag{15}$$

can be written for any sets of compatible quantities (starred) and of (unstarred) quantities pertaining to an actual static process, with no need of causality relation between the two sets. For $\boldsymbol{G} = \boldsymbol{0}$, (15) becomes a usual virtual work equation.

c. Since the following analytical condition plays a key role in the sequel,

$$\boldsymbol{A} \equiv \boldsymbol{H} - \tilde{\boldsymbol{N}}\boldsymbol{Z}\boldsymbol{N} \quad \text{symmetric positive semidefinite} \tag{16}$$

its mechanical meaning is pointed out below.

Symmetry of $\boldsymbol{A}$ implies symmetry of $\boldsymbol{H}$ and this excludes some hardening rules, [2]. Let a prime mean reference to only those yield planes which, at a certain instant, are potentially active, i.e. contain the current stress point. Then, whatever $\boldsymbol{H}$ may be, for any load rates $\dot{\boldsymbol{F}}$ and the consequent $\dot{\boldsymbol{\lambda}}'$, $\dot{\boldsymbol{p}}$ and $\dot{\boldsymbol{u}} = \dot{\boldsymbol{u}}^e + \dot{\boldsymbol{u}}^p$:

$$\tilde{\dot{\boldsymbol{\lambda}}}'\boldsymbol{A}'\dot{\boldsymbol{\lambda}}' = \tilde{\dot{\boldsymbol{Q}}}^e\dot{\boldsymbol{p}} = \tilde{\dot{\boldsymbol{F}}}\dot{\boldsymbol{u}}^p = \tilde{\dot{\boldsymbol{F}}}\dot{\boldsymbol{u}} - \tilde{\dot{\boldsymbol{u}}}^e\boldsymbol{S}\dot{\boldsymbol{u}}^e. \tag{17}$$

The first equality rests on Theorem IV of Ref. [5], the third on (13), the second on the two equations of type (15):

$$\tilde{\dot{\boldsymbol{Q}}}^e(\dot{\boldsymbol{p}}+\dot{\boldsymbol{e}}^p)+\tilde{\dot{\boldsymbol{u}}}^e\boldsymbol{G}\dot{\boldsymbol{u}}^p = \tilde{\dot{\boldsymbol{F}}}\dot{\boldsymbol{u}}^p, \qquad \tilde{\dot{\boldsymbol{Q}}}^p\dot{\boldsymbol{e}}^e+\tilde{\dot{\boldsymbol{u}}}^p\boldsymbol{G}\dot{\boldsymbol{u}}^e = 0. \tag{18}$$

Since $\boldsymbol{A}'$ is a principal minor of $\boldsymbol{A}$, Eqs. (17) show that, if the situation $\bar{\Sigma}$ is strictly stable ($\boldsymbol{S}$ positive definite), condition (16) guarantees overall stability ($\tilde{\dot{\boldsymbol{F}}}\dot{\boldsymbol{u}} > 0$ for any incremental process), whatever set of yielding modes may become active.

'Static' shakedown theorems

I. When (16) holds, shakedown occurs if a vector $\boldsymbol{\lambda}^*$ exists such that

$$\boldsymbol{\varphi}_t^* \equiv \tilde{\boldsymbol{N}}\boldsymbol{Q}_t^e - \boldsymbol{A}\boldsymbol{\lambda}^* - \boldsymbol{K} < \boldsymbol{0} \quad \text{for any} \quad t \geqq 0. \tag{19}$$

Proof. For the (unstarred) quantities pertaining to the actual evolution of the structure, (5) and (6) with $\boldsymbol{Q}_t = \boldsymbol{Q}_t^e + \boldsymbol{Z}\boldsymbol{p}_t$ imply that:

$$\boldsymbol{\varphi}_t = \tilde{\boldsymbol{N}}\boldsymbol{Q}_t^e - \boldsymbol{A}\boldsymbol{\lambda}_t - \boldsymbol{K} \leqq \boldsymbol{0} \quad \text{for any} \quad t \geqq 0. \tag{20}$$

Consider the function L_t ('Liapunov function' for the similarity to known procedures in dynamics) and its time derivative $\dot{L}_t$:

$$L_t \equiv \tfrac{1}{2}(\tilde{\boldsymbol{\lambda}}_t - \tilde{\boldsymbol{\lambda}}^*)\boldsymbol{A}(\boldsymbol{\lambda}_t - \boldsymbol{\lambda}^*), \tag{21}$$

$$\dot{L}_t = \tilde{\dot{\boldsymbol{\lambda}}}_t\boldsymbol{A}(\boldsymbol{\lambda}_t - \boldsymbol{\lambda}^*) = \tilde{\dot{\boldsymbol{\lambda}}}_t\boldsymbol{H}(\boldsymbol{\lambda}_t - \boldsymbol{\lambda}^*) - \tilde{\dot{\boldsymbol{\lambda}}}_t\tilde{\boldsymbol{N}}(\boldsymbol{Q}_t^p - \boldsymbol{Q}^{p*}). \tag{22}$$

In (22) use is made of the symmetry of $\boldsymbol{A}$ and of (14a). By virtue of (19) (20) (9) and (10):

$$\dot{L}_t = \tilde{\dot{\boldsymbol{\lambda}}}_t\boldsymbol{\varphi}_t^* \leqq 0 \qquad (=0 \text{ if and only if } \dot{\boldsymbol{\lambda}}_t = \boldsymbol{0}). \tag{23}$$

Inequality (19) implies that there is a number $\alpha > 1$ such that:

$$\tilde{\boldsymbol{N}}\boldsymbol{Q}_t^e - \boldsymbol{A}\boldsymbol{\lambda}^* \leqq \boldsymbol{K}\alpha^{-1} \quad \text{for any} \quad t \geqq 0. \tag{24}$$

Substitution of Eq. (20) into (10) yields:

$$-\tilde{\dot{\boldsymbol{\lambda}}}_t \boldsymbol{N} \boldsymbol{Q}_t^e + \tilde{\dot{\boldsymbol{\lambda}}}_t \boldsymbol{A} \boldsymbol{\lambda}_t = -\tilde{\boldsymbol{K}} \dot{\boldsymbol{\lambda}}_t. \tag{25}$$

By multiplying (24) by $\dot{\boldsymbol{\lambda}}_t$ and adding to (25), account taken of (22), one obtains:

$$\dot{L}_t \leqq (\alpha^{-1}-1)\tilde{\boldsymbol{K}}\dot{\boldsymbol{\lambda}}_t \tag{26}$$

whence, integrating from $t = 0(\boldsymbol{\lambda}_0 = \boldsymbol{0})$ to $t = \infty(\boldsymbol{\lambda}_\infty = \boldsymbol{\lambda}_s)$:

$$(1-\alpha^{-1})\tilde{\boldsymbol{K}}\boldsymbol{\lambda}_s \leqq L_0 - L_\infty. \tag{27}$$

Observe now that: $L_t \geqq 0$ when (16) holds; $L_\infty \leqq L_0$ due to (23); $L_0 = \frac{1}{2}\tilde{\boldsymbol{\lambda}}^* \boldsymbol{A} \boldsymbol{\lambda}^*$ is bounded; $\boldsymbol{K} > \boldsymbol{0}$. As a consequence, (27) means that not only $\dot{\boldsymbol{\lambda}}_t \to \boldsymbol{0}$ as $t \to \infty$, but also $\boldsymbol{\lambda}_s$, to which $\boldsymbol{\lambda}_t$ tends, is necessarily bounded (q.e.d.).

II. Shakedown cannot occur if there is no $\boldsymbol{\lambda}^*$ such that

$$\tilde{\boldsymbol{N}} \boldsymbol{Q}_t^e - \boldsymbol{A}\boldsymbol{\lambda}^* - \boldsymbol{K} \leqq \boldsymbol{0} \quad \text{for any} \quad t \geqq 0. \tag{28}$$

Proof. Since $\boldsymbol{\lambda}_s$ defines actual plastic strains, inequality (28) must be satisfied, as a consequence of (6) (7), by $\boldsymbol{\lambda}^* = \boldsymbol{\lambda}_s$ at all t subsequent to shakedown (in an unlimited time interval, in which loads attain any values within the given limits); hence, it will be so also for $t \geqq 0$ (q.e.d.).

III. When (16) holds, if there is a vector $\boldsymbol{\lambda}^*$ complying with (28), there is also a $\boldsymbol{\lambda}^* \geqq \boldsymbol{0}$ complying with (28).

Proof. Consider the maximum of the projection of the elastic stress vector $\boldsymbol{Q}_t^{ei}$ in element i, on the normal vector $\boldsymbol{N}_j^i$ relative to yield plane j of that element. Then, for all j and i, define the vector:

$$\tilde{\boldsymbol{M}} \equiv [\ldots M_j^i \ldots], \quad \text{with} \quad M_j^i = \max_t \tilde{\boldsymbol{N}}_j^i \boldsymbol{Q}_t^{ei}. \tag{29}$$

Inequality (28) can be replaced now by the equivalent one

$$\boldsymbol{A}\boldsymbol{\lambda}^* \geqq \boldsymbol{M} - \boldsymbol{K}. \tag{30}$$

Consider the convex quadratic programming problems:

$$\max_{\boldsymbol{\lambda}^*} \tilde{\boldsymbol{\lambda}}^*(\boldsymbol{M}-\boldsymbol{K}) - \tfrac{1}{2}\tilde{\boldsymbol{\lambda}}^* \boldsymbol{A}\boldsymbol{\lambda}^* \quad \text{subject to} \quad \boldsymbol{\lambda}^* \geqq \boldsymbol{0}, \tag{31}$$

$$\min_{\boldsymbol{\lambda}^*} \tfrac{1}{2}\tilde{\boldsymbol{\lambda}}^* \boldsymbol{A}\boldsymbol{\lambda}^* \quad \text{subject to} \quad \boldsymbol{A}\boldsymbol{\lambda}^* \geqq \boldsymbol{M}-\boldsymbol{K}. \tag{32}$$

Problem (32) is derived from (31) via formal dualization resting on condition

(16), [6][7]. If some $\boldsymbol{\lambda}^*$ satisfies (28), the constraints of (32) define a nonempty feasible domain. Therefore, as its objective is quadratic and bounded below, problem (32) is solvable and, hence also the primal (31) is so, [7]. Any $\boldsymbol{\lambda}^* \geqq \mathbf{0}$ which solves (31), solves the dual (32) as well, [7], and, hence, satisfies (30) and (28) (q.e.d.)

REMARKS. In contrast to (I), statement (II) is not subject to the restriction (16). Using Theor. (III) and a customary argument, [8], founded on a perturbation $\delta\boldsymbol{K} > \mathbf{0}$ of $\boldsymbol{K}$ in (19), it is clearly possible to combine (I) and (II), and state in mechanical terms the following sufficient and necessary shakedown criterion, which generalizes the classical one.

Under condition (16), shakedown occurs if, and only if, there exist plastic strains such that, should they be generated at $t = 0$ (with consequent changes in stresses, configuration and yield surfaces), the whole loading history would cause no further yielding.

Special cases. For $\boldsymbol{G} = \mathbf{0}$ (no geometric effects) the above statements reduce to those given by the author in [2]. For $\boldsymbol{G} = \mathbf{0}$ and $\boldsymbol{H} = \mathbf{0}$: the qualification (16) always holds; selfstresses (or redundants) can be used instead of $\boldsymbol{\lambda}^*$ as 'test variables'; (III) becomes immaterial; (I) (II) specialize to Prager's extension to dislocations of Melan-Bleich theorem, [9][10]. For $\boldsymbol{Q}_t^e = \boldsymbol{Q}^e$ constant, besides (but not without) $\boldsymbol{G} = \mathbf{0}$ and $\boldsymbol{H} = \mathbf{0}$, kinematic quantities (including $\boldsymbol{D}$) and elastic properties turn out to play no role, and (I) (II) reduce to the static theorem of limit analysis.

Determination of the safety factor

Let $k \geqq 0$ be a common multiplier of the loads $\boldsymbol{F}_t$, $\boldsymbol{D}_t$. 'Safety factor' is the value s such that for $k \leqq s$ the structure shakes down, for $k > s$ it does not.

IV. When (16) holds, s is the optimal value of any of the linear programming problems:

$$\max_{k,\, \boldsymbol{\lambda}^*} k \quad \text{subject to} \quad \boldsymbol{M}k - \boldsymbol{A}\boldsymbol{\lambda}^* \leqq \boldsymbol{K}, \tag{33}$$

$$\max_{k,\, \boldsymbol{\lambda}^*} k \quad \text{subject to} \quad \boldsymbol{M}k - \boldsymbol{A}\boldsymbol{\lambda}^* \leqq \boldsymbol{K},\ k \geqq 0,\ \boldsymbol{\lambda}^* \geqq \mathbf{0}, \tag{33a}$$

$$\min_{\bar{\boldsymbol{\lambda}}} \tilde{\boldsymbol{K}}\bar{\boldsymbol{\lambda}}\beta^{-1} \quad \text{subject to} \quad \tilde{\boldsymbol{M}}\bar{\boldsymbol{\lambda}} = \beta,\ \boldsymbol{A}\bar{\boldsymbol{\lambda}} = \mathbf{0},\ \bar{\boldsymbol{\lambda}} \geqq \mathbf{0}, \tag{34}$$

$$\min_{\bar{\boldsymbol{\lambda}}} \tilde{\boldsymbol{K}}\bar{\boldsymbol{\lambda}}\beta^{-1} \quad \text{subject to} \quad \tilde{\boldsymbol{M}}\bar{\boldsymbol{\lambda}} \geqq \beta,\ \boldsymbol{A}\bar{\boldsymbol{\lambda}} \leqq \mathbf{0},\ \bar{\boldsymbol{\lambda}} \geqq \mathbf{0} \tag{34a}$$

where β is an arbitrary positive constant with the dimensions of (generalized) stress $\times$ strain.

Proof. Program (33) follows straightforwardly from the definition of s, statements (I) (II) and the remark that the requirement $k \geqq 0$ is immaterial (as the constraints can be certainly satisfied with $k = 0$). The alternative, computationally advantageous formulation (33a) is justified by theorem (III). The formulations (34) and (34a) are obtained by dualizing (33) and (33a), respectively, and by interpreting the dual variables as plastic multipliers (this requires to introduce β) (q.e.d.)

V. Even if (16) fails to hold, the optimal value k^0 of the linear programs listed in (IV) represents an upper bound to s.

Proof. Any k such that $k^0 < k < s$, would imply shakedown by definition of s, non-shakedown by statement (II). Hence, $s \leqq k^0$ (q.e.d.).

VI. When $\boldsymbol{A}$ is symmetric, positive definite, shakedown occurs under any loading $\boldsymbol{F}_t$, $\boldsymbol{D}_t$.

Proof. Since (16) holds, statement (IV) applies. If det $\boldsymbol{A} \neq 0$, the constraints of problem (34) are inconsistent for any $\boldsymbol{M}$. But those of its dual (33) are not so, because, when $\boldsymbol{A}$ is positive definite, the quadratic program (31) admits for any $\boldsymbol{M}$ a solution, which solves problem (32) too and, hence, complies with its constraints. The only alternative is that k be unbounded above over the feasible domain of (33), [7], i.e. $s = \infty$ (q.e.d.).

REMARKS. Though the variables $\boldsymbol{\lambda}^*$ and $\boldsymbol{\lambda}$ refer to the same mechanical quantities, their role, meaning and values at the optimum, are different. Duality theory of linear programming supplies the following 'combinatorial' relation for the solutions of (33) and (34):

$$\tilde{\bar{\boldsymbol{\lambda}}}(\boldsymbol{M}s - \boldsymbol{A}\boldsymbol{\lambda}^* - \boldsymbol{K}) = 0. \tag{35}$$

For paucity of space, (35) and related points cannot be discussed here (some hints can be found in [11] for a narrower context); neither solutions based on (IV) are compared to those from continuous approaches.

If both $\boldsymbol{H}$ and $-\boldsymbol{Z}$ are symmetric and positive semidefinite ($-\boldsymbol{Z}$ is always so if $\boldsymbol{G} = \boldsymbol{0}$), then $\boldsymbol{A}\bar{\boldsymbol{\lambda}} = \boldsymbol{0}$ requires simultaneously:

$$\boldsymbol{H}\bar{\boldsymbol{\lambda}} = \boldsymbol{0}, \tag{36}$$

$$\boldsymbol{Z}\boldsymbol{N}\bar{\boldsymbol{\lambda}} = \boldsymbol{0}. \tag{37}$$

In fact $\tilde{\bar{\boldsymbol{\lambda}}}\boldsymbol{A}\bar{\boldsymbol{\lambda}} = 0$ implies both $\tilde{\bar{\boldsymbol{\lambda}}}\boldsymbol{H}\bar{\boldsymbol{\lambda}} = 0$ and $\tilde{\bar{\boldsymbol{p}}}\boldsymbol{Z}\bar{\boldsymbol{p}} = 0$, whence a known property of semidefinite quadratic forms, [7], leads to (36) (37). Eq. (36)

with (6) (7), and Eq. (37) with (14a), mean, respectively, that the plastic strains $\bar{\boldsymbol{p}}$ defined by $\bar{\boldsymbol{\lambda}}$ must correspond to unaltered yield surfaces and must cause no stresses (both requirements need not apply as plastic strains develop, but only at the end). When $\boldsymbol{G} = \boldsymbol{0}$, Eq. (37) imposes that $\bar{\boldsymbol{p}}$ be compatible, i.e. form a 'mechanism'; in fact it becomes equivalent to $\boldsymbol{N\lambda} = \boldsymbol{Cu}$, as easily seen from (14). For $\boldsymbol{G} = \boldsymbol{0}$ and also $\boldsymbol{H} = \boldsymbol{0}$, the primal and the dual linear programs in (IV), can be readily reduced to the static and, respectively, kinematic matrix formulations given in [11] for the classical shakedown analysis problem.

'Kinematic' shakedown theorems

A plastic flow process $\dot{\boldsymbol{\lambda}}_t \geqq \boldsymbol{0}$ over some time interval τ, characterized by the properties

$$\int_\tau \dot{\boldsymbol{\lambda}} \mathrm{d}t = \bar{\boldsymbol{\lambda}} \quad \text{such that} \quad \boldsymbol{A}\bar{\boldsymbol{\lambda}} = \boldsymbol{0}, \quad \bar{\boldsymbol{\lambda}} \geqq \boldsymbol{0}, \tag{38}$$

will be called 'admissible yielding cycle'.

VII. The structure does not shakedown, if

$$\int_\tau \tilde{\boldsymbol{F}}_t \dot{\boldsymbol{u}}_t^p \mathrm{d}t + \int_\tau \tilde{\boldsymbol{D}}_t \dot{\boldsymbol{Q}}_t^p \mathrm{d}t > \tilde{\boldsymbol{K}}\bar{\boldsymbol{\lambda}} \tag{39}$$

for some load path $\boldsymbol{F}_t$, $\boldsymbol{D}_t$ within the assigned limits, and for some admissible yielding cycle and its consequences $\dot{\boldsymbol{u}}_t^p$, $\dot{\boldsymbol{Q}}_t^p$.

Proof. By applying the virtual work relation (15):

$$\tilde{\boldsymbol{Q}}_t^e(\dot{\boldsymbol{p}}_t + \dot{\boldsymbol{e}}_t^p) + \tilde{\boldsymbol{u}}_t^e \boldsymbol{G} \dot{\boldsymbol{u}}^p = \tilde{\boldsymbol{F}}_t \dot{\boldsymbol{u}}_t^p, \tag{40}$$

$$\tilde{\dot{\boldsymbol{Q}}}_t^p(\boldsymbol{e}_t^e + \boldsymbol{D}_t) + \tilde{\dot{\boldsymbol{u}}}_t^p \boldsymbol{G} \boldsymbol{u}_t^e = 0. \tag{41}$$

Since $\tilde{\dot{\boldsymbol{Q}}}_t^p \boldsymbol{e}_t^e = \tilde{\dot{\boldsymbol{e}}}_t^p \boldsymbol{Q}_t^e$, Eqs. (40) and (41) combined yield:

$$\tilde{\boldsymbol{F}}_t \dot{\boldsymbol{u}}_t^p + \tilde{\boldsymbol{D}}_t \dot{\boldsymbol{Q}}_t^p = \tilde{\boldsymbol{Q}}_t^e \dot{\boldsymbol{p}}_t. \tag{42}$$

By the definition (29) of $\boldsymbol{M}$, since $\dot{\boldsymbol{\lambda}}_t \geqq \boldsymbol{0}$:

$$\tilde{\boldsymbol{Q}}_t^e \dot{\boldsymbol{p}}_t = \tilde{\boldsymbol{Q}}_t^e \boldsymbol{N} \dot{\boldsymbol{\lambda}}_t \leqq \tilde{\boldsymbol{M}} \dot{\boldsymbol{\lambda}}_t \tag{43}$$

whence, integrating over τ:

$$\int_\tau \tilde{\boldsymbol{F}}_t \dot{\boldsymbol{u}}_t^p \mathrm{d}t + \int_\tau \tilde{\boldsymbol{D}}_t \dot{\boldsymbol{Q}}_t^p \mathrm{d}t \leqq \tilde{\boldsymbol{M}}\bar{\boldsymbol{\lambda}}. \tag{44}$$

Through (44), inequality (39) implies

$$\tilde{K}\bar{\lambda} < \tilde{M}\bar{\lambda}. \tag{45}$$

A vector $\bar{\lambda}$ complying with both (38) and (45), fulfills the constraints of problem (34) (with suitable, always possible choice of β), and makes its objective less than 1. Hence, the minimum, k^0, of (34) and the safety factor $s \leqq k^0$ must be <1 as well. (q.e.d.)

VIII. When condition (16) holds, shakedown will occur if all F_t, D_t within the prescribed limits and all admissible yielding cycles satisfy the inequality:

$$\int_\tau \tilde{F}_t\dot{u}_t^p \mathrm{d}t + \int_\tau \tilde{D}_t\dot{Q}_t^p \mathrm{d}t \leqq \tilde{K}\bar{\lambda}. \tag{46}$$

Proof. In the class of all combinations of (nonrelated) loading and plastic flow paths, it is possible to single out those such that: while each component of $\dot{\lambda}_t$ is >0, all other components are zero and the corresponding component, of $\tilde{N}Q_t^e$ attains its maximum. For such a process, in view of the meaning (29) of M, inequality (43) and, hence, (44) are satisfied as equalities; therefore (46) implies:

$$\tilde{K}\bar{\lambda} \geqq \tilde{M}\bar{\lambda}. \tag{47}$$

This inequality leads to a value $\geqq 1$ in the objective function of problem (34). Since same conclusion can be drawn for any $\bar{\lambda}$ complying with (38) and, hence, with the constraints of program (34), the minimum k^0 of this must be $\geqq 1$. Hence, if (16) holds, $s \geqq 1$ by virtue of (IV) (q.e.d.).

REMARKS. The plastic work performed along a yielding cycle resulting in $\bar{\lambda}$, by means of the constitutive laws (5) (6) (10), turns out to be

$$\int_\tau \tilde{Q}_t\dot{p}_t \mathrm{d}t = \tilde{K}\bar{\lambda} + \tfrac{1}{2}\tilde{\bar{\lambda}}H\bar{\lambda}. \tag{48}$$

Therefore the r.h.s. of (39) and (46) represent the total dissipation only in the absence of hardening ($H = 0$). Hardening intervenes, through A, merely in definition (38).

By analogy and symmetry to the criterion derived from (I) and (II), it is clearly possible to combine theorems (VII) (VIII) into a sufficient and necessary 'kinematic' condition for non-shakedown centered on (46) and subjected to restriction (16). Note, however, the different levels of generality of (VII) and (VIII) with respect to (16), as long as hardening and/or geometric effects are present.

Special cases. When $\boldsymbol{G} = \mathbf{0}$ and $\boldsymbol{H} = \mathbf{0}$ (but still in the presence of variable dislocations $\boldsymbol{D}_t$), statements (VII) and (VIII) reduce to those established in [11] and, by a continuous approach, in [12]. For $\boldsymbol{G} = \mathbf{0}$, $\boldsymbol{H} = \mathbf{0}$ and $\boldsymbol{D}_t = \mathbf{0}$, they coincide with Koiter's theorems [10]. If, in addition, $\boldsymbol{F}$ and $\boldsymbol{D}$ are constant, the second terms in the l.h.s. of (39) and (46) vanish, and the kinematic theorem of limit analysis is thus arrived at.

Upper bounds for displacements, stresses and plastic strains

The safety factor s, as previously defined, often represents inadequate and sometimes meaningless information. In fact, even if shakedown is predicted to occur ($s > 1$), the development of plastic strains may involve one of the following events: (a) excessive displacements make the structure unserviceable or/and the theory unreliable; (b) stresses exceed the strength of possible brittle elements; (c) the deformation capacity of materials or structural elements is exceed; (d) strains increase beyond the validity range of the assumed idealization of material behavior.

Methods of checking these occurrences become particularly important when hardening and geometric effects are allowed for: indeed $s = \infty$ may be then a consistent and sometimes obvious result, cf. theorem (VI), which emphasizes the inadequacy of shakedown as a safety criterion, rather than invalidating the practical value of the theory. The upper bound procedures suggested below seem capable to meet the above requirements and are meant as a natural, often necessary continuation of the analysis after s has been found.

Let V_s denote the value assumed after shakedown by a scalar quantity x which depends linearly on $\boldsymbol{\lambda}$:

$$V_s = \tilde{\boldsymbol{a}}\boldsymbol{\lambda}_s. \tag{49}$$

When x represents the *h-th* component of $\boldsymbol{u}_t^p$ or $\boldsymbol{Q}_t^p$, vector $\boldsymbol{a}$ is the *h-th* row of matrix $\boldsymbol{S}^{-1}\tilde{\boldsymbol{C}}\boldsymbol{E}\boldsymbol{N}$, Eq. (13a), or of $\boldsymbol{Z}\boldsymbol{N}$, Eq. (14a), respectively. If $\boldsymbol{a}$ is formed by the above rows modified by replacing all negative entries with zero, then since $\boldsymbol{\lambda}_t \geqq \mathbf{0}$, (49) supplies an upper bound V_m to the maximum value attained by u_{ht}^p or Q_{ht}^p over $0 \leqq t \leqq \infty$. However in most practical cases, for meaningful components u_h^p, Q_h^p (e.g. mid-span deflections) it can be easily foreseen that the value V_s at shakedown will represent the maximum;

in these cases checks against possibilities (a) and (b) can be carried out simply by adding to V_s the readily calculated maxima of the elastic counterparts $\max_t u^e_{ht}$, $\max_t Q^e_{ht}$, without recourse to V_m.

When x is a component of $\boldsymbol{\lambda}_t$, say λ_{ht}, then obviously $V_m = V_s = \lambda_{hs}$. The value λ_{hs} at shakedown can be used for checks concerning (c) and (d). Rigorous criteria for (b) (c) and (d) would involve nonlinear functions of $\boldsymbol{\lambda}_s$. However, since rigorous criteria seem to be both unnecessary for the estimates needed in most cases, and are covered by easy generalizations, we will consider below only the central point of generating upper bounds to the values V_s defined by (49). To this purpose the following statement is central.

IX. When (16) holds, if $s > 1$, a finite upper bound to V_s is the optimal value $U(\boldsymbol{\lambda}^*)$ of the convex programming problem:

$$U(\boldsymbol{\lambda}^*) = \max_{\boldsymbol{\lambda}} \tilde{\boldsymbol{a}}\boldsymbol{\lambda} \tag{50}$$

subject to:

$$\boldsymbol{M} - \boldsymbol{A}\boldsymbol{\lambda} \leqq \boldsymbol{K} \tag{51}$$

$$\boldsymbol{\lambda} \geqq \boldsymbol{0} \tag{52}$$

$$\tilde{\boldsymbol{K}}\boldsymbol{\lambda} \leqq s(s-1)^{-1}\tfrac{1}{2}\tilde{\boldsymbol{\lambda}}^*\boldsymbol{A}\boldsymbol{\lambda}^* \tag{53}$$

$$\tfrac{1}{2}\tilde{\boldsymbol{\lambda}}\boldsymbol{A}\boldsymbol{\lambda} - \tilde{\boldsymbol{\lambda}}\left(\boldsymbol{A}\boldsymbol{\lambda}^* - \frac{s-1}{s}\boldsymbol{K}\right) \leqq \boldsymbol{0} \tag{54}$$

$\boldsymbol{\lambda}^*$ being any vector such that:

$$\boldsymbol{M} - \boldsymbol{A}\boldsymbol{\lambda}^* \leqq \boldsymbol{K}s^{-1}. \tag{55}$$

Proof. Let us show first that the constraints (51)–(55) define in the $\boldsymbol{\lambda}$-space a domain $\Lambda(\boldsymbol{\lambda}^*)$ to which the vector $\boldsymbol{\lambda}_s$ of the actual plastic multipliers at shakedown belongs. Because of (30) and (33), s can replace α in inequality (24), which thus becomes (55). Also, the consequence (27) of (24) holds for $\alpha = s$ and through (21) and some algebra, supplies the constraint (54). Since, by (21) and (23), $L_0 \geqq L_\infty \geqq 0$, inequality (27) for $\alpha = s$ remains valid if L_∞ is suppressed, and thus, through (21), yields (53). The inequalities (51) (52) follow from the constitutive laws (6) (7) (9) through the definition (29) of $\boldsymbol{M}$ and (16) of $\boldsymbol{A}$. The domain $\Lambda(\boldsymbol{\lambda}^*)$ is: nonempty, because, if $s > 1$, at least the feasible vector $\boldsymbol{\lambda} = \boldsymbol{\lambda}_s$ does exist; bounded, because (52) and (53) alone clearly define a bounded domain; convex, because the l.h.s. of (54) is a convex function as a consequence of (16) and all other constraints are linear. Since the objective function (50) is linear, the stated finiteness and convexity are proved (q.e.d.).

Inequality (54) can be easily seen to be more stringent than (53).

Let Λ' be a domain which contains the above defined domain Λ. The following ways of generating upper bounds U to V_s flow immediately from (IX):

(i) $U'(\boldsymbol{\lambda}^*) = \max_{\boldsymbol{\lambda}} \tilde{\boldsymbol{a}}\boldsymbol{\lambda}$ subject to $\boldsymbol{\lambda} \in \Lambda'(\boldsymbol{\lambda}^*)$, $\boldsymbol{\lambda}^*$ obeying inequality (55);

(ii) dualize problem (i) formulated with a convex Λ'; in the dual (minimization) problem, any value U^d the objective function assumes for a feasible vector is an upper bound to V_s (because $U^d \geqq U'$).

(iii) $U'_m = \min_{\boldsymbol{\lambda}^*} \max_{\boldsymbol{\lambda}} \tilde{\boldsymbol{a}}\boldsymbol{\lambda}$ subject to (55) and to $\boldsymbol{\lambda} \in \Lambda'(\boldsymbol{\lambda}^*)$.

Various combinations of the above procedures are possible and differ in computational effort and quality of results[1]). The most stringent upper bound resulting from (iii) for $\Lambda' = \Lambda$, is usually difficult to achieve. The method described below might be useful in many cases.

If the nonlinear constraint (54) is dropped from (IX), the relaxed problem (i) is the linear program (50)–(53). The dual to this reads:

$$\min_{\boldsymbol{\lambda}, \xi} \gamma\tilde{\boldsymbol{\lambda}}(\boldsymbol{K}-\boldsymbol{M})+\xi s(s-1)^{-1}\tfrac{1}{2}\tilde{\boldsymbol{\lambda}}^*\boldsymbol{A}\boldsymbol{\lambda}^* \tag{56}$$

subject to:

$$\boldsymbol{K}\xi-\gamma\boldsymbol{A}\boldsymbol{\lambda} \geqq \boldsymbol{a}, \quad \boldsymbol{\lambda} \geqq \boldsymbol{0}, \quad \xi \geqq 0 \tag{57}$$

where ξ is a new variable, γ an arbitrary positive constant with suitable dimensions to make $\boldsymbol{a}$ and $\gamma\boldsymbol{A}\boldsymbol{\lambda}$ homogeneous. According to (ii), it can be stated in general that:

X. When (16) holds, the linear form (56), with any $\boldsymbol{\lambda}^*$ obeying (55) and any $\gamma > 0$, supplies a (finite) upper bound to V_s for any $\boldsymbol{\lambda}$, ξ complying with the linear inequalities (57).

The search via (iii) for the best upper bound in this simplified context, reduces to minimize the r.h.s. of (53) with $\boldsymbol{\lambda}^*$ constrained by (55), i.e. to a convex quadratic program analogous to (32).

The corresponding primal problem, analogous to (31), is fully equivalent for the present purposes and computationally more convenient.

Special cases. When $\boldsymbol{G} = \boldsymbol{0}$ and $\boldsymbol{H} = \boldsymbol{0}$: the inequalities (53) (54) reduce to

[1]) It is worth noting that when an upper bound is found to be less than the relevant critical value, the check can often be regarded as accomplished with positive result.

Koiter's bounds for total plastic work [10]; the formulations (i) (iii) with $\Lambda' = \Lambda$, applied to bound components of $\boldsymbol{\lambda}_s$, become those established in [13] by Vitiello who also obtained encouraging numerical results. For $s = \infty$ all the above conclusions are seen to hold unaltered.

Concluding remarks

What precedes is intended as a contribution to broaden the shakedown theory rather than to improve the methods already available for solving shakedown problems. Some critical considerations on the results achieved seem appropriate in closing.

a. The essential role of geometric changes is obvious in many practical situations, and was recognized and studied previously in [14]. However these effects were ignored thus far in the basic shakedown theorems, to the author's knowledge. The limited extent ('second order') at which they are covered in the present theory is specified by the linear nature of the equations of compatibility (1) and equilibrium (2), where both the compatibility matrix $\boldsymbol{C}$ and the stiffness matrix $\boldsymbol{G}$ are assumed constant. Indeed the geometric effects are allowed for only by the presence of the 'geometric term' $\boldsymbol{G}(\overline{\boldsymbol{Q}})\boldsymbol{u}_t$ in the equilibrium equation (1). This is strictly valid in general for incremental processes starting from any loaded configuration to which $\boldsymbol{C}$ and $\boldsymbol{G}$ are referred. Beams subjected to constant axial loads and to variable transversal loads unquestionably admit the above representation (in the range of deflections for which reference to the undeformed geometry would be legitimate in the absence of axial loads).

An analogous distinction between constant stresses $\overline{\boldsymbol{Q}}$ with an essential role in the geometrical term of the equilibrium equations, and variable stresses $\boldsymbol{Q}_t$ with negligible role in it, can be often reasonably made for slender structures of various types; sometimes conservative estimate of $\overline{\boldsymbol{Q}}$ may be assumed (e.g. for tall frames under wind loads and heavy dead loads). Obviously such distinction becomes arbitrary and its accuracy unclear in many cases. Therefore in the presence of important geometric effects the present 'second order' theory must be used with caution and its inherent limitations in this regard kept in mind.

b. Piecewise linear approximation of yield loci is a customary practice in perfect plasticity. The representation of actual work-hardening behaviour by means of translating-interacting yield planes has not yet a comparable support of computational experience, but appears to be the most natural extension of the former approximation and equally flexible and practically convenient.

c. The use of the vector-matrix analytical dress provides compactness and simplicity to the preceding study and implicitly establishes clear connections between the theory and some efficient mathematical programming techniques for its applications. The discrete structural models referred to, include, besides some elementary structures like trusses, general continua discretized in constant-strain finite elements. This sort of discretization is clearly not always convenient computationally (it is certainly not so for plates and shells), but combines particularly well generality and simplicity; moreover its convergence on the continuous description has intuitive, computational [3] [4], and recently also theoretical [15] support.

The extension, not discussed here, to more refined finite element models, such those suitable for plates and shells, deserves further study. The 'lumped-parameter' models already successfully employed in elastoplastic analysis of plates [16], seem particularly promising for direct applications of the present results.

d. In the broadened theory, in contrast to the classical one, two ingredients play a central role: the distribution of the plastic multipliers ($\boldsymbol{\lambda}$) and the operator ($\boldsymbol{Z}$) which transforms plastic strains into the consequent stresses. The latter is far more tractable for discretized structures than in the tensor field description, where it assumes the form discussed in [17].

The dimension of the vector $\boldsymbol{\lambda}$ might be reduced for the numerical solution of practical problems (particularly for calculating upper bounds [13]) by neglecting those yield modes which can be conjectured 'a priori' not to be activated in the evolution of the structure.

Nomenclature

Boldface symbols denote matrices and column-vectors; $\mathbf{0}$ is a matrix or vector whose entries are zero. A tilde $\sim$ means transpose; a dot, time derivative; s = safety factor; $\in$ = belongs to. Loads = given forces and 'dislocations' (such as thermal strains).

Subscript t = time dependent. Superscripts e and p = elastic consequence of loads and of plastic strains, respectively.

$\boldsymbol{F}, \boldsymbol{u}$ = n-vectors of all components of external forces and displacements, respectively, at free nodes.

$\boldsymbol{q}, \boldsymbol{e}, \boldsymbol{p}, \boldsymbol{D}, \boldsymbol{Q}$ = m-vectors of *all* components of, respectively: total, elastic, plastic generalized strains; dislocations; generalized stresses. They are partitioned into subvectors, each relevant to each of the finite elements, taken in a fixed order.

$\boldsymbol{\lambda}$, $\boldsymbol{\varphi}$, $\boldsymbol{K}$ = y-vectors of plastic multipliers, plastic potentials (assumed linear) and positive constants, respectively, for *all y* yielding modes in the structure, elementwise partitioned in the same order as before.

$\boldsymbol{G} = n \times n$-geometric stiffness matrix of the structure.

$\boldsymbol{E} = \text{diag}\, \boldsymbol{E}^i = m \times m$-block diagonal matrix, the *i-th* block $\boldsymbol{E}^i$ being the natural elastic stiffness matrix of the finite element i, index i running over the ordered set of all elements.

$\boldsymbol{N} = \text{diag}\, \boldsymbol{N}^i = m \times y$-block diagonal matrix, $\boldsymbol{N}^i$ being the matrix containing as columns the outward normal vectors to the yield planes in the stress space of the *i-th* finite element.

$\boldsymbol{H} = \text{diag}\, \boldsymbol{H}^i = y \times y$-block diagonal matrix, $\boldsymbol{H}^i$ being the matrix of the hardening coefficients which govern the translations due to plastic flow, of the yield planes of element i.

$\boldsymbol{Z} = m \times m$-matrix which transforms dislocation vectors into consequent stress vectors.

Other symbols are defined where they are used first.

Acknowledgement

This study was carried out while the writer was a Visiting Professor at the University of Illinois, Urbana, U.S.A. The writer wishes to thank D. C. Drucker, Dean of the College of Engineering, and N. M. Newmark, Head of the Department of Civil Engineering.

References

[1] Sawczuk, A., On incremental collapse of shells under cyclic loading. *Theory of Thin Shells, Proc. IUTAM Symp.*, Springer-Verlag, Berlin, (1969) 328.

[2] Maier, G., A matrix structural theory of piecewise linear elastoplasticy with interacting yield planes. *Meccanica*, *5* (1970) 54.

[3] Zienkiewicz, O. C., *The finite element method in engineering sciences.* McGraw-Hill Co., New York, (1971).

[4] Argyris, J. H., *Continua and discontinua.* Proc. Conf. matrix methods in structural mechanics. Wright-Patterson A.F.B., Ohio, (1967) 12.

[5] Maier, G., Incremental plastic analysis in the presence of large displacements and physical instabilizing effects. *Int. J. Solids Structures*, *7* (1971) 34.

[6] Zangwill, W. S., *Nonlinear programming: a unified approach.* Prentice Hall Inc., Englewood Cliffs, N. J., (1969).

[7] Künzi, H. P. and W. Krelle, *Nonlinear programming.* Blaisdell Publ. Co., Waltham, Mass., (1966).

[8] Hodge, P. G., *Plastic analysis of structures.* McGraw-Hill Co., New York, (1959).

[9] Prager, W., Shakedown in elastoplastic media subjected to cycles of load and temperature. *Proc. Symp. Plasticità in Sc. d. Costr., Varenna,* Zanichelli, Bologna, (1956) 239.

[10] Koiter, W. T., General theorems for elastic-plastic solids. *Progress in Solid Mechanics,* Vol. I, North Holland Publ., (1960) 167.

[11] Maier, G., Shakedown theory in perfect elastoplasticity with associated and non-associated flow-laws: a finite element linear programming approach. *Meccanica,* No. 3, *14* (1969) 250.

[12] De Donato, O., Second shakedown theorem allowing for cylces of both loads and temperatures. *Ist. Lombardo, Rendic. Sci., A 104* (1970) 265.

[13] Vitiello, E., Upper bounds to plastic strains in shakedown of structures subjected to cyclic loads. To appear in *Meccanica.*

[14] Davies, J. M., Collapse and shakedown loads of plane frames. *J. Structural Div., Proc. ASCE, 93* (1967) 35.

[15] Havner, K. S., On convergence of a discrete aggregate model in polycrystalline plasticity. To appear in *Int. J. Solids and Structures,* (1971).

[16] Ang, A. H. and L. A. Lopez, Discrete model analysis of elastoplastic plates. *J. Eng. Mech. Div., Proc. ASCE, 94* (1968).

[17] Maier, G., A minimum principle for incremental elastoplasticity with non-associated flow-laws. *J. Mech. Phys. Solids, 18* (1970) 319.

Methods of solving problems in the shakedown theory of continua

D. A. Gokhfeld and O. F. Cherniavsky
Polytechnic Institute, Cheliabinsk, USSR

The fundamental theorems of shakedown are transformed such that a time-dependent analysis of stress and strain may be separated from the solution of the fundamental extremum problem. Analysis of incremental collapse is reduced to the limit analysis problem for a nonhomogeneous body in which fictitious yield surfaces are constructed beforehand for each point of the body subjected to a variable stress state. Generalized variables are suitably introduced.
The theory permits one to employ the classical procedures of limit analysis in shakedown analysis of plates and shells as well as the present-day methods of solving non-classical variational problems. Some examples of the applications of the proposed methods are given.

Introduction

Two fundamental theorems of shakedown of ideally elastic-plastic bodies, formulated by Melan and Koiter [1] permit one to bound values of repeatedly alternating loading parameters from below and above. The first theorem states that the body will shake down if there exists a time independent distribution of residual stresses ρ_{ij} such that the elastic stresses $\sigma_{ij}^{(e)}(\tau)$ generated by external loads, does not violate the yield conditions at any point of the body at any instant of time, i.e.

$$f(\sigma_{ij}^{(e)} + \rho_{ij}) < 0, \tag{1}$$

$$\rho_{ij,\,j} = 0, \qquad \rho_{ij}n_j = 0 \tag{2}$$

where $f(\sigma_{ij}) = 0$ is the equation of the yield surface.

In accordance with the second theorem, shakedown never takes place if any kinematically admissible cycle of the plastic strain rate $\dot{\varepsilon}''_{ijo}(\tau)$ can be found such that the following inequality is satisfied

$$\int_0^T \left\{ \int X_i \dot{u}_{i0}\, dv + \int_{Sp} p_i \dot{u}_{i0}\, dS \right\} d\tau > \int_0^T d\tau \int \sigma_{ij} \dot{\varepsilon}''_{ij0}\, dv, \tag{3}$$

$$\Delta\varepsilon''_{ij0} = \tfrac{1}{2}(\Delta u_{i0,j} + \Delta u_{j0,i}), \quad \left(\Delta\varepsilon''_{ij0} = \int_0^T \dot{\varepsilon}''_{ij0}\, d\tau, \quad \Delta u_{i0} = \int_0^T \dot{u}_{i0}\, d\tau \right) \tag{4}$$

T is a cycle period, X_i and p_i are the body forces and surface tractions respectively, and $\dot{u}_{i0}$ indicate the residual displacement rates.

The plastic strain rates $\dot{\varepsilon}''_{ij0}$ are assumed to be related to the yield condition by means of the associated flow law. In the general case, some terms should be added to the inequality (3) to allow for a discontinuous rate field.

For a limiting cycle correspondent to the boundary of a purely elastic response of a body the inequalities (1), (3) become equalities.

Shakedown theorems are widely applied in the analysis of beams, trusses and so on (Rhzanitsyn, Neal, Hodge and others). However, the analysis of spatial bodies has proven to be rather complicated, mainly due to the fact that the collapse mechanisms are not easyly to assess and that in the solutions of the extremum problems the variables such as stresses and strain rates depend not only on the coordinates of a point considered but also on the current time. As will be seen later on, these difficulties can be overcome to a considerable extent if the classical theorems (1)–(4) are restated in such a way that the analysis of time dependent stresses and strains is separated from the analysis of the non-classical variational problem.

Restatement of the fundamental theorems

It appears convenient to divide the external loads into time-dependent loads and constant loads. Associating the elastic stresses generated by constant loads X_i^0, p_i^0 with the self-equilibrating residual stresses ρ_{ij}^0, the conditions (1), (2) can be expressed in the form

$$f(\sigma_{ij\tau}^{(e)} + \sigma_{ij}^0) < 0, \tag{5}$$

$$\sigma_{ij,j}^0 + X_i^0 = 0, \qquad \sigma_{ij}^0 n_j = p_i^0. \tag{6}$$

It is necessary and sufficient that the maximum value of the yield function (within the cycle of $\sigma_{ij\tau}^{(e)}(\tau)$) remain negative, in order that the inequality (5) is satisfied at any instant of time, i.e.

$$\max_\tau f[\sigma_{ij\tau}^{(e)}(\tau) + \sigma_{ij}^0] < 0. \tag{7}$$

The stresses $\sigma_{ij\tau}^{(e)}$ generated by time dependent loads and evaluated under the assumption of ideal elasticity, are considered to be known – at least to within the external loading parameters[1]). Thus, on account of the inequality (7), the unique (the 'determining') value of the variable elastic stresses $\sigma_{ij*}^{(e)}$ can be found for each point of the yield surface together with the corresponding limiting value of the constant part of stresses such that the inequality holds (for a given infinitesimal volume).

The equality

$$f(\sigma_{ij*}^{(e)} + \sigma_{ij}^{0}) = \max_{\tau} f[\sigma_{ij\tau}^{(e)}(\tau) + \sigma_{ij}^{0}] = 0 \tag{8}$$

can be looked upon as the equation of a surface in the nine-dimensional stress space bounding the region of admissible, time-independent residual stresses since on the left-hand side of (8) no time-dependent terms exist. This surface will be shortly referred to as the fictitious yield surface.

Thus the lower bound (statical) theorem can be stated as follows: The body shakes down if there exists a field of time-independent stresses σ_{ij}^{0} which are in equilibrium with the constant loads X_i^0, p_i^0, and belong to the region bounded by the fictitious yield surface, i.e. the region of constant stresses which constitute an admissible stress state if added to fluctuating elastic stresses at any instant of time. The fictitious yield surface should be determined for each point of the body.

In order to construct, a fictitious yield surface in plane stress, let the directions x, y of the principal stresses be fixed at a certain point of the body and the stresses generated by variable load be linearly related to two time-dependent, but mutually independent, parameters:

$$\sigma_{x\tau}/\sigma_0 = 0.2p_1(\tau) - 0.4p_2(\tau), \quad \sigma_{y\tau}/\sigma_0 = 0.5p_1(\tau) - 0.1p_2(\tau), \tag{9}$$

$$-p_1 \leqq p_1(\tau) \leqq p_1, \qquad 0 \leqq p_2(\tau) \leqq p_2 \tag{10}$$

where σ_0 stands for the yield point.

Adopting the Tresca-Saint Venant yield condition

$$\max(|\sigma_x|, |\sigma_y|, |\sigma_x - \sigma_y|) - \sigma_0 = 0. \tag{11}$$

the equation of the fictitious yield surface takes on account of (8) the form

$$\max_{\tau} \{\max [|\sigma_{x*}^0 + \sigma_{x\tau}|, |\sigma_{y*}^0 + \sigma_{y\tau}|, |\sigma_{x*}^0 - \sigma_{y*}^0 + \sigma_{x\tau} - \sigma_{y\tau}|]\} - \sigma_0 = 0 \tag{12}$$

where σ_{x*}^0, σ_{y*}^0 are the stresses belonging to that surface.

[1]) Necessity of their predetermination appears to be a particular feature of the shakedown problems.

On substituting the values of variable stresses (9) into (12) and considering (10), the parameters $p_1(\tau)$, $p_2(\tau)$ can be found that compare with previous proof the left-hand side of equation (12) be maximum. Thus the equation of the fictitious yield surface assumes the form

$$\max [\sigma^0_{x*} + 0.2p_1\sigma_0,\ -\sigma^0_{x*} + (0.2p_1 + 0.4p_2)\sigma_0,\ \sigma^0_{y*} + 0.5p_1\sigma_0,\ -\sigma^0_{y*} + (0.5p_1 + 0.1p_2)\sigma_0,\ \sigma^0_{x*} - \sigma^0_{y*} + 0.3p_1\sigma_0,\ -\sigma^0_{x*} + \sigma^0_{y*} + (0.3p_1 + 0.3p_2)\sigma_0] - \sigma_0 = 0. \tag{13}$$

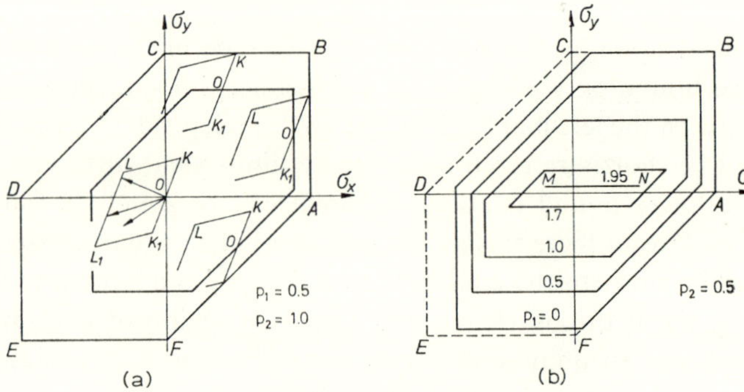

Fig. 1. Construction of fictitious yield surface.

This is illustrated by Fig. 1a where in the centre a bounding hodograph KLL_1K_1 of varying elastic stresses (9) is shown. The starting point of the hodograph plots the fictitious yield surface (13) by sliding 'without rotation' the parallelogram KLL_1K_1 over the yield surface $ABCDEFA$. The results for different values of the parameters p_1 and p_2 are shown in Fig. 1b.

Some general properties of the fictitious yield surface will now be indicated without proof:

a). The fictitious yield surface is convex if the actual yield surface is convex,

b). If, for an elementary volume, several flow profiles are possible in the limiting cycle at different instances, a corner forms on the fictitious yield surface,

c). The plastic strain increments $\Delta\varepsilon''_{ij0}$ over the limiting cycle are related to the stresses on the fictitious yield surface by means of the associated flow law

$$\Delta\Sigma''_{ij} = \sum_m \mu_m \partial f_m(\sigma^*_{ij\tau} + \sigma^0_{ij*})/\partial(\sigma^*_{ij\tau} + \sigma^0_{ij*}) \tag{14}$$

where σ^0_{ij*} are the stresses belonging to the fictitious yield surface.

In the case where either the actual yield surface or the hodograph of variable stresses is piecewise linear, relation (14) takes the form

$$\Delta\varepsilon''_{ij} = \sum_m \mu_m \partial f_m(\sigma^*_{ij\tau} + \sigma^0_{ij*})/\partial\sigma^0_{ij*}. \tag{15}$$

Consider now the kinematical (upper bound) theorem. Isolating, as in the transformation of the statical theorem, the constant components of loading, using the principle of virtual work, i.e. the inequality (3) can be given the form [2]

$$\int X^0_i \Delta u_{i0}\,\mathrm{d}v + \int_{Sp} p^0_i \Delta u_{i0}\,\mathrm{d}S > \int_0^T \mathrm{d}\tau \int (\sigma_{ij} - \sigma^{(e)}_{ij\tau})\dot{\varepsilon}''_{ij0}\,\mathrm{d}v. \tag{16}$$

In accordance with Koiter's theorem, a necessary and sufficient condition for shakedown not to occur is that the inequality (16) be satisfied when the product $(\sigma_{ij} - \sigma^{(e)}_{ij\tau})\dot{\varepsilon}''_{ij0}$, dependent upon the fluctuating elastic stresses, attains its minimum value at every point of the body. In this situation the inequality (16) can be rewritten in the form

$$\int X^0_i \Delta u_{i0}\,\mathrm{d}v + \int_{Sp} p^0_i \Delta u_{i0}\,\mathrm{d}S > \int \sigma^0_{ij*}\Delta\varepsilon''_{ij0}\,\mathrm{d}v \tag{17}$$

where σ^0_{ij*} are the stresses belonging to the fictitious yield surface (8).

Thus Koiter's theorem can be reformulated as follows: The body never shakes down if there exists per cycle any kinematically admissible distribution of plastic strain increments, for which the increment of work done by constant external loads over corresponding displacements exceeds the increment of work done by the stresses corresponding to the fictitious yield surface, which relates the stresses to strains by means of (14) or (15).

The above restatement of the shakedown theorems permits one to resolve the solution to the problem into two consecutive stages. Firstly, the fictitious yield surface is constructed for every point of the body and secondly the limit analysis problem is solved for a body of fictitious plastic non-homogeneity.

Local and global collapse

When the conditions for shakedown are violated, two different situations may arise.

The limit state can be reached if in accordance with the lower bound theorem, at least at one point of the body the fictitious yield surface shrinks to a line or to a point (Fig. 1b, line *MN*). The role of the residual stresses is restricted to the alteration of the characteristics of stress cycle in volumes which happen to be in a 'dangerous' state. In accordance with Melan's theorem, the required residual stresses must clearly appear provided the boundary conditions are still satisfied. It is important to note that distribution of residual stress need not be unique at all points except at the dangerous points.

A kinematical interpretation can be given to the indicated situation by means of Koiter's theorem. Analysis has shown that the degeneration of the fictitious yield surface into a line or a point is associated with alternating plastic flow in which the plastic strain increments per cycle vanish at all points of the body ($\Delta\varepsilon''_{ij} \equiv 0$, $\Delta u_{i0} \equiv 0$) although at the corresponding points the plastic strain rates during the cycle differ from zero.

For the determination of the limiting values of loading parameters it is sufficient to analyse the variation of elastic stresses at each point of the body. Direct use of the shakedown theorems appears not to be necessary in the solution of specific problems. It can be proven that in general the alternating flow must take place if the stress variation per cycle at any point of the body exceeds the double value of the yield point. All known solutions reduce to this sufficient condition through various procedures used for example in [3], [4]. Under complex (multiparameter) loading the alternating plasticity can also take place at lower absolute values of the stress amplitudes. The necessary condition for this case may be established in every problem.

In the situation when the fictitious yield surface is nowhere degenerated, shakedown can take place only if, in the region of admissible constant stresses (i.e. inside the fictitious yield surface), a statically admissible distribution (6) can be found. In this case plastic deformation of the same sign builds up leading to incremental collapse.

It should be pointed out that a clear distinction between the two types of cyclic deformation is of primary importance for the applications of the shakedown theory [5].

Generalized variables in the shakedown analysis of shells and plates

Introduction of generalized variables is necessarily based on certain assumptions regarding the distribution of strains or stresses across the thickness of a shell, or a plate or across the depth of a beam. When a repeated loading gives rise to local alternating flow no such hypothesis, is justified. However, in such a situation generalized variables are hardly necessary to be employed since the limiting cycle parameters may be calculated in terms of stresses alone which reader the fictitious yield surface degenerate.

It appears useful to introduce generalized variables in order to determine the conditions for incremental collapse of shells and plates. This is, however, by no means trivial since the stresses at points belonging to the same normal of the middle surface of a shell do not necessarily have to reach the yield surface at the same time. As for example when strains from tension and bending are combined. On the other hand, there is a number of problems in which either the interaction of those two types of strains is absent or not taken into account under the assumptions adopted [6].

The shakedown problem can be conveniently formulated in terms of generalized variables by means of the restated kinematical theorem (17). In case of incremental collapse the distribution of kinematically admissible plastic strain increments across the thickness of the shell or plate per cycle can be written down on the basis of Kirchhoff's hypothesis as follows

$$\Delta\varepsilon''_{\alpha\beta 0} = z\Delta x_{\alpha\beta} + \Delta e_{\alpha\beta}, \qquad (\alpha, \beta = 1, 2) \tag{18}$$

where $\Delta x_{\alpha\beta}$, $\Delta e_{\alpha\beta}$ are the generalized strain increments.

On extracting from the right-hand side of the inequality (17) the integral over the thickness and accounting for (18) ,

$$\int_{-h}^{+h} \sigma^0_{\alpha\beta *}(z\Delta x_{\alpha\beta} + \Delta e_{\alpha\beta})\mathrm{d}z = M^0_{\alpha\beta *}\Delta x_{\alpha\beta} + N^0_{\alpha\beta *}\Delta e_{\alpha\beta}. \tag{19}$$

The equation (19) appears clearly analogous to and, in the absence of alternating loading, fully coinciding with the corresponding equation of the limit analysis theory. The relationship between $M^0_{\alpha\beta *}$ and $N^0_{\alpha\beta *}$ i.e. the equation of the fictitious yield surface in terms of generalized stresses can be found in the usual way: The stresses $\sigma^0_{\alpha\beta *}$ are determined by means of the relationships between the components of the generalized strain vector $\Delta x_{\alpha\beta}$, $\Delta e_{\alpha\beta}$ and of the associated flow law in the form (14) or (15). The magnitudes of the generalized stresses are then found from equation (19).

Since the introduction of generalized stresses is connected with the kinematical approach and based on the kinematical theorem of the shakedown theory, the formulation of the statical theorem in terms of the generalized stresses can be arrived at along similar lines as in the general proof of theorem [1] to ensure that the incremental collapse will not take place if under a given external load there exists an equilibrium distribution of generalized stresses belonging to the region bounded by the fictitious yield surface.

Exact methods of solving the shakedown problems

Evaluation of an incremental collapse of a continuum in analogy with the related problem of limit analysis requires a certain variational technique which involves the solution of the differential equations of equilibrium or of compatibility, using restrictions on the magnitudes of stresses or strain rates and the optimization of loading parameter. In accordance with any of the theorems the differential equations describing the behaviour of a structure admit an infinite number of solutions. A tool for solving such extremum problems is provided by the mathematical theory of optimum control [7].

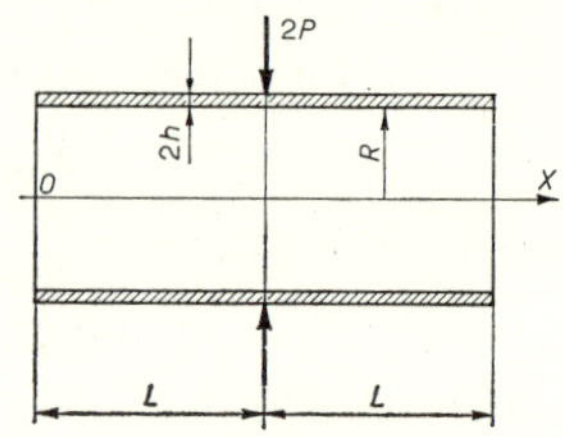

Fig. 2. Cylindrical shell.

Consider, as an example, a cylindrical shell of the length $2L$ (Fig. 2) subjected to a time independent ring load $2P$ and to a cyclic variation of temperature linearly distributed across the thickness of the shell

$$T(\tau, z) = T_0(\tau) + \Delta T(\tau)z/h, \qquad (-h \leqq z \leqq h), \tag{20}$$

$$-\Delta T \leqq \Delta T(\tau) \leqq \Delta T. \tag{21}$$

By means of the restated Melan's theorem, the limiting values of load P may be found as dependent upon the temperature difference ΔT.

Due to the temperature field (20) the following, axial and hoop, thermoelastic stresses develop

$$\begin{aligned} &\sigma_{x\tau} = \sigma_{\varphi\tau} = t(\tau)z\sigma_0/2h, \\ &t(\tau) = \alpha E \Delta T(\tau)/(1 - \nu)\sigma_0, \qquad -t \leqq t(\tau) \leqq t \end{aligned} \tag{22}$$

The edge effect is ignored for reason of simplicity. In (22) E stands for Young's modulus, ν for Poisson's ratio and α denotes the coefficient of linear expansion.

Adopting the yield condition (11) the fictitious yield surface (8) can be expressed in terms of variable stresses (22) as follows

$$\max(|\sigma^0_{x*}| + t|z|\sigma_0/2h, |\sigma^0_{\varphi*}| + t|z|\sigma_0/2h, |\sigma^0_{x*} - \sigma^0_{\varphi*}|) - \sigma_0 = 0. \quad (23)$$

The equation (19), assumes the form

$$M^0_{x*}\Delta x_x + N^0_{x*}\Delta e_x + N^0_{\varphi*}\Delta e_\varphi = \int_{-h}^{+h} [\sigma^0_{x*}(\Delta e_x + z\Delta x_x) + \sigma^0_{\varphi*}\Delta e_\varphi]\,\mathrm{d}z. \quad (24)$$

due to the rotational symmetry of the cylindrical shell considered.

Considering all possible relations between the increments Δx_x and $\Delta e_\varphi(\Delta x_x > 0, \Delta e_\varphi > 0; \Delta x_x > 0, \Delta e_\varphi < 0$ et) the corresponding stresses on the fictitious yield surface (23) will be found by means of (15) followed by the calculations of the generalized stresses. The magnitude of the axial elongation increment Δe_x, referred to the middle surface of the shell, can be found from the condition $N^0_{x*} = 0$. The results of the computations are shown in Fig. 3. with solid lines

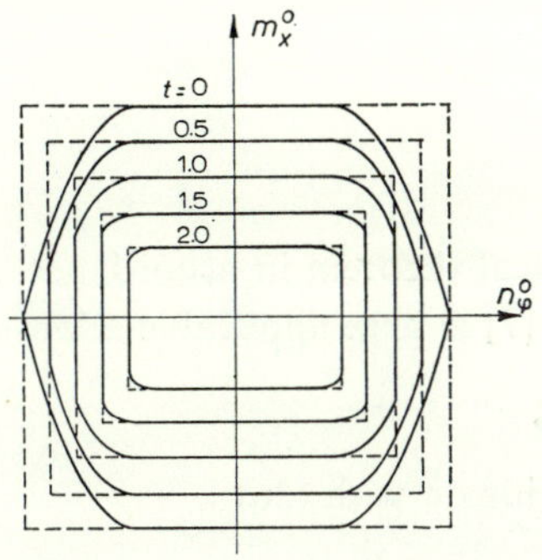

Fig. 3. Fictitious yield surface for cylindrical shell.

Let the fictitious yield surface be approximated by means of

$$\begin{aligned} &\max(|m^0_{x*}| + t/3,\ |n^0_{\varphi*}| + t/4) = 1, \\ &(m^0_{x*} = M^0_{x*}/\sigma_0 h^2,\ n^0_{\varphi*} = N^0_{\varphi*}/2\sigma_0 h) \end{aligned} \quad (25)$$

shown in Fig. 3 with dashed lines.

In accordance with the statical theorem, such a distribution of time-independent, dimensionless, generalized stresses m_x^0, n_φ^0 must be selected which belongs to the region bounded by the surface (25) and satisfies equilibrium of the shell

$$\mathrm{d}m_x^0/dx = q_x^0, \quad \mathrm{d}q_x^0/dx = -n_\varphi^0, \tag{26}$$

$$m_x^0(0) = 0, \quad q_x^0(0) = 0, \quad q_x(l) = p, \tag{27}$$

$$(x = X\sqrt{2}/\sqrt{Rh}, \quad l = L\sqrt{2}/\sqrt{Rh}, \quad p = P\sqrt{Rh}/\sigma_0 h^2\sqrt{2})$$

the condition for the maximum of the load parameter p being met at the same time.

Eqs. (26) admits an infinite number of solutions, but can be treated as a description of a 'controlled process' (given, for example, the function $n_\varphi^0(x)$, different solutions to the system (26) can be found). Following the terminology adopted in the theory of system control, the function $n_\varphi^0(x)$ will be called a control and its first order discontinuities will be admitted. The continuous functions $m_x^0(x)$, $q_x^0(x)$ will be referred to as phase coordinates. The load parameter p constitutes an optimization criterion. The problem consists in finding such an optimum control n_φ^0 which leads to the maximum of p, simultaneously satisfying (26) and (27) as well as the constraints on the magnitudes m_x^0, n_φ^0

$$|m_x^0| + t/3 \leqq 1, \tag{28}$$

$$|n_\varphi^0| + t/4 \leqq 1, \tag{29}$$

resulting from the statical theorem in accordance with (25). The maximum principle of Pontragin [7] is here applicable. Consider a Hamiltonian

$$H = \psi_1 q_x^0 - \psi_2 n_\varphi^0 \tag{30}$$

and system of eqs. conjugate with (26)

$$d\psi_1/\mathrm{d}x = 0, \qquad d\psi_2/\mathrm{d}x = -\psi_1. \tag{31}$$

According to the maximum principle, the optimum control n_x^0 associated with the maximization of the load parameter should render the Hamiltonian (30) an absolute minimum. Thus, on account of the constraint (29)

$$n_\varphi^0 = (1 - t/4)\,\mathrm{sign}\,\psi_2. \tag{32}$$

Firstly a solution may be found disobeying the constraint (28) and the limits of its applicability may be determined. Substituting (32) into (26) and using (31), a system of four differential equations with the boundary conditions (27) is arrived at and the continuity conditions for the phase coordinates q_x^0, m_x^0 is obtained. The solution yields

$$n_\varphi^0 = -(1 - t/4), \quad p = (1 - t/4)l \tag{33}$$

for $l < \sqrt{2b}$ where

$$b = \frac{1 - t/3}{1 - t/4}.$$

In this range of length of the shell the condition (28) is satisfied as an inequality and does not affect the obtained results. The mechanism of incremental collapse is that of uniform compression. As the length of the shell increases the condition (28) must be taken into account as a constraint on the phase coordinate $m_x^0(x)$. The Pontragin principle for the system with constrained phase coordinates leads in this case to the results:

$$\text{for } \sqrt{2b} \leqq l \leqq 4\sqrt{b} \qquad p = (1 - t/4)(2\sqrt{0.5l^2 + b} - l) \tag{34}$$

$$n_\varphi^0 = 1 - t/4 \quad \text{for} \quad 0 \leqq x \leqq a, \; n_\varphi^0 = -(1 - t/4) \quad \text{for} \quad a < x \leqq l,$$

$$a = l - \sqrt{0.5l^2 + b}.$$

In this situation a plastic hinge is found to form in the middle of the shell ($|m_x^0| = 1$) and for $x = a$ the hoop strain increments per cycle are changing sign.

Finally, for $l > 4\sqrt{b}$

$$p = 2\sqrt{(1 - t/4)(1 - t/3)}. \tag{35}$$

Here three plastic hinges are formed along the shell, the circumferential strain increments being negative between every two of them.

Results of the above calculations are shown in Fig. 4 with solid lines. The same results were obtained applying the kinematical theorem of shakedown. A similar problem for an infinitely long cylinder was solved in [2] in which the upper bound to the load parameter p was found by the kinematical approach. This value coincides with the exact result (35).

For low intensities of alternating load the fictitious yield surface shown in Fig. 3 with solid lines can be better approximated by the inequalities

$$|m_x^0| \leqq 1 - 2t/3, \; |n_\varphi^0| + 0.5|m_x^0| \leqq 1 - t/4. \tag{37}$$

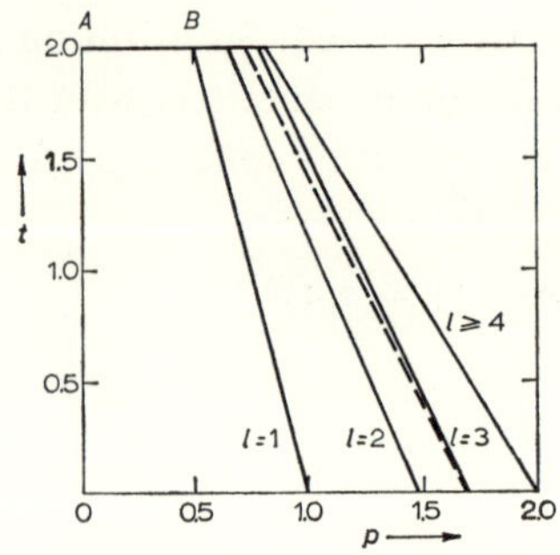

Fig. 4. Shakedown diagram for shell.

The latter indicates that the constraints on the magnitudes of the control and the phase coordinates are not independent of each other. In this case the optimum system theory leads to a system of nonlinear differential equations [8] the solution of which is very time consuming. Hence the methods of linear or convex programming appears to be convenient for the study of the incremental collapse. Those methods are being widely applied to the problems of limit analysis and the appropriate procedures will not be discussed in detail. It will only be mentioned that the mathematical programming employs a discretized description of the behaviour of a structure which entails some difficulties in the assessment of the accuracy of calculations, unlike in the mathematical theory of optimum processes. Specifically, in the above considered example the conditions for the incremental collapse can be found by means of linear programming provided the constraints (28), (29) or (37) are imposed only in a finite number of the cross-sections of the shell, $x = x_1, x_2, \ldots, x_n$, and the differential equilibrium equations (27) are replaced by a corresponding set of linear algebraic equations. The results of calculations made by means of the simplex programme are shown for the shell with the length $l > 4\sqrt{b}$ in Fig. 4 by a dashed line, the constraints (37) being employed.

To complete the shakedown analysis of the cylindrical shell, conditions necessary for alternating plasticity may be determined. The fictitious yield surfaces (23) are not degenerated at any point of the shell, i.e. no alternating flow sets in, if $t < 2$. For $t = 2$ fictitious yield surfaces do degenerate at the inner and outer surface of the shell, $|z| = h$.

Approximate methods of solving the shakedown problems

Similarly as in the problems of limit analysis, the statical and the kinematical theorems of the shakedown theory can be employed to bound the magnitudes of the load multipliers. The results appear to be fully applicable to a large number of practical problems leading often to the exact, complete solution by satisfying both the statical and the kinematical requirements.

Some of the methods of shakedown analysis will now be illustrated by means of an example. Consider a circular plate with constant thickness (Fig. 5) subjected to a time-independent, concentrated force P_1 and a variable, radial edge load

$$-P_2 \leqq P_2(\tau) \leqq P_2. \tag{38}$$

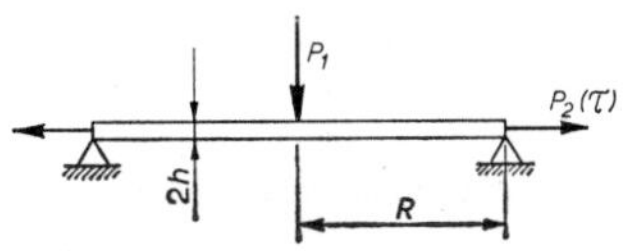

Fig. 5. Circular plate.

The problem is assumed to be geometrically linear and the Tresca yield condition (11) is adopted. On account of (8), the fictitious yield surfaces – identical for all points throughout the plate – are expressed by

$$\max\left(|\sigma^0_{r*}| + p_2\sigma_0, |\sigma^0_{\varphi*}| + p_2\sigma_0, |\sigma^0_{r*} - \sigma^0_{\varphi*}|\right) - \sigma_0 = 0,$$

$$(p_2 = P_2/2\sigma_0 h). \tag{39}$$

To formulate conditions for the incremental collapse, the statical method will be used first.

To begin with, the fictitious yield surface will be found in the space of generalized stresses. On assuming that no work per cycle is done by the load $P_2(\tau)$ (which amounts to saying that no plastic deformation develops in the middle surface of the plate), the equation (19) together with the conditions (15) and (38) results in

$$\max\left(|M^0_{r*}| + p_2\sigma_0 h^2, |M^0_{\varphi*}| + p_2\sigma_0 h^2, |M^0_{r*} - M^0_{\varphi*}|\right) - \sigma_0 h^2 = 0 \tag{40}$$

where M^0_{r*}, $M^0_{\varphi*}$ denote the radial and the circumferential bending moments, respectively, belonging to the fictitious yield surface. Now, such a relation between the limiting values of the parameters of loads applied should be found at which the distribution of the bending moments M_r^0, M_φ^0 belongs to the region bounded by the surface (40) and the equilibrium equations

$$\frac{\mathrm{d}}{\mathrm{d}r}(rM_r^0) - M_\varphi^0 + p_1\sigma_0 h^2 = 0, \quad M_r^0(R) = 0, \quad \left(p_1 = \frac{P_1}{2\pi\sigma_0 h^2}\right) \tag{41}$$

have to be satisfied at the same time.

Bearing in mind (40), assume the circumferential bending moments to obey

$$M_\varphi^0 = \sigma_0 h^2(1 - p_2). \tag{42}$$

Equilibrium condition (41) supplies the radial bending moment

$$M_r^0 = [p_1 - (1 - p_2)]\frac{R - r}{r}\sigma_0 h^2. \tag{43}$$

It is readily seen that the bending moments will belong to the region (40) only if

$$p_1 = 1 - p_2. \tag{44}$$

Thus the required lower bound relationship between the load parameters is found.

To find an upper bound, the kinematical method will now be applied. Consistent with the above, the collapse mechanism is assumed to be of a pure bending type, i.e. the plastic elongations of the middle surface being neglected. Thus, as in limit analysis of a plate with no in-plane loading [9]

$$\Delta w = \Delta a\left(1 - \frac{r}{R}\right). \tag{45}$$

In accordance with (18), the increments of radial and circumferential strains are related to the curvature increments

$$\Delta\varepsilon_r'' = z\Delta x_r, \qquad \Delta\varepsilon_\varphi'' = z\Delta x_\varphi. \tag{46}$$

The formulae of the type (4) have here the form

$$\Delta x_r = -\frac{\mathrm{d}^2(\Delta w)}{\mathrm{d}r^2}, \qquad \Delta x_\varphi = -\frac{1}{r}\frac{\mathrm{d}(\Delta w)}{\mathrm{d}r} \tag{47}$$

hence

$$\Delta\varepsilon_r'' = 0, \qquad \Delta\varepsilon_\varphi'' = z\frac{\Delta a}{Rr}. \tag{48}$$

On account of the associated flow law (15), the strain increment vector (48) corresponds to the stresses on the fictitious yield surface (39)

$$\begin{aligned} \sigma^0_{\varphi *} &= \sigma_0(1 - p_2) \quad \text{for} \quad z > 0; \\ \sigma^0_{\varphi *} &= -\sigma_0(1 - p_2) \quad \text{for} \quad z < 0. \end{aligned} \tag{49}$$

Thus the inequality (16) enables the limiting condition to be found in the form

$$P_1 \Delta a = 4\pi \int_0^R r\,\mathrm{d}r \int_0^h \sigma_0(1 - p_2) z \frac{\Delta a}{Rr}\,\mathrm{d}z. \tag{50}$$

The coincidence of both upper and lower bound means that the obtained solution is complete. Note that in the application of the kinematical method to the plate it was not necessary to introduce the generalized stresses.

The obtained solution holds only in the absence of local alternating flow. Suitable constraint can be found from the condition for degeneration of the fictitious yield surface (39); this yields $p_2 = 1$.

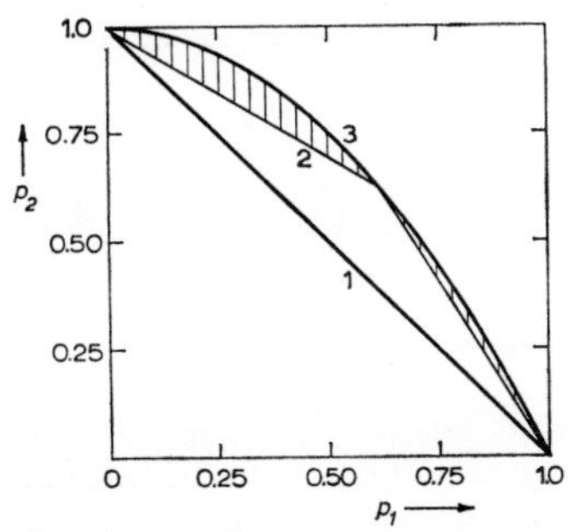

Fig. 6. Shakedown diagram for plate.

The diagram of shakedown of the plate considered is shown in Fig. 6. The straight line 1 corresponds to (44). It seperates the shakedown region from the region of incremental collapse. The lines 2 and 3, taken from [9], show the lower and upper bound results for an 'instantaneous' collapse.

A decrease in the limit load of the plate is clearly seen as a result of the variation of one of the loads.

References

[1] Koiter, W. T., General theorems for elastic-plastic solids. Progress in solid mechanics, *1*, I. N. Sneddon and R. Hill Editors, North-Holland, Amsterdam, 1960.

[2] Gokhfeld, D. A., Some problems of the shakedown theory for plates and shells, in Russian. Proceedings of VI All-Union Conference on theory of shells and plates, (Baku 1966) pp. 284–292, Nauka, Moscow 1966.

[3] König, J. A., Shakedown theory of plates. Archiwum Mechaniki Stosowanej, *2* (1969) 623–637.

[4] Rosenblum, V. I., On shakedown analysis of nonuniformly heated elastic-plastic bodies, in Russian. Zh. Prikl. Mekh. Tekh. Fiz. (PMTF), No. 5 (1965) 98–102.

[5] Gokhfeld, D. A., Carrying capacity of structures under variable thermal conditions, in Russian. Mashinostroenje, Moscow, 1970.

[6] Sawczuk, A., Evaluation of upper bounds to shakedown loads for shells. *Journal Mech. Phys. Solids*, *17* (1969) 291–301.

[7] Pontragin, L. S., V. G. Boltiansky, P. M. Gamkrelidze, E. F. Mischenko, Mathematical theory of optimum processes, (in Russian). Nauka, Moscow 1965.

[8] Cherniavsky, O. F., Solutions of limit equilibrium and shakedown problems by means of Pontragin's maximum principle, (in Russian). Mekh. Tverd. Tela (MTT), No. 4 (1970) 99–105.

[9] Hodge, Ph. G., Limit analysis of rotationally symmetric plates and shells, Prentice-Hall, Englewood Cliffs, N.J. 1963.

Some experimental investigations of the elastic-plastic wave propagation in bars

J. Klepaczko

Institute of Fundamental Technological Research, Warsaw, Poland

Introduction

Although a considerable amount of experimental information was reported in the literature on the subject of plastic wave propagation in metallic bars, there are still some uncertainties left as to the use of rate-dependent or rate-independent constitutive equations in the analysis. The simplest analytical approach assumes a rate-independent dynamic strain hardening curve which lies above the static one. Experimental and theoretical work has been done to obtain a shape of the wave front by assuming certain constitutive equations for the material. Few results have been published as the material behaviour as deduced from the measurements of the wave front.

The purpose of this paper is to describe some experimental results and methods leading to a dynamic strain-hardening curve based on observations of the wave front. A procedure of algorithms combined with error analysis is proposed, which appears to uncover some additional facts unknown so far in the elastic-plastic stress wave analysis.

There are essentially two groups of wave experiments. In the first group no initial preloading is applied to a specimen whereas in the second group, which is more frequently reported in recent papers, incremental loading is employed such that the incremental wave is superimposed on the initial plastic state of a specimen. The technique of incremental loading appears to be a valuable tool in the study of the dynamic plasticity of materials. There are three types of incremental experiments, each of them differs from the others by the initial state of specimen at the instant of incremental loading, which can be specified as follows:

i. creep state, $\sigma_w = \text{const.}$, $\varepsilon_w \neq 0$, $\dot{\varepsilon}_w \neq 0$, $\ddot{\varepsilon}_w < 0$,
ii. relaxation state, $\varepsilon_w = \text{const.}$, $\sigma_w \neq 0$, $\dot{\sigma}_w \neq 0$, $\ddot{\sigma}_w < 0$,
iii. slow active loading, $\dot{\varepsilon}_w = \text{const.}$

where σ and ε denote stress and strain, respectively, subscript w refers to the initial state of specimen prior to incremental loading, and a dot implies the time derivative. If the specimen material is strain rate sensitive, the precursor wave travelling along the specimen at the elastic wave speed would be different for each of the initial states. The incremental experiments may be performed for three types of one-dimensional stress state, i.e. tension, compression or pure shear (torsion of a tubular specimen). Recent experimental results [1, 2] have shown that the lateral inertia effects associated with the stress wave propagation in bars are negligible. This statement is true for the relatively long rise times which are usually encountered in experimental conditions, i.e. duration of about 20 to 40 μs.

On the other hand, it was also recently shown, [1] and [3], that a good theoretical description of the wave propagation process in a bar may be obtained, when the TKR[1])-theory is applied, using the strain rate as a parameter in the rate sensitive constitutive equation. The analysis of earlier results of plastic wave propagation tests showed that the strain rate variations at the wave front do not differ very much in the $\{\varepsilon, \log \dot{\varepsilon}\}$ coordinates from the maximum value of the strain rate $\dot{\varepsilon}_{max}$, [3]. It was also found experimentally that the maximum strain rate $\dot{\varepsilon}_{max}$ at the wave front changes less than one order of magnitude during its travel along the bar [3]. These facts confirm the first simplifying assumption that the strain rate may be a parameter in the constitutive equation.

Analysis

Serious difficulties are encountered in recording dynamic stress-strain measurements. Many experimental methods have to been proposed. Relatively early the possibility was shown that the elastic-plastic wave propagation experiment can be applied to this purpose, [5] and [6]. All experimental methods associated with wave propagation experiments may be divided into two groups. The first one is based on a distribution of the plastic deformation along a specimen which is measured after the test and makes it possible to derive a dynamic strain-hardening curve. The second method is based on immediate measurements of the elastic-plastic wave front as it travels along a specimen.

[1]) TKR: Taylor – Kármán – Rakhmatulin elastic-plastic stress wave propagation theory, see for example review in Ref. [4].

Theoretical considerations as well as experimental practice indicate which quantities can be measured experimentally or obtained in the analytical way for the wave propagation process. The following quantities are basically involved:

i. displacement U,

$$U(x, t), \qquad \text{for} \quad x = \text{const} \qquad U(x, t)_x; \tag{1}$$

ii. mass velocity v,

$$v(x, t), \qquad \text{for} \quad x = \text{const} \qquad v(x, t)_x \tag{2}$$

where

$$v(x, t)_x = \left(\frac{\partial U}{\partial t}\right)_x;$$

iii. acceleration a,

$$a(x, t), \qquad \text{for} \quad x = \text{const} \qquad a(x, t)_x \tag{3}$$

where

$$a(x, t)_x = \left(\frac{\partial v}{\partial t}\right)_x = \left(\frac{\partial^2 U}{\partial t^2}\right)_x;$$

iv. strain ε,

$$\varepsilon(x, t), \qquad \text{for} \quad x = \text{const} \qquad \varepsilon(x, t)_x \tag{4}$$

where

$$\varepsilon(x, t)_x = \left(\frac{\partial U}{\partial x}\right)_{\bar{x}}$$

and $\bar{x}$ denotes an average coordinate at the strain measurements

$$\bar{x} = \frac{x_1 + x_2}{2} = x_1 + \frac{\Delta x}{2}$$

where, in turn, Δx denotes the base of an extensometer or a strain gage,

$$\Delta x = x_2 - x_1;$$

v. strain rate $\dot{\varepsilon}$,

$$\dot{\varepsilon}(x, t), \qquad \text{for} \quad x = \text{const} \qquad \dot{\varepsilon}(x, t)_x \tag{5}$$

where

$$\dot{\varepsilon}(x, t)_x = \left(\frac{\partial^2 U}{\partial x \partial t}\right)_{\bar{x}}.$$

In addition, a basic measurable quantity is the wave propagation velocity

$$C = \frac{dx}{dt}. \tag{6}$$

It can be seen from the above listed quantities that it is the displacement $U(x, t)$ which is the most important measurable quantity. It is measured at constant distances x from the impact end of a bar, $U(x, t)_x$. According to the TKR-description, the displacement $U(x, t)$ is identified with the solution of the wave equation

$$\frac{\partial^2 U}{\partial t^2} = C^2(\varepsilon) \frac{\partial^2 U}{\partial x^2} \tag{7}$$

which is solved by the method of characteristics,

$$C(\varepsilon) = \left(\frac{1}{\rho} \frac{d\sigma}{d\varepsilon}\right)^{\frac{1}{2}} \quad \text{and} \quad d\left[\int_0^{\varepsilon_0} C(\varepsilon)\, d\varepsilon \pm v\right] = 0. \tag{8}$$

However, in many cases it is the mass velocity $v(x, t)_x$ or the deformation $\varepsilon(x, t)_x$ at different distances x from the impact end which are measured.

Deformation measurements $\varepsilon(x, t)_x$ at different distances x (case iv) lead to a relatively simple way of determining the dynamic strain-hardening curve of a metal. Such a method was put forward in [1] and [5], and can be applied both in normal and incremental tests. Having recorded the $\varepsilon(x, t)_x$ at several positions, the current tangent modulus can be obtained numerically, and equation (8) can be integrated to get σ. Although the method was proposed as early as in 1952 it has not so far been applied, probably due to measuring difficulties. The results obtained in [5] for copper are not conclusive because of substantial errors in measurements which proved to be much greater than a possible dynamic overstress in copper.

If strain gages are applied at two distances from the impact end $\Delta x = x_2 - x_1$, and time elapsing for the wave passage of the particular strain level ε is $\Delta t = t_2 - t_1$, then

$$C_\varepsilon = \left(\frac{x_2 - x_1}{t_2 - t_1}\right)_\varepsilon, \qquad \Delta x = x_2 - x_1 = \text{const.} \tag{9}$$

The method discussed above can be summarized to form the following algorithm

$$\varepsilon(x,t)_x\Big|_e \xrightarrow[①]{m} C(\varepsilon)_\varepsilon = \left(\frac{x_n - x_{n-1}}{t_n - t_{n-1}}\right)_\varepsilon \xrightarrow[②]{n} C^2(\varepsilon)_\varepsilon \xrightarrow[④]{n} \sigma = \rho \int_0^{\varepsilon_0} C^2(\varepsilon)\,\mathrm{d}\varepsilon$$

$$\hookrightarrow \xrightarrow[③]{n} \varepsilon = C_\varepsilon^{-1}\left(\frac{x}{t}\right) \tag{10}$$

where $|_e$ means the experimental data, i.e. $\varepsilon(x, t)_x$, an arrow with an encircled number denotes a successive operation, m and n denote measuring and numerical operations, respectively. Since the TKR – wave propagation theory is assumed to hold, operation ③ is possible, and the deformation can be calculated from the equation $(x/t) = C(\varepsilon)_\varepsilon$ or by inversion

$$\varepsilon = C^{-1}\left(\frac{x}{t}\right).$$

In this way the dynamic stress-strain relations can be determined.

A similar method has been proposed in [7], but with the measurements of the mass velocity $v(x, t)_x$. Experiments were performed on annealed aluminium with little or even negligible magnitudes of the dynamic overstress. In other words, the difference between static and dynamic strain-hardening curves was very small. Since from the TKR-theory the following expressions can be derived

$$v = \int_0^{\varepsilon_0} C(\varepsilon)\,\mathrm{d}\varepsilon \quad \text{or} \quad \frac{\mathrm{d}v}{\mathrm{d}\varepsilon} = C(\varepsilon),$$

$$\varepsilon = \int_0^{v_0} \frac{\mathrm{d}v}{C(v)}, \tag{11}$$

the respective algorithm takes the form

$$v(x,t)_x\Big|_e \xrightarrow[①]{m} C(v)_v = \left(\frac{x_n - x_{n-1}}{t_n - t_{n-1}}\right)_v \xrightarrow[②]{n} \varepsilon(v) =$$

$$= \int_0^{v_0} \frac{\mathrm{d}v}{C(v)} \xrightarrow[③]{n} C(\varepsilon) \xrightarrow[⑤]{n} C^2(\varepsilon) \xrightarrow[⑥]{n} \sigma = \rho \int_0^{\varepsilon_0} C^2(\varepsilon)\,\mathrm{d}\varepsilon$$

$$\hookrightarrow \xrightarrow[④]{n} \varepsilon = C^{-1}\left(\frac{x}{t}\right). \tag{12}$$

Finally, the third method, based on the displacement measurements $U(x, t)_x$, can also be shown in the algorithmic form. In this case more than one method can be proposed, but the simplest solution is to differentiate the $U(s, t)_x$ data, i.e.

$$U(x, t)_x \Big|_e \xrightarrow[①]{n} v(x, t)_x = \frac{\partial U(x, t)_x}{\partial t} \xrightarrow[②]{n} \tag{13}$$

and to proceed according to formula (12). Clearly a numerical differentiation is necessary.

Since the accuracy of the wave measurements is usually insufficient for several mathematical operations to follow in the derivation of algorithms and for the calculation of reliable dynamic stress-strain relations, error analysis seems to be of importance. Error analysis [8] for resistance strain gage technique has shown that for algorithm (10) the relative stress error can reach a value from 20 to 38 per cent. The analysis was limited to the base of a strain gage $l_0 = 3$ mm and to the maximum strain level $\varepsilon = 2$ per cent. No initial preloading was considered. It can be shown that smaller relative errors can be expected for the incremental experiments. This is the reason why only the experiments of the incremental type were performed. The algorithm (10) can be applied also for the incremental experiments, but at an instant of dynamic loading the initial point of the static strain-hardening curve $(\sigma_w, \varepsilon_w)_{\varepsilon_w}$ on the $\{\sigma, \varepsilon\}$ plane must be known, and in this case

$$\varepsilon = \varepsilon_w + \varepsilon_d \quad \text{or} \quad \varepsilon = \varepsilon_w + C^{-1}\left(\frac{x}{t}\right) \tag{14}$$

and

$$\sigma = \sigma_w + \sigma_d \quad \text{or} \quad \sigma = \sigma_w + \rho \int_0^{\varepsilon_d} C^2(\varepsilon)\, \mathrm{d}\varepsilon$$

the subscript d denotes the incremental loading process.

Results of experiments

The results of experiments on the wave propagation process in long specimens exposed to preloading at a small constant strain rate were obtained for annealed aluminium, annealed copper and deep drawing steel. The general arrangement as well as the detailed description of the apparatus employed is given elsewhere, [8]. It may be mentioned here that the specimen about one meter in length was loaded from one end at a constant static velocity. The

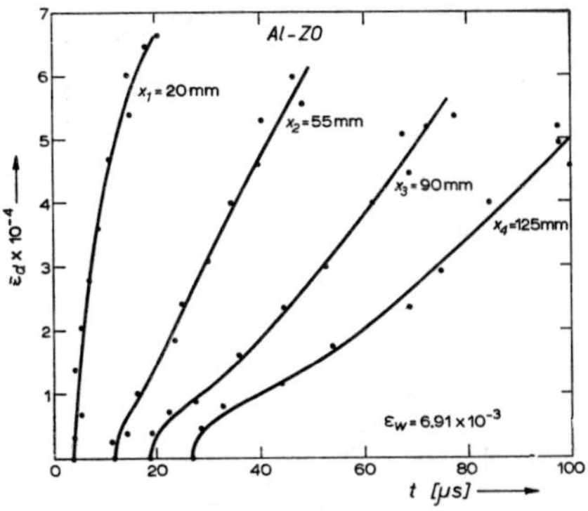

Fig. 1. The incremental wave profiles for annealed aluminium specimen.

other end of the specimen was loaded dynamically by means of the rotating hammer. Specimens used had a rectangular cross section of 10 mm in width and about 1 mm in thickness. The strains increments as a function of time were measured at four places along the specimen by recording the output of four strain gages. The wave profiles $\varepsilon_d(x, t)_x$ for four strain gage positions and prestrain $\varepsilon_w \approx 6{,}91 \times 10^{-3}$ in annealed aluminium specimen are shown in Fig. 1. Similar $\varepsilon_d(x, t)_x$ experimental data for annealed copper and deep drawing steel have been obtained. In all cases considered the strain rate of initial static preloading was $\dot{\varepsilon} \approx 10^{-3}\, s^{-1}$. According to the algorithm (10) the values of the incremental wave speed $C(\varepsilon_d)$ for constant values of ε_d were obtained. The following relations were used

$$C_n = \frac{x_{n+1} - x_n}{t_{n+1} - t_n}$$

where $n = 1, 2, 3$, $\varepsilon_d = $ const., and the average value of C at a constant level of strain were found

$$C = \frac{1}{n} \sum_{i=1}^{n} \frac{x_{i+1} - x_i}{t_{i+1} - x_i}$$

where $n = 1, 2, 3$, $\varepsilon_d = $ const., and $n + 1$ denotes the position of the strain gage as regarded from the impact end of the specimen. Tests on annealed aluminium at three initial strains gave similar results. For two values of prestraining the results are shown in Fig. 2. The elastic wave speed C_0 is shown by broken lines. In addition, values of the wave speed which are derived from the TKR-theory with no preloading, $\varepsilon_w = 0$, are shown in the

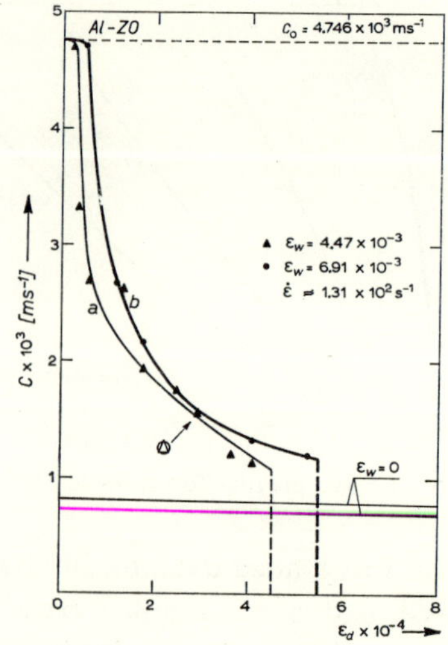

Fig. 2. The incremental wave velocity C for aluminium as a function of the incremental strain ε_d; a) initial strain $\varepsilon_w = 4.47 \times 10^{-3}$, average strain rate at the wave front $\dot{\varepsilon} \approx 1.31 \times 10^2\ s^{-1}$; b) initial strain $\varepsilon_w = 6.91 \times 10^{-3}$, average strain rate at the wave front $\dot{\varepsilon} \approx 1.31 \times 10^2\ s^{-1}$.

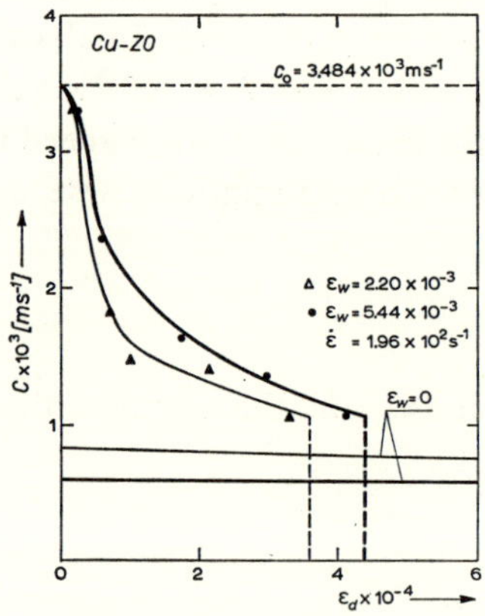

Fig. 3. The incremental wave velocity C for copper; a) $\varepsilon_w = 2.20 \times 10^{-3}$, $\dot{\varepsilon} \approx 1.98 \times 10^2\ s^{-1}$; b) $\varepsilon_w = 5.44 \times 10^{-3}$, $\dot{\varepsilon} \approx 1.98 \times 10^2\ s^{-1}$.

figure. It is seen that over some time the material responds elastically and precursor travels at the elastic wave speed C_0. For higher values of additional strains the wave speed diminishes rapidly, but these speeds are still higher than those obtained from the TKR-theory with no preloading, $\varepsilon_w = 0$.

Quite similar results we obtained for annealed copper, as shown in Fig. 3. Fig. 4 shows the incremental wave velocities $C(\varepsilon_d)$ obtained in tests on the deep drawing steel specimens. In comparison with the previous results on aluminium and copper, the shape of the $C(\varepsilon_d)$ function is seen to be quite different. The material responsc is mostly elastic, i.e. precursor magnitude is relatively high. However, the speed of precursor is a little lower than the elastic wave speed C_0 for the annealed material. This phenomenon is associated with a decrease of the Young modulus due to initial plastic deformation, as it was observed during the static tests, [8], [9]. For higher strains a very sharp decrease of the incremental wave speed is observed. In Fig. 4 the wave speed predicted by the TKR-theory can also be recognized for the delivery state of the material, $\varepsilon_w = 0$. These values are very low in comparison with the elastic wave speed C_0.

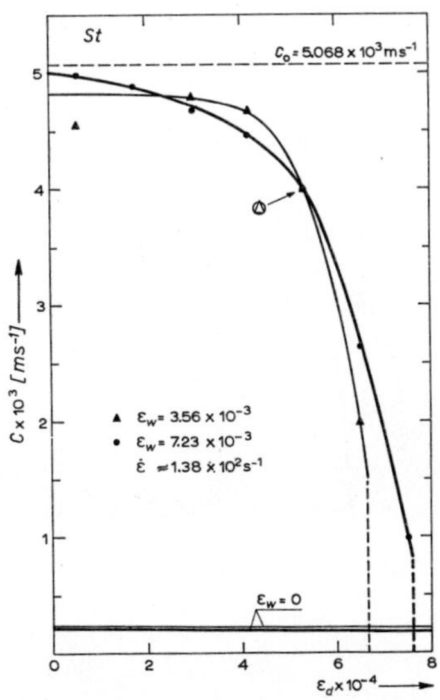

Fig. 4. The incremental wave velocity C for deep drawing steel; a) $\varepsilon_w = 3.56 \times 10^{-3}$, $\dot{\varepsilon} \approx 1.38 \times 10^2\ s^{-1}$; b) $\varepsilon_w = 7.23 \times 10^{-3}$, $\dot{\varepsilon} \approx 1.38 \times 10^2\ s^{-1}$.

Using the results $C(\varepsilon_d)$, the numerical operation from ② to ④ was performed following the algorithm (10) and the dynamic incremental stress-strain curves were obtained for all of the three materials tested. The results presented in Figs. 5, 6 and 7 show the incremental stress-strain relations for aluminium, copper and deep drawing steel which are superimposed on the

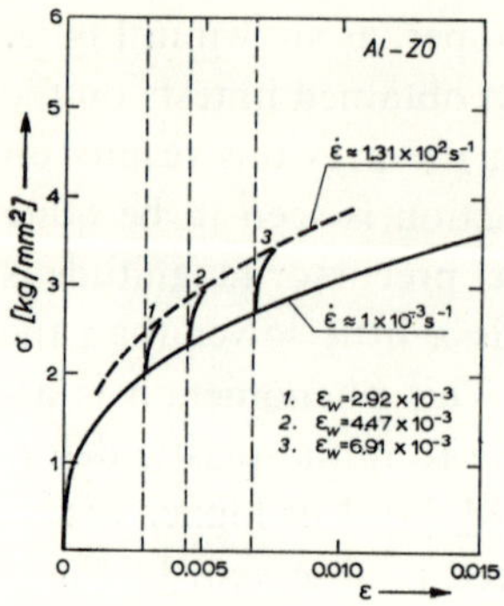

Fig. 5.

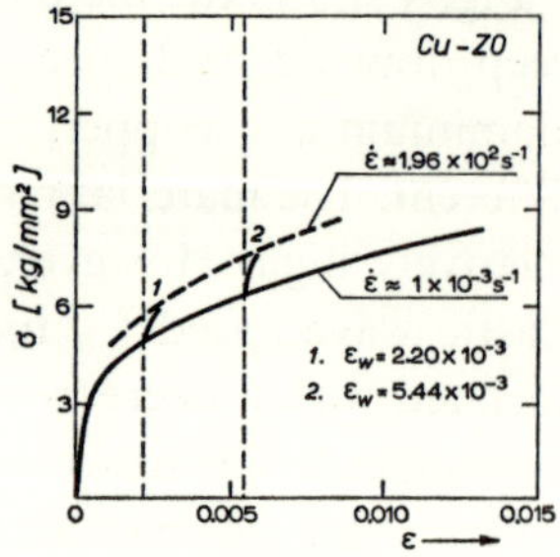

Fig. 6.

Fig. 5. The incremental stress-strain relations for annealed aluminium, superimposed on the quasi-static strain hardening curve.

Fig. 6. The incremental stress-strain relations for annealed copper, superimposed on the quasi-static strain hardening curve.

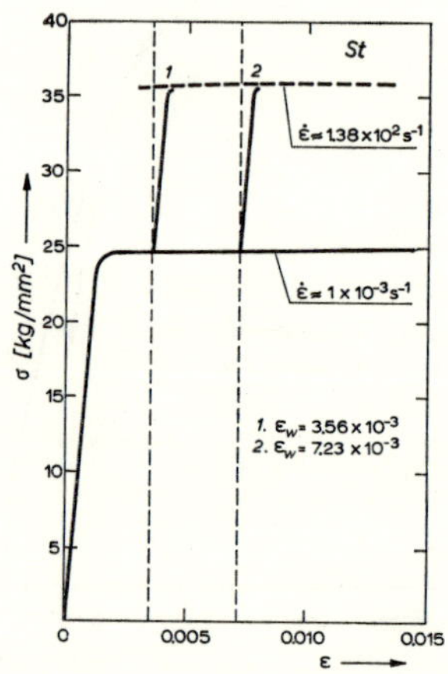

Fig. 7. Two incremental stress-strain relations for deep drawing steel, curves derived from the algorithm (10) and experimental data of Fig. 4 .The incremental curves are superimposed on the quasi-static relationship.

quasi-static curves. For aluminium and copper a gradual transition can be seen to take place from a quasi-static to dynamic curve. This is not observed for deep drawing steel. For aluminium and copper the initial slope of the incremental stress-strain curve was found to be roughly equal to that corresponding to the Young modulus. For steel the transition from the quasi-static to dynamic stress-strain relation appears rather abrupt.

Discussion and conclusions

Observation that the initial part of the incremental wave front, i.e. the precursor, travels along a specimen with the elastic wave speed C_0 confirms the existence of the elastic material response. This fact was reported as early as in 1953, [10] and encountered several times in tension or compression as well as in torsion tests (shear incremental waves). Thus, it cannot be attributed to the lateral inertia. Recently, this feature was observed for aluminium, copper and mild steel at the shear conditions [11]. Only a negligible decrease in the elastic wave speed occurred in copper as the prestrain was increased. In this situation the constitutive laws of materials tested so far in the wave conditions, mostly aluminium, copper and mild steels, are strain rate dependent.

The proposed method to apply the measured wave velocities for obtaining the incremental stress-strain relations works rather well, in the sense that the expected errors are relatively small. It was found that in aluminium and copper the transition from the quasi-static stress-strain relation to a dynamic is asymptonic. A similar result has recently been obtained for copper through application of the algorithm (10), but incremental experiments were performed from the creep state at the initial loading [12].

The above mentioned behaviour indicates that the strain rate history is significant in the elastic-plastic wave propagation in aluminium or copper bars. These effects were reported previously on the basis of quasi-static experiments, see [13] and [14]. A further discussion of this effect, in the light of the incremental wave experiment, can be found in [15].

The following conclusions can be drawn:

i. The incremental wave experiments must be performed exclusively from the initial active loading, and neither creep nor relaxation state.

ii. The proposed method to obtain incremental stress-strain dynamic relations from the wave speed measurements appears suitable for the purpose and expected measuring errors do not affect results appreciably.

iii. The incremental wave experiments for annealed aluminium and copper as well as for deep drawing steel have shown that the precursor travels at the elastic wave speed C_0.

iv. The strain rate history effects are significant in the elastic-plastic wave propagation in bars made of aluminium and copper.

References

[1] Convery, E. and H. L. D. Pugh, Velocity of torsional waves in metals stressed statically into plastic range. *Journ. Mech. Engng. Science 10* (1968) 153.

[2] Yew, C. H. and H. A. Richardson, The strain-rate effect and the incremental plastic wave in copper. *Exp. Mech. 9* (1968) 366.

[3] Klepaczko, J., Analysis of the elasto-plastic wave front in a bar (in Russian). *Mekhanika, 22* (1971) 112.

[4] Simmons, J. A., F. Hauser and J. E. Dorn, *Mathematical theories of plastic deformation under impulsive loading*. Univ. of Calif. Publ. in Engineering, Berkeley, 1962.

[5] Campbell, W. R., Determination of dynamic stress-strain curves from strain waves in long bars. *Proc. SESA, 10* (1952) 113.

[6] Schestopalov, L. Ml., *Deformation of metals and plastic waves* (in Russian), AN-SSSR, Moscow, 1958.

[7] Efron, L. and L. E. Malvern, Electromagnetic velocity transducer studies of plastic waves in aluminium bars. *Exp. Mech. 9* (1969) 255.

[8] Klepaczko, J., Properties of metals and the acceleration waves at the dynamic loadings (in Polish). *Institute of Fundamental Technological Research Reports*, No. 61, 1970.

[9] Litoński, J. and J. Klepaczko, Influence of the initial plastic deformation on the young modulus in brass and low carbon steel (in Polish). *Rozprawy Inzynierskie 12* (1964) 251.

[10] Sternglass, E. J. and D. A. Stuart, A experimental study of the propagation of transient longitudinal deformations in elastoplastic media. *Journ. Appl. Mech. 20* (1953) 427.

[11] Campbell, J. D. and A. R. Dowling, The dynamic behaviour of materials subjected to dynamic incremental shear loading. *Journ. Mech. Phys. of Solids 18* (1970) 43.

[12] Santosham, T. V. and H. Ramsey, Small plastic strain wave propagation in prestressed soft copper rods. *Int. Journ. Mech. Sci. 12* (1970) 447.

[13] Klepaczko, J., Effects of strain-rate history on the strain hardening curve of aluminium. *Arch. Mech. Stos. 19* (1967) 211.

[14] Klepaczko, J., Strain-rate history effects for polycrystalline aluminium and theory of intersections. *Journ. Mech. Phys. of Solids 16* (1968) 255.

[15] Klepaczko, J., Discussion of effects developed by the incremental wave in a bar. *Archives of Applied Mech. 24* (1972) 187.

Deformation and fracture of structural metals under complex stress at low temperatures

A. A. Lebedyev and N. V. Novikov
Institute for Problems of Strength, Kiev, USSR

This work represents an experimental and theoretical investigation of theorems governing deformation and fracture of steel, grey pig iron, and aluminium alloy under conditions of plane stress at both normal temperatures and sub-zero temperatures down to −180°C. Tubular specimen were subjected to compressive or tensile axial load and internal pressure. The temperature was lowered using nitrogen. Low temperatures appear to cause a change in the shape of strain hardening curves $\sigma_i(\varepsilon_i)$ and a termination of proportionality between stress and strain deviators. In addition, bulging of the yield surface is accompanied by a change in its shape. The low-temperature behaviour of aluminium alloy and iron appears to be qualitatively the same as that at room temperature.

Introduction

Machine parts are subjected to mechanical loads and temperatures which can cause a complex stress state in the material.

Information about materials is, generally, confined to mechanical properties estimated from results of tests conducted in uniaxial tension, uniaxial compression or in pure shear. Because of this, three major problems arise which are studied in the present paper.

The first problem is associated with the development of criteria of equivalence for any arbitrary stress system, and with experimental verification of those criteria for a wide temperature range.

The second problem is linked to the formulation of a relationship between stress and strain. At room temperature, this problem is solved with some accuracy, by employing existent plasticity theories. In view of new experimental data it is necessary to show to what extend stress-strain relations remain valid under sub-zero temperature loading conditions.

The third problem arises in connection with brittle fracture under 'rigid' loading condition at small ratios[1]) of τ_{max}/σ_{max}. Even relatively ductile materials are known to fail in a brittle manner under uniform biaxial tension. Brittle fracture, associated with the above stress state, may increase with decreasing temperature.

Very few reliable experimental data exist regarding the effect of low temperature on deformation and fracture of metals under combined stress. Due to experimental difficulties, investigations of stress-strain relationships, as well as strain and strength of structural materials exposed to both temperature changes and mechanical loading are virtually nonexistent above the yield point.

A qualitative comparison of test results for carbon steels in uniaxial tension with those obtianed from tubular specimens and Charpy specimens exposed by biaxial and triaxial stress states respectively was made by Kirsch [1, 2]. The results show that the critical temperature of brittleness determined from results of Charpy specimens is higher than of uniaxial and biaxial specimens. On the other hand, the critical temperature for biaxial tension was found to be higher than that for uniaxial tension.

Interesting data on low-temperature strength of structural materials under combined stress were obtained by Kurkin [3] and Arone [4]. However, their experiments were conducted primarily to solve some particular problems, which predetermined the test procedures employed.

The data represented in this paper have been obtained in a study of hardening laws and yield criteria for carbon steel (C = 0,37%), grey pig iron of type Cч12–28 and a malleable aluminium alloy D16T (3,62%Si, 0,72%Mn) at sub-zero temperature down to $-180°C$.

Testing-technique

The experimental study was conducted using a set-up [5, 6] designed at the Institute for Problems of Strength. The set-up consisted of a hydraulic testing machine, a high-pressure unit, a vacuum chamber; a cooling and temperature-control systems and a strain measuring system.

A tubular specimen was subjected to an axial load, internal pressure and a torque. Load control was performed by hydromechanical governors.

The high-pressure unit consisted of a multipiston pump, a hydraulic amplifier, a separating chamber and control-measuring instruments. In order

[1]) Notation is given in Appendix.

to produce high pressure in a specimen at low temperatures petroleum ether and isopentane were used.

The hydraulic system made it possible to measure the axial load within an accuracy of 1,5 per cent. The accuracy of internal pressure measurements was 0,5 kG/cm^2.

The cooling system consisted of circulation cooler type placed inside of a tubular specimen and of a device controling circulation of liquid nitrogen to maintain the preserved temperature. A thermocouple fastened onto the sample was employed as a feed-back sensor. In order to obtain a uniform temperature distribution over the gauge length of the specimen, elastic plates were soldered to the cooler mantle free to slide on the internal surface of the specimen in the neibourhood of its end portions [7]. The number and location of plates were adjusted such that there was practically no temperature gradient over the gauge length of the specimen in excess of 2°C.

Longitudinal and lateral elongations as well as shear strains were measured by means of electro-mechanical strain gauges [8] with 25 mm measuring length and were automatically recorded. The order to increase the accancy of the measurements, the deformation range between 0 and 30 per cent was split up into three parts. Within each part the strains were registered with different sensitivity in order to ensure the required accuracy of $\pm 0{,}002$ mm.

The vacuum chamber kept the heat supply to the specimen at a low level and increased the accuracy of measurements of strain gauges fixed to the outer surface of the specimen.

Test results were processed in terms of true stresses and strains. The radial strain of the specimen was determined in an analytical way accounting for elastic volume changes.

Results and discussion

The most significant data on mechanical behavior of a material are the stress-strain curves. These curves represent an overall description of all processes which take place in a material subjected to external forces and thermal loading.

Over 100 curves were obtained for various ratios of the principal stresses at temperatures ranging from ambient temperature in to -180°C. The most typical results obtained in the first quadrant of the stress plane $\{\sigma_1, \sigma_2\}$ will be represented only here.

Primary stress-strain curves for the tested metals are shown in Fig. 1.

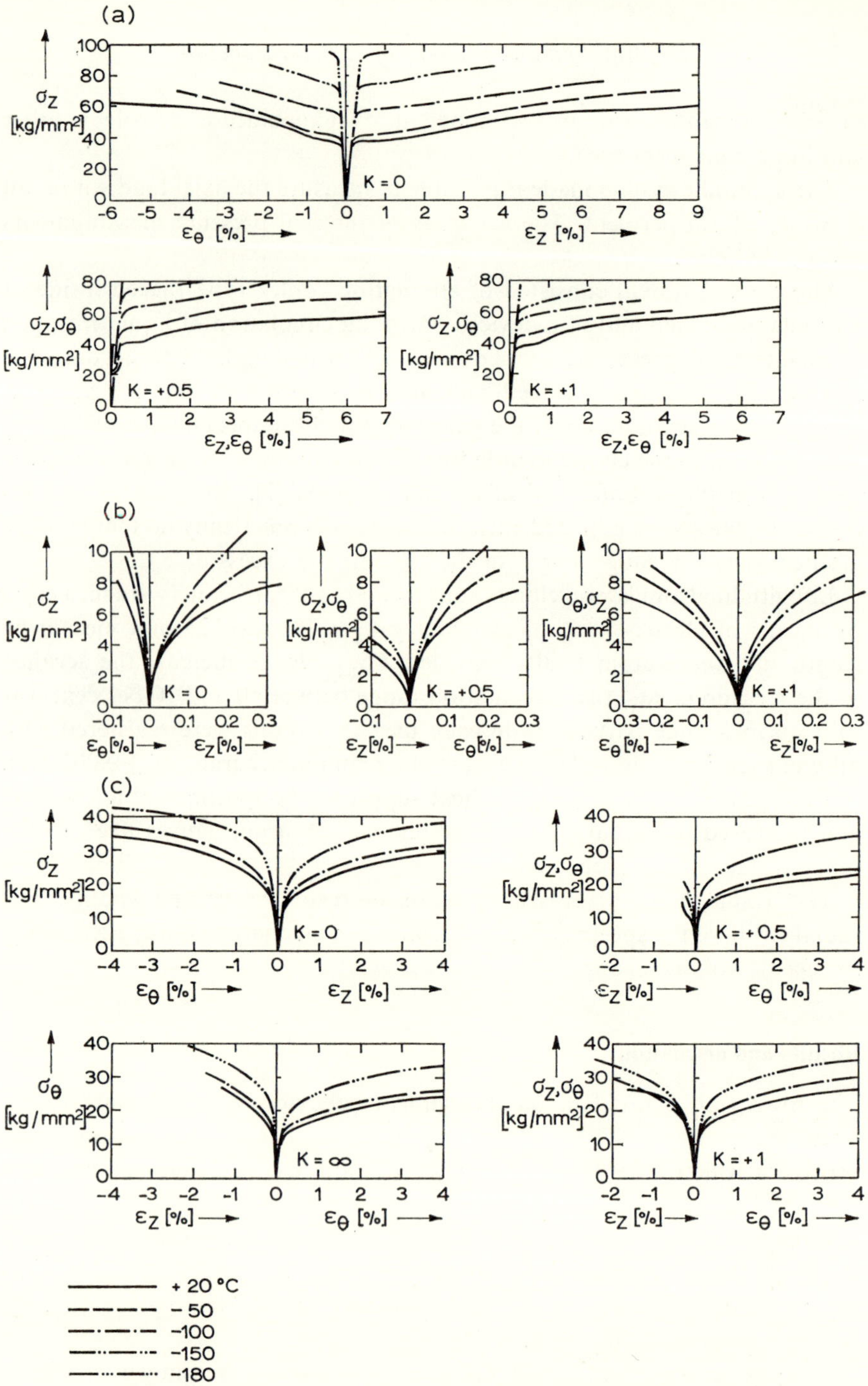

Fig. 1. Stress-strain curves for steel (a), cast iron (b), and alloy (c).

The majority of curves are plotted from test results obtained from two or three specimens.

Deformability and strength of carbon steel reveal a strong dependence on both temperature and type of the stress state as shown in Fig. 1a. As temperature decreases, yield stress and rupture stress show a substantial increase, whereas ductility decreases. Such a behaviour at lower temperature is observed for all considered principal stress ratios. The decrease in ductility at lower temperatures is more pronounced in biaxial tension. It is also found that the resistance to rupture of steel increases more appreciably with decreasing temperature in uniaxial compression than in uniaxial tension.

Stress-strain relationships for cast iron are practically nonlinear over the entire range given in Fig. 1b. The nonlinearity is maintained at sub-zero temperature, such that the portion of relatively small curvature is extended into the region of higher stresses. In addition, low-temperature hardening depends on the type of stress state, i.e. for an increase in the ratio $K^1 = \sigma_2/\sigma_1$ a 'fan' of curves is obtained at different temperatures at lower stresses. Moreover, the initial portions of stress-strain curves coincide in uniaxial tension, but not in biaxial tension.

Fig. 1c shows stress-strain curves for aluminium alloys. Here, a decrease in temperature is accompanied not only by an increase in strength, but also by an increase in the ductility of the alloy. Some change in anisotropy also takes place. This has been confirmed by results from tests conducted in uniform biaxial tension where higher rupture resistance and lower ductility are observed in the axial direction.

The effect of both temperature and stress state on metal ductility are shown in Fig. 2. The maximum value of effective strain is plotted versus temperature. The relations $\varepsilon_i = f(\times T°C)$ have been found to be almost linear at all principal stress ratios. For the metals investigated minimum ductility was observed in biaxial tension at the principal stress ratio $K = +0{,}5$, irrespective of temperature. The materials reveal higher ductility under combined tension-compression than in uniaxial tension.

For carbon steel and cast iron, the value ε_i for $K = +0{,}5$ is equal to about half the value of the corresponding strain in uniaxial tension at relatively high sub-zero temperatures. At very low temperatures a second tensile component may produce a brittle rupture of the metal. For instance, carbon steel tested in uniaxial tension at $-180°C$ ruptured with an appreciable amount of plastic strain i.e. 0,4%, while in biaxial tension the same steel failed in a quasi-brittle manner.

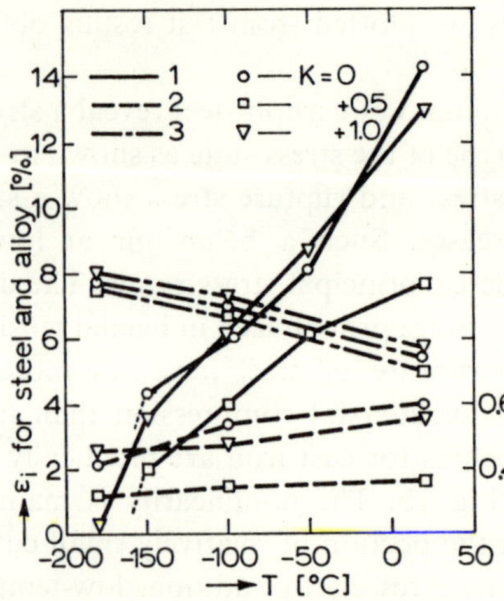

Fig. 2. Influence of temperature on the deformability of steel (1), cast iron (2), and aluminium alloy (3).

Ductility of aluminium alloy increases with decreasing temperature. The slopes of the curves have been found to be virtually identical for both uniaxial and biaxial tensions.

Strain hardening at low temperatures depends on the type of metal. Low-temperature deformability of cast iron and aluminium alloys remains qualitatively the same as at normal temperature. Carbon steels, however, are particularly sensitive to conditions of mechanical and thermal loading. It appears that during plastic straining the proportionality of stress and strain deviators is not maintained, and a remarkable dispersion between curves of stress intensity versus strain intensity is observed.

Temperature dependency of the Lode-Nadai parameter

$$\mu_\varepsilon = \frac{2\varepsilon_2 - \varepsilon_1 - \varepsilon_3}{\varepsilon_1 - \varepsilon_3}$$

at various values of the parameter

$$\mu_\sigma = \frac{2\sigma_2 - \sigma_1 - \sigma_3}{\sigma_1 - \sigma_3}$$

are shown in Fig. 3. A deviation from the proportionality condition $\mu_\varepsilon = \mu_\sigma$ is seen to increase at lower temperatures. The largest deviations are observed in uniaxial compression or tension. At first sight, this result appears to be

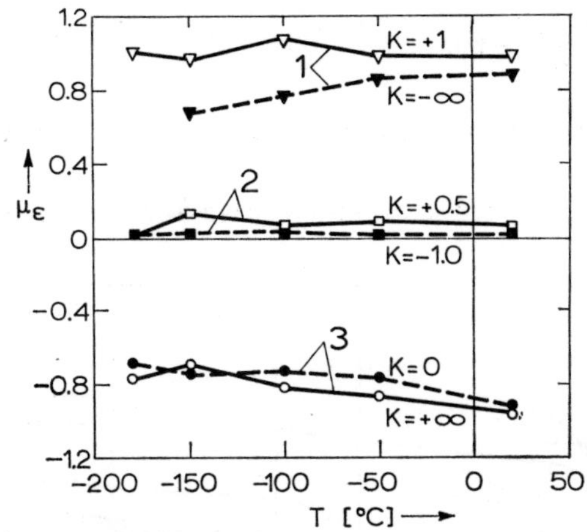

Fig. 3. Temperature dependence of the Lode-Nadai parameter μ_ε for steel. $1 - \mu_\sigma = +1$; $2 - \mu_\sigma = 0$; $3 - \mu_\sigma = -1$.

exeptional since for the above types of loading applied to an isotropic material, the condition $\mu_\varepsilon = \mu_\sigma$ must be satisfied due to symmetry.

A detailed analysis of this contradiction is beyond the scope of the present paper. However, some deviations from symmetry condition are possible during variation of temperature the only independent parameter and the obtained results may be explained by specific internal processes taking place in the material strained at low temperatures. There is no doubt that at lower temperatures, stresses of the second kind, i.e. micro stresses would develop due to the difference in both the elastic constants and the coefficients of linear expansion of individual grains and the phases within each grain. Stress concentrations may arise in the areas of grain contact, increasing the non-homogeneity of the stress field and resulting the formation of microcracks.

Due to phase transformations occurring in the metal, the kinetics of low-temperature straining becomes more complicated. The Bauschinger effect is present in microvolumes of the metal. Kinking, and twinning, begin to play a substantal role in addition to slip. As a result, disorientation of micro-plastic strains takes place with respect to the stresses, leading to a deviation of principal directions of strain and stress deviators. The symmetry of the deviators ceases to hold true.

The obtained data are in agreement with an analysis of variation in lateral deformation ratio for steel, i.e. the deviation of experimental data from the theoretical curve

$$\nu = \frac{1}{2} - \frac{\sigma}{E\varepsilon}\left(\tfrac{1}{2} - \nu'\right),$$

is increasing with decreasing temperature. The above equation follows from the condition $\mu_\varepsilon = \mu_\sigma$.

Fig. 4 shows generalized stress-strain curves for steel. It can be seen that the

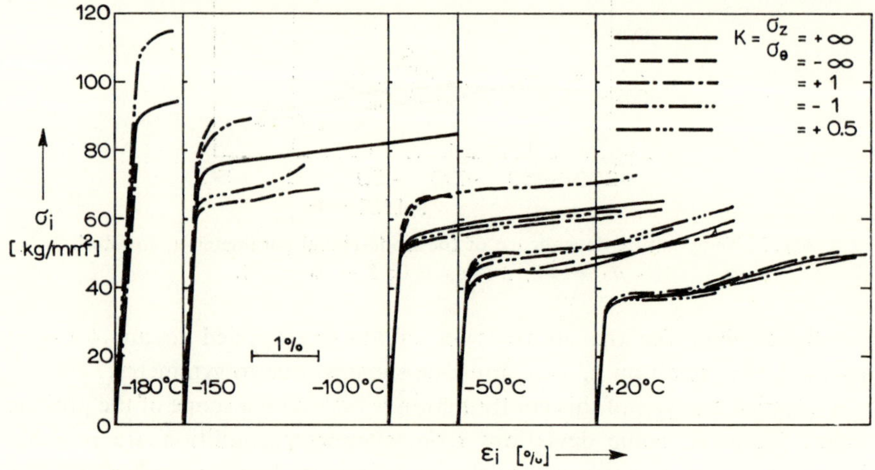

Fig. 4. Effective stress-strain curves for steel.

generalized curves are unvariant with respect to the stress state. The scatter of the curves can be attributed to errors in recording and processing of experimental data. The stress state affects, in this case, only the magnitude of the effective plastic strain at rupture.

As the temperature is lowered, the scatter of the curves increases and, amounts to ± 15–$20\,\%$ at strains $\varepsilon_i = 0{,}2 + 1{,}0\,\%$ using $-50°C$ as a reference temperature. A remarkable feature is observed in diagrams corresponding to the same temperature. The stress-strain curves obtained in stress states where both stress components are positive fall below the curves for uniaxial tension, whereas for stress states in which the principal stresses have different signs or in case of uniaxial compression, the curves are located higher. The curves $\tau_{max}(\gamma_{max})$ were plotted using the test results. Dispersions of these curves expressed by relative deviations of $\Delta\tau/\tau_{max}$ were of the same or higher order than those of the curves $\sigma_i(\varepsilon_i)$, where the relative deviation was expressed by $\Delta\sigma_i/\sigma_i$.

Yield and rupture curves for the materials tested are shown in Fig. 5. At

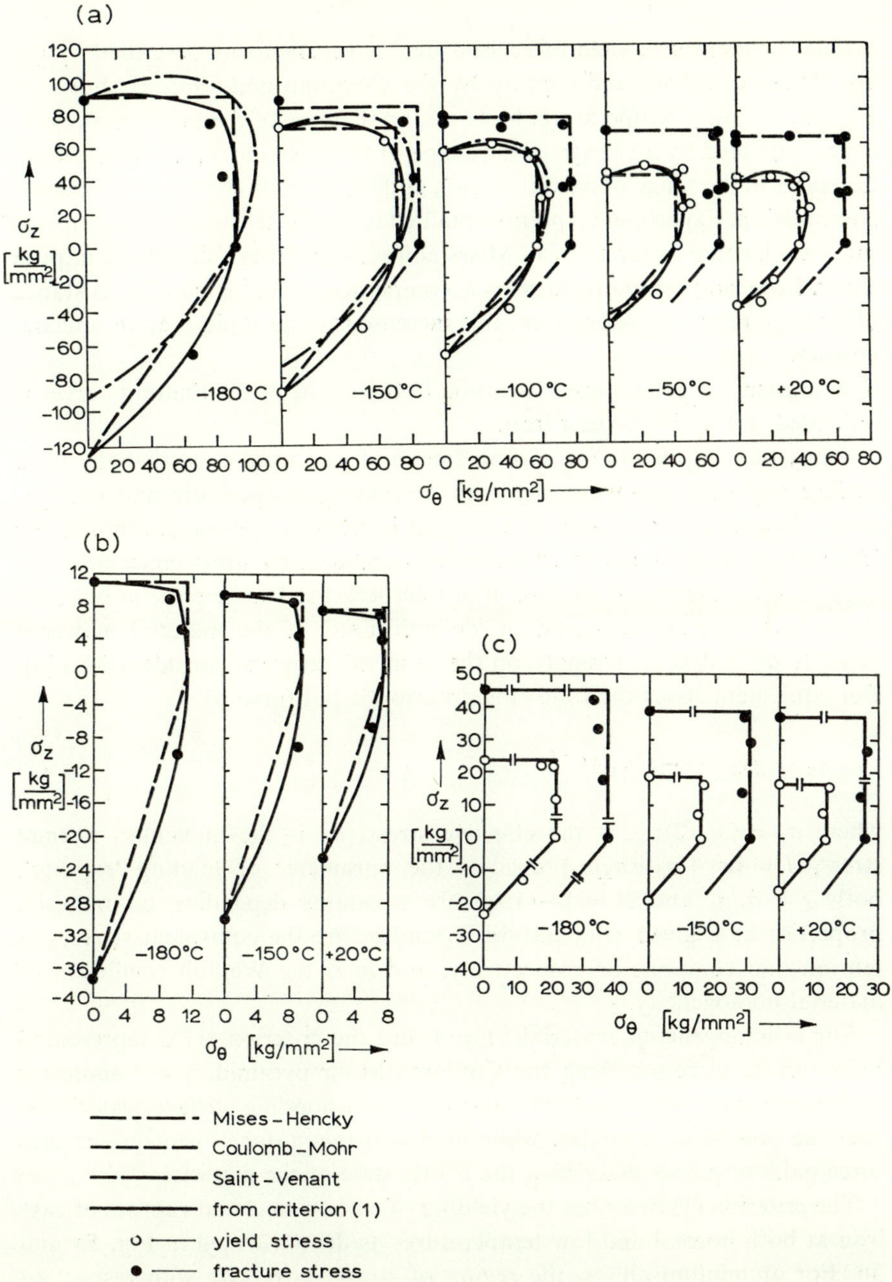

Fig. 5. Conditions of yielding and rupture for carbon steel (a), cast iron (b), and aluminium alloy (c).

normal temperature, yield of carbon steel is satisfactorily described by the von Mises criterion, and rupture by the Coulomb condition as shown in Fig. 5a. At lower temperature, however, an increase of the safe stress region is accompanied by a change in the shape of the yield curve, i.e. the strength decreases in the range of biaxial tension and increases in the range of tension-compression. Experimental points obtained in biaxial tension at low temperatures are located inside the von Mises ellipse, while they fall outside in pure shear. Lowering the temperature is accompanied by an increase in resistance of steel to uniaxial compression, this increase is more rapid than in uniaxial tension.

A similar less pronounced variation in the shape of the limiting curves is indicated in Fig. 5b for cast iron.

Thus, the test results for carbon steel and cast iron show that the yield surface expands anisotropically with decreasing temperature and changes shape. This is probably due to a change in both the plastic strain and the fracture mechanism caused at lower temperatures, by stress concentrations at grain boundaries and formation of microcracks. The increase in both the stress field heterogeneity and in 'defectiveness' of the material makes it possible to analyse its strength on the basis of statistical considerations [9]. For equivalent stress the following criterion is put forward

$$\bar{\eta}_{\Sigma} = \chi\sigma_{i} + (1-\chi)\sigma_{1}A^{1-J} \quad (1)$$

where $\sigma_i = (3/\sqrt{2})\tau_{oct}$ is the effective stress, σ_1 is the maximum normal stress, $J = (\sigma_1+\sigma_2+\sigma_3/\sigma_i)$ denotes the parameter of loading 'rigidity', both $\chi = \sigma_t/\sigma_c$ and $A = 1-(1/m)$ are constants depending on material properties at a given temperature, σ_t and σ_c are the equivalent stresses in tension and compression respectively, and m is the Weibull coefficient of material homogeneity.

For a homogeneous material, $A = 1$ and the criterion (1) is represented by a surface circumscribing the Coulomb-Mohr pyramid. $\chi = 1$ indicates the plastic state of the material and the corresponding surface transforms into the von Mises cylinder, while at $\chi = 0$ the surface disintegrates into three pairs of planes describing the brittle state of the material.

The criterion (1) describes the yielding of carbon steel and rupture of cast iron at both normal and low temperatures as demonstrated in Fig. 5a and 5b. For aluminium alloys, the region of stress states safe with respect to yielding extends as the temperature is lowered as shown in Fig. 5c, but the shape of the yield curve does not change and is consistent with the Saint-Venant criterion for an anisotropic body.

The experimental data obtained are usefull to assess the critical temperature of brittleness for metals with different principal stress ratios, this critical temperature is usually considered as the temperature corresponding to a sharp decrease in ductility. Since the definition is somewhat ambiguous the authors prefer to define the critical temperature of brittleness as the temperature at which the residual strain at rupture does not exceed the value of strain at the yield point, i.e. $\varepsilon_{res} \leqq 0{,}2\%$ and, therefore, $\sigma_{02} \geqq \sigma_b$.

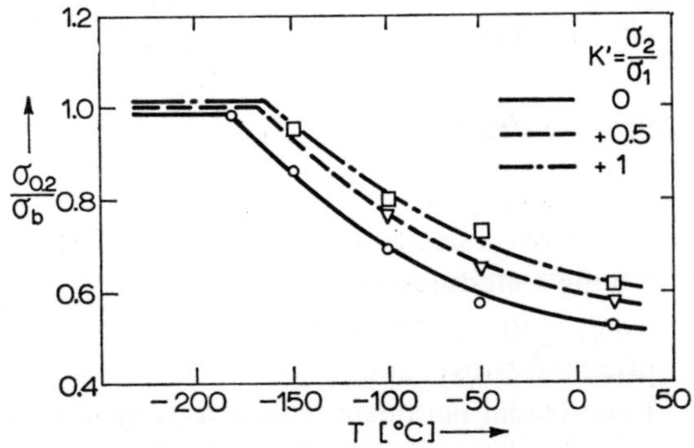

Fig. 6. Temperature dependence of the ratio of proof stress to ultimate strength for steel.

Cast iron and aluminium alloys showed no increase in brittleness in the investigated sub-zero temperature range. Therefore, Fig. 6 presents a temperature dependence of the ratio σ_{02}/σ_b only for steel. As it is seen from the diagram, the temperature at which the ratio σ_{02}/σ_b attains 1 depends on the principal stress ratio, and in biaxial tension the transition into the brittle state occurs at the temperature, which is 25–30° higher than the corresponding one in uniaxial tension.

Tests on anisotropic aluminium alloys and on prestrained carbon steels showed that lowering the temperature causes a more rapid increase in strength in that direction which exhibits lower strength properties at normal temperatures. As a result, the anisotropy of the metal diminishes with decreasing temperature. The similarity of results obtained for different materials i.e. aluminium alloy and carbon steel, with different anisotropy indicates that the increase in isotropy of mechanical properties at lower temperatures appears to be a general feature for a large class of metals.

Appendix

Notation

$\sigma_1, \sigma_2, \sigma_3$	principal stresses,
$\varepsilon_1, \varepsilon_2, \varepsilon_3$	principal elongations
τ_{max}	maximum shear stress
σ_{max}	maximum normal stress
γ_{max}	maximum shear
ε_{max}	maximum elongation
$\sigma_i = \frac{3}{\sqrt{2}} \tau_{oct}$	effective stress
$\varepsilon_i = \frac{1}{\sqrt{2}} \gamma_{oct}$	effective strain
σ_b	tensile strength,
σ_{02}	yield stress referred to 0,2 per cent offset
E	Young's modulus
$K = \sigma_z/\sigma_\theta$	axial – –to – hoop stress ratio
$K' = \sigma_2/\sigma_1$	principal stress ratio
μ_σ	Lode-Nadai parameter characterising stress deviator
μ_ε	Lode-Nadai parameter for deformation
$\bar{\eta}_\varepsilon$	criterion of limit state of material
$\chi = \sigma_t/\sigma_c$	constant, where σ_t and σ_c denote equivalent tensile and compressive stresses, respectively
ν'	Poisson's ratio
ν	lateral deformation ratio in plastic range
ε_{res}	residual deformation after fracture

References

[1] Krisch, A., Archiv für Eisenhüttenwesen, *31* (1960) 113–119.

[2] Krisch, A., Archiv für Eisenhüttenwesen, *38* (1967) 337–343.

[3] Kurkin, S. A. and D. I. Umarov, Izvestai vysshikh uchebnykh zavedenii. Mashinostroyenie, *No 2*, (1968) 186–192.

[4] Arone, R. G., Zavosdkaya laboratoria, *32* (1966) 1254–1257.

[5] Koval'chuk, B. I., A. A. Lebedyev and N. V. Novikov, Soviet Author's Certificate N 274439, Bulletin 'Otkrytia, isobretenia, problmyshennye obraztsy, tovarnye znaki', *No 21* (1970) 111.

[6] Koval'chuk, B. I., A. A. Lebedyev and N. V. Novikov, Zavodskaya laboratoria, *35* (1969) 1116–1118.
[7] Lebedyev, A. A. and V. P. Lamashevskii, Soviet Author's Certificate 300806, Bulletin 'Otkrytia, izobretenia, promyshlennye obraztsy, tovarnye znaki', No 13, (1971) 185.
[8] Lebedyev, A. A. and V. P. Lamashevskii, Soviet Author's Certificate 24265, Bulletin 'Otkrytia, izobretenia, promyshlennye obraztsy, tovarnye znaki', No 15, (1969) 90.
[9] Lebedyev, A. A. and B. I. Koval'chuk, Problemy prochnosti, No 8, (1970) 40–42.

Subsequent yield surface determined in consideration of the Bauschinger effect

Eiryo Shiratori, Kozo Ikegami and Kenji Kaneko
Tokyo Institute of Technology, Tokyo, Japan

Summary

Plastic behavior of metal under complex loading is experimentally investigated by subjecting thin-walled brass specimens to combined axial load, internal (or external) pressure and torsion. After some radial preloadings the subsequent yield surfaces, which are dependent on the Bauschinger effect, are established in the axial-circumferential-shear stress space. Experimental results are represented in the deviatoric plane, and the influence of the Bauschinger effect on the subsequent yield surface is studied. It is shown that subsequent yield surfaces after radial preloadings are congruent with one another irrespective of the directions in preloading, if they correspond to the same amount of prestraining and are represented in the deviatoric plane.

1. Introduction

Subsequent yield surfaces after preloading are generally of in stress space complicated shape. An extensive and systematic study of the yield surface for various preloading paths is essential in the theory of anisotropic plasticity, specifically in the theory of plastic flow in which the effect of strain history is taken into consideration.

Experimental results on subsequent yield surfaces have been reported in considerable number of papers [1]–[10][1]). However, in most of those papers, the yield loci are obtained only in specified planes of stress space. No three-dimensional representation of subsequent yield surfaces are given showing the correlation between the yield loci obtained in the two principal stress plane and those in the normal stress-shear stress plane. Knowledge on three-dimensional subsequent yield surfaces seems to be insufficient.

This paper represents the shape of three dimensional subsequent yield

[1]) Numbers in brackets designate References at end of paper.

surface obtained by subjecting thin-walled tubular specimens to combined axial load, internal (or external) pressure and torsion. The tests results are represented in the cylindrical coordinate system constituted by the deviatoric plane and the rotation angle of the principal stress axes. The shape of subsequent yield surfaces is discussed on the basis of those experimental results.

2. Experimental procedure

The tests were conducted with an apparatus which is capable of subjecting a thin-walled specimen to combined axial load, internal (or external) pressure and torsion simultaneously. The transducer of the strain gauge type was used for measuring the loads applied to the specimen.

Ninety thin-walled tubular specimens with flanges were machined from brass bars (BsBM-2 according to the Japanese standard) 70 mm in diameter, which were annealed at 550° C for 3 hours. The length of the cylindrical part of a specimen was 100 mm, the nominal inside diameter was 20 mm and the nominal wall thickness was 1.5 mm. The wall thickness measured at eight locations varied from the nominal value by less than 0.01 mm. Thirty one specimens were used to determine the initial and subsequent yield surface. The axial, circumferential and '45 degree' strains were measured by two strain gauges which were located diametrically opposite each other on the outer surface in the middle part of the specimen. The nominal values of axial, circumferential and shear stress were calculated on the basis of the original dimensions of the specimen.

The specimen were prestrained by radial loading under combined axial, circumferential and shear stresses. The amounts of the prestrain were chosen as 0.8, 1.8, 3.4 and 5.1 per cent of the equivalent plastic strain. The directions of the preloading paths are shown in Fig. 1. The axial, circumferential and shear stresses are denoted as σ_x, σ_y and τ_{xy}, respectively, see Fig. 1. The ratios of σ_x, σ_y and τ_{xy} are indicated in the parentheses in Fig. 1.

The subsequent yield point is determined by the reloading straight path such as $O'A$ in Fig. 4(b). The point O' is chosen within the yield surface in such a way that the elastic behavior of the specimen is expected to take place in the neighborhood of this point. The yield point is determined by the proof strain of 0.02 per cent in the equivalent stress-strain diagram by reloading after prestraining.

Only one specimen was used to determine one point on the initial yield surface. All points for the subsequent yield surface under a given prestrain

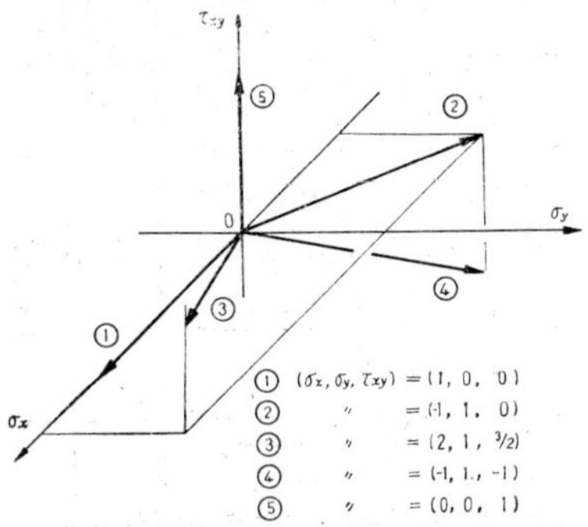

Fig. 1. Directions of preloading paths.

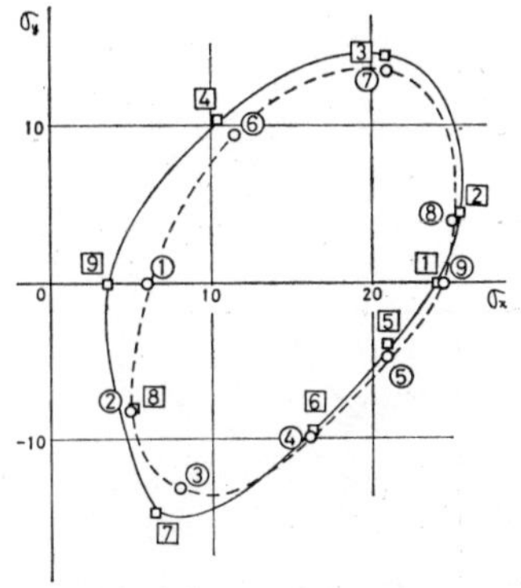

Fig. 2. Effect of the sequence of determining the yield point on the subsequent yield locus.

were obtained by means of a single specimen. In the latter case, the sequence of determining the yield point might affect the shape of the yield surface because of the effect of strain hardening. To estimate this effect, two yield surfaces were determined under the same prestraining by using two different sequences as shown in Fig. 2. Although the effect of the sequence was negligible in the vicinity of the prestress point, it was found to exist in its opposite side. To reduce this effect on the subsequent yield surface, the sequence of obtaining the yield point was started from the opposite side of the prestress point.

3. Experimental results

Some of the experimental results regarding the initial and the subsequent yield surfaces are shown in Figs. 3 to 5.[1]) Figs. 3(a) and (b) show the subsequent yield loci established in the $\{\sigma_x, \sigma_y\}$ plane for preloading along the paths (1) and (2) in Fig. 1. The subsequent yield loci after preloading along the paths (3) and (4), Fig. 1, are shown in Figs. 3(c) and (d). The principal stress in the radial preloading obtained by combined σ_x, σ_y and τ_{xy} are denoted as σ_1 and σ_2 in Figs. 3(c) and (d). The yield points in the yield loci of Figs. 3(c) and (d) are determined by the coaxial stress tests in which principal stress axes do not rotate with respect to those in preloading.

[1]) The yield point in the preloading direction was determined by reloading after preloading.

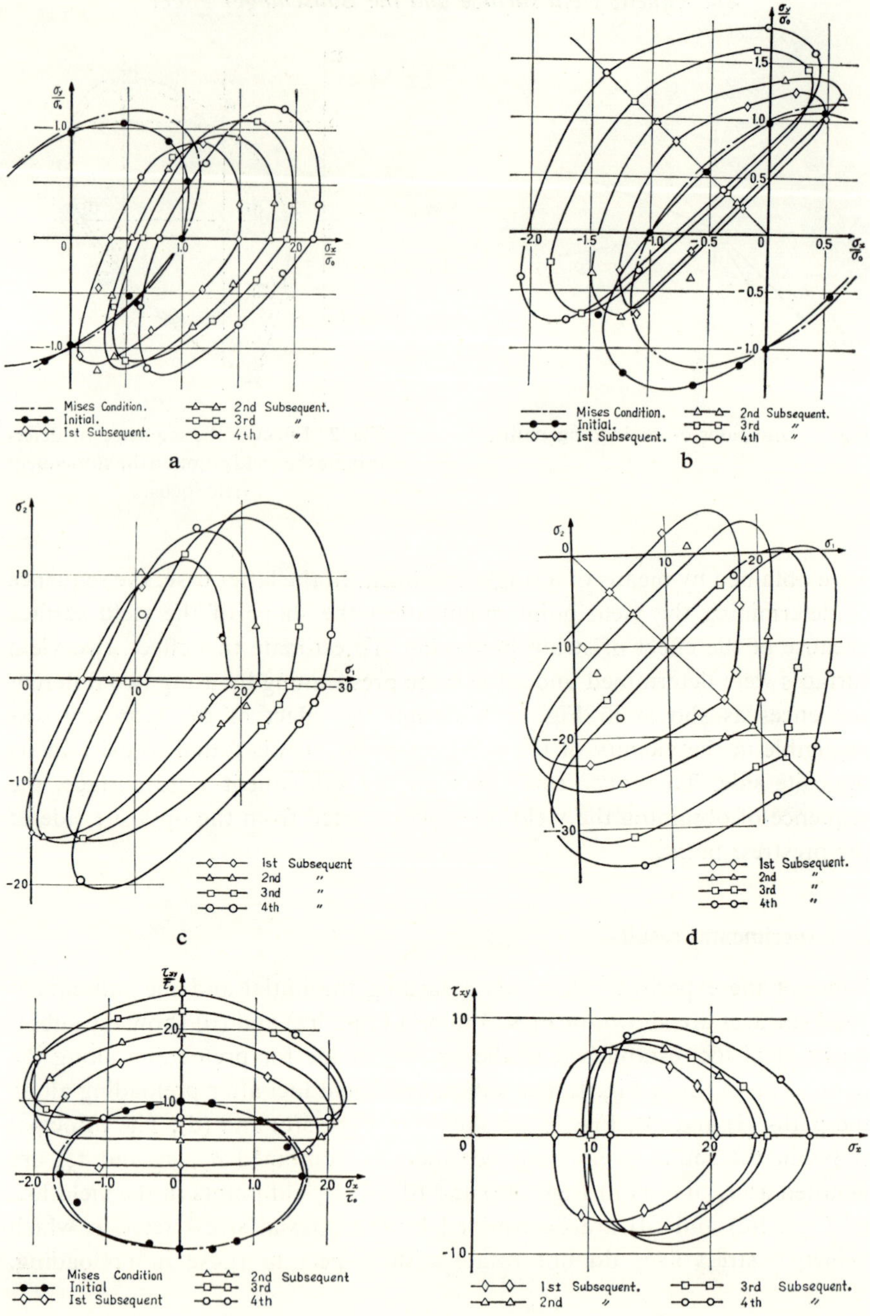

Fig. 3. Experimental results on the subsequent yield loci. Preloadings along the paths indicated in Fig. 1. (a) preloading along (1); (b) along (2); (c) along (3); (d) along (4); (e) along (5); (f) along (1).

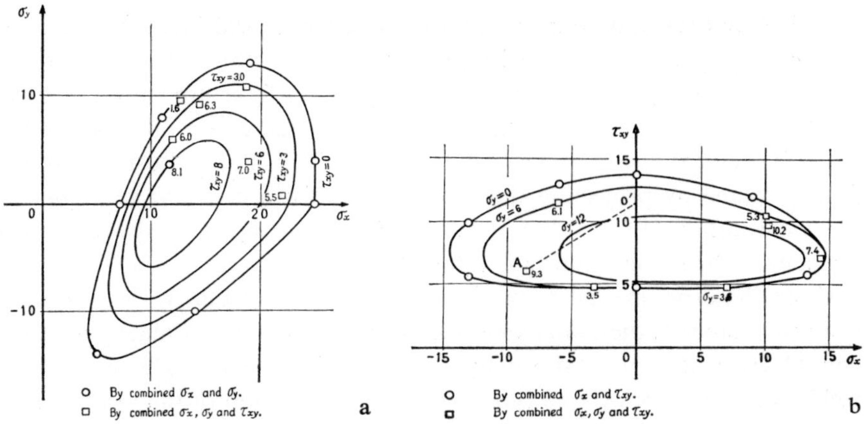

Fig. 4. Experimental results on the subsequent yield surfaces. (a) preloading along the path (1) in Fig. 1; (b) preloading along the path (5).

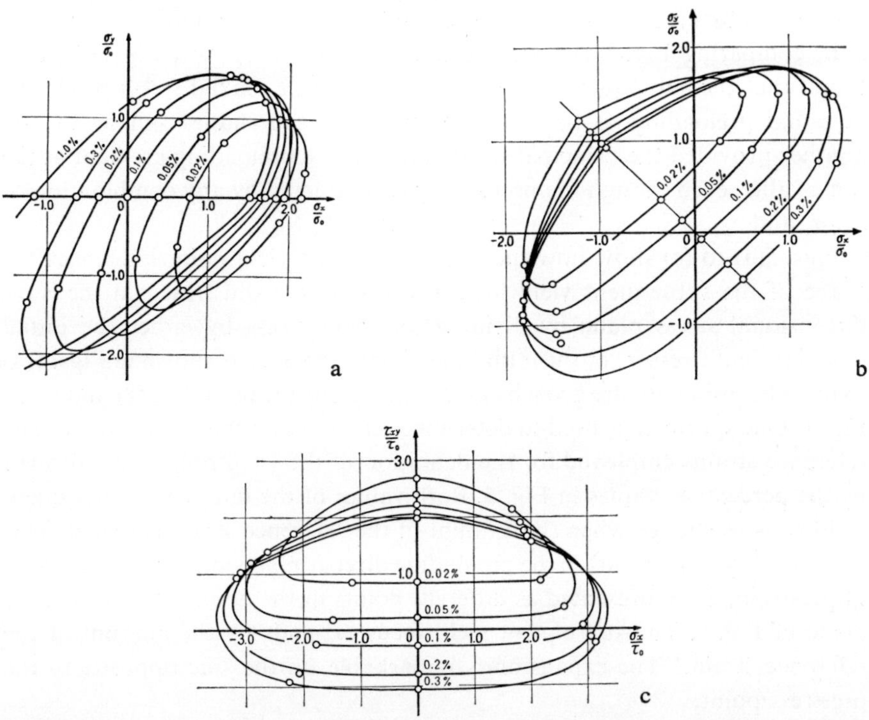

Fig. 5. Effect of the definition of the yield point on the yield locus after 5.1 per cent prestrain. (a) uniaxial tensile preloading; (b) pure shear preloading; (c) simple shear preloading.

Figs. 3(e) and (f) show respectively the subsequent yield loci determined in the $\{\sigma_x, \tau_{xy}\}$ plane for preloading along the paths (5) and (1), shown in Fig. 1. The subsequent three dimensional yield surfaces are shown in Figs. 4(a) and (b). The preloading directions in these figures are taken along the paths (1) and (5) in Fig. 1. The yield surface of Fig. 4(a) is illustrated in the $\{\sigma_x, \sigma_y\}$ plane by projecting the contour lines for a constant value of τ_{xy}. The yield surface of Fig. 4(b) is illustrated in the $\{\sigma_x, \tau_{xy}\}$ plane by projecting the contour lines for a constant value of σ_y.

From the experimental results shown in Fig. 3 and Fig. 4, the following can be concluded: the initial yield surface coincides approximately with the von Mises condition as shown in Figs. 3(a), (b) and (e). The subsequent yield locus does not expand isotropically. A translation of the yield surface is observed in the preloading direction. The subsequent yield surface has a rounded corner with large curvature at the prestress point. The corner is not as sharp as predicted by the slip theory [13]. The flattening of the yield surface is observed at the opposite side of the prestress point.

In comparing Fig. 3(a) with Fig. 3(c), and Fig. 3(b) with Fig. 3(d), it is found that the subsequent yield loci determined in coaxial stress programs at radial preloading of the same magnitude, have shapes similar to one another provided the principal stress ratios in the preloading are kept at the same value, even though the preloading is conducted by any combination of σ_x, σ_y and τ_{xy}.

Figs. 5(a) to (c) show how the definition of the yield point influences the shape of the subsequent yield loci. The results are illustrated in the non-dimensional stress plane by dividing the yield stress by either the initial uniaxial yield stress σ_0 or the initial shear yield stress τ_0 as shown in Figs. 5(a) to (c). The prestrains are given by preloading along the paths (1), (2) and (5) in Fig. 1. One specimen is used to determine one point on the yield surface. The reference strains employed for the definition of the yield point are indicated by the percentage values in Fig. 5. Both values of the initial and subsequent yield stresses change when the amount of the reference strain is varied. This explains why yield points in the preloading direction, having the same amount of prestrain, are represented as different points in the nondimensional stress plane of Fig. 5. The subsequent yield locus expands by the amount of the reference strain.[1] The expansion is remarkable on the side opposite to the prestress point.

[1]) The initial yield locus in such a non-dimensional stress plane as shown in Fig. 5 intersects the points (± 1, 0) and (0, ± 1) independently of the amount of the reference strain.

4. Discussion on the shape of the subsequent yield surface

4.1 *Deviatoric stress space*

The yield locus is usually plotted in the Cartesian coordinate system constituted by two or three stresses used in the test as shown in Figs. 3 and 4. In such a representation the variation of the shape of the yield surface is extremely complex since it depends on both the magnitude and the direction of preloading. This makes it difficult to draw a definite conclusion about the effect of preloading on the yield locus from experimental result of the subsequent yield loci after different radial preloadings. The authors represent the yield surface in the deviatoric plane due to the fact that the plastic deformation is characterized by incompressibility and independence of the mean hydrostatic stress.

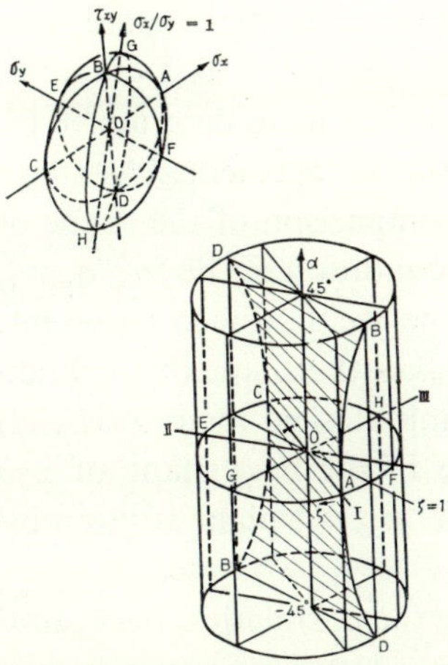

Fig. 6. Initial yield surface in the deviatoric stress space.

The stress space shown in Fig. 6 is used to represent the yield surface. The principal axes of the specimen is considered to be a fixed coordinate system. This space will hereafter be referred to as the 'deviatoric stress space'. The base plane in this stress space, denoted as *AFHCEG* in Fig. 6, corresponds to the deviatoric plane in the principal stress space whose axes coincide with the axial, circumferential and radial directions of the specimen. A vertical axis α indicates the rotation angle of the principal stress axes with respect to the radial direction of the specimen. The rotation angle is specified by

$$\tan 2\alpha = \frac{2\tau_{xy}}{\sigma_x - \sigma_y} \quad (|\alpha| \leqq 45°). \tag{1}$$

Thus, the deviatoric plane in the principal stress space specified by a rotation angle α, is represented as the plane at a vertical distance of α from the origin 0 in Fig. 6. The following indices are used for representation of the magnitude and direction of the deviatoric stress vector [14]:

$$\xi = \frac{1}{\sqrt{2}\sigma_0} [(\sigma_1 - \sigma_2)^2 + (\sigma_2 - \sigma_3)^2 + (\sigma_3 - \sigma_1)^2]^{\frac{1}{2}}, \tag{2}$$

$$\zeta = \frac{6}{\pi} \arctan \frac{(\sigma_2 - \sigma_3)\sqrt{3}}{2\sigma_1 - \sigma_2 - \sigma_3} + 3\left(1 - \frac{\sigma_2 - \sigma_3}{|\sigma_2 - \sigma_3|}\right), \tag{3}$$

where σ_1, σ_2, σ_3 are the principal stresses, and σ_0 is the initial yield stress in uniaxial loading. In the deviatoric stress space, a yield locus determined by coaxial stress program is represented as a closed curve in a plane with the value of α kept constant. A yield locus determined by tests causing rotation of principal directions of stress is represented as a space curve.

Fig. 6 also shows the comparison of the shape of the initial yield surface represented by the Mises condition in the $\{\sigma_x, \sigma_y, \tau_{xy}\}$ stress space and in the deviatoric stress space. The yield surface represented as an ellipsoid in the $\{\sigma_x, \sigma_y, \tau_{xy}\}$ stress space takes the shape of a cylinder in the deviatoric stress space. For example, the initial yield locus $AGECHFA$ in the $\{\sigma_x, \sigma_y\}$ plane is represented by a circle in the base plane of $\alpha = 0°$, whereas the initial yield locus $DABCD$ in the $\{\sigma_x, \sigma_y\}$ plane is given by two space curves in the curved hatched surface.

Figs. 7(a) and (b) show the correlation between the shapes of subsequent yield surfaces in the above mentioned two stress spaces. The subsequent yield surfaces after uniaxial ($\zeta = 0, \alpha = 0°$) preloading are shown in Fig. 7(a), where the loci S_0 in the plane $\tau_{xy} = 0$ and S_c in the plane $\sigma_y = 0$ correspond to the loci $\bar{S}_0$ in the plane $\alpha = 0°$ and $\bar{S}_c$ in the curved hatched surface respectively. Subsequent yield surfaces after simple shear preloading are shown in Fig. 7(b),[1] where the loci S_1 in the plane $\sigma_x - \sigma_y = 0$ and S_b

[1]) When the yield surface is divided into two parts by the plane $\sigma_x - \sigma_y = 0$, one part will lie in the semi space $\alpha > 0°$ and the other in $\alpha < 0°$. Only the yield surface in the semi space $\alpha > 0°$ is illustrated in Fig. 7(b).

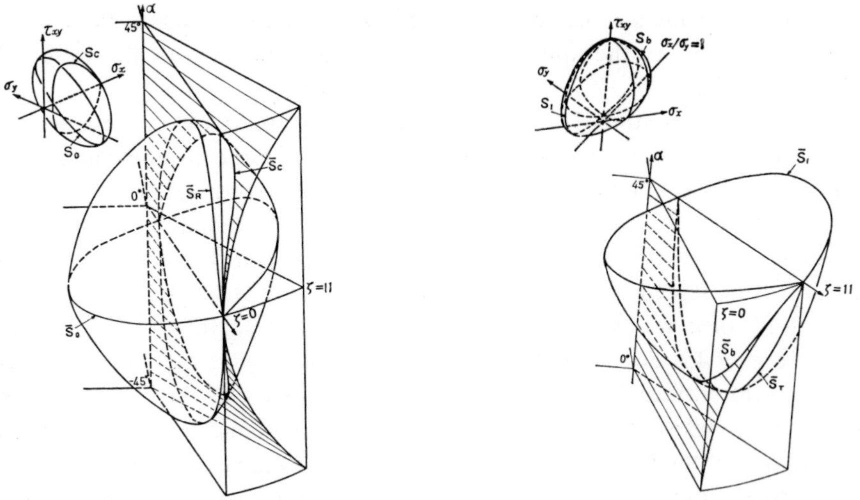

a b

Fig. 7. Subsequent yield surface in the deviatoric stress space. (a) uniaxial tensile preloading; (b) simple shear preloading.

in the plane $\sigma_y = 0$ are represented as the loci $\bar{S}_1$ in the plane $\alpha = 45°$ and $\bar{S}_b$ in the curved hatched surface respectively.

The shape of the subsequent yield surfaces in the deviatoric stress space will now be discussed. The subsequent yield surface is expected to be symmetric after radial preloading with respect to the plane generated by the vertical axis of α and the preloading path, since the initial yield surface takes a cylindrical shape. For example, the subsequent yield surfaces shown in Figs. 7(a) and (b) will be symmetric with respect to the planes $\bar{S}_R$ and $\bar{S}_T$, respectively. Although the yield surfaces for radial preloading in the planes corresponding to $\alpha = 0°$ and $\alpha = 45°$ are also symmetric with respect to the plane $\alpha = 0°$,[1]) this symmetry is not generally observed in the subsequent yield surfaces for radial preloading, since the yield surface corresponding to preloading path near $\alpha = 45°$ or $\alpha = -45°$, may have the shape of a surface whose upper or lower portion is cut by the plane $\alpha = 45°$ or $\alpha = -45°$. The dependence of the symmetry on the α-value of preloading path is closely related to the adopted range ($|\alpha| \leqq 45°$) of α-value in the deviatoric stress space. Even the subsequent yield surface, whose upper or lower portion

[1]) In Fig. 7(b) the two surfaces divided by the plane $\sigma_x - \sigma_y = 0$ will be symmetrically situated with respect to the plane $\alpha = 0°$.

is cut by the plane $\alpha = 45°$ or $\alpha = -45°$, will be a continuous surface when it is represented in the appropriate range of α-value such as the range of $0° \leqq \alpha \leqq 90°$.

In the previous study [11] the authors have proposed a hypothesis on the congruency of the shape of the subsequent yield loci in the same plane if the value of α is kept constant. This hypothesis on the congruency will also be applicable to the yield loci in the planes where α is constant.

As previously mentioned, the upper or lower part of the subsequent yield surface represented in the deviatoric stress space may be cut by the plane $\alpha = +45°$ or $\alpha = -45°$, when the α-value of the preloading path is chosen close to $\alpha = 45°$ or $\alpha = -45°$. In such cases, the congruency of the subsequent yield surfaces will hold for the common parts of the yield surfaces obtained by translating vertically their preloading planes to the base plane. For example, the yield surface in Fig. 7(b) will be made to coincide with the lower part of the yield surface in Fig. 7(a) by translating the former vertically to the plane $\alpha = 0°$ and then rotating its preloading direction to $\zeta = 0$.[1])

4.2 *Experimental results*

The initial yield points determined by radial loading in various directions of α are shown in Fig. 8(a) by projecting the yield points in the plane $\alpha \neq 0°$ onto the plane $\alpha = 0°$. The initial yield points in different α-planes are found to fall approximately on a circle. This shows that the initial yield surface takes a cylindrical shape in the deviatoric stress space (cf. Fig. 6).

The congruency of the subsequent yield loci after radial preloading is examined in Fig. 9[2]) by replotting the third subsequent yield loci of Figs. 3(a) to (d) in the deviatoric stress space.

Fig. 10 shows the congruency of the three-dimensional yield surface by representing two subsequent yield surfaces of Figs. 4(a) and (b) in the deviatoric stress space. In Fig. 10 the preloading plane and direction of the yield surface of Fig. 4(b) are translated to the plane $\alpha = 0°$ and then rotated to the direction $\zeta = 0$, respectively. The shape of the coupled surface obtained by that translation and rotation is shown in Fig. 10 by drawing the contour lines of $\alpha = 0°$, 20° and 30°.

[1]) If the subsequent yield surface for simple shear preloading is represented in the range $0° \leqq \alpha \leqq 90°$, it becomes a continuous yield surface. The surface will be congruent with that shown in Fig. 7(a).

[2]) ξ in Fig. 9 indicates the non-dimensional value of the subsequent yield stress divided by the value of the yield stress in the preloading direction.

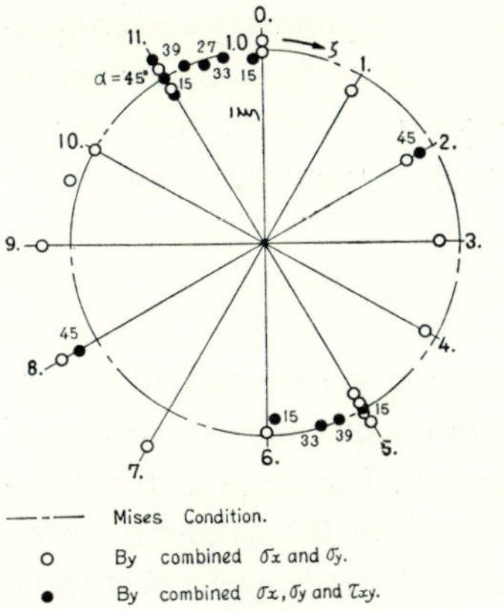

Fig. 8.

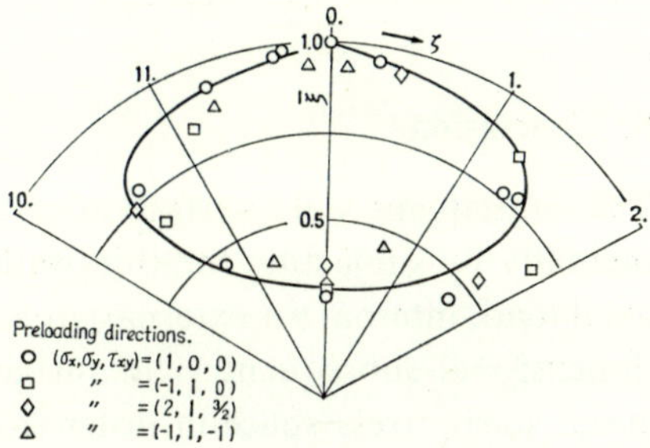

Fig. 9.

Fig. 8. Representation of experimental results of the initial yield surface in the deviatoric stress space.

Fig. 9. Symmetry and congruency of the subsequent yield loci after radial preloading. Data of Figs. 3.

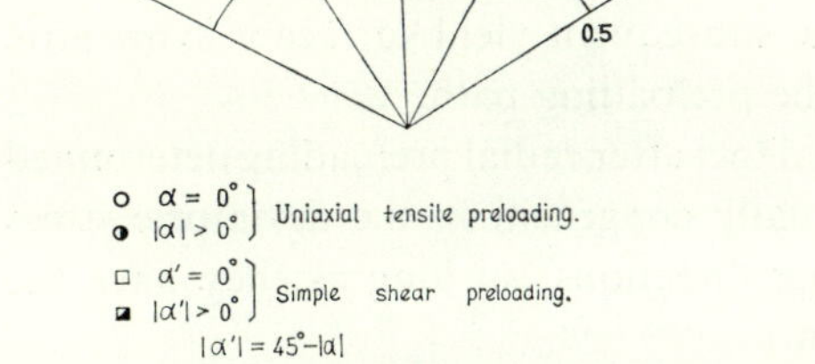

Fig. 10.

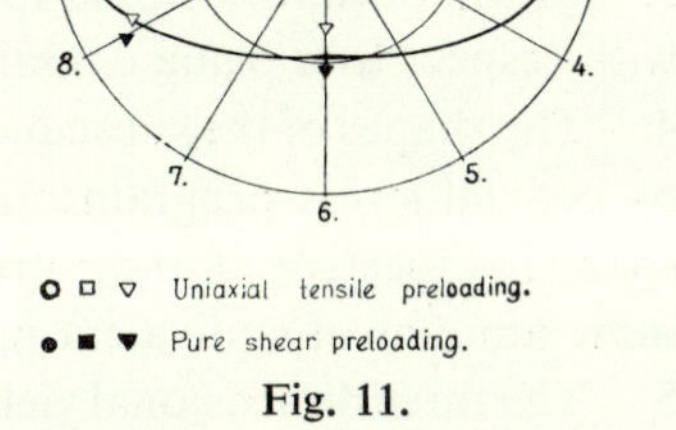

Fig. 11.

Fig. 10. Symmetry and congruency of the subsequent yield surface after radial preloading. Data of Figs. 4.

Fig. 11. Effect of the definition of the yield point on the symmetry and congruency of the yield loci.

The effect of varying the yield point definition on the symmetry and congruency of the subsequent yield loci is investigated in Fig. 11 by using the results of Figs. 5(a) and (b). The results show that the symmetry and congruency of the yield loci will be independent of the definition of the yield point.

5. Conclusion

The subsequent yield surface after radial preloading is investigated experimentally by subjecting the thin-walled tubular brass specimen to combined axial load, internal (or external) pressure and torsion. The shape of the three-dimensional subsequent yield surface is determined in the axial-circumferential-shear stress space in order to show how the Bauschinger effect influences the subsequent yield surface.

It has been shown that, for a given preloading, the correlation between the yield locus determined in the $\{\sigma_x, \sigma_y\}$ plane and that in the $\{\sigma_x, \tau_{xy}\}$ plane can be obtained by representing the yield locus in the deviatoric stress space. This method of representing the yield surface is also useful in the experimental investigation of the congruency of the yield surface suggested theoretically by Ilyushin [12] and Sayir [15].

As to the shape of the subsequent yield surface, the following conclusions can be drawn.

1. The subsequent yield surface after radial preloading has a rounded corner of large curvature in the vicinity of the prestress point. The flattening of the surface is observed on the opposite side of the prestress point.
2. The effect of varying the definition of the yield point on the shape of the yield surface is more pronounced on the opposite side of the prestress point than in the vicinity of this point.
3. In the deviatoric stress space, the subsequent yield surface is symmetric with respect to a plane containing the preloading path.
4. The shapes of the subsequent yield loci after radial preloading determined by coaxial stress programs are mutually congruent in the deviatoric stress space irrespective of their preloading directions, so long as they have the same amount of equivalent prestrain.
5. The three dimensional yield surface for simple shear preloading is congruent with the lower (or upper) part of the yield surface for uniaxial preloading with respect to the preloading plane.

References

[1] Yagn, Yu. I. and O. A. Shishmarev, Some results of an investigation of elastic state limit of plastically drawn nickel samples (in Russian). *Doklady Akademii Nauk, SSSR, 119*, 431–433.

[2] Naghdi, P. M., F. Essenburg and W. Koff, An experimental study of initial and subsequent yield surfaces in plasticity. *Journal of Applied Mechanics 25* (1958) 201–209.

[3] Hu, L. W. and J. F. Bratt, Effect of tensile plastic deformation on yield conditions. *Journal of Applied Mechanics, 25* (1958) 411.

[4] McComb, H. G., Jr., Some experiments concerning subsequent yield surfaces in plasticity. *NASA TN* D-396, 1960.

[5] Ivey, H. J., Plastic stress-strain relations and yield surfaces for aluminium alloys. *Journal of Mechanical Engineering Science, 6* (1961) 15–31.

[6] Mair, W. M. and H. Ll. D. Pugh, The effect of torsional pre-strain on the yield surface in copper. *NEL Report* No. 90, 1963.

[7] Parker, J. and M. B. Bassett, Plastic stress-strain relationships – some experiments to derive a subsequent yield surface. *Journal of Applied Mechanics 31* (1964) 676–682.

[8] Theocaris, P. S. and C. R. Hazell, Experimental investigation of subsequent yield surfaces using the moiré method. *Journal of Mechanics and Physics of Solids 13* (1965) 281–294.

[9] Szczepinski, W. and J. Miastkowski, An experimental study of the effect of the prestraining history on the yield surfaces of an aluminium alloy. *Journal of Mechanics and Physics of Solids 16* (1968) 153–162.

[10] Shiratori, E. and K. Ikegami, Experimental study of the subsequent yield surface by using cross-shaped specimens. *Journal of Mechanics and Physics of Solids 16* (1968) 373–394.

[11] Shiratori, E. and K. Ikegami, Studies of the anisotropic yield condition. *Journal of Mechanics and Physics of Solids, 17* (1969) 473–491.

[12] Ilyshin, A. A., *Plasticity* (in Russian). Izdatel'stvo Akademii Nauk, Moscow, 1963. of also V. S. Lensky, *Analysis of plastic behavior of metals under complex loading.* Proceedings of 2nd Symposium of Naval Structural Mechanics (Providence 1960) Pergamon Press, Oxford (1960) 259–278.

[13] Batdorf, S. B. and B. Budiansky, A mathematical theory of plasticity based on the concept of slip. *NASA TN* 1871, 1949.

[14] Hsu, T. C., The effect of the rotation of the stress axes on the yield criterion of prestrained materials. *Journal of Basic Engineering 88* (1966) 61–70.

[15] Sayir, M., Zur Fliessbedingung der Plastizitätstheorie. *Ingenieur-Archiv 39* (1970) 414–432.

Experimental study on one type of plastic anisotropy not considered in simplified flow theories

O. A. Shishmarev

Riazan, U.S.S.R.

Anisotropy due to plastic deformation was studied on tubular steel specimens. This type of anisotropy is found to influence the plastic strain increment vector as the stress point moves over the yield surface associated with a prescribed prestraining. The length of the plastic strain increment vector is found to decrease as the stress point moves from the point associated with the prestraining program. An appropriate measure of this effect is introduced in the form of a modulus of the effective plastic strain increment. Variation of the modulus over the yield surface is compared with the results furnished by the existing flow theories.

Introduction

Standard theories of plastic flow are formulated to calculate deformations along a given loading path only. Material anisotropy due to plastic deformation is either not accounted for, as it is the case in the theory of isotropic hardening, or is considered only as regards the change of shape and position of subsequent yield surfaces. In all such theories the following ratio

$$K = \frac{\delta \ni^p}{\delta S_{\delta \ni}} \tag{1}$$

is assumed to be constant on each subsequent yield locus. This ratio – we shall call it the modulus of the effective plastic strain increment – involves $\delta \ni^p$ which is the length of the plastic strain increment vector and $\delta S_{\delta \ni}$, which stands for projection of the stress increment vector on the direction of $\delta \ni^p$. According to this assumption all the vectors δS with equal $\delta S_{\delta \ni}$ at any point of the yield surface which corresponds to an actually prestrained material produce the same $\delta \ni^p$, [1].

The results of experiments [2, 3], suggest however, that the deformation induced anisotropy manifests itself not only through the change of shape and position of subsequent yield surfaces but also makes the modulus K to vary with the stress profile on these surfaces.

The aim of the present paper is to examine how the modulus (1) varies with the loading path and with stress trajectory on the yield surface.

Outfit, material and specimens

The experiments have been carried out in plane stress under combined tension and torsion. A suitable experimental outfit has been developed, resembling that described in [4]. The deformations were measured by means of mirror extensometers.

Thin-walled, tubular specimens were manufactured from drawn mild steel tubes. The outer diameter and the wall thickness of the specimens were 12 mm and 0.5 mm respectively. Initial anisotropy was removed by annealing.

Experimental procedure and discussion of results

The test results were interpreted in stress and strain spaces referred to the coordinate systems

$$S_1 = \bar{\sigma}, \qquad S_2 = \bar{\tau}\sqrt{3} \quad \text{and} \quad \ni_1^p = \varepsilon^p. \qquad \ni_2 = \frac{\gamma^p}{\sqrt{3}},$$

respectively. The applied stresses are denoted by $\bar{\sigma}$ and $\bar{\tau}$ whereas ε^p and γ^p stand for plastic strains recorded in the program. The lengths of vectors S_i and $\ni_i^b$ correspond to the equivalent stress and strain.

In order to establish the magnitude of the modulus K corresponding to the prescribed value of equivalent plastic strain, specimens without prior plastic prestraining were tested under proportional loading. For this purpose straight paths of loading associated with different values of the Lode parameter μ were chosen and on each of such paths the magnitude of the modulus K was established at $\ni_i^p$ equal to 3 per cent[1]). Two specimens were tested for each loading path associated with a prescribed value of μ. The results showed that the scatter of K was irregular but did not exceed 20 per cent.

The essential part of the testing procedure was as follows. After the specimens were subjected to equal initial plastic strain in tension or in torsion, they were unloaded and reloaded at different angles to the direction of

[1]) The material used had a long platform of yielding. Such an extensive deformation was necessary in order to ensure entering into the hardening range.

prestraining well into the plastic range. Values of the modulus K were determined for each loading path when the yield surface associated with the prestraining was reached and plastic strains were evaluated by subtracting elastic strains from the measured strains by disregarding variations in the elastic moduli E and G with plastic deformation. In order to eliminate the influence of viscous effects, specimens were loaded in steps of $\delta S_i = 100$ kg/cm^2 with an adequate time interval between the consecutive steps. The results are presented in Figs. 1 and 2 where the abscissa specifies the angle α between the initial and the consecutive path of loading and the ordinate represents the ratio $n = K_\alpha / K_0$. K_0 denotes the modulus of plastic strain increment at the end of prestraining and K_α the corresponding value for the second loading. The magnitudes of initial plastic strain are marked on the corresponding curves.

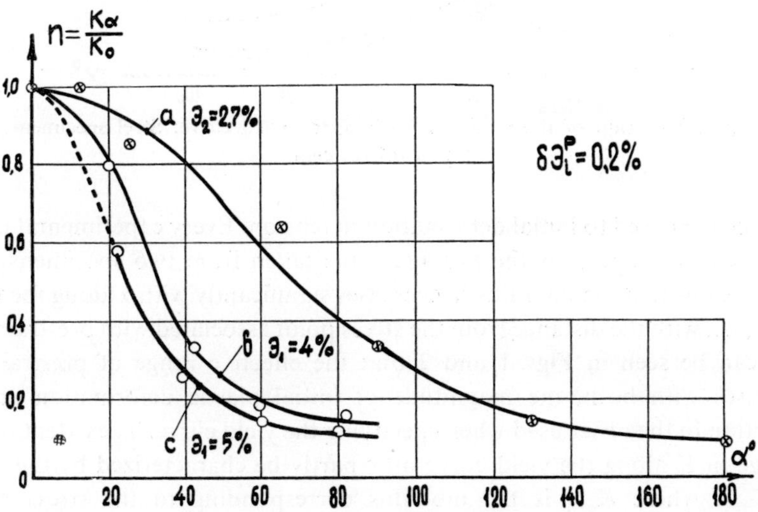

Fig. 1. Variation of the modulus K along the yield curve. Steel specimens, $\delta \ni_i^p = 0.2$ per cent.

Fig. 1 specifies the variation of K along the yield curve. The value of plastic strain increment at which K defined in (1) was computed is $\delta \ni_i^p = 0{,}2$ per cent. Similar curves relating to $\delta \ni_i^p = 0.1$ per cent are plotted in Fig. 2. In these figures the lines labelled a) correspond with results of specimens prestrained in torsion whereas those marked b) and c) represent results of

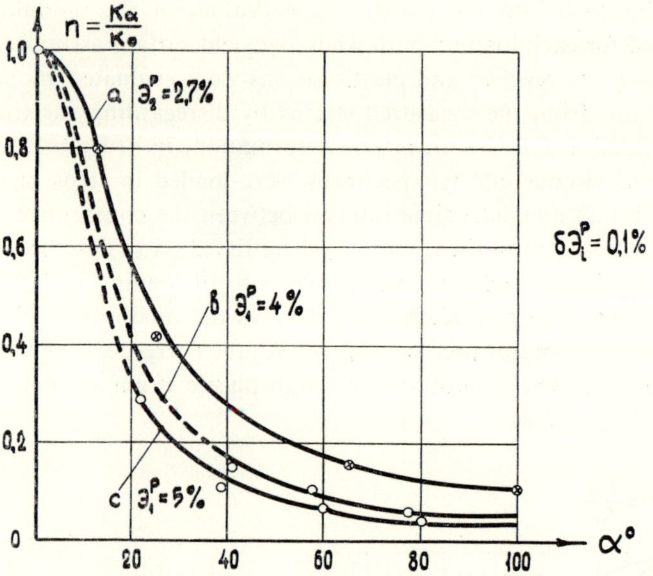

Fig. 2. Variation of the modulus K along the yield curve. Steel specimens, $\delta Э_i^p = 0.1$ per cent.

specimen exposed to initial deformation in tension. Every experimental point marked in the figures is the average value taken from two specimens. The figures show that the modulus K decreases significantly with α along the yield curve, i.e. with the distance from the stress point associated with prestraining.

It can be seen in Figs. 1 and 2 that the effective range of prestraining increases with both, the magnitude of initial plastic deformation and a reduction in the offset used when specifying the yield curve. The extent of the change in K along the yield curve may partly be characterized by the ratio $K_0/K_{\pi/2}$, where $K_{\pi/2}$ is the modulus corresponding to the stress point specified by the radius-vector which makes the angle $\alpha = \pi/2$ with the direction of prestraining. In Fig. 3 the plot of $K_0/K_{\pi/2}$ against initial effective plastic strain is given. It can be readily seen that this relationship is well approximated by a straight line for a comparatively narrow range of initial strains considered in the experiments.

In order to eliminate a possible conjecture that the observed variation of the modulus K is a property of the steel tested, similar experiments were conducted on nickel tubes (outer diameter 5 mm, wall thickness 0.2 mm). The specimens were subjected to initial plastic prestraining of 1.1 per cent

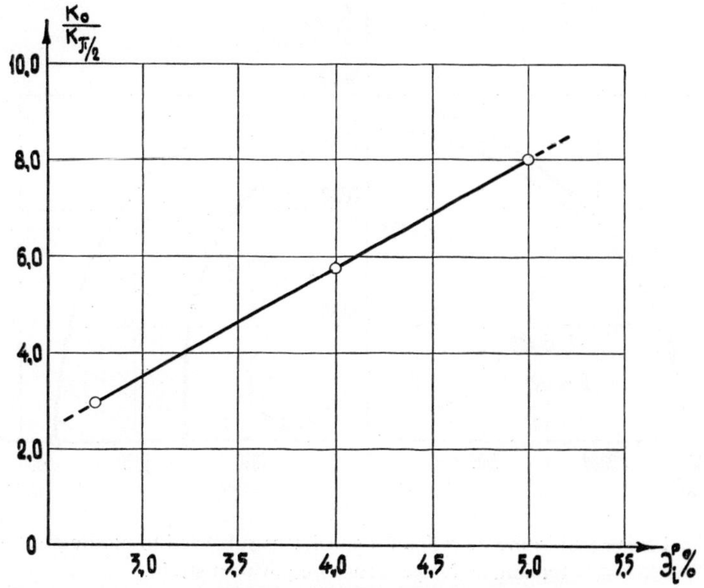

Fig. 3. Rate of change of the modulus K along the yield curve depending on the value of initial plastic strain.

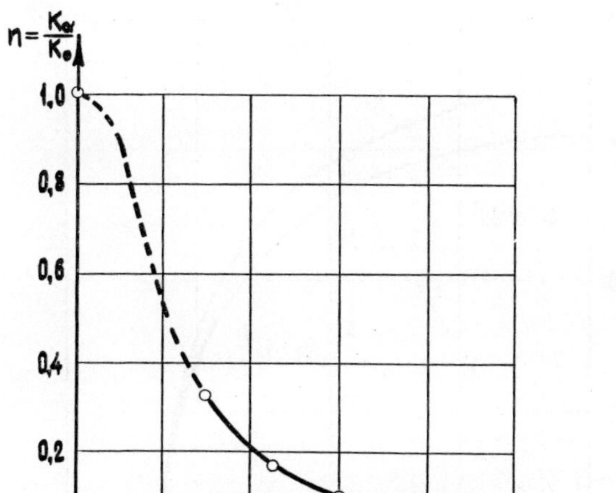

Fig. 4. Variation of the modulus K along the yield curve. Nickel specimens, $\delta Э_i^p = 0.02$ per cent.

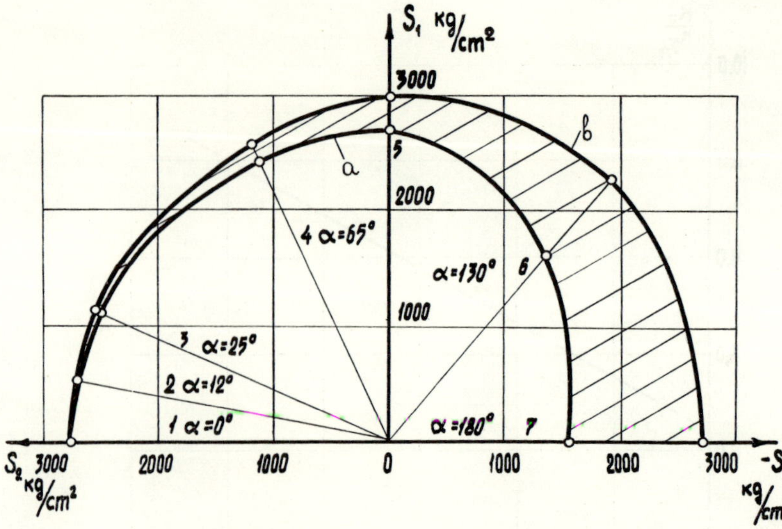

Fig. 5. Yield curve and isomodular line for specimens subjected to plastic prestraining in torsion, at 2.7 per cent of equivalent strain.

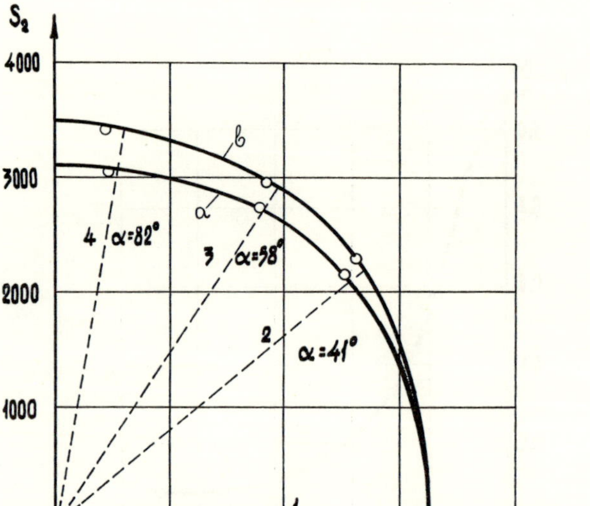

Fig. 6. Yield curve and isomodular line for specimens subjected to plastic prestraining in tension at $\ni_i^p = 4.0$ per cent.

TABLE 1

Variation of the modulus K with the length of the stress profile on the yield curve and with the length of the secondary loading path

α \ ΔS_i	0	150	300	450	600	750	850	950	1050	1150	1250
0°	$3.7\cdot10^{-3}$	$4.3\cdot10^{-3}$	$5.0\cdot10^{-3}$	$6.4\cdot10^{-3}$	$8.6\cdot10^{-3}$						
25°	$3.2\cdot10^{-3}$	$3.8\cdot10^{-3}$	$4.4\cdot10^{-3}$	$5.5\cdot10^{-3}$	$6.4\cdot10^{-3}$	$7.9\cdot10^{-3}$					
65°	$2.2\cdot10^{-3}$	$3.1\cdot10^{-3}$	$4.0\cdot10^{-3}$	$5.0\cdot10^{-3}$	$6.2\cdot10^{-3}$	$7.6\cdot10^{-3}$	$9.1\cdot10^{-3}$				
90°	$1.2\cdot10^{-3}$	$2.2\cdot10^{-3}$	$3.2\cdot10^{-3}$	$4.3\cdot10^{-3}$	$5.3\cdot10^{-3}$	$6.6\cdot10^{-3}$	$7.5\cdot10^{-3}$	$8.5\cdot10^{-3}$			
130°	$5.2\cdot10^{-4}$	$6.0\cdot10^{-4}$	$8.0\cdot10^{-4}$	$1.2\cdot10^{-3}$	$1.8\cdot10^{-3}$	$2.4\cdot10^{-3}$	$3.2\cdot10^{-3}$	$3.9\cdot10^{-3}$	$4.6\cdot10^{-3}$	$5.6\cdot10^{-3}$	$7.2\cdot10^{-3}$
180°	$3.5\cdot10^{-4}$	$4.1\cdot10^{-4}$	$6.0\cdot10^{-4}$	$7.0\cdot10^{-4}$	$1.0\cdot10^{-3}$	$1.5\cdot10^{-3}$	$1.8\cdot10^{-3}$	$2.2\cdot10^{-3}$	$2.7\cdot10^{-3}$	$3.1\cdot10^{-3}$	$3.7\cdot10^{-3}$

and variations of K were recorded along the yield curve corresponding to an offset of 0.02 per cent. These results are represented in Fig. 4, but cannot be compared quantitatively with the curves shown in Figs. 1 and 2. However, it may be concluded that changes along yield curves in the modulus K introduced in this work prove to be a common property of metals.

As already mentioned, the modulus K was measured for the secondary straight paths of loading along the yield curve established by using a standard offset of 0.2 per cent. A certain regularity of change in K was observed which appears to depend on how far secondary paths of loading ΔS_i advanced into the plastic range. It was found that at increasing ΔS_i the difference between the K's for different paths of secondary loading decreased and vanished at a definite value ΔS_{i0} at which the magnitudes of the modulus reached the stable value K_0. The quantity ΔS_{i0} is called 'the length of restoration path of K' and should not be confused with the so called trace of delay of scalar properties [5, 6]. Loci corresponding to stationary values of K are called 'isomodular lines'.

In Figs. 5–7 the yield curves obtained using the standard offset of 0.2 per cent (curves a) and the corresponding isomodular lines (curves b) are presented for specimens subjected to plastic prestrain of 2,7, 4 and 5 per cent respectively. In Fig. 5 the results of the most complete analysis of the curves are represented. The specimens of the relevant experiment were subjected to plastic prestraining in torsion, and the angles of secondary loading paths were $0 \leqq \alpha \leqq \pi$. The shaded area of the diagram corresponds to the states associated with 'slight' plastic strains, i.e. to the restoration path of K. Numerical values of K are given in mm^2/kg on table 1. On the basis of both the tabulated data and the curves presented in Fig. 5 it is evident that the length ΔS_{i0} of the restoration path of K depends on the angle α defined before the length increases as the angle grows. In Figs. 6 and 7 the curves corresponding to the specimens subjected to prestraining in tension are given. The curves are shown only in the first quadrant of the $\{S_1, S_2\}$ plane. It is easy to see that in these cases also ΔS_{i0} increases as α grows. The comparison of the curves presented in figures leads to the conclusion that the restoration path length ΔS_{i0} of the modulus depends also on the value of initial plastic strain. The curves in Fig. 8 show the relationship between ΔS_{i0} and α for different initial plastic prestraining.

Using the isomodular line as a datum, the plastic deformation process under secondary loading becomes analogous to the deformation at pre-

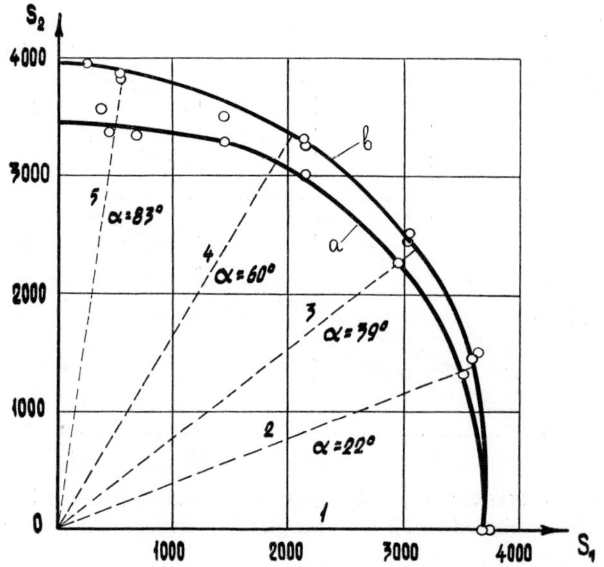

Fig. 7. Yield curve and isomodular line for specimens subjected to plastic prestraining in tension at $\ni_i^p = 5.0$ per cent.

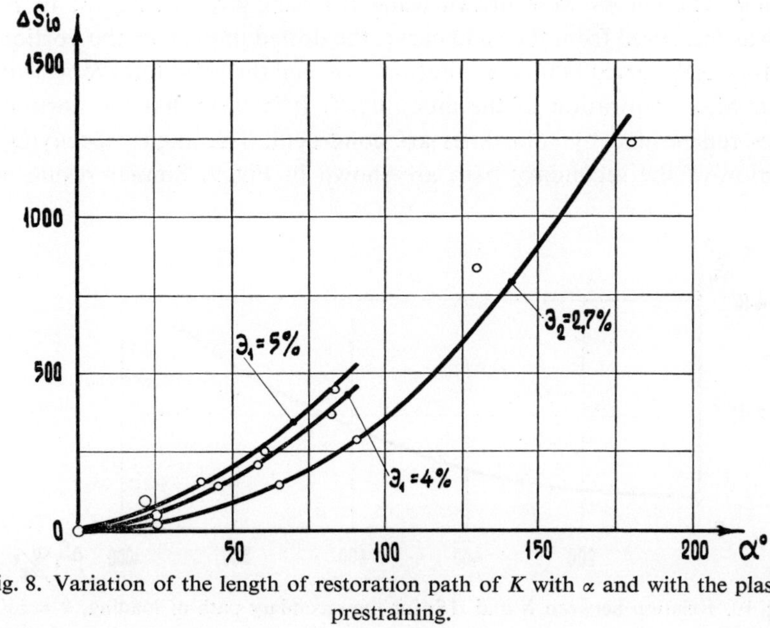

Fig. 8. Variation of the length of restoration path of K with α and with the plastic prestraining.

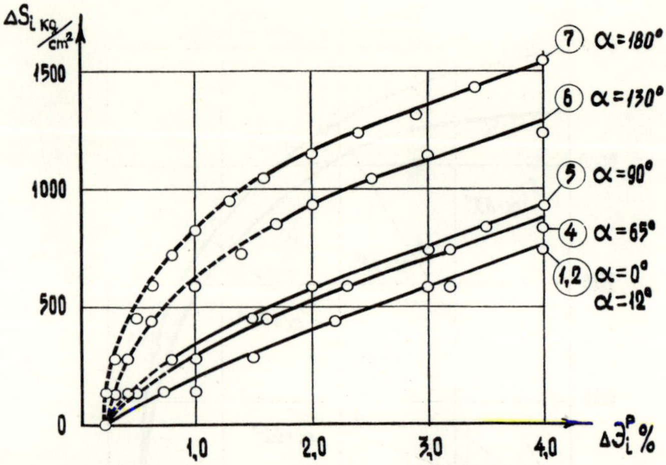

Fig. 9. Relationship between $\Delta Э_i^p$ and ΔS_i for the secondary paths of loading for specimens subjected to 2.7 per cent plastic prestraining in torsion.

straining. This follows from the data given in table 1 and also from Fig. 9 in which the relationship $\Delta Э_i^p - \Delta S_i$ is represented for the secondary paths of loading. The curves were drawn using the data given in Fig. 5. In Fig. 9 ΔS_i was measured from the yield curve, the dotted lines show the portions of the $\Delta Э_i^p - \Delta S_i$ curves before restoration of K and the solid lines represent the results after restoration of the modulus K. It is clear that the portions of curves represented by solid lines are congruent. The angles specifying the direction of the secondary path are shown in Fig. 9. Similar results were

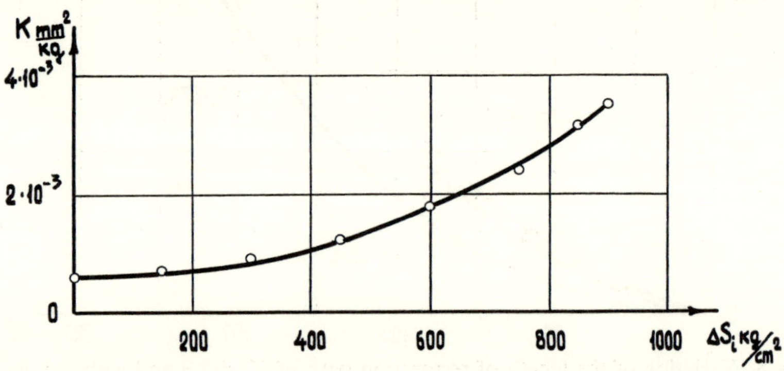

Fig. 10. Relation between K and ΔS_i for the secondary path of loading, $\alpha = 130°$.

obtained in experiments when specimens were subjected to plastic prestrain of 4 and 5 per cent.

The plot in Fig. 10 can be useful for the evaluation of the change of K along the path of its restoration. Since it is drawn for the secondary path making a 130° angle with the initial path for prestraining up to $\ni_i^p = 2.7$ per cent. The $K-\Delta S_i$ relationship for other paths of secondary loading were almost the same but differed a little in curvature.

It is thus evident that anisotropy in metals caused by the initial plastic straining may be associated with the change of modulus K along the yield curve.

Proposed method

Using the obtained data, an attempt was made to develop an approximate method to estimate plastic strains by a simplified flow theory such that deformation anisotropy was appropriately accounted for. The method is not universal and can be applied to a definite combined loading only, i.e. when total or partial unloading takes place after preliminary proportional loading[1]) and secondary straight loading trajectories are followed. But as indicated in [7], the greatest discrepancy between predictions of the flow theory and experiments is observed under such loading.

The main point of the method is as follows. Once the loading path is specified, the angle α is known and the magnitudes of plastic prestrain and of modulus K_0 corresponding to the loading point (yield stress) can easily be estimated from the curve $\sigma-\varepsilon^p$ valid for uniaxial tension. The appropriate K_α and ΔS_{i0} can be estimated from Figs. 1 and 8. When initial strains differ from those for which the plots have been drawn, approximate interpolations may be used. Values of K along the restoration path may be evaluated using the formula

$$K = K_\alpha + \frac{K_0 - K_\alpha}{\Delta S_{i0}} \Delta S_i \qquad (2)$$

which was obtained by linearizing the $K-\Delta S_i$ relationship shown in Fig. 10 and by making use of the relationship (1) in the form[2])

$$\delta \ni^p = K \delta S_n, \qquad (3)$$

[1]) The specimens should not be deformed more than by 5–6 per cent.

[2]) In flow theories the associate flow law is employed. The vector $\delta\ni^p$ is then directed

$\delta \ni^p$ can be found for small increments of δS_n along the restoration path of K. The next step is to determine graphically the increments of deformation, i.e. $\delta \ni_1^p$ and $\delta \ni_2^b$ once the shape, the theoretical yield curve and the plastic strain $\delta \ni^p$ are known. At the end of the restoration path of K, subsequent values of $\delta \ni^p$ are determined from the curve $\sigma - \varepsilon^p$ as it is usually done in incremental theories.

Comparison of theoretical and experimental results

In order to estimate errors that are caused by neglecting the anisotropy considered in the paper, experimental data and theoretical predictions obtained by the theory of kinematical hardening and by the theory accounting for the Bauschinger effect [9] were compared, by considering and disregarding a change of modulus K. The comparison was made for two combined loading paths which caused the greatest discrepancy between theoretical results and experimental data.

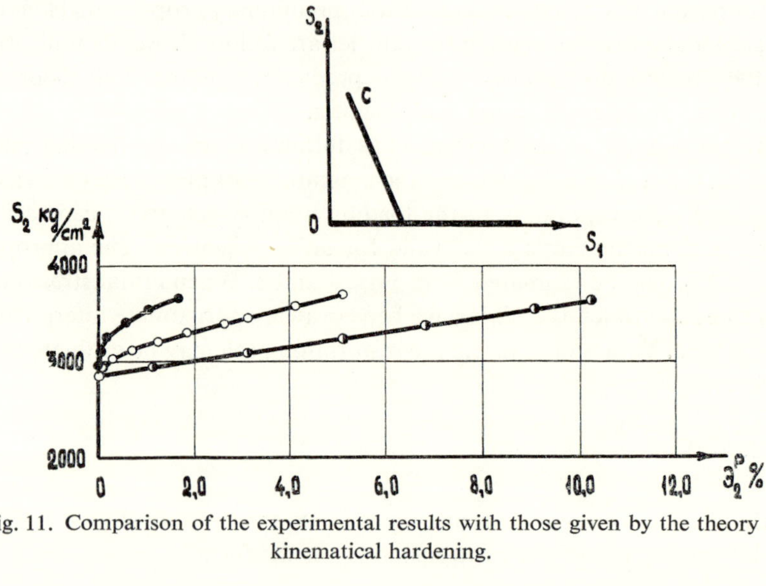

Fig. 11. Comparison of the experimental results with those given by the theory of kinematical hardening.

along the exterior normal to the yield surface. This postulate is confirmed experimentally, [8], if elastic anisotropy due to plastic strain is ignored. Under these conditions $\delta S_{\delta \ni}$ is proportional to δS_n which stands for projection of the stress vector increment on the normal to the yield curve.

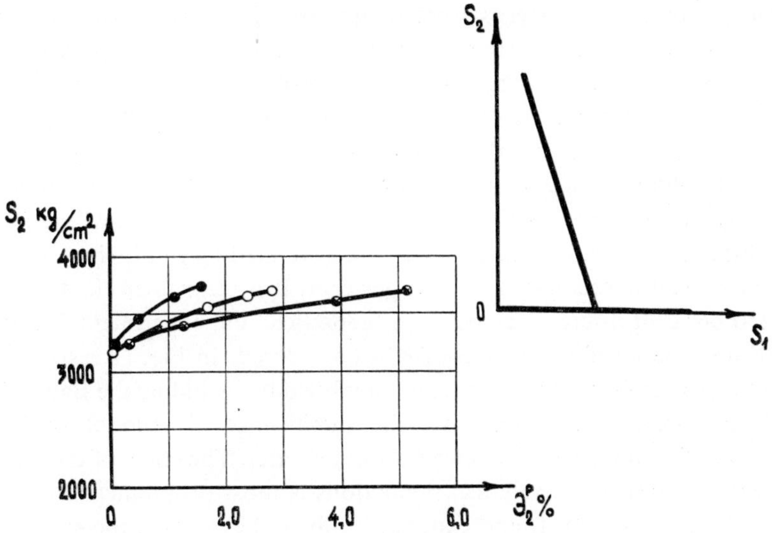

Fig. 12. Comparison of the experimental results with those given by the theory accounting for the Bauschinger effect.

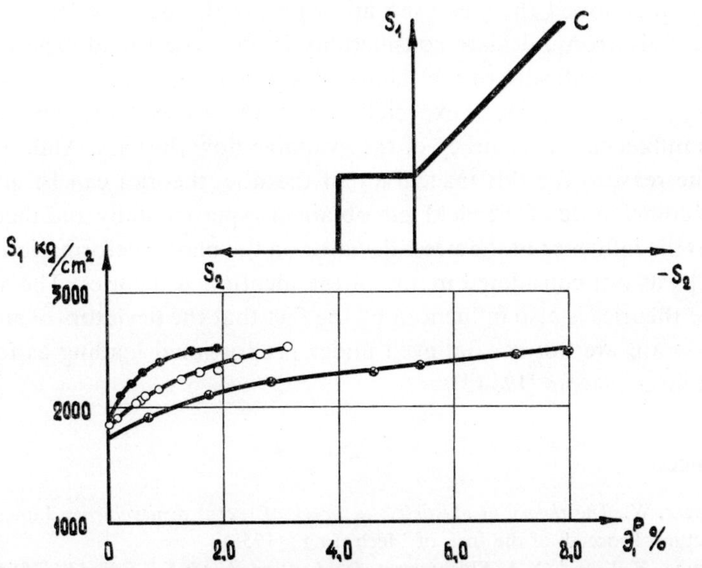

Fig. 13. Comparison of the experimental results for significantly differing secondary loading path with the values given by the theory accounting for the Bauschinger effect.

Plastic strain components were determined in graphic form using the curve $\sigma - \varepsilon^p$ obtained experimentally from data of ten tensile specimens. The results are presented in Figs. 11–13 where black dots correspond to experimental data, half-filled dots indicated results obtained by flow theories assuming that K is constant along the yield curve and empty dots correspond to results obtained according to flow theories taking variations of K into account. In the upper parts of the figures the loading programs are shown.

In Fig. 11 data are compared which are evaluated by employing the theory of kinematical hardening at $\ni_1^p = 5$ per cent. It is clear form the figure that the 450 per cent discrepancy between theoretical and experimental results is two times smaller if a variation of K is considered. In Fig. 12 results corresponding to the same loading path but analysed by including the Bauschinger effect are presented. In this case the significance of accounting for the variation of K along the yield curve is even greater. The effect of the variable K on the accuracy of analytical predictions is most pronounced in Fig. 13, in which plastic strains found experimentally and evaluated theoretically by accounting for the Bauschinger effect are compared for a specimen subjected to initial shearing strain $\ni_2^p = 2.7$ per cent.

It has to be remarked that the strain components obtained in accordance with the mentioned theories even allowing for the described plastic strain induced anisotropy, deviate considerably from those found experimentally (this applies specifically to the theory of kinematical hardening). No better agreement can, however, be expected as we take into account, only one of the factors influencing inaccuracy of the available flow theories. And in view of [7] some reasons for this inaccuracy of the flow theories can be given, i.e. lack of coincidence of the yield loci obtained experimentally and theoretically. Also the influence of the stress deviator on the plastic deformation process [10, 11] was not considered in any of the mentioned theories. The accuracy of these theories is also influenced by the fact that the deviators of stress and plastic strain are not coaxial even under proportional loading as found by several investigators [12, 13].

References

[1] Prager, W., *The theory of plasticity: a survey of recent achievements*. James Clayton Lecture, Proceed. of the Inst. of Mech. Eng., (1955).

[2] Yougn, Y. I. and O. A. Shishmarev, *Dokl. Akad. Nauk S.S.S.R. 119* (1958) 46–48.

[3] Isotov, I. N. and Y. I. Yougn, *Dokl. Akad. Nauk S.S.S.R. 139* (1961) 576–579.

[4] Yougn, Y. I. and O. A. Shishmarev, *Zavodskaya Laboratoriya, 10* (1958) 1243–1245.

[5] Ilyushin, A. A., Plastichnost (in Russian), *Izd. AN S.S.S.R.* Moscow (1963).
[6] Lensky, V. S., Voprosi teorii plastichnosti, *Izd. AN S.S.S.R.* Moscow (1961) 58.
[7] Shishmarev, O. A., *Izvestiya vissihkh uchebnikh zavedeni. Mashinostroenie 1* (1971) 33–37.
[8] Shishmarev, O. A., *Inzh. Zhurn. 3* (1963) 740–747.
[9] Talypov, G. B., *Inzh. Zhurn., Mekh. Tverd. Tela, 6* (1966) 81–88.
[10] Mitrokhin, N. M. and Y. I. Yougn, *Dokl. Akad. Nauk S.S.S.R. 135* (1960) No. 4.
[11] Shishmarev, O. A., *Inzh. Zhurn., Mekh. Tverd. Tela* (1966) No. 5, 112–118.
[12] Taylor, G. J. and H. Quinney, *Phil. Trans. Roy. Soc.*, London (1931) 323.
[13] Davis, E., *Journ. Appl. Mech. 10* (1943) 187–196.

On the concept of influence surfaces of plastic prestraining in the analysis of mechanical performance of metals

W. Szczepiński

Institute of Fundamental Technological Research, Warsaw, Poland

1. Introduction

Over the last twenty years a considerable number of experimental investigations has been devoted to the analysis of influence of plastic prestraining on the yield condition for metals. The main purpose of that work was to verify various theoretical concepts of strain-hardening. It is very convenient to represent results of such experiments in stress space.

Stress space has been found to be suitable for interpretation of recent experiments on the behaviour of yield surfaces for metals prestrained along some complex loading paths. In Section 2 a short description is presented of former investigations of the yield surfaces for different modes of plastic prestraining.

Section 3 contains remarks on the possible use of stress space for graphical presentation of results of experimental tests regarding the effect of plastic prestraining on other properties of metals such as fatigue strength or various types of rheological performance.

In Section 4 a new concept is introduced usefull for presentation of results of experimental investigations on plastic, rheological and fatigue-type properties of metals in stress space. This concept consists in using the so-called surfaces of influence of plastic prestraining.

2. Recent investigations of yield surfaces for plastically prestrained metals

In this section recent results will be described of experimental studies of yield surfaces for metals prestrained plastically along complex loading paths.

The purpose of these investigations was to obtain more information on plastic behaviour of cold-worked metals. Yield surfaces of prestrained and then fully unloaded material are analysed for two types of prestraining histories. The first type of investigations consists in a comparison of yield surfaces for two prestraining paths which are different at the beginning but have a common part at the end. In the second type of experiments the prestraining paths consist of three parts; the first and third parts are the same while the second part differs by the numbers of loading cycles.

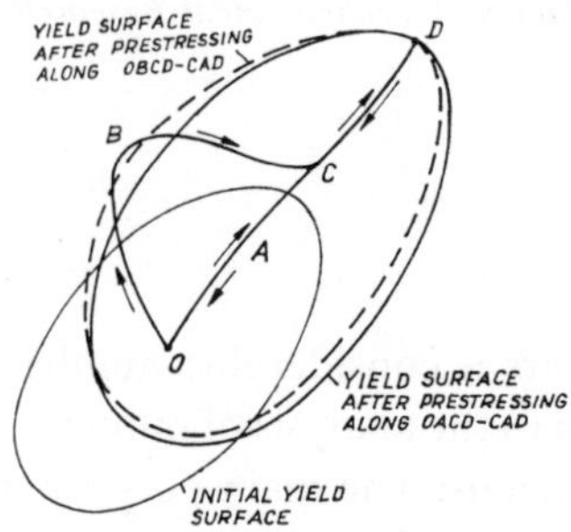

Fig. 1.

The first type of experiments is shown in Fig. 1. Two sets of specimens are tested. Each specimen of the first set is loaded beyond the initial yield surface along the path *OACD* and then undergoes unloading along *DCAO*. Specimens of the second set are loaded along *OBCD* and then unloaded along *DCAO*. Thus the unloading path is the same in both cases. The two loading paths have the common end section *CD*. Experimental results obtained for an aluminium alloy [1] and for the *M*-63 brass [2] indicate that if the common element *CD* is sufficiently long (roughly speaking, if it is larger than the portion *OBC*), yield surfaces for the two prestraining hisotries coincide practically. It is interesting to note that a similar result is obtained from the theoretical analysis, if the kinematic hardening rule is assumed to be valid. However, the shapes of theoretical surfaces are quite different from the shapes of the actual ones.

Fig. 2 presents an example of such experimental results [1]. Tests have been performed using flat specimens cut in different directions from a prestressed sheet. Two prestraining programmes are schematically shown in the insert to Fig. 2. The first of the two sheets was prestressed by uniaxial tension in the *y*-direction to point *A*, and then after unloading to the point *O* prestressed by

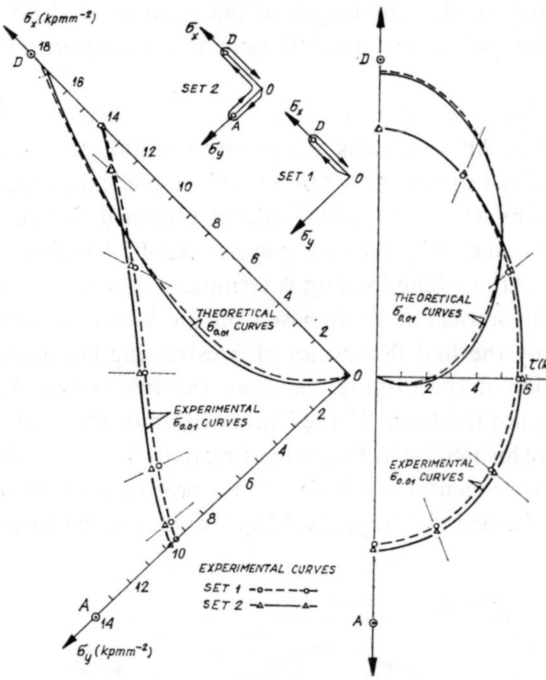

Fig. 2.

uniaxial tension in the x-direction to point D. The second sheet has been prestressed by uniaxial tension to point D only, the section AO being omitted. Experimental points corresponding to the conventional yield loci $\sigma_{0.01}$ (yield stress for 0.01 per cent offset) obtained during simple tension tests of small specimens cut from the two prestressed sheets are presented in the $\{\sigma_x, \sigma_y, \tau_{xy}\}$ stress space. Spatial curves plotted through these points differ only slightly for the two prestraining programmes. For the sake of comparison theoretical curves obtained in accordance with the kinematic hardening rule [3] are also shown in the figure. The difference between theoretical and experimental curves is so large that it is evident that the kinematic hardening rule departs qualitatively from the real behaviour of cold-worked metals. However, the distance between theoretical curves for both prestraining programmes is practically the same as the distance between experimental curves. Similar results for other loading paths were obtained by Miastkowski [2] using tubular specimens. This means that the kinematic hardening rule may predict the length of the common end portion of prestraining paths along which the metal 'forgets' the initial portion from O to C (Fig. 1).

An example of the second type of experiments is schematically shown in the insert to Fig. 3. All specimens are prestressed along a complex loading path composed of an initial section *OA* and of the cyclically repeated portion *ABACABA*... Compared are yield surfaces obtained for several sets of specimens. The number of prestressing cycles *ABACA* is different for each set of specimens. After cyclic loading is terminated specimens are unloaded following *AO*. Theoretical analysis based on the kinematic hardening rule indicates that only the first few cycles of prestressing are accompanied by plastic deformation if the distance between the two points *B*, *C* and the central point *A* is not too large. In the final position of the yield surface both points B and C are located on it. Entire loading path lies inside the translated yield surface. Experimental study for cyclic prestraining histories of this type has been performed by Marjanović [4]. Tubular specimens of the *M*-63 brass were used.

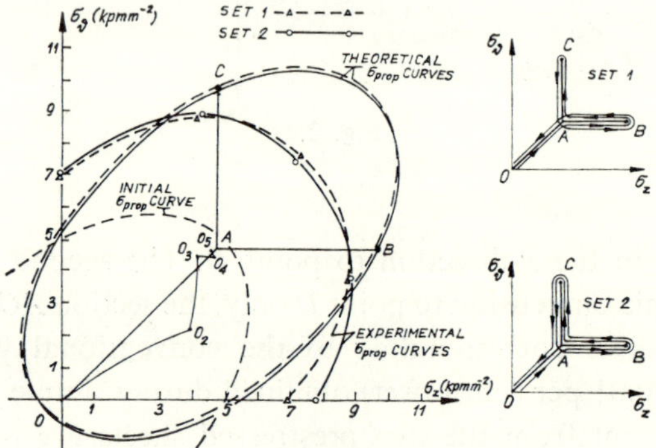

Fig. 3.

Fig. 3 shows an example of results of his experiments in which tubular specimens were loaded by axial force and internal pressure. Two prestraining programmes are shown in the insert. Prestraining history of specimens of set No 1 consists of one and a half of the cycle *ABACA*, while specimens of the set No 2 were prestressed in two full cycles. Shifted positions of the theoretical proportional limit curve were found for the two programmes if a kinematic hardening rule was assumed. The line $OO_2..O_5$ represents the path of the central point of the shifted Huber-Mises ellipse. The central point for the programme 1 is marked as O_4, and for the programme 2 as O_5. Two curves plotted through experimental points represent actual yield surfaces for both prestraining programmes. As before the difference between actual and theoretical yield curves is very large. Using a kinematic hardening rule one may predict only the distance between yield surfaces of metals prestrained along loading paths with different number of stress cycles.

3. Fatigue strength surfaces and rheological performance surfaces of plastically deformed metals

The effect of previous plastic deformation on certain other properties of metals such as fatigue strength or creep appears to be remarkable. However, in the papers available on fatigue performance, only such test results are reported in which solid cylindrical specimens prestressed by uniaxial tension undergo simple loading such as, reversed bending or tension-compression. Similarly, creep tests on uniaxially prestressed specimens are usually performed under simple tension (e.g., see [5]). No fatigue or creep test under combined stress with prestressed specimens is known to the author. Experimental studies of this kind would lead to a natural generalization of the well known studies of yield surfaces of plastically deformed metals. Suppose that for a given undisturbed material the fatigue strength surface is known in stress space (such surfaces are presented, for example, in [6] and [7]). Consider a set of identical tubular specimens made from this material, and prestressed beyond the initial yield locus along the same prescribed loading path. After unloading all specimens undergo fatigue-type tests, each of them under a different state of stress and the same type of stress cycles (symmetrical or non-symmetrical). The surface of fatigue strength will be different from the surface for undisturbed material. Unfortunately, such experimental studies are not available and even an approximate shape cannot be given for the fatigue strength surface of prestrained metals. Also experimental surfaces of the fatigue strength of prestrained metals could be found for a prescribed number of cycles.

Similar surfaces in the stress space may be experimentally obtained also for other quantities characterizing fatigue-type performance of prestrained metals. Suppose, that each specimen of a prestrained set is loaded under different cyclic stress. The type of cycles is the same in each case (for example all stress components change symmetrically). Assume the effective stress amplitude $A_{ef} = \sqrt{\frac{3}{2}B_{ij}B_{ij}}$ to be constant; B_{ij} denotes the deviator of the amplitude tensor A_{ij}. Each of the tested specimens will fracture after different number of cycles which may be represented in stress space as vectors of the same directions as the corresponding stress amplitude vectors. End points of these vectors form a surface of life-time of prestrained metal for various states of stress of the fatigue-type loading with constant effective amplitude.

Surfaces of this kind may also be constructed for various rheological properties of prestrained metals. It is known, for example [8], that the strain rate of secondary creep of nickel specimens prestrained beyond the yield locus in simple tension and then annealed at the temperature 700° C depends significantly on the extent of previous plastic deformation. Therefore one can expect that if, various combined stress states σ_{ij}^* are applied during creep tests of identically prestrained specimens instead of a simple tensile stress the effective strain rates of secondary creep $\dot{\varepsilon}_{ef}^*$ will depend not only on the extent of previous plastic deformation but also on the stress tensor σ_{ij}^*, even when the effective stress $\sigma_{ef}^* = \sqrt{\frac{3}{2}s_{ij}^* s_{ij}^*}$ remains the same in all creep tests. Representing experimentally found effective strain rates $\dot{\varepsilon}_{ef}^*$ in stress space as vectors with directions of the corresponding stress vectors σ_{ij}^*, the surface $\dot{\varepsilon}_{ef}^*(\sigma_{ij}^*)$ for prestrained metal may be obtained. For nonprestrained metals such a surface has a constant radius $\dot{\varepsilon}_{ef}^* =$ const. if the simplest creep hypothesis $\dot{\varepsilon}_{ef}^* = B(\sigma_{ef}^*)^n$ is assumed.

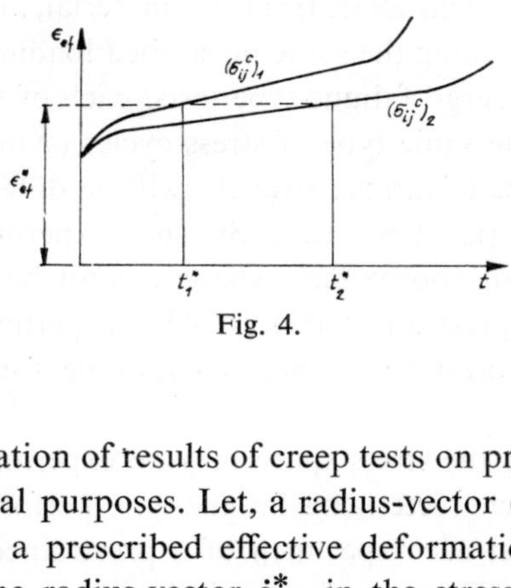

Fig. 4.

Another representation of results of creep tests on prestrained metals may be useful for practical purposes. Let, a radius-vector corresponding to the time t^* after which a prescribed effective deformation ε_{ef}^* is reached be plotted instead of the radius-vector $\dot{\varepsilon}_{ef}^*$, in the stress space for each test (Fig. 4). End points of all vectors t^* form the surface $t^*(\sigma_{ij}^*)$ for a chosen value of ε_{ef}^*.

4. Surfaces of influence of plastic deformation on mechanical performance of metals

Experimental analysis of the effect of plastic prestraining on the mechanical performance of metals discussed in previous sections are characterized by a procedure which will be called *the loading scheme* 1.

i. All specimens of the set are plastically deformed up to the same strain level under identical state of stress.

ii. Each specimen of the prestrained set is loaded by a different state of stress (static loading, cyclic loading, creep test etc.)

iii. Measured property (yield stress, fatigue strength, strain rate of secondary creep etc.) of prestrained metal is represented in stress space by a radius vector oriented in the same manner as the stress vector in *i.* End points of these vectors form a surface of the considered property.

Experiments performed according to the above procedure furnish data that enable the following question to be answered:

How will a metal previously plastically deformed along a prescribed loading path perform under service loadings characterized by different stress states?

From the point of view of engineering applications, the following question of no less importance can be formulated:

How will a metal plastically deformed to the same level with the use of different stress states perform under a strictly prescribed service loading?

If an answer to such a question is to be given, data must be collected from experiments employing the following procedure called *the loading scheme* 2:

i. Each specimen of the set is plastically prestrained with the use of different stress states along proportional loading path until a prescribed effective stress $\sigma_{ef} = \sqrt{\frac{3}{2}s_{ij}s_{ij}}$ is reached.

ii. All specimens of the prestrained set are loaded (static loading, cyclic loading, creep test etc.) in the same manner by the same state of stress.

iii. Measured property (yield stress, fatigue strength, strain rate of secondary creep etc.) of the prestrained metal is represented in stress space by a radius-vector oriented in the same manner as the stress vector from the point *i.*

End points of all vectors mentioned in *iii* form a certain surface in the stress space. Such surfaces obtained according to the *loading scheme* 2 will be called *surfaces of influence* of arbitrary prestraining on the measured property of metal.

4.1 *Surfaces of influence for the effective yield stress*

The concept of surfaces of influence of arbitrary plastic prestraining on the effective yield stress by subsequent loading along a prescribed path has been proposed in [9] in which experimental results for a particular case are also presented. More extensive experimental study of surfaces of influence for *M*-63 brass is presented in [10].

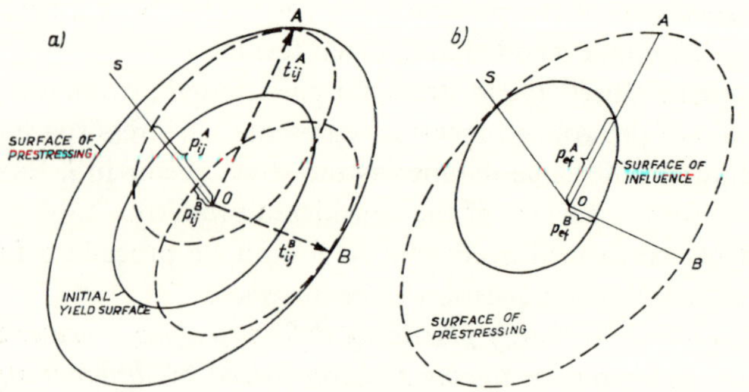

Fig. 5.

The principle of constructing surfaces of influence is shown in Fig. 5 for equivalent yield stress in the general case of six-dimensional stress space. Specimens of a material for which the initial yield surface $\sigma_{ef} = \sigma_0$ is known have been plastically deformed, each by a different stress t_{ij}. All points representing stresses t_{ij} are located on the common surface $t_{ef} = \alpha\sigma_0$ where $t_{ef} = \sqrt{\frac{3}{2}\tau_{ij}\tau_{ij}}$ (τ_{ij} is the stress deviator). Such surface will be called the surface of prestressing. The quantity $\alpha(\alpha > 1)$ is a scalar multiplier. Thus the surface of prestressing is formed by isotropic expansion of the initial yield surface. By loading along each of the chosen prestress paths, the initial yield surface is deformed and translated; two paths of prestressing represented by points *A* and *B* have been shown in Fig. 5a. Assume that each of previously deformed specimens is subsequently loaded along a common path *OS*. For these subsequent loadings the effective yield stress $p_{ef}^{(n)} = \sqrt{\frac{3}{2}\tau_{ij}^{(n)}\tau_{ij}^{(n)}}$ will be different for particular specimens. Here $r_{ij}^{(n)} = p_{ij}^{(n)} - \frac{1}{3}\delta_{ij}p_{kk}^{(n)}$ is the stress deviator, *n* is the number of a consecutive specimen and $p_{ij}^{(n)}$ is the plastic stress state defined by intersection of the subsequent loading path with the corresponding actual yield surface. For the two prestressing paths shown in Fig. 5a (points *A* and *B*) plastic stress states for subsequent loading along *OS* are defined by stress vectors p_{ij}^A and p_{ij}^B, respectively. Effective yield stresses $p_{ef}^{(n)}$ are then represented in stress space as vectors with the directions coinciding with the directions of corresponding prestress vectors $t_{ij}^{(n)}$. For the two prestress modes *A* and *B* both vectors p_{ef}^A and p_{ef}^B are shown in Fig. 5b. Through end points *A*, *B*, ... of all vectors $p_{ef}^{(n)}$ a surface of influence is plotted for the prescribed path of subsequent loading *OS*.

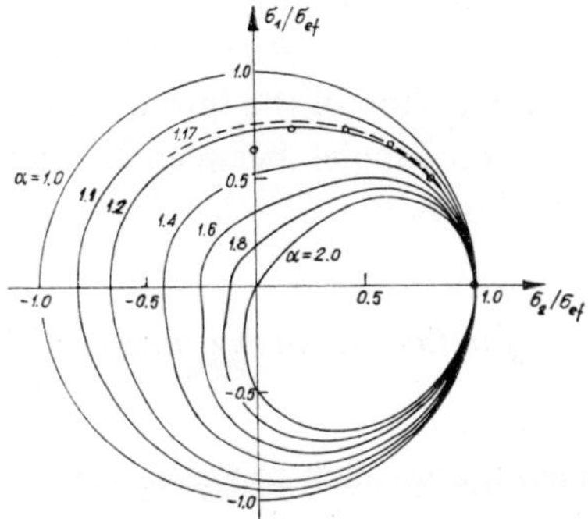

Fig. 6.

Surfaces of influence for effective yield stress may be constructed analytically if any theoretical hardening rule is assumed. In Fig. 6 an example is presented of a family of theoretical surfaces of influence for different values of α [9]. The kinematic hardening rule has been assumed. Construction of surfaces has been performed graphically for plane stress state with fixed principal directions and for the particular case of subsequent loading by tensile principal stress σ_2 only. As an initial yield surface, the Huber-Mises ellipse

$$\sigma_1^2 - \sigma_1\sigma_2 + \sigma_2^2 = \sigma_0^2$$

was assumed and, therefore, surfaces of prestress were represented by ellipses

$$t_1^2 - t_1 t_2 + t_2^2 = \alpha^2\sigma_0^2.$$

In Fig. 6 experimental points are also shown obtained with the use of tubular thin-walled specimens of the *M*-63 brass prestressed by different combinations of the axial force and internal pressure [9]. The value of the coefficient α was equal to 1.17 with respect to the conventional $\sigma_{0\cdot01}$ yield locus (yield stress for 0.01 per cent offset). After prestressing all specimens were loaded by axial tensile stress σ_2. In the vicinity of the axis of circumferential stress σ_1 experimental points depart from the theoretical curve for $\alpha = 1.17$ (dashed line).

Further experimental study [10] shows that the theoretical surfaces of influence calculated in accordance with either kinematic or isotropic hypothesis of hardening depart qualitatively even from the actual surfaces. Tubular specimens of the M-63 brass were used. Much better coincidence of experimental and actual surfaces of influence was obtained if in the theoretical analysis the mixed hardening hypothesis was used which assumes simultaneous translation and expanxion of the initial yield surface. However, even in this case remarkable discrepancies between theoretical and experimental surfaces were sometimes observed.

4.2 *Surfaces of influence for fatigue strength and rheological performance of metals*

In Section 3 a more general approach to the concept of surfaces in the stress space was presented for plastically deformed metals. It was indicated that, using such surfaces, experimental results can be represented for various mechanical properties of prestrained metals. In a similar way the concept of the surfaces of influence of previous plastic deformation generated by various stress states can be generalized for properties of metals others than effective yield stress.

Assume that in the point *iii* of *the loading scheme* 2 fatigue strength of all prestrained specimens is tested under identical conditions. If combined stress state with symmetrical cycle of all stress components is used in fatigue tests, and if for each specimen of the set the fatigue strength is defined by the tensor $A^{(n)}_{ij}$ of the stress amplitude, values of effective amplitude $A^{(n)}_{ef}$ may be plotted as vectors forming the surface of influence. Similarly surface of influence for the life-time of specimens tested under identical stress amplitude level may be obtained.

Rheological performance of prestrained metal can be described, for example, by the surface of influence representing values of the effective strain rate of secondary creep under combined stress. Surfaces of influence for the time after which a prescribed effective creep deformation is reached, may also be useful in engineering applications.

References

[1] Szczepiński, W. and J. Miastkowski, *Journ. Mech. Phys. Solids 16* (1968) 153–162.
[2] Miastkowski, J., *Arch. Mech. Stos. 20* (1968) 261–277.
[3] Shield, R. T. and H. Ziegler, *Zeitschr. Angew. Math. Physik IXa* (1958) 260–275.
[4] Marjanović, R., *Panašanje površina tečenja pri složenom cikličnom naprezanju*, Dissertation, Beograd University, 1970.
[5] Dieter, G. E., *Mechanical metalurgy*, McGraw-Hill, 1961, p. 364.
[6] Roš, M. and A. Eichinger, *Die Bruchgefahr Fester Körper bei wiederholter Beanspruchung - Ermüdung*, EMPA-Bericht 173, Zurich 1950.
[7] Rotvel, F., *Int. Journ. Mech. Sci. 12* (1970) 597–613.
[8] Parker, E. R., *Trans. ASM 50* (1958).
[9] Szczepiński, W. and K. Turski, *Arch. Mech. Stos. 21* (1969) 49–57.
[10] Turski, K., *Mech. Teoret. Stos. 9* (1971) 155–199.

The method of local variations for numerical solutions of elastic-plastic problems

N. V. Banichuk, and F. L. Chernous'ko

Institute of Mechanics, USSR Academy of Sciences Moscow

1. Introduction

The method of local variations as a tool for numerical solutions of variational problems was originally suggested in (1) and then further developed and extended in (2–5). The method is applicable to both conventional variational problems and to problems with all types of constraints on the desired functions. Over many years the method was used for numerical solutions of various nonlinear problems in solid mechanics, particularly in elastic-plastic analysis (3, 6, 7) and has been wholly standardized in the of ALGOL-60 programmes.

This paper will review the results of the method of local variations and its applications to elastic-plastic problems. The basic idea is described and some estimates of the convergence are given. The feasibility of a numerical solution of various boundary value problems for elastic-plastic solids via the method of local variations, is shown for the following cases:

(1) torsion of a bar;

(2) plane problems for the Prandtl-Reuss elastic-plastic medium with different types of boundary conditions;

(3) axially symmetric indentation problems for an elastic-plastic half-space;

(4) steady motion of visco-plastic medium.

Results of numerical solutions are given and mechanical effects disclosed in numerical solutions of specific elastic-plastic problems are discussed.

2. The method

The algorithm for the method of local variations will be discussed using the following problem: Find a function $u(x, y)$ determined in the region D of the plane xy under the conditions

$$(x, y, u) \in G \qquad \text{in } D$$

and minimizing the functional

$$J = F(J_1, \dots J_\tau) \tag{2}$$

where

$$J_i = \int_D f_i(x, y, u, u_x, u_y)\mathrm{d}x\mathrm{d}y, \qquad i = 1, \dots \tau. \tag{3}$$

f_i, F are specified functions; G is a set specified in the three-dimensional space x, y, u and the subscripts x, y denote partial derivatives. Decompose the region D into equal cells by straight lines

$$x = x_j = x_0 + j\Delta x, y = y_k = y_0 + k\Delta y, \Delta x > 0, \Delta y > 0;$$
$$j = 0, \dots m; \qquad k = 0, \dots n,$$

where m, n are integers and find approximate values of the function u at the nodes of the grid $(x_j, y_k) \in D + C$, or values $u_{jk} = u(x_j, y_k)$. C is the contour of D. The constraint (1) for these values takes the form

$$(x_j^*, y_k^*, u_{jk}^*) \in G. \tag{4}$$

The integrals (3) will be approximately replaced by sums of integrals for cells completely belonging to the region $D + C$,

$$J_i = \sum f[x_j^*, y_k^*, u_{jk}^*, (u_x)_{jk}^*, (u_y)_{jk}^*]\Delta x\Delta y. \tag{5}$$

The following notation for mean values of functions in a cell is introduced

$$\begin{aligned} x_j^* &= (x_j + x_{j+1})/2, \qquad y_k^* = (y_k + y_{k+1})/2, \\ u_{jk}^* &= (u_{jk} + u_{j+1,k} + u_{j,k+1} + u_{j+1,k+1})/4, \\ (u_x)_{jk}^* &= (u_{j+1,k} + u_{j+1,k+1} - u_{jk} - u_{j,k+1})/2\Delta x, \\ (u_y)_{jk}^* &= (u_{j,k+1} + u_{j+1,k+1} - u_{jk} - u_{j+1,k})/2\Delta y. \end{aligned} \tag{6}$$

The initial variational problem (1)–(3) may be restated as the problem of finding the values of u_{jk} that would satisfy the constraints (4) and minimize the function (2) where J_i's are determined by the formula (5).

In order to start operations according to the algorithm a certain initial approximation is needed, viz. the table of numbers for u_{jk}. The algorithm first screens this approximation checking the condition (4) and evaluating the functionals (5), (2). If the constraint (4) is satisfied, the values of u_{jk} are varied otherwise the information is printed out that the constraint does not hold and the initial approximation has to be modified.

The variation algorithm consists of successive iterations. Over one iteration the algorithm looks through all nodes of the grid in a certain order checking each node just once. For each node a new value of u_{jk} is sought, such that the algorithm tries the old value of u_{jk} and the values of $u_{jk}+h$, $u_{jk}-h$. In order to compute the functional J for each of these possible values, in each sum (5) for J_i no more than four addenda should be calculated containing u_{jk}. These addenda are associated with cells lying completely in the region $D+C$ and with the point (x_j, y_k) as the node. If the replacement of u_{jk} by $u_{jk}+h$ or $u_{jk}-h$ does not violate the constraint (4) and reduces the functional J, then in the solution table the quantity u_{jk} is replaced by $u_{jk}+h$ or $u_{jk}-h$ and the algorithm goes to the next node, otherwise the old value of u_{jk} is retained in the table. Note that in each node the variation in that direction is first tried (viz either $u_{jk}+h$ or $u_{jk}-h$ is taken) for which the functional was reduced in the previous node. Thus a number of operations is considerably reduced (see (1)).

Following the completion of one iteration, another iteration is started, etc. until complete convergence is obtained at the given Δx, Δy, h, in other words, until none of the numbers u_{jk} is found to change as a result of the iteration. Then the variation step h is reduced, e.g. halved, and the iterations start a new, etc. An approximate solution on a specific grid is obtained when complete convergence of iterations has been achieved at $h = O\ (\min/(\Delta x)^2, (\Delta y)^2/)$. A solution should be obtained on a grid with finer mesh by increasing the numbers m, n and reducing accordingly the numbers Δx, Δy while the initial approximation for the new grid is interpolated from the solution for the old grid. A check and a variation take place again where the step h is again reduced gradually from a certain initial value as described above. When complete convergence of iterations has been achieved at the minimum Δx, Δy, h $[h = O(\min/(\Delta x)^2, (\Delta y)^2/)]$, the numerical solution may be considered to be finished.

3. Some extensions

The simplest version of the method was described above consider now some possible modifications. First of all, the integral (3) can be approximated in different ways by using, more accurate difference-quadratic formulae instead of (5), (6). In some specific problems non-uniform, triangular mesh as well as other types of grids can be used. The variation step h should not necessarily be the same for all nodes. Finally, at each iteration only those nodes can be checked for which the values of u_{jk} have changed at the preceding iteration. This would require a greater storage capacity but reduce the computing time. Such a modification for variational problems with one independent variable programmed in ALGOL-60 was reported in (5).

Let us consider now certain modifications and extensions of the statement of the problem (1)–(3) where the above alogrithm of local variations is applicable.

(1) The desired function $u(x, y)$ in the relations (1), (3) can be a vector-valued function with the components $u^1, \ldots, u^k$. In this case the variation is implemented sequentially for each of the components u^i $(i = 1, \ldots, K)$ at fixed values of the remaining ones (see (7)).

(2) The method is applicable to variational problems in which the integrals (3) depend on higher order derivatives of the desired function. Then the expression (5) will include difference approximations of these derivatives, such that the variations of the function u at one node of the grid would require the calculation of more than four addenda in the sum (5). Problems with higher order derivatives were solved by the method of local variations in (8).

(3) The method is essentially applicable to problems in which the desired function depends on three and more variables while the integrals (3) are taken for regions of appropriate dimensionality. The formulae (5), (6) and the solution algorithm change in an obvious way: the elementary cells will be the parallelepipeds of the appropriate dimensionality and the variation of the function u in one cell would entail changes of integrals for at least eight cells in the vicinity.

(4) There are numerous variational problems in which the functional (2) contains integrals over the boundary C of the region D. Assume, for instance, that the integral J_1 of eq. (2) is

$$J_1 = \int_C \varphi(x, y, u, u_x, u_y)\mathrm{d}s \tag{7}$$

where φ is the desired function, ds is an arcelement of the boundary C. In this case it appears convenient first to transform the independent variables x, y so that the region D becomes a block or another region with boundaries parallel to the coordinate axes. Then the integral (7) is replaced by a sum of integrals over the regions between the neighbouring boundary points. The variation proceeds as before with the only difference that in variations of u_{jk} at the boundary points the changes of addenda corresponding to the integral (7) are recognized.

4. Estimates of convergence

Certain results of the convergence analysis of the methods will be formulated without proof.

(1) Let J^*, u^* be the magnitude of the integral functional minimum and the function which implements that minimum, respectively, for the variational problem

$$J = \int_D [au_x^2 + bu_y^2 + Cu^2 + 2gu]\mathrm{d}x, \tag{8}$$

$$u = \psi(x, y) \quad \text{on} \quad C \tag{9}$$

where $\psi(x, y)$ is the specified continuous function. Then if the conditions $a > 0$, $b > 0$, $a+b \geqq q \geqq 0$, $\frac{1}{2}q \geqq c \geqq 0$ are met, the values of J, u obtained by the method of local variations (at fixed Δx, Δy and $h = 0(\delta^2)$, $\delta = \min(\Delta x, \Delta y)$) converge to J^*, u^* with $\Delta x \to 0$, $\Delta y \to 0$.

(2) Let us consider a particular case of the problem (1)–(3) in which the functional (2) has the form

$$J = J_1/J_2 = \int_D f_1(x, y, u, u_x, u_y)\mathrm{d}x\mathrm{d}y \Big/ \int_D f_2(x, y, u, u_x, u_y)\mathrm{d}x\mathrm{d}y \tag{10}$$

while the constraint (1) includes boundary conditions only. The minimization of the functionals of the form (10) results from certain eigenvalue problems for which J_1, J_2 are quadratic functionals. In a particular case f = const. a conventional variational problem with an integral functional exists. For variational problems with functionals of the form (8) the solution u_{jk} obtained by the method of local variations with finite values of h and $\delta = \min(\Delta x, \Delta y)$ may be shown to satisfy finite difference approximation of the Euler equation

$$\frac{\partial f_1}{\partial u} - \frac{\partial}{\partial x}\left(\frac{\partial f_1}{\partial u_x}\right) - \frac{\partial}{\partial y}\left(\frac{\partial f_1}{\partial u_y}\right) - \\ -\lambda\left[\frac{\partial f_2}{\partial u} - \frac{\partial}{\partial x}\left(\frac{\partial f_2}{\partial u_x}\right) - \frac{\partial}{\partial y}\left(\frac{\partial f_2}{\partial u_y}\right)\right] = 0, \qquad \lambda = \frac{J_1}{J_2} \tag{11}$$

to within the accuracy of the order of magnitude max $(h, h/\delta, h/\delta^2)$. It follows, in particular, that for the termination of variation on a specific grid, $h = 0(\delta^2)$.

(3) Consider a non-conventional variation of the minimum of the functional

$$J = \int_a^b f(x, u, u_x)\mathrm{d}x \tag{12}$$

under the condition

$$u^-(x) \leqq u(x) \leqq u^+(x) \tag{13}$$

where u^+ and u^- are functions given in the range (a, b) while $u^+ \geqq u^-$. One or both constraints may be absent for some points of the range (a, b); in this case it is assumed that $u^+ = \infty$ or $u^- = -\infty$ at these points. The boundary conditions $u(a) = c_1$, $u(b) = c_2$ are also contained in (13). At these points $u^+(a) = u^-(a) = c_1$ and $u^+(b) = u^-(b) = c_2$. The function f is assumed to be twice continuously differentiable over u and u_x. Divide the stretch (a, b) into N equal parts by the points $x_i = i\Delta x = i(b-a)/N$. Replace the continuous problem (10), (13) by a nfiite difference analog

$$J = F = \sum_{i=0}^{N-1} f\left(x_i + \frac{\Delta x}{2}, \frac{u_i + u_{i+1}}{2}, \frac{u_{i+1} - u_i}{\Delta x}\right), \tag{14}$$

$$u^-(x_i) \leqq u_i \leqq u^+(x_i), \qquad 0 \leqq i \leqq N. \tag{15}$$

The following theorem is valid: If the conditions

$$f_{u_x u_x} f_{uu} - f_{uu_x}^2 \geqq 0, f_{u_x u_x} \geqq 0, f_{uu} \geqq 0$$

hold true, then in solving the problem (14), (15) by the method of local variations the functional (12) is convergent of the order $h/(\Delta x)^2$. When stricter constraints are imposed on the integrand, there is a quadratic convergence in terms of the variation step $h^2/(\Delta x)^4$ for the functional and linear convergence of the order $h/(\Delta x)^{\frac{3}{2}}$ for the desired function. Under analogous assumptions such estimates of convergence are valid for two-dimensional variational problems in which the desired function u depends on two variables, or $u = u(x, y)$.

5. Torsion of elastic-plastic bar

The study of elastic-plastic torsion of a bar may be reduced to the solution of a variational problem,

$$J = \int_D [\tfrac{1}{2}(u_x^2+u_y^2)-2G\theta u]\mathrm{d}x\mathrm{d}y \to \min, \tag{16}$$

$$u = 0 \qquad \text{on} \quad C, \tag{17}$$

$$u_x^2+u_y^2 \leqq \tau_0^2 \qquad \text{in} \quad D. \tag{18}$$

G is the shear modulus, τ_0 is the yield point in pure shear (const), θ is the angle of twist per unit length of the bar, D is the cross-section area, and C is the boundary of the region D. $u(x, y)$ denotes the shearing stress function such that $\tau_{xz} = u_y$, $\tau_{yz} = -u_x$. The minimum of the functional (16) is sought in a class of functions which satisfy the conditions (17) and (18). The strict inequality in (18) applies in the elastic region while equality is true in the plastic zone. If the solution to the problem (16)–(18) is found, the torque can be obtained by the formula

$$M = 2\int_D u(x, y)\mathrm{d}x\mathrm{d}y.$$

The problem thus formulated was expressed in the dimensionless form by the formulae $x' = x/d$, $y' = y/d$, $u' = u/2G\theta d^2$, $\alpha = 2G\theta d/\tau_0$, $\mu = M/2\tau_0 d^3$ (d is the reference length in the region D) and solved numerically on a computer by the use of algorithm developed in of Sec. 2 (see [3, 6]) for a

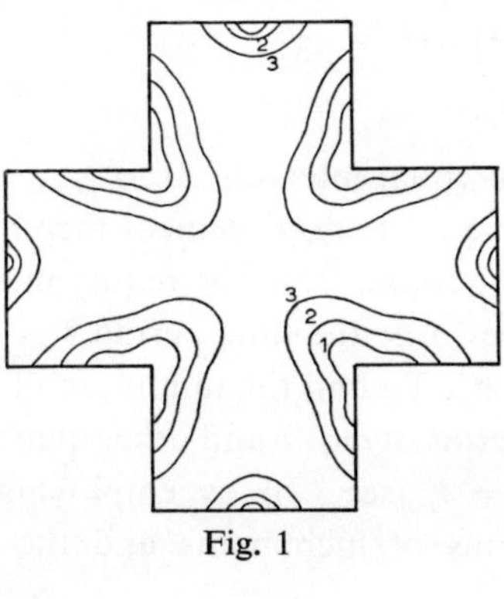

Fig. 1

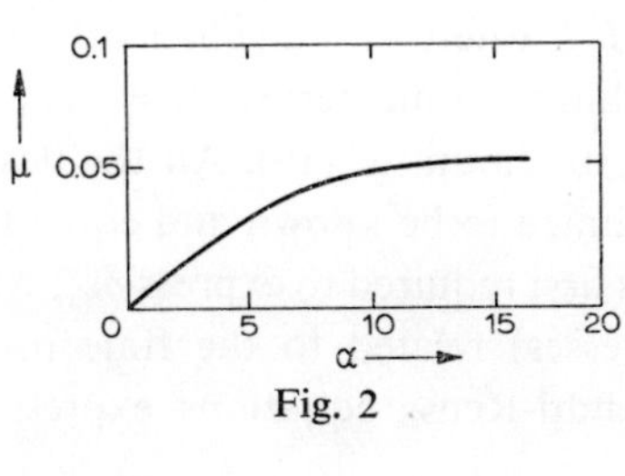

Fig. 2

number of regions D with different values of the parameter $\alpha = 2G\theta d/\tau_0$. Note that, unlike (1), the constraint (18) includes derivatives of the desired function. This required an additional modification of the method described in Sec. 2. The boundary between the elastic and the plastic zone was determined following the solution to the problem by using the relation (18). Computations were performed for many shapes of the cross-section; the available programmes permit a standard solution to twisting problems for practically all shapes. The results for an X-shaped cross-section are given in Figs. 1 and 2. Curves 1, 2, 3 in Fig. 1 represent the elastic-plastic boundaries at $\alpha = 20/3$, 10, 20. Fig. 2 shows the calculated relationship between the torque and the angle of twist.

6. Plane elastic-plastic problems

A quasistatic loading of an elastic-plastic body can be studied using the Prandtl-Reuss theory by considering sufficiently close time intervalls and sequentially solving the problems of a transition from one equilibrium state of the medium into another. The variational principle of the plastic flow theory (9, 10) is used whereby the determination of displacement increments $\delta u(x, y)$, $\delta v(x, y)$ reduces to minimization of the functional

$$J(\delta u, \delta v) = \int_D (\delta\sigma_x\delta\varepsilon_x + 2\delta\tau_{xy}\delta\varepsilon_{xy} + \delta\sigma_y\delta\varepsilon_y)\mathrm{d}x\mathrm{d}y - \int_{C_2} (\delta p\delta u + \delta q\delta u)\mathrm{d}S \quad (19)$$

which is taken in a class of functions δu, δv satisfying the boundary conditions for δu, δv given on the part C_1 of C bounding the region D. On the part C_2 of the circumference the surface load increments δp, δq are prescribed. The Mises yield condition is employed

$$g = \tfrac{1}{6}[(\sigma_x - \sigma_y)^2 + (\sigma_y - \sigma_z)^2 + (\sigma_z - \sigma_x)^2] + \tau_{xy}^2 = \tau_0^2 \quad (20)$$

where τ_0 is the yield point in pure shear.

Divide the range $(0, T)$ by points t_l into equal intervals $\Delta t = t_{l+1} - t_l = T/L$ where L is a sufficiently large number. The displacement increments within the time period $\Delta t = t_{l+1} - t_l$ will be found from the requirement of the minimum of (19). All the desired values over the time period $t = t$ are assumed to be known and equal to $u^1, \ldots, \sigma_y^1$. To find the minimum of (19), it is first required to express $\delta\varepsilon_x, \ldots, \delta\varepsilon_y$ in terms of δu, δv and other quantities (stresses) related to the time instances $t = t_l$ (see (7)) by employing the Prandtl-Reuss equations expressed in terms of increments and the geometrical relations between deformations and displacements. For the sake of convenience the following dimensionless variables $x' = x/d$,

$$u' = Eu/d\tau_0, \varepsilon'_x = E\varepsilon_x/\tau_0, \sigma'_x = \sigma_x/\tau_0$$

e, t, c are introduced. The resulting functional has the form

$$J(\varphi, \psi) = \tfrac{1}{2}\int_D [a(\varphi_x^2+\psi_y^2)+2b\varphi_x\psi_y+\tfrac{1}{2}(\varphi_y+\psi_x)^2-2C\Lambda^2]\mathrm{d}x\mathrm{d}y-$$

$$-\int(\delta p\varphi+\delta q\psi)\mathrm{d}S,$$

$$\Lambda = \begin{cases}\tfrac{1}{2}[(\sigma_x^l-\sigma^l)\varphi_x+(\sigma_y^l-\sigma^l)\psi_y+\tau_{xy}^l(\varphi_y+\psi_x)], q = 1, \\ 0, \quad \text{at} \quad g < 1,\end{cases} \tag{21}$$

$$a = (1-\nu)/(1+\nu)(1-2\nu), b = \nu/(1+\nu)(1-2\nu), c = 1/(1+\nu),$$

where $\delta u = \varphi$, $d\delta = \psi$.

If a solution to the variational problem (21) is obtained for a given C_1 in other words, $\delta u = \varphi(x, y)$, $\delta v = \psi(x, y)$ are determined, then the deformation and stress increments are readily computed from both the relations between deformations and displacements and Prandtl-Reuss equations (7). The values of displacements, deformation, and stresses for $t = t$ are given by the formula $u^{l+1} = u^l+\delta u, \ldots, \sigma_z^{l+1} = \sigma_z^l+\delta\sigma_z$.

In (7) the following problem of extrusion was solved. A rectangular block

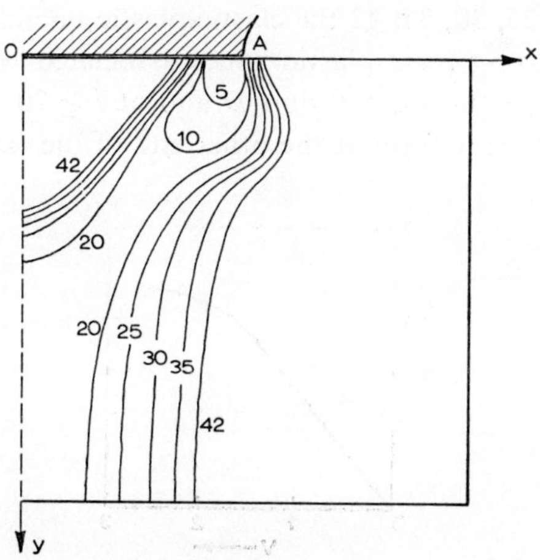

Fig. 3

with edges parallel to the coordinate axes x, y, z is compressed between two perfectly rigid plates (extruders), $y = 0, y = 1$, in the direction of the axis y. The lower extruder (Fig. 3) covers the entire side surface $y = 1$ while the upper extruder makes contact with the surface of the bar in the interval $y = 0, |x| \leqq 0.5$. The contact surfaces were assumed to be ideally smooth. One half of the region of the bar cross-section is shown in Fig. 3 in the xy-plane. The deformation in the z-direction is prevented. Let V_* denote the depth of the extruder penetration into the block such that yielding appears first in one of the cells into which the region D is decomposed for the purpose of numerical solution. Initially, the lagorithm of Sec. 2 was used to solve the variational problem (21) with the boundary condition on the contact surface $\delta v = v_1 < v_*$. The calculated equivalent stress is at its maximum in the vicinity of the point A (Fig. 3) with the coordinates $x = 0.5, y = 0$. Considering the yield condition (20) and using the stress-strain relation, the depth of the extruder penetration

$$v_* = v_1 (\max_{xy})^{-\frac{1}{2}} \qquad xy \in D + C$$

was found at which in a cell adjacent to the point A the plastic state sets in consequently the solution corresponding to that instant was determined. Note that in a strict formulation of the problem the plastic zones develop near the edge of the extruder at any infinitesimal penetration. The quantity v^* depends on the size of the mesh and tends to zero when these sizes tend to zero. Further computations were performed for $v_l = v_* + l\Delta v, l = 1, \ldots, 42$. For $l = 5, 10, 20, 25, 30, 35, 42$ the elastic-plastic boundaries are shown in Fig. 3. With increasing l the plastic zones associated with these numbers penetrate even deeper into the region D and at $l \geqq 20$ merge with plastic zones that symmetrically form at the other end of the extruder ($x = -0.5$,

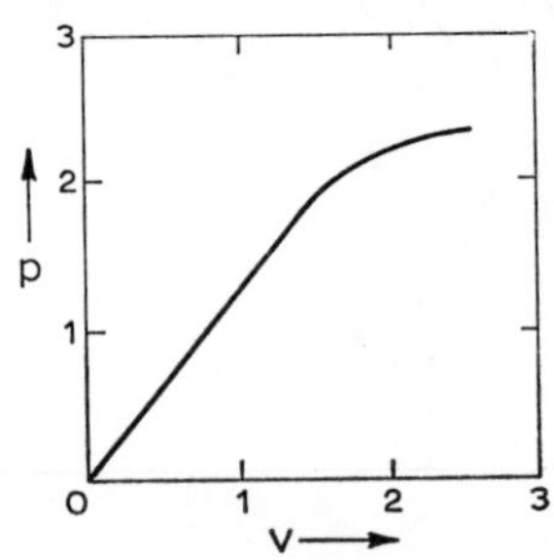

Fig. 4

$y = 0$) and pierce the region D. The total load on the extruder is given by the formula

$$P = 2\int_0^1 \sigma_y \mathrm{d}x$$

which is shown in Fig. 4 versus the penetration depth.

7. Axially symmetric indenter penetration of an into an elaxtic-plastic half-space

Penetration of a rigid, ideally smooth cylinder into an elastic-plastic half space (Fig. 5) using cylindrical system or coordinates r, v, z is an extension of the plane problem treated above. Since the region $D(r \geqq 0, z \geqq 0)$ is

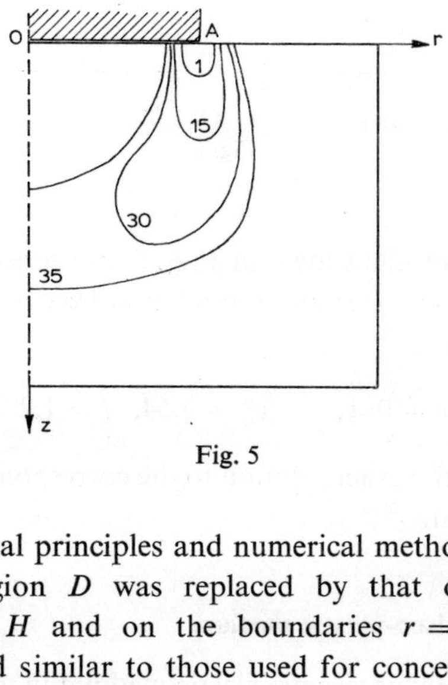

Fig. 5

not bounded, extremal principles and numerical methods may not be used.

Therefore the region D was replaced by that of finite dimensions $0 \leqq r \leqq R, 0 \leqq z \leqq H$ and on the boundaries $r = R, z = H$ the conditions were imposed similar to those used for concentrated forces in the Boussines problem.

A justification of such an approach follows from the fact that at sufficiently large distances from the indenter the displacements u, v asymptotically tend to the values given by the Boussinesq solution. The boundary conditions for these displacements at $z = H$, $r = R$ contain as a multiplier the magnitude of the force P exerted by the medium on the extruder. This force is unknown in advance but can be found simultaneously with the displacements, deformations and stresses. Consequently, in using the procedure of Sec. 2 an auxiliary iteration process had to be employed for evaluation of P. Another specific feature of the numerical solution to this problem is the use of a triangular non-uniform mesh r_j, z_k which expands linearly with the distance from the contacting surface.

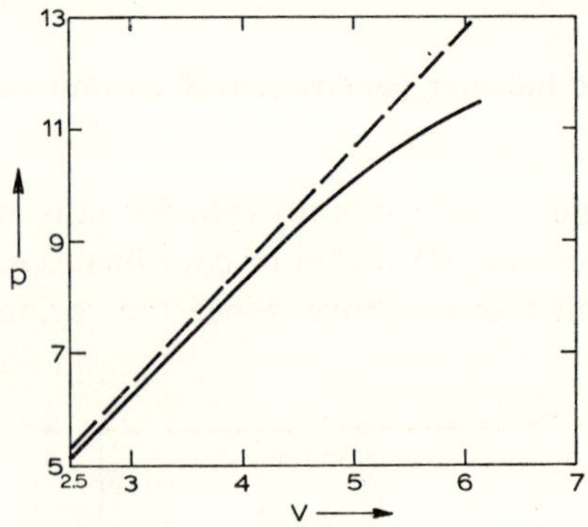

Fig. 6

The computation results shown in Figs. 5 and 6 permit one to show the development of plastic zones and the relation between the force P and the indenter penetration,

$$v = v^* + l\Delta v, \quad \Delta v = 0.1, \qquad v^* = 2.54, \quad l = 1,2$$

The dash line shows the exact solution to the corresponding contact problem for an elastic medium.

8. Steady flow of a visco-plastic medium

Analysis of steady flow of viscous-plastic medium in a cylindrical tube with cross-section D and boundary C may be reduced to a variational problem (11) with a minimum

$$J = \int_D \left\{ \frac{\mu}{2} (u_x^2 + u_y^2) + \tau_0 (u_x^2 + u_y^2)^{\frac{1}{2}} - u\Delta p \right\} \mathrm{d}x\mathrm{d}y, \tag{22}$$

$$u = 0, \quad \text{on} \quad C \tag{23}$$

whereas the flux of the medium is given by the formula

$$Q = \int_D u(x, y)\mathrm{d}x\mathrm{d}y$$

$u = u(x, y)$ is the flow velocity, Δp is the constant pressure drop in the direction of the tube axis, and μ describes the medium viscosity.

It is shown in (11) that the variational problem does not to belong to the category of conventional variational problems which may be reduced directly to a boundary value problem for the Euler equations. Nevertheless, the method of local variations provides a numerical solution. In (12) the problem was thus solved for a number of regions D and a wide range of variations of the parameter $\alpha = \Delta p/\tau_0$. Results for the square region D and certain specific features of the motion will be given. The computed velocities vary from zero on the boundary C to a certain value in the central part where the velocity is constant. The regions in which the viscoplastic medium moves at a constant velocity are the kernels of fluxes. Closed curves 1, 2, 3,

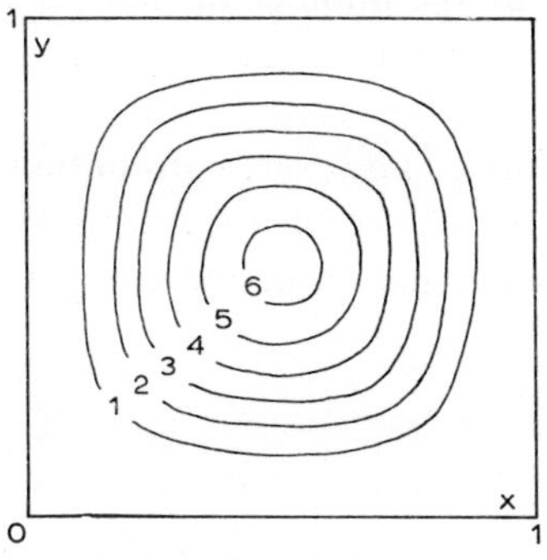

Fig. 7

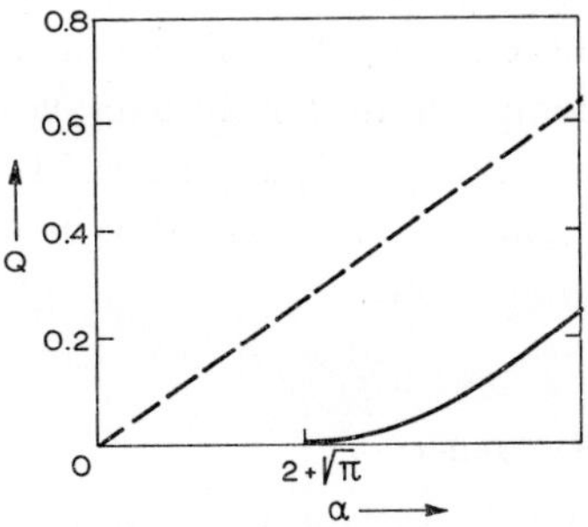

Fig. 8

4, 5, 6 in Fig. 7 show the boundaries of kernels for $\alpha = 6, 7, 8, 10, 14, 20$, respectively. With increasing α the sizes of kernels diminish while the distribution of velocities approaches that of a viscous liquid. The dependence of the dimensionless flux $Q' = Q\mu/\tau_0$ on the parameter α is shown in Fig. 8 as a solid curve. The region $\alpha \leqq \alpha_*$ where the flux $Q = 0$ represents the case in which there is no flow of the visco-plastic medium at pressure drops below the critical value. At $\alpha \geqq \alpha_*$ the function $Q'(\alpha)$ is nonlinear but tends to be linear increases further. For comparison in Fig. 8 the dash straight line shows the relationship between a viscous liquid flux and the pressure drop for a tube of the same cross-sectional area. In (12) the problem was also solved of a stationary motion of a visco-plastic medium in an inclined channel.

9. Further problems

Another class of problems in which the method of local variations can be used concerns elastic and elastic-plastic stability. In a variational statement these problems reduce to the minimization of functionals of the type (10) where f_1 and f_2 depend also on higher order derivatives of the desired functions. An extension of the method for this case is described in Sec. 3. Note that the problems of elastic stability were treated by the method of local variations in (2).

We shall give now an information on the order of magnitude of the computer time required for a typical two-dimensional problem (for instance, for the torsion problem, Sec. 5 or the flow problem of Sec. 8; for problems of Secs. 6 and 7 this time must be multiplied by the number of time steps employed.) For the BESM-6 computer (addition time 10^{-6} sec), for $30 \times 30 = 900$ cells mesh and for the final variation step $h \sim 10^{-7}$ the computer time is in average 0.5 minute. It differs, however, substantially depending on parameters of the problem, on the choice of initial approximation, and on the way of programming and translation of the computer programme.

References

[1] Chernous'ko, F.L., The method of local variations for numerical solutions to variational problems. (in Russian) *Zh. vychisl. matem. fiz.*, *5* (1965) No. 4.

[2] Krylov, I. A. and F. L. Chernous'ko, Solution of optimal control problems by the method of local variations (in Russian). *Zh. vychisl. matem. i matem. fiz.*, *6* (1966) No. 2.

[3] Banichuk, N. V., V. M. Petrov and F. L. Chernous'ko, Numerical solution of variational and boundary value problems by the method of local variations. (in Russian) *Zh. vychisl. matem. i matem. fiz.*, *6* (1966) No. 6.

[4] Banichuk, N. V., V. M. Petrov and F. L. Chernous'ko, The method of local variations for variational problems with non-additive functionals. (in Russian) *Zh. vychisl. matem. i matem. fiz.*, *3* (1969) No. 3.

[5] Banichuk, N. V., I. A. Krylov, V. M. Petrov and F. L. Chernous'ko, An algorithm of the method of local variations for solution to variational problems with one

independent variable. Algorithms and Algorithmic Languages (in Russian). USSR Academy of Sciences Computing Center, (1969), Issue 4.

[6] Banichuk, N. V., The calculation of elastic-plastic twisting of rods by the method of local variations. (in Russian) *Inzh. zh. MTT*, (1967) No. 1.

[7] Banichuk, N. V., Calculating the load of an elastic-plastic body, (in Russian) *Izv. AN SSSR, MTT*, (1969) No. 1.

[8] Banichuk, N. V., A numerical solution to the problem of deflection for an elastic plate with restrictions. (in Russian) *Inzh. zh. MTT*, (1967) No. 4.

[9] Hodge, P. and W. Prager, A variational principle for plastic materials with strain hardening. *J. Math. and Phys.*, *27* (1948) p. 1.

[10] Greenberg, H. J., Complementary minimum principles for an elastic-plastic material. *Quart. Appl. Math.*, *7* (1949) p. 85.

[11] Mosolov, P. P. and V. P. Myasnikov, Variational methods in the theory of viscous plastic medium fluxes. (in Russian) PPM, *29* (1965) No. 3.

[12] Banichuk, N. V., Calculating the flux of a viscous-plastic medium in tubes by the method of local variations. (in Russian) *Izv. AN SSSR, MZhG*, (1966) No. 6.

Constitutive equations of nonisothermal elastic-plastic deformation and methods of their practical application

A. G. Ugodchikov and Yu. G. Korotkikh
Gorky State University, Gorky, USSR

In order to investigate stress and strain states in structural elements exposed to complex temperature and deformation rate conditions the improvement of the constitutive equations, describing the processes of nonisothermal elastic-plastic deformation is necessary. Approximate methods of solutions of boundary value problem are required. A theory of thermoplasticity based on the concept of kinematic, isotropic hardening and accounting for the effect of strain rate and hydrostatic pressure is developed in the paper. Solutions of some specific problems are given.

Introduction

The basic factors determining the response of a material subjected to elastic-plastic deformation are: temperature, hydrostatic pressure, strain and strain rate.

The temperature and the deformation rate affect strain hardening and work softening of metals. The higher the temperature, the higher is the rate of work softening. Above the temperature of recrystallization (0.3 to $0.4T_p$ where T_p is the melting point) the rate of accumulation of lattice imperfections due to plastic deformation may be lower than their rate of decay due to recrystallization [1].

Existence of physical limits on both the dislocation velocity and the rate of dislocation density is responsible for the experimentally observed fact that under certain temperature-strain rate conditions plastic flow is checked so that resistance to deformation increases with the strain rate [2–4]. Under blast loading where the pressure reaches tens and hundreds of thousand bars, mechanical properties of the material are influenced by the hydrostatic pressure causing phase transitions and producing an increase in the material strength, shear and bulk moduli, yield stress, etc. [5–7].

Elastic-plastic deformation of materials results in general, in plastic anisotropy, which may be in the form of the 'spatial' Bauschinger effect. A second essential feature of elastic-plastic straining of metals is generation of the so-called latent (implicit) potential energy. This is mainly due to non-uniform strain distribution caused by slips in crystal as well as to the lattice imperfections which strongly effect the process of elastic-plastic deformation of metals [1, 8]. Below the recrystallization temperature the latent energy may be as high as 10% to 15% of the total energy absorbed [1]. Above the temperature of recrystallization the stored latent potential energy may be released.

A large number of mechanical models concerned with various aspects of elastic-plastic deformation of metals exists [2, 10–23]. The models relate mainly to the deformation theory of plasticity, to the flow (incremental) theory, and to the statical theory. The flow theory employing the concept of kinematic hardening describes the experimental results well enough even for quite complex loading programmes [11, 13, 23, 24, 25].

The present article develops a nonisothermal theory of plasticity with kinematic and isotropic strain hardening. The theory takes into account the strain rate and hydrostatic pressure effects. An algorithm for solving the static and dynamic problems of thermoplasticity is presented, and specific results of numerical computations are given.

Basic equations

To derive fundamental relationships for an initially isotropic medium, the stress σ_{ij} and strain e_{ij}-tensors and their increments are resolved into spherical σ, $\Delta\sigma$, e, Δe and deviatoric p_{ij}, Δp_{ij}, ε_{ij}, $\Delta\varepsilon_{ij}$ components

$$\Delta e_{ij} = \Delta e\delta_{ij} + \Delta\varepsilon'_{ij} + \Delta\varepsilon''_{ij} \tag{1}$$

$$\Delta\sigma_{ij} = \Delta\sigma\delta_{ij} + \Delta p_{ij} \tag{2}$$

where prime and double prime denotes respectively the reversible and irreversible part of the strain increment deviator. Separate relations between spherical and deviatoric components will be established. Their interaction is taken into account through certain scalar parameters.

The relationship between increments of the stress and the strain deviators is taken in the form of generalized Hooke's law

$$\Delta p_{ij} = 2G^*\Delta\varepsilon'_{ij} + \frac{\Delta G^*}{G^*} p_{ij} \tag{3}$$

where G^* is the variable shear modulus depending on both the stress state and temperature.

The yield condition is assumed in the form, [2, 13]

$$S_{ij}S_{ij} - \tfrac{2}{3}C^2 = 0 \tag{4}$$

$$S_{ij} = p_{ij} - \rho_{ij} \tag{5}$$

where S_{ij} is the 'active' stress tensor, ρ_{ij} is the 'residual microstress' tensor, $\sqrt{\tfrac{2}{3}}C$ denotes the radius of a current yield surface, a hypersphere in the stress deviator space. C depends on the hardening parameter x, the temperature T, the second invariant of the strain rate tensor $I_{2\varepsilon}$ and on hydrostatic pressure σ. The dependence of C on the above parameters is determined experimentally and may be given in the form of tables or formulae. For example, the dependence of C upon strain rate at constant temperature and constant hydrostatic pressure can be expressed as follows, [2]

$$C = C(x, T)\left\{1 + \left[\frac{I_{2\varepsilon}}{D(x, T)}\right]^{1/n(T,x)}\right\} \tag{6}$$

or

$$C = C(x, T)\left\{1 + \ln\left[1 + \frac{I_{2\varepsilon}}{D(T, x)}\right]\right\} \tag{7}$$

where C, D and n are the experimentally determined parameters. The dependence of C on hydrostatic pressure at constant temperature and strain rate may be described by, [6]:

$$C = C_0 + C_1(\sigma - \sigma_0) \tag{8}$$

or

$$C = C_0 + C_1 h + C_2 h^2 \tag{9}$$

where C_0, C_1, C_2 are the experimentally determined constants, h is the ratio of the current material density to the initial density.

The relation between the residual microstress tensor increment $\Delta\rho_{ij}$ and the plastic strain tensor increment $\Delta\varepsilon''_{ij}$ is

$$\Delta\rho_{ij} = g\Delta\varepsilon''_{ij} + \theta\rho_{ij} \tag{10}$$

where $g = g[I_{2\rho}, I_{2\varepsilon}, T, x, \sigma, \operatorname{sign}(\rho_{ij}\Delta\varepsilon''_{ij})]$ is some scalar function, and $I_{2\rho}$ is the second invariant of the microstress tensor ρ_{ij}. The first term on the right hand side of (10) species the process of generating the oriented micro-inhomogeneity [15, 16]. The dependence of g upon the above parameters must be determined experimentally and can be given in the form of either tables or formulae. The shape and the behaviour of the hysteresis loop under cyclic loading is affected by the form of the function g. The material response under cyclic loading can easily be described by assuming that g depends on the sign of $\rho_{ij}\Delta\varepsilon''_{ij}$. The second term in (10) characterizes the recrystallization process; θ is a function of temperature, strain hardening, strain rate, hydrostatic pressure, etc. and may be taken in the following form

$$\theta = \frac{\Delta\psi}{\psi} \quad \Delta\psi = \int_{t^n}^{t^{n+1}} \dot{\psi}\, dt; \quad \psi = \frac{I_{2\rho}}{I_{2\rho}^0} \tag{11}$$

where $I_{2\rho}^0$ is the magnitude of the second invariant of the micro-stress tensor evaluated at a certain initial temperature T_0, a certain hydrostatic pressure σ_0, a strain rate $I_{2\varepsilon}^0, \ldots$, and at the same value of the strain hardening parameter x as that used when computing $I_{2\rho}$. Under the assumption that the recrystallization is an instantaneous process depending only upon the current value of the homologous temperature $\xi = T/T_p$, the expression (11) can be written in the form [26] (see fig. 1):

$$\theta = \begin{cases} \dfrac{\psi(\xi_2) - \psi(\xi_1)}{\psi(\xi_1)} & \text{for } \Delta\xi \geqq 0 \\ 0 & \text{for } \Delta\xi < 0 \end{cases} \tag{12}$$

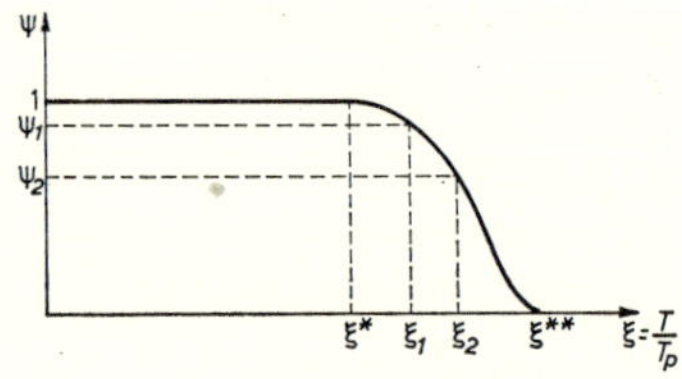

Fig. 1.

Eqs. (10 to 12) describe the process of initiation of plastic anisotropy which depends upon ρ_{ij} and on the latent energy $\ni_H = \int_0^t \rho_{ij}\dot{\varepsilon}''_{ij}dt$. The process of removing the plastic anisotropy by recrystallization as well as the process of the latent potential energy release at temperatures exceeding the temperature of recrystallization are also taken care of by these formulae [1].

The plastic strain rate tensor $\dot{e}''_{ij}$ and the 'active' stress S_{ij} are related by the plastic potential flow law associated with the instantaneous ('dynamic') yield surface [2, 10, 11, 15, 16, 22, 23]. The instantaneous yield surface given by Eq. (4) represents a hypersphere in the space of deviatoric stress. The center of the sphere is specified by ρ_{ij}. Eventually the following constitutive relation is obtained

$$\dot{e}''_{ij} = \lambda S_{ij} \tag{13}$$

If $\rho_{ij} \equiv 0$, $C \equiv$ const. the relations (3) to (13) describe an ideally plastic medium.

For $\rho_{ij} \equiv 0$ and $C \equiv C(x, T)$ these relations reduce to the equations of the nonisothermal theory of plasticity developed by Prager [10, 23] in [17].

At $g =$ const, $C =$ const, $\theta \equiv 0$ the relations (3) to (13) specify the isothermal theory of plastic deformation accounting for linear strain hardening, [13].

At $g = g(T)$, $C = C(T)$, $\theta \equiv 0$ they represent the theory of residual microstresses under nonisothermal deformation [16].

Finally, if $g = g(x, I_{2\varepsilon})$, $C = C(x, I_{2\varepsilon})$ the relations (3) to (13) concern the rate-dependent theory of plasticity, [2].

The relationship between the volumetric strain and the isotropic stress at low hydrostatic pressure can be written as follows

$$\Delta\sigma = 3K^*[\Delta e - \Delta(\alpha T)] + \frac{\Delta K^*}{K^*}\sigma \tag{14}$$

where K^* is the variable bulk modulus depending on both the stress and temperature, and α denotes the coefficient of linear expansion.

At high hydrodynamic pressures ($\sigma \lesssim 50$ kbar), and in order to obtain a relation between isotropic parts of stress and strain tensors, the constitutive equations for solids under high pressures and temperatures must be applied. We shall use the Mie-Grüneisen [27] equations for the purpose,

$$-\sigma = \sigma^*(v) + \gamma\rho[E - E(v)] \tag{15}$$

where $\sigma^*(v)$ is the 'cold' part of the pressure specified in [27], as

$$\sigma^*(v) = a(h-1) + b(h-1)^2 + c(h-1)^3 \tag{16}$$

The coefficients a, b, c are obtained from experiments, $h = \rho/\rho_0$, ρ is the current density, and $v = 1/\rho$ is the specific volume. Again

$$\gamma = \gamma_0 + A(h-1) + B(h-1)^2 + C(h-1)^3 \tag{17}$$

is the Grüneisen parameter, whereas γ_0 A, B, C have to be determined experimentally. As for the second term on the right hand side of (15) E is the internal energy, and $E(v)$ is the potential energy density. They are specified as follows

$$\Delta E = -\sigma^* \Delta v + p_{ij}\Delta\varepsilon_{ij} + q_{ij}\Delta\varepsilon_{ij} + \Delta E_B \tag{18}$$

$$\Delta E(v) = -\sigma\Delta v + p_{ij}\Delta\varepsilon'_{ij} + \rho_{ij}\Delta\varepsilon''_{ij} \tag{19}$$

where q_{ij} denotes the viscosity tensor regarded as function of the strain rate, ΔE_B is the internal energy increment due to internal thermal sources and thermal conductivity. The term $\rho_{ij}\Delta\varepsilon''_{ij}$ involved in (19) corresponds to the latent potential energy increment.

To justify the incremental form of the theory we shall make the following remarks. In order to determine the yield surface experimentally, a certain offset of plastic deformation is usually established, $e'' = 0.002$ to 0.001. The decrease of this offset results in the fact that the fundamental requirements of the phenomenological theory of plasticity are no longer satisfied [14]. That is, on the one hand, the experimentally determined stress and strain increments cannot be considered as differentials, while, on the other hand, for practical purposes it is meaningless to evaluate the plastic deformation to within an accuracy less than the established offset. Therefore one cannot consider the increments as infinitesimal quantities. Moreover, in solving dynamic problems, it is difficult to control stress and strain increments, as they may vary rapidly from one element to another. That is the reason why in the set of equations of the proposed theory differentials are replaced by the increments.

The 'physical' algorithm for application of the developed thermoplasticity

Let us assume that at a given instant n at a given material point the residual microstress tensor ρ^n_{ij} (vector $\boldsymbol{OO^n}$), the stress deviator p^n_{ij} (vector $\boldsymbol{OA}$), and the yield surface radius $\sigma^n_T = \sqrt{\frac{2}{3}}C_n$ are prescribed (Fig. 2). Assume that at the next instant $(n + 1)$ at this given point the temperature increment ΔT, the hydrostatic pressure and strain deviator increments $\Delta\varepsilon_{ij}$ as well as the mean strain rate for the given time interval ε_{ij} are known. When solving a dynamic problem by explicit methods the above increments $\Delta\varepsilon_{ij}$ are found once the displacement vector increments are known at the instant $(n + 1)$. In quasi-static problems the displacement vector increments in the first approximation may be determined considering the prescribed load increment for the elastic response of the material.

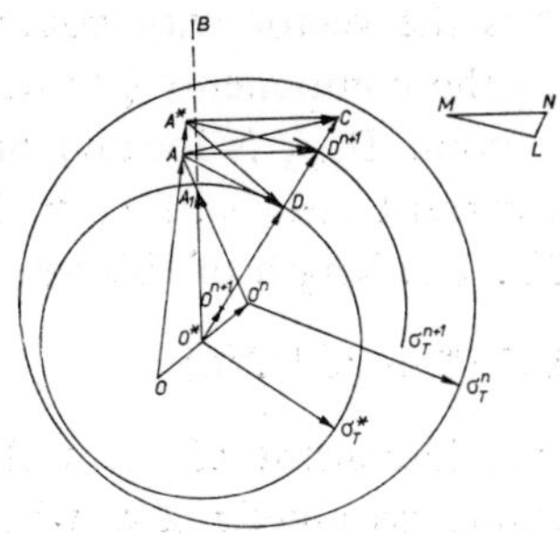

Fig. 2.

It is therefore possible to specify the yield surface variation due to temperature, strain rate and pressure changes as follows

$$\rho^*_{ij} = \rho^n_{ij}(1 + \theta) \tag{20}$$

$$\sigma^*_T = \sigma_T(T^{n+1}, x^n, I^{n+1}_{2\dot{\varepsilon}}, \sigma, \ldots) \tag{21}$$

ρ^*_{ij} are the components of the vector $\boldsymbol{OO^*}$ (see fig. 2). Further, $\boldsymbol{AC}$ with the components Δp^*_{ij} is determined assuming the elastic response of the material

$$\Delta p^*_{ij} = 2G^*\Delta\varepsilon_{ij} + \frac{\Delta G^*}{G^*} p^n_{ij} \tag{22}$$

In the case of nonlinear-elastic behaviour the vector $\boldsymbol{AC}$ is the required deviator increment sought. Moreover

$$|\boldsymbol{O^*C}| = |\boldsymbol{OA} + \boldsymbol{AC} - \boldsymbol{OO^*}| \leqq \sqrt{\tfrac{2}{3}}C^* \tag{23}$$

In the case of plastic yielding, $|\boldsymbol{O^*C}| > \sigma^*_T$, the required stress deviator increment is represented by the vector $\boldsymbol{AD}_{n+1}$. Subtracting (3) from (22) one obtains

$$\{\boldsymbol{D}_{n+1}\boldsymbol{C}\} = \{\boldsymbol{AC} - \boldsymbol{AD}_{n+1}\} = 2G^*\Delta\varepsilon''_{ij} \tag{24}$$

It means that the vector representing the difference between the elastic and actual increments of stress deviator is colinear with the vector of plastic strain increment and is related to the plastic strain increment through a law similar to that of Hooke.

Thus, the direction of the vector $\boldsymbol{CD}_{n+1}$ is fully specified by the direction of $\Delta\varepsilon''_{ij}$. The triangle A^*CD_{n+1} is similar to the triangle MNL (to within the multiplier $2G^*$). $\boldsymbol{AA^*}$ is the vector with components $(\Delta G^*/G^*)\cdot\{p_{ij}\}$, $\boldsymbol{MN}$, $\boldsymbol{ML}$ whereas $\boldsymbol{LN}$ has the components $\{\Delta\varepsilon_{ij}\}$, $\{\Delta\varepsilon'_{ij}\{$, $\{\Delta\varepsilon''_{ij}\}$.

One concludes that the point D_{n+1} is located on the straight line O^*C, joining the centre of the current yield surface with the end of the elastic stress increment vector $\boldsymbol{AC}$. The length of this vector is

$$|\boldsymbol{CD}| = g|\boldsymbol{NL}| + \sqrt{\tfrac{2}{3}}\Delta C^{**} + 2G^*|\boldsymbol{NL}| \tag{25}$$

where $\sqrt{\frac{2}{3}}\Delta C^{**}$ denotes the increment of the yield surface radius due to an increase of the hardening parameter Δx, whereas $\boldsymbol{CD}$ represents the difference between the perfectly elastic and perfectly plastic solutions (the perfectly plastic solution corresponds to $\boldsymbol{AD}$).

It is obvious that in the case of linear hardening when $C = C(T, I_{2\varepsilon}, \sigma)$ and $g = g(T, I_{2\varepsilon}, \sigma)$ the point D_{n+1} is to be found from the expression:

$$|\boldsymbol{CD}_{n+1}| = 2G^*|\boldsymbol{NL}| = 2G^*\frac{|\boldsymbol{CD}|}{g + 2G^*} \tag{26}$$

If both C and g depend on the hardening parameter, the iteration procedure according to the rule

$$|\Delta\varepsilon''_{ij}|^{k+1} = \frac{|CD| - \sqrt{\dfrac{2}{3}}\dfrac{\Delta C^{**(k)} + \Delta C^{**(k-1)}}{2}}{\dfrac{g(k) + g(k+1)}{2} + 2G^*} \tag{27}$$

must be applied in order to find $\boldsymbol{AD}_{n+1}$.

The iteration is considered to be completed when

$$\frac{2G^*(|\Delta\varepsilon''_{ij}|^{(k+1)} - |\Delta\varepsilon''_{ij}|^{(k)})}{\sqrt{\frac{2}{3}}C^{n+1}} \leqq \delta \tag{28}$$

The iteration procedure is found to be rapidly convergent (2–5 iterations for $\delta = 10^{-8}$).

If the effect of delayed yielding occurs the plastic deformation can start only if the following condition is satisfied, [4]

$$\int_0^{t_3}\frac{dt_k}{t_k} \geqq 1 \tag{29}$$

where dt_k is the time interval of the current intensity I_{2S} of active stresses, and t_k is the yield retardation time corresponding to the current relative increase of I_{2S} over the current value of yield surface radius σ_T:

$$\beta = (I_{2S} - \sigma_T)/\sigma_T$$

The curves $t_k = t_k(\beta, T)$ are considered to be known from experiments.

Numerical solutions of problems of thermoplasticity

Variational-difference methods belong to the most effective methods of solving quasi-static boundary value problems of thermoplasticity. Finite-difference equations are derived employing appropriate discretisations of the energy functionals. Such values of the required functions are looked for at the nodes of the mesh which yield the extrema of the respective functional [28, 29].

When the 'deformational' theory of plasticity is used the total energy of the system Π is the appropriate functional if the problem is described in terms of 'displacements'. Analogously the complementary energy A of the system is the pertinent functional if solution 'in terms of stresses' are attempted, [28, 32]. The respective variations are

$$\delta\Pi = \delta\left\{\int_V\left[\int_0^{e_u}\sigma_u de_u + 3\int_0^{e}\sigma de\right]dV - \int_{S'} P_i U_i dS\right\} = 0; \tag{30}$$

$$\delta A = \delta\left\{\int_V\left[\int_0^{\sigma_u} e_u d\sigma_u + 3\int_0^{\sigma} e\, d\sigma\right] dV - \int_{S''} P_i U_i dS\right\} = 0; \tag{31}$$

where V is the volume occupied by the body, S' is the part of the boundary surface of the body on which external loads P_i $(i = 1, 2, 3)$ are prescribed, S'' is the part of the boundary surface on which the displacements U_i are prescribed, e_u, σ_u are the effective strain and stress, respectively.

In solving boundary value problems employing incremental theories, the combined variational principle formulated in [33–36] may be used:

$$\delta K = \delta\left\{\tfrac{1}{2}\int_V \dot\sigma_{ij}(\dot e_{ij} - \dot e^*_{ij})dV - \int_{S''}\dot P_i(\dot U_i - \dot U^e_i)dS - \int_{S'}\dot P^e_i \dot U_i dS\right\} = 0; \tag{32}$$

where $\dot e^*_{ij}$ is the sum of plastic, creep and temperature strain rates.

The discretization of the above-mentioned variational problems may be achieved either by the finite element method [37–40] or by an appropriate approximation of the derivatives for actually chosen multiangular mesh [41].

The most effective methods for solving dynamic problems of thermoplasticity are at present the finite difference methods employing the explicit integration procedures for the equations of motion [42].

Numerical results

We shall consider some examples, solved using the presented model. All examples concern the aluminum alloy D16T, having mechanical properties as shown in fig. 3, 4 and 5. In fig. 3 the initial curves $\sigma \sim e^p$ are given, and in fig. 5 and 4 – the curves $C \cup g$, correspond to these initial curves. Algorithms, used for the examples considered below, are described in detail in [43].

Example 1. In fig. 6 deflections along the generator of a shell with clamped ends are given. The shell length is 260 mm, its diameter 300 mm and the wall thickness 3 mm. All three diagrams correspond to the same loading intensity $q = 8$ newton/mm^2 and temperature field as indicated in fig. 7. The curve O gives the final deflection due to proportional increase of temperature and load q at each cross section of the shell (proportional loading). The loading history, associated with 1 consists of heating the shell from 0° up to the final temperature, and then applying the pressure. The curve 2 corresponds to the case when the shell is loaded first from $q = 0$ up to $q = 8$ newton/mm^2, and then heated. The solution prescribed was obtained by the variational-difference method using the variational principle (32). The shell generatrix was divided into 20 elements. It is worthwhile to note that the deflection diagram obtained for the final state employing the deformation theory of plasticity when the variational principle (30) applies is close to the curve 0 and occupies the intermediate position with respect to the curves 1 and 2.

Example 2. A square plate 2,4 × 2,4 cm with a central circular hole of radius $R = 0.8$ cm was subjected to uniform biaxial tension. Two loading histories, resulting in the same final state were considered.

1. Uniform loading in directions OX and OY up to 20 kilonewton/cm^2 in both directions simultaneously.

2. Successive loading first in the X-direction up to 20 kilonewton/cm^2 and then perpendicularly up to the same intensity.

The problem was solved by employing the variational principle (32) and triangular finite elements assuming uniform strain and stress distribution over the element. The mesh used is shown in fig. 8 and 9 for a quarter of the plate. The load step was 2 kilonewton/cm^2. Plastic zones under uniform

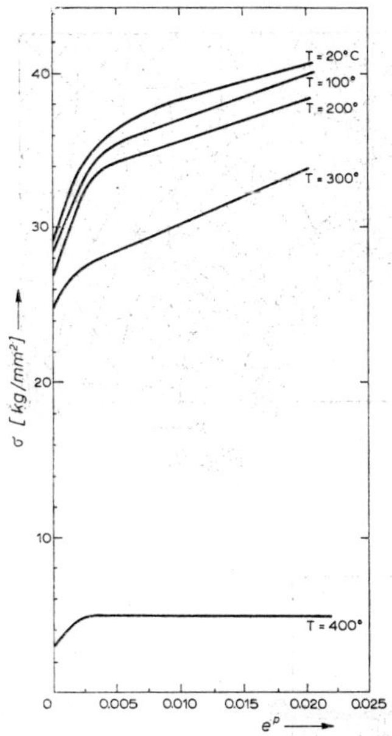

Fig. 3.

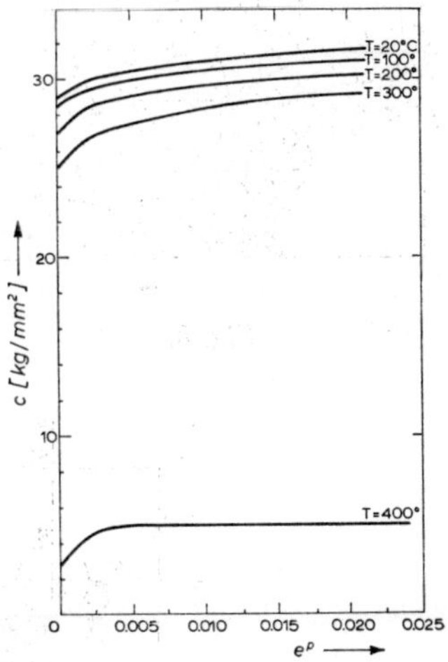

Fig. 4.

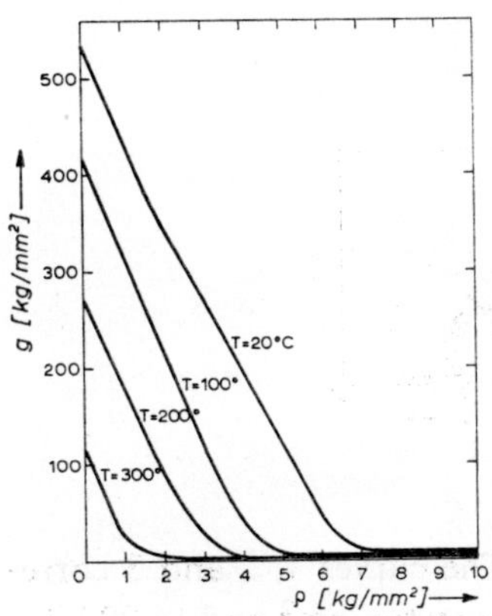

Fig. 5.

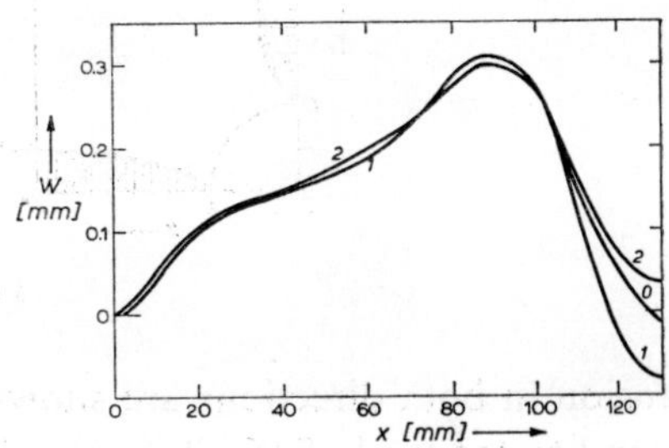

Fig. 6.

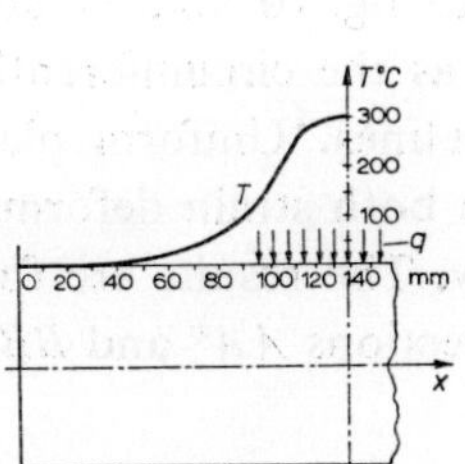

Fig. 7.

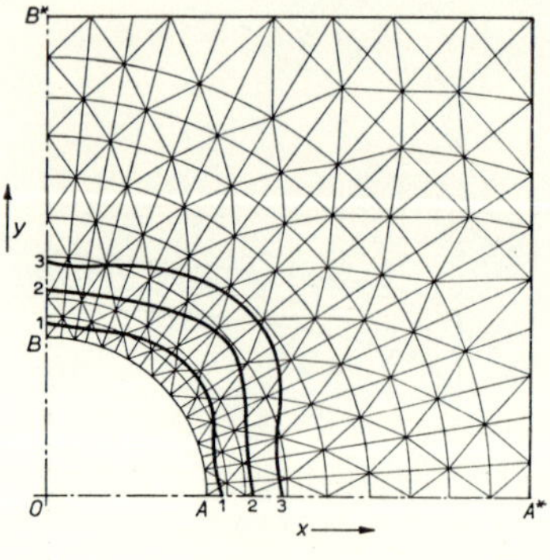

Fig. 8.

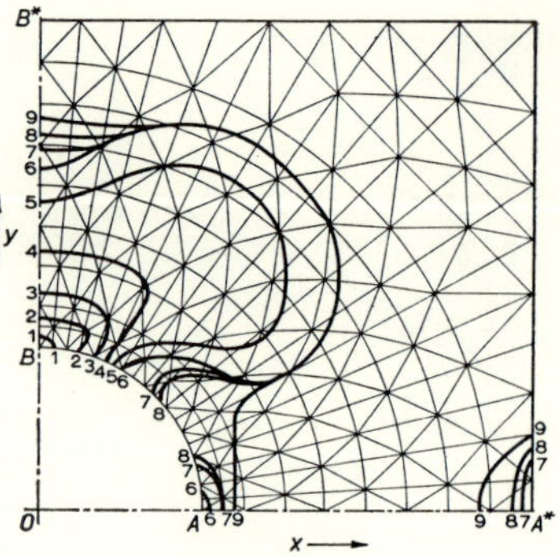

Fig. 9.

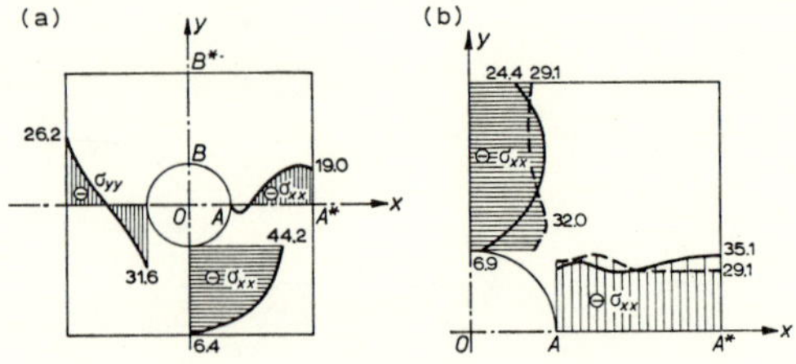

Fig. 10.

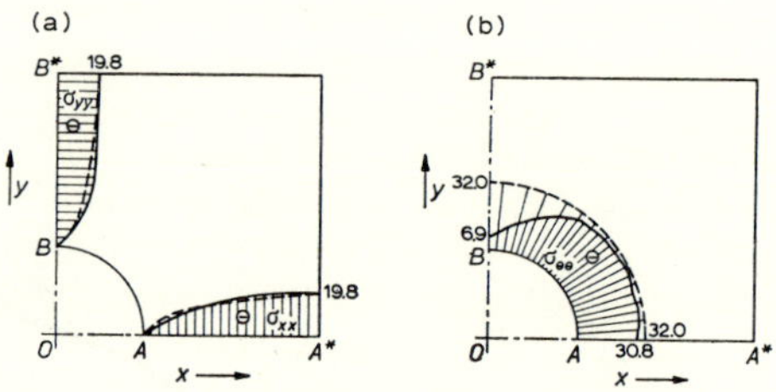

Fig. 11.

tension in both directions are shown in fig. 8. The curves 1, 2 and 3 correspond to the load of 16 kilonewton/cm^2, 18 kilonewton/cm^2 and to 20 kilonewton/cm^2, respectively. In fig. 10 and 11 stresses σ_{xx} and σ_{yy} in cross sections AA^*, BB^* as well as the circumferential stress $\sigma_{\theta\theta}$ along the hole contour are shown in broken lines. 'Uniform' plate loading in both directions was calculated according to both strain deformation and differential incremental theories of plasticity. The results are found to coincide. The stress distribution over the cross sections AA^* and BB^*, corresponding to 20 kilonewton/cm^2 load the X-direction is shown in fig. 10a. Whereas the plastic zone boundaries are seen in fig. 9. The curves 1 to 5 correspond to the load intensities 10, 12, 14, 16, 18 and 20 kilonewton/cm^2 respectively. The consecutive tension in Y-direction leads to development of new plastic zones, while within the originally yielded zones elastic unloading takes place.

In fig. 9 the curves 7, 8 and 9 correspond to the loading in the y-direction of 16, 18 and 20 kilonewton/cm^2 respectively. The distribution of stresses σ_{xx}, σ_{yy} and $\sigma_{\theta\theta}$ over the cross sections AA^*, BB^* and along the hole contour are given in fig. 10 and 11 in heavy lines. They correspond to the second loading programme. The comparison of the results shows the great influence of the loading history.

References

[1] Rodzivonchik, V. F., Rapid plastic deformation of metals, (in Russian). Kharkov, 1967.

[2] Perzyna, P., Fundamental problems in viscoplasticity. Advances in Applied Mechanics, *9*, Academic Press, New York, 1966.

[3] Gilman, J. J., Dislocation dynamics and the response of materials to impact. *Appl. Mech. Rev.*, *21* (1968) 767–783.

[4] Voloshenko-Klimovitsky, Yu. Ya., Dynamic yield stress (in Russian). Nauka, Moscow, 1965.

[5] Altschuler, L. V., Application of shock waves in high pressure physics (in Russian). Uspekhi fiz. nauk Nauka, Moscow, 1965.

[6] Erkman, J., Elastoplastic effects in the attenuation of shock waves. Proc. 4th Intern. Symp. on Detonation (White Oak, 1965), Washington, 1965.

[7] Novikov, S. A. and L. M. Sinitsina, Shock compression effect on critical shear stresses in metals. Prikl. Mat. Tekh. Fiz. (PMTF) No. 6, 1970.

[8] Grace, F. I., Shock-wave strengthening of copper and nickel. *J. Appl. Phys.*, *40* (1969).

[9] Volshanina, M. A. and V. E. Panin, Larent strain energy (in Russian). Collected articles 'Research in solid-state physics'. Publishing House of the Academy of Sciences, USSR, Moscow, 1957.

[10] Naghdi, P. M., Stress-strain relations in plasticity and thermoplasticity. Proc. of the Second Symp. on Naval Structural Mech. (Providence 1960), Pergamon Press, London 1960, 121–167.

[11] Olszak, W., Z. Mróz and P. Perzyna, Recent trends in the development of the theory of plasticity, Pergamon Press, Oxford 1963.

[12] Ilyushin, A. A., Plasticity (in Russian). Publishing House of the Academy of Sciences, USSR, Moscow, 1963.

[13] Kadashevich, Yu. I. and V. V. Novozhilov, The theory of plasticity, taking into account residual microstresses (in Russian). Prikl. Mat. Mekh. (PMM), *22* (1958).

[14] Novozhilov, V. V., On complex loading and perspectives of the phenomenological

approach to the investigation of microstresses (in Russian). Prikl. Mat. Mekh. (PMM), *28* (1964).

[15] Vakulenko, A. A., On relation between stresses and strains in isotropic non-elastic media (in Russian). Collected articles 'Research in elasticity and plasticity', vol. 1, Leningrad State University Press, Leningrad, 1969.

[16] Pallej, I. Z., Application of residual microstresses theory to low-temperature deformation (in Russian). In AN USSR. Mechanics, N 2, 1965.

[17] Birger, I. A. and I. V. Demjanushko, The theory of plasticity under non-isothermal loading (in Russian). Mekh. Tverd. Tekv., No. 6, (1968).

[18] Zarubin, V. S. and A. A. Polyakov, Non-isothermal plastic deformation model of polycrystal material (in Russian). Collected articles 'Thermal stresses in structural elements', vol. *9* 'Naukova dumka', Kiev, 1970.

[19] Batdorf, S. B. and B. Budiansky, A mathematical theory of plasticity based on the concept of slip. NACA, Technical Note, N 1871, April 1949.

[20] Malmeister, A. K., The foundations of local strain theory (in Russian). Mekhanika polymerov, No. 4, 1964.

[21] Korotkikh, Yu. G., and S. M. Belevich, On thermoplastic plane problem solution under complex loading (in Russian). Collected articles. 'Thermal stresses in structural elements'. Vol. 9, Naukova dumka, Kiev, 1970.

[22] Koiter, W. T., General theorems for elastic-plastic solids. Progress in solid mechanics, vol. 1, edited by I. N. Sneddon and R. Hill. North-Holland, Amsterdam, 1960.

[23] Knets, I. V., On basic current trends in mathematical theory of plasticity (in Russian). 'Zinatne', Riga, 1971.

[24] Miastkowski, J., Analysis of the memory effect of plastically prestrained material. Arch. Mech. Stosowanej, *20* (1968).

[25] Szczepiński, W. and J. Miastkowski, An experimental study of the effect of the prestraining history on the yield surfaces of an aluminium alloy. J. Mec. Phys. Solids, *16* (1968) 153–162.

[26] Korotkikh, Yu. G. and S. M. Belevich, The equations of state for static and dynamic problems of thermoplasticity under complex loading (in Russian). Collected articles 'The methods of solving of elasticity and plasticity problems. M., Memoirs of Gorky State University, Mechanics No. 2, vol. 108, Gorky, 1970.

[27] Rice, M. H., R. G. McQueen and J. M. Walsh, Solid State Physics. Adv. in Res. and Appl., vol. 6, Academic Press, New York, 1958.

[28] Mikhlin, S. G., The numerical realization of variational methods (in Russian). 'Nauka', Moscow, 1966.

[29] Ugodchikov, A. G. and Yu. G. Korotkikh, Some methods of solving physically nonlinear problems of the theory of plates and shells by means of electronic digital computer (in Russian). 'Naukova dumka', Kiev, 1971.

[30] Kachanov, L. M., The variational methods of solving problems of the theory of plasticity (in Russian). Prikl. Mat. Mekh. *23* (1959).

[31] Kachanov, L. M., Some variational principles and methods in the theory of plasticity (in Russian). Proceedings of the All-Union Congress on Theoretical and Applied Mechanics (January 27–February 3, 1960). Publishing House of the Academy of Science USSR, Moscow, 1962.

[32] Kauderer, H., Nichtlineare Mechanik. Berlin, Springer, 1958.

[33] Sanders, J. L., H. C. McComb and F. R. Schechte, A variational theorem for creep with applications to plates and columns. NACA, Rep., N 1342, 1957.
[34] Reissner, E., On variational theorem for finite elastic deformations. J. Math. and Phys., *32* (1953).
[35] Wang, A. J. and W. Prager, Thermal and creep effects in work hardening elastic-plastic solids. Journ. Aeronaut. Sci., *21* (1954).
[36] Rabotnov, Yu. I., The creep of structural elements (in Russian). 'Nauka', Moscow, 1966.
[37] Argyris, J. H., Recent advances in matrix methods of structural analysis. Pergamon Press, Oxford, 1964.
[38] Pian, T. H. H., Derivation of element stiffness matrices by assumed stress distributions. AIAA Journal, *2* (1964).
[39] Zienkiewicz, O. C., The finite element method: from intuition to generality. Appl. Mech. Rev. (1970) 249–156.
[40] Actem, H., G. Isakson and A. Pifko, Discrete element methods for the plastic analysis of structures subjected to cyclic loading. Intern. J. for Numer. Meth. in Eng., *2* (1970) 189–206.
[41] Noh, N. F., Methods in computational physics advances in research and applications. Ed. by B. Alder, S. Fernbach, M. Rotenberg. Vol. 3. Fundamental methods in hydrodynamics. Academic Press, New York, 1964.
[42] Wilkins, M. L., Methods in computational physics advances in research and application. Ed. by B. Alder, S. Fernbach, M. Rotenberg. Vol. 3, Fundamental methods in hydrodynamics. Academic Press, New York, 1964.
[43] Collected articles. The methods of solving the problems of elasticity and plasticity (in Russian). Memoirs of the Gorky State University, Mechanics, *1–5*, Gorky, 1970–1971.

A description of workhardening of metals with application to variable loading

Z. Mróz

Institute of Fundamental Technical Research, Warsaw

A mathematical formulation of time-independent theory of plasticity using internal state parameters is discussed. During the plastic deformation process, besides yield condition, flow rule, and rate equations for state parameters, a set of additional scalar equalities and consistency relations must be satisfied. Restrictive thermodynamic and stability conditions are considered. Some particular forms of composite work-hardening models are presented and a mechanical interpretation of the state parameters is given. Application of these models is discussed in two cases: 1. for a hardening curve approximated by three linear segments, 2. for any stable hardening curve with Masing unloading rule.

Introduction

Simple theories of plasticity, such as flow rules associated with a single yield surface, are sufficiently accurate to predict the plastic behaviour of a material for monotonically increasing loads. On the other hand, for alternating loads, and in particular for cyclic loads, such flow rules cannot accurately describe complex plastic behaviour. In order to formulate constitutive relationships applicable to alternating loading conditions, one has to depart from the notion of a single yield surface completely defining the state of the material and introduce a set of additional internal state parameters considering the internal material structure in a more realistic way.

In this paper, a mathematical theory of plasticity formulated in terms of internal state parameters is discussed and some of its applications to plastic behaviour under arbitrarily varying loads is demonstrated. Isothermal behaviour is emphasized and creep effects are neglected. For small strains in the order of several percent, the anisotropic hardening is primarily due to nonhomogeneity of plastic flow and the effect of residual microstresses within a polycrystalline aggregate. For the uniaxial stress state, this anisotropy is exhibited as a familiar Bauschinger effect whereas for a complex stress state it is manifested as a directional change of the initial stiffness of the element after plastic prestrain.

There are several possible ways to approach the formulation of more general constitutive laws. First, it can be assumed that a representative macro-element constitutes a composition of sub-elements with specified plastic properties, for instance randomly distributed monocrystals with definite hardening rules on a system of slip planes. Such theories have been considered in numerous works [6–9] to mention a few. Depending on the degree of idealization, more or less complex constitutive relations were obtained which combined the piecewise linear yield condition and simple hardening rules for monocrystals with the effect of constraint hardening due to non-homogeneity of deformation. General inequalities for elasto-plastic composites expressed in terms of surface tractions and displacement rates have been analyzed in [28]. In particular, by postulating series or parallel arrangement of micro-elements, the problem of averaging the stress and strain within the macro-element becomes simple and bounds on the actual hardening curve can be obtained. Such composite models have been discussed in several works [10–13] by assuming the plastic body to be an assemblage of weak (perfectly plastic or hardening) and strong (elastic) subelements. The hardening curve and the hysteresis loop under uniaxial cyclic loading can be simulated in this manner for any material.

The concept of internal state parameters is similar to that of composite structure approach, since stresses and strains in particular phases can be regarded as state parameters. It differs, however, in some important aspects since these parameters do not enter equilibrium equations nor kinematic relations but occur in the free energy or dissipation functions or directly in constitutive relationships. Thus, no averaging procedure is to be carried out since the macro-element is regarded as uniform and the theory is constructed in a more formal manner. The real problem, which remains open, lies in proper selection of such parameters and formulation of their law of variation in the process of plastic straining. However, it will be shown that the general mathematical structure of this plasticity theory can be discussed without detailed knowledge of these parameters, and some interpretation may be ascribed by considering only macroscopic behaviour.

The possibility of constructing a theory of plasticity with internal state variables was already discussed by Vakulenko [14] and Kröner [15] for isothermal processes and by Kratochvíl and Dillon [16] for temperature effect. However, the emphasis of previous works was on the thermodynamical aspect of the problem [14, 16], or on the relation of the internal parameters to the dislocation arrangement within the polycrystalline material [15]. Our aim is to discuss concrete relations that could be used in order to describe plastic behaviour phenomenologically. Moreover, we shall consider a general structure of constitutive relations which, so far, have not been discussed in the literature.

Description of workhardening by means of internal state parameters

Let us first discuss the mathematical structure of the theory, restricting the analysis to small strains. Assume that the state of material is defined by the elastic strain component $\boldsymbol{\varepsilon}^e$, temperature θ, and by $N+M$ state parameters which are assumed to be scalars $\gamma^{(r)}$, $(r = 1, 2, \ldots, N)$, and second order tensors $\boldsymbol{\alpha}^{(s)}$, $(s = 1, 2, \ldots, M)$. Since elastic strain is related by Hooke's law to stress $\boldsymbol{\sigma}$, it is equivalent to assume the stress $\boldsymbol{\sigma}$ as the state parameter instead of $\boldsymbol{\varepsilon}^e$. Let the yield condition be a function of stress, temperature and some or all state parameters, that is

$$f^{(0)}(\boldsymbol{\sigma}, \gamma^{(r)}, \boldsymbol{\alpha}^{(s)}, \theta) = 0. \tag{1}$$

During deformation in the elastic range, there is no variation of the state parameters, thus

$$\dot{\gamma}^{(r)} = \dot{\boldsymbol{\alpha}}^{(s)} = 0, \quad \text{when} \quad \dot{\boldsymbol{\varepsilon}}^p = 0$$

where a dot over a symbol denotes the rate of variation of respective quantity. Assume that besides the scalar relationship (1), the internal parameters satisfy $K+L$ additional scalar relations which define *regions of internal irreversibility*, i.e.

$$\begin{aligned} f_\alpha^{(q)}(\boldsymbol{\sigma}, \boldsymbol{\alpha}^{(s)}, \gamma^{(r)}, \theta) &= 0, \qquad q = 1, 2, \ldots, K, \\ f_\gamma^{(v)}(\boldsymbol{\sigma}, \boldsymbol{\alpha}^{(s)}, \gamma^{(r)}, \theta) &= 0, \qquad v = 1, 2, \ldots, L, \end{aligned} \tag{3}$$

where the first set of equations (3) pertains to the parameters $\boldsymbol{\alpha}^{(q)}$ and the second to parameters $\gamma^{(v)}$. For instance, when the equality $f_\alpha^{(q)} = 0$ is satisfied, the corresponding parameter $\boldsymbol{\alpha}^{(q)}$ varies during plastic deformation according to some specified law; on the other hand, when $f_\alpha^{(q)} < 0$, then $\dot{\boldsymbol{\alpha}}^{(q)} = 0$, whereas the other parameters can vary. It may occur that a single equality of the type of Eq. (3) governs the variability of several parameters. This case can be obtained from (3) by requiring the respective equalities to be identical. Assume further that rates of internal parameters are linearly related to plastic strain increments, that is

$$\begin{aligned} &\dot{\alpha}_{ij}^{(s)} = A_{ijkl}^{(s)} \dot{\varepsilon}_{kl}^{p} = b_{ij}^{(s)} \dot{\lambda}, \quad \text{when} \quad f_{\alpha}^{(s)} = 0, \\ &\dot{\alpha}_{ij}^{(s)} = 0, \quad \text{when} \quad f_{\alpha}^{(s)} < 0, \end{aligned} \tag{4}$$

and

$$\begin{aligned} &\dot{\gamma}^{(r)} = c_{ij}^{(r)} \dot{\varepsilon}_{ij}^{p} = d^{(r)} \dot{\lambda}, \quad \text{when} \quad f_{\gamma}^{(r)} = 0, \\ &\dot{\gamma}^{(r)} = 0, \quad \text{when} \quad f_{\gamma}^{(r)} < 0, \end{aligned} \tag{5}$$

where, $A_{ijkl}^{(s)}$, $b_{ij}^{(s)}$, $c_{ij}^{(r)}$, $d^{(r)}$, are functions of stress, temperature and internal parameters, and $\dot{\lambda} = (\dot{\boldsymbol{\varepsilon}}^{p} \cdot \dot{\boldsymbol{\varepsilon}}^{p})^{\frac{1}{2}} = \dot{\varepsilon}^{p}$. Let us use the flow law in the form

$$\dot{\boldsymbol{\varepsilon}}^{p} = \frac{1}{K} \boldsymbol{n}^{(0)} (\dot{\boldsymbol{\sigma}} \cdot \boldsymbol{n}^{(0)}) = \frac{1}{K} \boldsymbol{n}^{(0)} \dot{\sigma}_{f}, \tag{6}$$

where $\boldsymbol{n}^{(0)}$ denotes the unit normal vector to the yield surface and the dot between two vectors denotes the scalar product. Using the flow rule (6) we have

$$A_{ijkl}^{(s)} n_{kl}^{(0)} = b_{ij}^{(s)}, \qquad c_{ij}^{(r)} n_{kl}^{(0)} = d^{(r)}. \tag{7}$$

In general, the number of scalar relationships (3) can be smaller than that of internal parameters.

Assume that the yield condition (1) and some scalar relationships (3) are satisfied. The corresponding state parameters thus vary in the course of plastic deformation. Since the relationships (3) are satisfied for any increments of $\boldsymbol{\sigma}$, $\boldsymbol{\alpha}^{(s)}$, $\gamma^{(r)}$ and θ, we have

$$\begin{aligned} &\frac{\partial f^{(0)}}{\partial \boldsymbol{\sigma}} \cdot \dot{\boldsymbol{\sigma}} + \sum_{s=1}^{M} \frac{\partial f^{(0)}}{\partial \boldsymbol{\alpha}^{(s)}} \cdot \dot{\boldsymbol{\alpha}}^{(s)} + \sum_{r=1}^{N} \frac{\partial f^{(0)}}{\partial \gamma^{(r)}} \dot{\gamma}^{(r)} + \frac{\partial f^{(0)}}{\partial \theta} \dot{\theta} = 0, \\ &\cdots\cdots\cdots\cdots\cdots\cdots, \\ &\frac{\partial f_{\gamma}^{(v)}}{\partial \boldsymbol{\sigma}} \cdot \dot{\boldsymbol{\sigma}} + \sum_{s=1}^{M} \frac{\partial f_{\gamma}^{(v)}}{\partial \boldsymbol{\alpha}^{(s)}} \cdot \dot{\boldsymbol{\alpha}}^{(s)} + \sum_{r=1}^{N} \frac{\partial f_{\gamma}^{(v)}}{\partial \gamma^{(r)}} \dot{\gamma}^{(r)} + \frac{\partial f_{\gamma}^{(v)}}{\partial \theta} \dot{\theta} = 0. \end{aligned} \tag{8}$$

Using the flow rule (6) and the incremental relationships (4), (5), expressions in (8) take the form

$$\begin{aligned} &K \left| \frac{\partial f^{(0)}}{\partial \boldsymbol{\sigma}} \right| + \sum_{s=1}^{M} \frac{\partial f^{(0)}}{\partial \boldsymbol{\alpha}^{(s)}} \cdot \boldsymbol{b}^{(s)} + \sum_{r=1}^{N} \frac{\partial f^{(0)}}{\partial \gamma^{(r)}} d^{(r)} + \frac{\partial f^{(0)}}{\partial \theta} \left(\frac{\dot{\theta}}{\dot{\sigma}_{f}} \right) = 0, \\ &\cdots\cdots\cdots\cdots\cdots\cdots, \\ &K \left| \frac{\partial f_{\gamma}^{(v)}}{\partial \boldsymbol{\sigma}} \right| + \sum_{s=1}^{M} \frac{\partial f_{\gamma}^{(v)}}{\partial \boldsymbol{\alpha}^{(s)}} \cdot \boldsymbol{b}^{(s)} + \sum_{r=1}^{N} \frac{\partial f_{\gamma}^{(v)}}{\partial \gamma^{(r)}} d^{(r)} + \frac{\partial f_{\gamma}^{(v)}}{\partial \theta} \left(\frac{\dot{\theta}}{\dot{\sigma}_{f}} \right) = 0 \end{aligned} \tag{9}$$

whenever

$$\frac{\partial f^{(0)}}{\partial \boldsymbol{\sigma}} \cdot \dot{\boldsymbol{\sigma}} + \frac{\partial f^{(0)}}{\partial \theta} \dot{\theta} > 0, \ldots, \frac{\partial f_{\gamma}^{(v)}}{\partial \boldsymbol{\sigma}} \cdot \dot{\boldsymbol{\sigma}} + \frac{\partial f_{\gamma}^{(v)}}{\partial \theta} \dot{\theta} > 0. \tag{10}$$

and summation pertains to such s and v for which the scalar equalities (3) are satisfied.

Equations (9) are the *consistency relationships* on the set (3), (4) and (6) which impose restrictions on the coefficients K, $\boldsymbol{b}^{(s)}$, and $d^{(r)}$ occurring in (4), (5), (6). Moreover, when the number of consistency relationships changes, the coefficients occurring in (4) and (5) should also change discontinously in order to satisfy all restrictive equalities (9). In other words, when a new state parameter starts to vary, the rate equations of other parameters may change. It can therefore alternatively be postulated that scalar equalities (3) define regions where the law of variation of internal parameters changes, for instance

$$\begin{aligned} \dot{\alpha}_{ij}^{(s)} &= b_{ij}^{(s)} \dot{\lambda}, \qquad \text{for} \quad f_{\alpha}^{(s)} = 0, \\ \dot{\alpha}_{ij}^{(s)} &= \beta_{ij}^{(s)} \dot{\lambda}, \qquad \text{for} \quad f_{\alpha}^{(s)} < 0, \end{aligned} \tag{11}$$

We have thus constructed the general form of constitutive relations for time-independent thermoplasticity which can be regarded as generalization of the simpler relations proposed by Prager [17]. To complete the description, further restrictive conditions following from the second law of thermodynamics or from the requirements of stability have to be derived.

Assume that specific thermodynamic functions per unit mass depend on elastic strain, temperature and internal state parameters, thus

$$\begin{aligned} e &= e(\boldsymbol{\varepsilon}^{e}, \gamma^{(r)}, \boldsymbol{\alpha}^{(s)}, \theta) \\ \psi &= e - \theta\eta = \psi(\boldsymbol{\varepsilon}^{e}, \gamma^{(r)}, \boldsymbol{\alpha}^{(s)}, \theta), \\ \eta &= \eta(\boldsymbol{\varepsilon}^{e}, \gamma^{(r)}, \boldsymbol{\alpha}^{(s)}, \theta) \end{aligned} \tag{12}$$

where e, ψ, and η respectively are internal energy, free energy and entropy. The local energy balance equation takes the form

$$\rho \dot{e} = \rho \dot{r} + \sigma_{ij} \dot{\varepsilon}_{ij} = q_{ii} \tag{13}$$

and the local entropy balance equation is expressed as follows

$$\rho \dot{\eta} = \frac{\rho \dot{r}}{\theta} - \left(\frac{q_i}{\theta} \right)_{,i} + \rho \dot{\eta}_i \tag{14}$$

where q_i is the heat flux vector and $\dot{r}$ denotes the heat supply per unit time and unit mass; the spatial derivative is denoted by an index proceded by a comma. The irreversible entropy production is given by the formula, [18]

$$\rho\dot{\eta}_i = q_i\left(\frac{1}{\theta}\right)_{,i} + \rho\dot{\eta}_i^d > 0 \tag{15}$$

where $\rho\dot{\eta}_i^d$ denotes the rate of entropy production due to internal dissipation. Using (13), (14), and (4), (5)

$$\rho\theta\dot{\eta}_i^d = -\rho\dot{\psi} - \rho\dot{\theta}\eta + \sigma_{ij}\dot{\varepsilon}_{ij} = \left(\sigma_{ij} - \rho\frac{\partial\psi}{\partial\varepsilon_{ij}^e}\right)\dot{\varepsilon}_{ij}^e - \rho\left(\eta + \frac{\partial\psi}{\partial\theta}\right)\dot{\theta} +$$
$$+\left[\sigma_{ij} - \rho\sum_r \frac{\partial\psi}{\partial\gamma^{(r)}} c_{ij}^{(r)} - \rho\sum_s \frac{\partial\psi}{\partial\alpha_{kl}^{(s)}} A_{klij}\right]\dot{\varepsilon}_{ij}^p. \tag{16}$$

Since within the elastic region $\dot{\eta}_i^d = 0$, it follows from (16) that

$$\sigma_{ij} = \rho\frac{\partial\psi}{\partial\varepsilon_{ij}^e}, \qquad \eta = -\frac{\partial\psi}{\partial\theta}. \tag{17}$$

Assuming that elastic properties are not affected by plastic deformation, relationships (17) will be valid in both elastic and plastic regions. The rate of entropy production due to dissipation is thus expressed as follows

$$\rho\theta\dot{\eta}_i^d = (\sigma_{ij} - \sigma_{ij}^*)\dot{\varepsilon}_{ij}^p = \sigma_{ij}^d\dot{\varepsilon}_{ij}^p \tag{18}$$

where σ_{ij}^d denotes the dissipative part of stress, expressed by the formula

$$\sigma_{ij}^d = \sigma_{ij} - \gamma\sum_r \frac{\partial\psi}{\partial\gamma^{(r)}} c_{ij}^{(r)} - \gamma\sum_s \frac{\partial\psi}{\partial\alpha_{kl}^{(s)}} A_{klij}^{(s)}. \tag{19}$$

The dissipation function has thus the form $D = D(\boldsymbol{\varepsilon}^p, \gamma^{(r)}, \boldsymbol{\alpha}^{(s)})$ and can be used to generate the dissipative stress as for a perfectly plastic body,

$$\sigma_{ij}^d = \frac{\partial D}{\partial\dot{\varepsilon}_{ij}^p}. \tag{20}$$

In the general case, the second law of thermodynamics requires the inequality (15) to be satisfied; in the isothermal case, however, the familiar inequality derived already by Vakulenko [14][1]) is obtained using (16)

$$D = \boldsymbol{\sigma}^d \cdot \dot{\boldsymbol{\varepsilon}}^p = (\boldsymbol{\sigma} - \boldsymbol{\sigma}^*) \cdot \dot{\boldsymbol{\varepsilon}}^p > 0. \tag{21}$$

[1]) The necessity of introducing dissipative stress $\boldsymbol{\sigma}^d$ rather than total stress $\boldsymbol{\sigma}$ in the expression for plastic dissipation rate has not been recognized by numerous authors. Thus,

Inequality (20) is not very restrictive since it requires the scalar product of σ^d and $\dot{\varepsilon}^p$ to be positive.

Stability conditions will be formulated by following the general result obtained for composition of elastic, plastic and viscous elements with convex specific elastic energy and dissipation functions [22, 20, 23]. For any specified loading or deformation programme, all processes starting from different initial states tend to each other until the stress distribution within plastic or viscous phases attains the same values. In other words, this composition possesses the property of fading memory since previously set internal stresses can be wiped out in plastic regions by subsequent deformation process. In particular, the form of the steady plastic cycle is independent of the previous deformation history or fixed deformation superposed upon cyclically varying loads [23]. Whereas the concept of asymptotic stability cannot be valid on a microscale when the details of dislocation arrangement are considered, it may prove useful in a phenomenological description of plastic behaviour since numerous metals subjected to cyclic loading exhibit a fading memory effect such that previous plastic straining has negligible effect on a form of the steady cycle.

For two processes starting from different initial values of state parameters and undergoing the same prescribed stress or strain history, it can bepostulated that the state parameters will approach each other until a pair of states are reached in which the macroscopic behaviour for all processes is indistinguishable; this means that subsequent variation of internal parameters for all processes is the same although their absolute values may differ. To express a sufficient condition of this property in a more formal manner, let us assume that the free energy can be decomposed into two terms,

$$\psi = \psi_0(\varepsilon^e, \theta) + \psi_1(\gamma^{(r)}, \boldsymbol{\alpha}^{(s)}), \tag{22}$$

where ψ_0 denotes the free energy due to elastic straining and temperature changes whereas ψ_1 is the hidden free energy expressed in terms of internal state parameters. The rate of dissipation is now expressed as follows

$$D(\dot{\varepsilon}^p) = \boldsymbol{\sigma}^d \cdot \dot{\boldsymbol{\varepsilon}}^p = \boldsymbol{\sigma} \cdot \dot{\boldsymbol{\varepsilon}}^p - \rho \dot{\psi}_1. \tag{23}$$

Consider the case when the hidden free energy is a positive definite quadratic

for instance, Ziegler [19] using the inequality $\boldsymbol{\sigma} \cdot \dot{\boldsymbol{\varepsilon}}^p > 0$ arrives at the conclusion that the yield surface for kinematic hardening cannot move outside the origin. However, starting from (19), it can easily be shown that $\boldsymbol{\sigma}^d = \boldsymbol{\sigma} - \boldsymbol{\alpha}$, where $\boldsymbol{\alpha}$ denotes the position of the yield surface centre, and inequality (21) is satisfied for any position of the yield surface.

function of state parameters, that is

$$\rho\psi_1 = \tfrac{1}{2}\sum_r L^{(r)}(\gamma^{(r)})^2 + \tfrac{1}{2}\sum_s M^{(s)}_{ijkl}\alpha^{(s)}_{ij}\alpha^{(s)}_{kl} \tag{24}$$

where $L^{(r)}$, $M^{(s)}_{ijkl} = M^{(s)}_{klij}$ are constants; let the rate of dissipation be a convex function of $\dot{\varepsilon}^p$, thus

$$(\boldsymbol{\sigma}^d_2 - \boldsymbol{\sigma}^d_1)\cdot(\dot{\boldsymbol{\varepsilon}}^p_2 - \dot{\boldsymbol{\varepsilon}}^p_1) > 0. \tag{25}$$

Consider two processes differing only in the initial values of the internal state parameters, whereas macroscopic stress or strain is identical. The dissipative stresses corresponding to these two processes are $\boldsymbol{\sigma}^d_1$, $\boldsymbol{\sigma}^d_2$ and the plastic strain rates are $\dot{\boldsymbol{\varepsilon}}^p_1$, $\dot{\boldsymbol{\varepsilon}}^p_2$. As a measure of distance between the two processes, consider the function

$$E = \tfrac{1}{2}\sum_r L^{(r)}(\Delta\gamma^{(r)})^2 + \tfrac{1}{2}\sum_s M_{ijkl}\Delta\alpha_{ij}\Delta_{kl} \tag{26}$$

where $\Delta\gamma^{(r)}$ and $\Delta\alpha^{(s)}_{ij}$ denote differences of internal state parameters corresponding to the same value of stress $\boldsymbol{\sigma}$. Using (19) it follows in analogy to (23),

$$\dot{E} = \sum_r L^{(r)}\Delta\gamma^{(r)}\Delta\dot{\gamma}^{(r)} + \sum_s M^{(s)}_{ijkl}\Delta\alpha^{(s)}_{ij}\Delta\dot{\alpha}^{(s)}_{kl} =$$
$$= -(\boldsymbol{\sigma}^d_2 - \boldsymbol{\sigma}^d_1)\cdot(\dot{\boldsymbol{\varepsilon}}^p_2 - \dot{\boldsymbol{\varepsilon}}^p_1) < 0. \tag{27}$$

Since E is a positive-definite measure of distance and its rate is negative in view of (27), any two processes will approach each other until a state is reached for which $\dot{E} = 0$ when $\dot{\boldsymbol{\varepsilon}}^p_2 = \dot{\boldsymbol{\varepsilon}}^p_1$ and $\boldsymbol{\sigma}^d_2 = \boldsymbol{\sigma}^d_1$. When $L^{(r)}$ and $M^{(s)}_{ijkl}$ in (24) are not constants but depend on the state parameters, the stability can be shown only for infinitesimal departures from the considered process.

In view of the preceding analysis materials may be classified into two groups: (i) asymptotically stable or exhibiting a fading memory, and (ii) asymptotically unstable or sensitive to initial state or previous deformation history. In the next section, some models representing stable behaviour during plastic deformation will be discussed.

Examples of composite models of hardening

1. Consider an isothermal case and assume that the state of material is defined by $n+1$ tensor and $n+1$ scalar parameters $\boldsymbol{\alpha}^{(0)}, \boldsymbol{\alpha}^{(1)}, \ldots \alpha^{(n)}, \gamma^{(0)}, \gamma^{(1)}, \ldots \gamma^{(n)}$. The yield condition has the form

$$f^{(0)}(\boldsymbol{\sigma} - \boldsymbol{\alpha}^{(0)}) - \gamma^{(0)2} = 0 \tag{28}$$

and scalar equalities determining regions of internal irreversibility are

$$f_\alpha^{(1)}(\boldsymbol{\alpha}^{(0)}-\boldsymbol{\alpha}^{(1)})-\gamma^{(1)2}=0$$
$$\cdots\cdots\cdots\cdots\cdots\cdots \tag{29}$$
$$f_\alpha^{(k)}(\boldsymbol{\alpha}^{(k-1)}-\boldsymbol{\alpha}^{(k)})-\gamma^{(k)2}=0$$

where, for simplicity, it is assumed that the scalar parameters $\gamma^{(r)}$ are constant and that only $k+1$ successive parameters $\boldsymbol{\alpha}^{(i)}$ vary during deformation. According to (4), the rate of variation of these parameters is given by

$$\dot{\boldsymbol{\alpha}}^{(0)}=\boldsymbol{b}^{(0)}\dot{\lambda},\ \dot{\boldsymbol{\alpha}}^{(1)}=\boldsymbol{b}^{(1)}\dot{\lambda},\ldots,\dot{\boldsymbol{\alpha}}^{(k)}=\boldsymbol{b}^{(k)}\dot{\lambda} \tag{30}$$

provided that equations (29) are satisfied. Differentiating (29) and using the flow rule (6), the following consistency relations exist

$$K_0=\boldsymbol{b}^{(0)}\cdot\boldsymbol{n}^{(0)},\ \boldsymbol{n}^{(1)}\cdot\boldsymbol{b}^{(0)}=\boldsymbol{n}^{(1)}\cdot\boldsymbol{b}^{(1)},\ldots\ \boldsymbol{n}^{(k)}\cdot\boldsymbol{b}^{(k-1)}=$$
$$=\boldsymbol{n}^{(k)}\cdot\boldsymbol{b}^{(k)} \tag{31}$$

where $\boldsymbol{n}^{(0)}, \boldsymbol{n}^{(1)}, \ldots \boldsymbol{n}^{(k)}$ are unit normal vectors to the yield surface (28) and to the surfaces (29), respectively, at points defined by the stress $\boldsymbol{\sigma}$ and by the parameters $\boldsymbol{\alpha}^{(0)}, \boldsymbol{\alpha}^{(1)}, \ldots \boldsymbol{\alpha}^{(k-1)}$.

The set of relationships (28–31) with the flow rule (6) is complete and can be used to describe mechanical behaviour for any deformation process. It follows from (31) that the tangent modulus of the hardening curve will vary discontinuously during proportional loading. In fact, assume that in the initial state $\boldsymbol{\alpha}^{(0)}=\boldsymbol{\alpha}^{(1)}=\ldots=\boldsymbol{\alpha}^{(n)}=0$. When only yield condition (28) is satisfied, the parameter $\boldsymbol{\alpha}^{(0)}$ starts to vary, and the tangent modulus is defined by the first relationship (31). However, when the first equality (29) is satisfied, the parameter $\boldsymbol{\alpha}^{(1)}$ starts varying and the two relationships (31) are to be satisfied. Now, the first relationship (31) provides a new value of the tangent modulus whereas the second defines $\boldsymbol{b}^{(0)}$ in terms of $\boldsymbol{b}^{(1)}$. In the particular cases when $\boldsymbol{n}^{(0)}=\boldsymbol{n}^{(1)}=\ldots=\boldsymbol{n}^{(k)}$, it follows that $\boldsymbol{b}^{(0)}=\boldsymbol{b}^{(1)}=\ldots=\boldsymbol{b}^{(k)}$. Thus, when one of successive conditions (29) is satisfied, say $f^{(i)}=0$, all parameters $\boldsymbol{b}^{(i)}$, $i=0, 1, \ldots l-1$ undergo discontinuous changes.

One may ascribe a mechanical interpretation to parameters $\boldsymbol{b}^{(i)}$ by considering the model represented in Fig. 1a. This model has already been discussed by Ivlev [24] and Prager [25]. Consider a system composed of rigid blocks connected by elastic springs and resting on a frictional foundation with prescribed limit friction forces $\gamma^{(0)}, \gamma^{(1)}, \ldots \gamma^{(n)}$. The displacement of block 0 and the force applied to it correspond to plastic strains and macroscopic

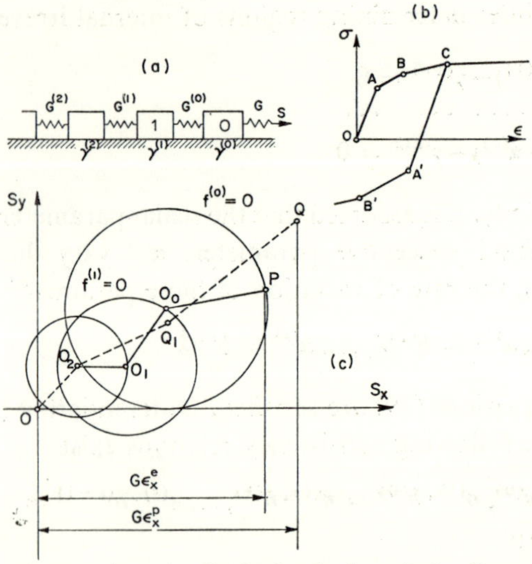

Fig. 1. A model simulating piecewise linear hardening a), loading and reverse loading curves approximated by linear portions, b), and representation of the model on the stress plane, c).

stress whereas forces in elastic elements $\boldsymbol{\alpha}^{(0)}$, $\boldsymbol{\alpha}^{(1)}$, ... $\boldsymbol{\alpha}^{(n)}$ and permanent displacements of blocks 1, 2, ... n simulate microstrains which will be denoted by $\boldsymbol{\varepsilon}^{(1)}$, $\boldsymbol{\varepsilon}^{(2)}$, ... $\boldsymbol{\varepsilon}^{(n)}$. Let the system under acting stress be assumed to reach a state for which yield condition (28) and the two first equalities (29) are satisfied. Rates of plastic microstrains are given by

$$\dot{\boldsymbol{\varepsilon}}^{(1)} = \frac{1}{K^{(1)}}\, \boldsymbol{n}^{(1)}(\boldsymbol{\alpha}^{(0)}\cdot \boldsymbol{n}^{(1)}),\ \dot{\boldsymbol{\varepsilon}}^{(2)} = \frac{1}{K^{(2)}}\, \boldsymbol{n}^{(2)}(\boldsymbol{\alpha}^{(1)}\cdot \boldsymbol{n}^{(2)}),$$

$$\dot{\boldsymbol{\varepsilon}}^{(3)} = 0, \tag{32}$$

and rates of micro-stresses are defined as follows

$$\dot{\boldsymbol{\alpha}}^{(0)} = G^{(0)}(\dot{\boldsymbol{\varepsilon}}^{(0)} - \dot{\boldsymbol{\varepsilon}}^{(1)}),\ \dot{\boldsymbol{\alpha}}^{(1)} = G^{(1)}(\dot{\boldsymbol{\varepsilon}}^{(1)} - \dot{\boldsymbol{\varepsilon}}^{(2)}),$$

$$\dot{\boldsymbol{\alpha}}^{(2)} = G^{(2)}\dot{\boldsymbol{\varepsilon}}^{(2)} \tag{33}$$

where $G^{(0)}$, $G^{(1)}$, $G^{(2)}$ denote constant coefficients. Using the consistency relationships in the form

$$\boldsymbol{n}^{(0)}\cdot\dot{\boldsymbol{\sigma}} = \boldsymbol{n}^{(0)}\cdot\dot{\boldsymbol{\alpha}}^{(0)},\ \boldsymbol{n}^{(1)}\cdot\dot{\boldsymbol{\alpha}}^{(0)} = \boldsymbol{n}^{(1)}\cdot\dot{\boldsymbol{\alpha}}^{(1)},\ \boldsymbol{n}^{(2)}\cdot\dot{\boldsymbol{\alpha}}^{(1)} =$$

$$= \boldsymbol{n}^{(2)}\cdot\dot{\boldsymbol{\alpha}}^{(2)} \tag{34}$$

and Eqs. (6), (32), (34), the following relationships between $G^{(i)}$ and $K^{(i)}$ is established

$$K^{(2)} = G^{(2)}, K^{(1)} = G^{(1)} - G^{(1)2} \frac{\boldsymbol{n}^{(1)} \cdot \boldsymbol{n}^{(2)}}{G^{(1)} + G^{(2)}},$$

$$K^{(0)} = G^{(0)} - G^{(0)2} \frac{\boldsymbol{n}^{(1)} \cdot \boldsymbol{n}^{(0)}}{G^{(0)} + G^{(1)}}, \tag{35}$$

and eliminating $\boldsymbol{\varepsilon}^{(1)}$, $\boldsymbol{\varepsilon}^{(2)}$, a set of relationships analogous to (30) are obtained

$$\dot{\boldsymbol{\alpha}}^{(0)} = \left[G^{(0)}\boldsymbol{n}^{(0)} - \frac{G^{(0)} - K^{(0)}}{\boldsymbol{n}^{(0)} \cdot \boldsymbol{n}^{(1)}} \boldsymbol{n}^{(1)} \right] \dot{\lambda} = \boldsymbol{b}^{(0)}\dot{\lambda},$$

$$\dot{\boldsymbol{\alpha}}^{(1)} = \left[G^{(1)}\boldsymbol{n}^{(1)} - \frac{G^{(1)} - K^{(1)}}{\boldsymbol{n}^{(1)} \cdot \boldsymbol{n}^{(2)}} \boldsymbol{n}^{(2)} \right] \frac{G^{(0)}(\boldsymbol{n}^{(0)} \cdot \boldsymbol{n}^{(1)})}{G^{(0)} + K^{(1)}} \dot{\lambda}$$

$$= \boldsymbol{b}^{(1)}\dot{\lambda},$$

$$\dot{\boldsymbol{\alpha}}^{(2)} = \frac{G^{(0)}G^{(2)}(G^{(1)} - K^{(1)})}{G^{(1)}(G^{(0)} + K^{(1)})} \frac{(\boldsymbol{n}^{(0)} \cdot \boldsymbol{n}^{(1)})}{(\boldsymbol{n}^{(1)} \cdot \boldsymbol{n}^{(2)})} \boldsymbol{n}^{(2)}\dot{\lambda} = \boldsymbol{b}^{(2)}\dot{\lambda}. \tag{36}$$

These relationships can be generalized for an arbitrary number of elements. Using the model of Fig. 1a, the tensor parameters $\boldsymbol{b}^{(i)}$ occurring in (30) are defined in terms of scalar quantities $\gamma^{(i)}$ and $G^{(i)}$.

Fig. 1b represents the hardening curve corresponding to this model and Fig. 1c shows a two-dimensional representation in the stress plane. The motion of particular blocks is governed by respective yield conditions which can be expressed in the forms

$$f^{(0)} = (s_x - \alpha_x^{(0)})^2 + (s_y - \alpha_y)^2 - \gamma^{(0)2} = 0,$$

$$f^{(1)} = (\alpha_x^{(0)} - \alpha_x^{(1)})^2 + (\alpha_y^{(0)} - \alpha_y^{(1)})^2 - \gamma^{(1)2} = 0. \tag{37}$$

These conditions represent circles on the plane (s_x, s_y). The vectors $OO_2/G^{(2)}$, $OO_1/G^{(1)}$, $OO_0/G^{(0)}$ indicate relative block displacements; thus the total displacement of O representing plastic strain equals $OO_2/G^{(2)} + OO_1/G^{(1)} + OO_0/G^{(0)}$. The stress is given by OP and OP/G corresponds to elastic strain.

To calculate the hidden free energy, the orthogonality of the set of residual stresses is postulated, thus

$$\rho\psi_1 = \tfrac{1}{2}\rho \sum_s \frac{1}{2G^{(s)}} \boldsymbol{\alpha}^{(s)2} \tag{38}$$

2. As previously, the internal parameters are denoted by $\boldsymbol{\alpha}^{(1)}, \boldsymbol{\alpha}^{(2)}, \ldots \boldsymbol{\alpha}^{(n)}$, $\gamma^{(1)}, \ldots \gamma^{(n)}$ and the yield condition is assumed in the form

$$f^{(0)}(\boldsymbol{\sigma}-\sum_{i=1}^{n}\boldsymbol{\alpha}^{(i)})-\gamma^{(0)2}=0, \tag{39}$$

whereas scalar equalities

$$f^{(i)}(\alpha^{(i)})-\gamma^{(i)2}=0. \tag{40}$$

Let $\gamma^{(i)}$ be constant during plastic deformation. It follows from (9) that the consistency relationships are

$$\boldsymbol{n}^{(i)}\cdot\dot{\boldsymbol{\alpha}}^{(i)}=0, \tag{41}$$

where $\boldsymbol{n}^{(i)}$ denote unit normal vectors to surfaces $f^{(i)}=0$.

Assume that the l equalities (40) are satisfied. Then rate equations for the parameters $\boldsymbol{\alpha}^{(i)}$, satisfying (40) can be expressed as

$$\begin{aligned}\dot{\boldsymbol{\alpha}}^{(i)} &= G^{(i)}[\dot{\boldsymbol{\varepsilon}}^{p}-(\dot{\boldsymbol{\varepsilon}}^{p}\cdot\boldsymbol{n}^{(i)})\boldsymbol{n}^{(i)}], \qquad f^{(i)}=0, i=1,2,\ldots l\\ \dot{\boldsymbol{\alpha}}^{i} &= G^{(i)}\dot{\boldsymbol{\varepsilon}}^{p}, f^{(i)}<0, \qquad i=l+1,\ldots n.\end{aligned} \tag{42}$$

The tangent modulus K of the hardening curve is obtained by using the flow rule (6) and (40), thus

$$\left|\frac{\partial f^{(0)}}{\partial\boldsymbol{\sigma}^{(0)}}\right|K=\sum_{i=1}^{l}G^{(i)}\left[\frac{\partial f^{(0)}}{\partial\boldsymbol{\alpha}^{(i)}}\cdot\boldsymbol{n}^{(0)}-\frac{\partial f^{(i)}}{\partial\boldsymbol{\alpha}^{(i)}}\cdot\boldsymbol{n}^{(i)}(\boldsymbol{n}^{(0)}\cdot\boldsymbol{n}^{(i)}\right]+\sum_{k=1}^{n}G^{(k)}\frac{\partial f^{(0)}}{\partial\boldsymbol{\alpha}^{(k)}}\cdot\boldsymbol{n}^{(0)}. \tag{43}$$

The mechanical interpretation of this model can be obtained by considering a system similar to that in Fig. 1a. A rigid block O whose displacement corresponds to plastic strain rests on a foundation with the limit friction force $\gamma^{(0)}$. A set of parallel blocks is connected to O by means of elastic springs and the friction forces between blocks and the foundation are $\gamma^{(1)}$, $\gamma^{(2)}, \ldots \gamma^{(n)}$.

3. Assume as state parameters $n+1$ tensor quantities $\boldsymbol{\alpha}^{(0)}, \ldots \boldsymbol{\alpha}^{(n)}$ and $2(n+1)$ scalar quantities $\gamma^{(0)}, \ldots \gamma^{(n)}, K^{(0)}, \ldots K^{(n)}$. The yield condition has the form

$$f^{(0)}(\boldsymbol{\sigma}-\boldsymbol{\alpha}^{(0)})-\gamma^{(0)2}=0 \tag{44}$$

and scalar relationships defining regions of variability of parameters $\boldsymbol{\alpha}^{(i)}$ are

$$f_{\alpha}^{(i)}(\boldsymbol{\sigma}-\boldsymbol{\alpha}^{(i)})-\gamma^{(i)2} = 0. \tag{45}$$

It is assumed that all surfaces for which (44) or (45) are satisfied are tangential to each other at the point defined by the current stress $\boldsymbol{\sigma}$. To each surface (45) or (46) there corresponds a tangent modulus $K^{(i)}$; if l surfaces are tangential at the stress point $P^{(l)}$, the hardening modulus has the value $K^{(l)}$ and in the flow rule (6) $K = K^{(l)}$. A configuration of surfaces (45) and (46) defines the field of hardening moduli in stress space; its variation can be observed by integrating the rate equations for a specified loading programme. This model has been discussed in detail by Mróz [20, 21] and applied to some cases of cyclic loading. A similar representation was also used by Haythornthwaite [29] in discussing his test results.

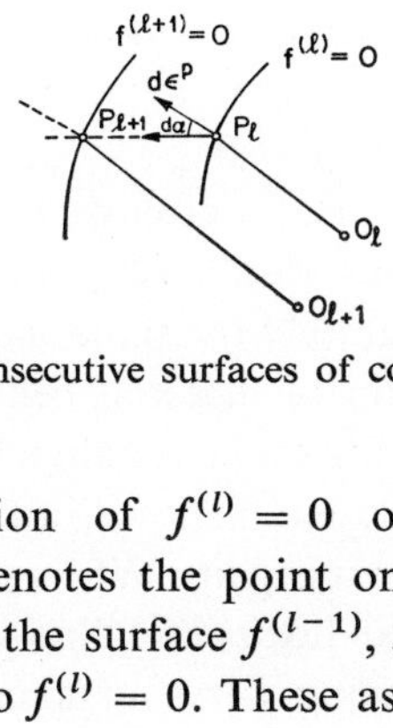

Fig. 2. Translation of consecutive surfaces of constant hardening moduli.

The instantaneous motion of $f^{(l)} = 0$ occurs along the direction $\boldsymbol{\sigma}^{(l+1)}-\boldsymbol{\sigma}^{(l)}$ where $\boldsymbol{\sigma}^{(l+1)}$ denotes the point on the surface with the same normal as at $P^{(l)}$, whereas the surface $f^{(l-1)}, \ldots f^{(0)}$ move with the point $P^{(l)}$, remaining tangential to $f^{(l)} = 0$. These assumptions can be expressed as follows (see Fig. 2):

$$\dot{\boldsymbol{\alpha}}^{(l)} = \dot{\mu}(\boldsymbol{\sigma}^{(l+1)}-\boldsymbol{\sigma}^{(l)}),\ \boldsymbol{\alpha}^{(l-1)} = \boldsymbol{\sigma}^{l} - \frac{\gamma^{(l-1)2}}{\gamma^{(l)2}}(\boldsymbol{\sigma}^{(l)}-\boldsymbol{\alpha}^{(l)}),$$

$$\dot{\boldsymbol{\alpha}}^{(l+1)} = \ldots = \dot{\boldsymbol{\alpha}}^{(n)} = 0, \tag{46}$$

$$\dot{\gamma}^{(i)} = C^{(i)}(\lambda)\dot{\lambda}. \tag{46a}$$

Using the consistency relationships it follows that

$$\dot{\mu} = \frac{|\partial f^{(l)}/\partial\boldsymbol{\sigma}|K^{(l)}-C^{(l)}}{|\,\partial f^{(l)}/\partial\boldsymbol{\sigma}|(\boldsymbol{\sigma}^{(l+1)}-\boldsymbol{\sigma}^{(l)})\cdot\boldsymbol{n}^{(0)}}\,\dot{\lambda}. \tag{47}$$

Whereas the present model is constructed more formally and has no direct mechanical interpretation, it may prove more useful since the field of hardening moduli exhibits a directional variation of material stiffness after plastic prestrain. The state parameters can therefore be measured in simple macroscopic tests.

It is obvious that many other models can be constructed in a similar manner by following the general discussion in the previous section. In addition to the flow rule and the yield condition, a set of scalar equalities and discontinuously changing rate equations for the state parameters have been introduced. Consistency relations restrict the arbitrary choice of material functions, so that material behaviour can be predicted by knowing a set of scalar functions.

Some applications to variable loading

Starting from a hardening curve for monotonic loading followed by unloading with subsequent reloading in a reverse direction, (Fig. 1b), and approximating it by a piecewise linear portion, the material parameters can easily be incorporated into models discussed in the previous section. This problem has been discussed in detail in [26]. It is not clear, however, whether such models describing accurately the same uniaxial stress strain curve can provide a rational description of material behaviour for non-proportional loading paths, in particular for cyclic loading. To illustrate this, consider a hardening curve under pure torsion shown in Fig. 3a and approximate it in the plastic range by two linear segments. Whereas the first segment represents linear hardening, the second horizontal line is a perfect plasticity upper bound on the actual hardening curve of small slope. Consider now the loading programme when upon initial shear $\tau_{xy} = \tau$ the alternating tension $\sigma_x = \pm\sigma_a$ is superposed, whereas the other stress components vanish. For Mises' yield condition, the yield curve is represented by a circle $\sigma_x^2 + t_{xy}^2 = \sigma_0^2$ on the (σ_x, t_{xy}) plane where $t_{xy} = \sqrt{3}\tau_{xy}$; the limit curve bounding elastic-plastic states is also represented by a circle $\sigma_x^2 + t_{xy}^2 = \sigma_c^2$. Let the cyclic loading programme under consideration be confined within the elastic-plastic region without reaching the limit curve. Fig. 3b shows the plastic shear strain accumulation during cyclic loading as predicted by three different models: i) the model shown in Fig. 1a and described by a set of relationships (32–38) (curves 'b') ii) the model based upon the concept of field of hardening moduli, Eqs. (45–48), (curves 'a') and iii) the kinematic hardening model, (curves 'c'). It is seen that whereas the first two models predict a steady shear

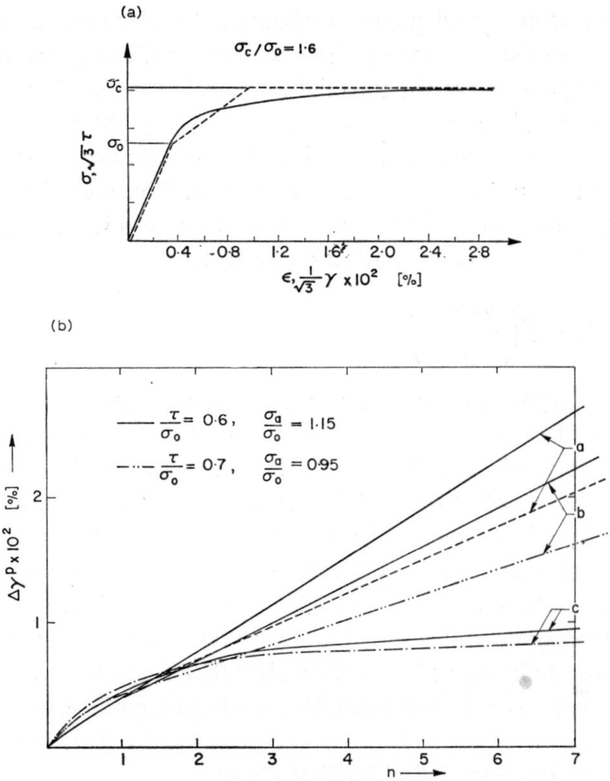

Fig. 3. Approximation of hardening curve by three linear segments, a), and predictions of three hardening models for torsional strain accumulation when alternating tensile and compressive stress is superposed upon constant shear stress, b), curves '*a*' correspond to model of field of hardening moduli, curves '*b*' to model of Fig. 1a, whereas curves '*c*' correspond to kinematic hardening model.

strain accumulation, this is not the case for the third model since the steady state corresponds to alternating plastic cycles in tension and compression with no progressive flow. Similar conclusions may be reached if more piece-wise linear segments were used [26]. In actuality, plastic strain accumulation has been observed in numerous tests performed on metallic tubular specimens subjected to alternating tensile or shearing stresses superposed upon a fixed stress state.[1]) It is apparent that models corresponding to the same

[1]) It can generally be stated that though models based upon the idea of series or parallel connection of strong (elastic) and weak (plastic) elements, such as those considered in

hardening curve may yield qualitatively different results for complex loadings.

Consider now the case when an infinite number of state parameters is used and the hardening curve for uniaxial loading is described by the relation $\varepsilon^p = f(\sigma)$ where ε^p denotes the plastic strain. Using the concept of field of hardening moduli, a simple analytical description can be provided for multiaxial stress state and non-proportional loading. Consider, for instance, the Tresca yield condition and the stress strain curve shown in Fig. 4d. During unloading, let the Masing equation be valid, that is,

$$\tfrac{1}{2}(\varepsilon_B^p - \varepsilon^p) = f\left(\frac{\sigma_B - \sigma}{2}\right) \tag{48}$$

where σ_B and ε_B^p denotes the stress and the plastic strain at B.

Let the field of hardening moduli be defined by an infinite number of Tresca hexagons of variable size which during deformation undergo only translation. Existence of the initial yield locus is not essential here and such materials may be considered for which plastic strain starts for arbitrarily small values of stress and for infinitesimal unloading from any state. Upon initial loading along OB, the field of hardening moduli is that depicted in Fig. 4a; all hexagons engaged by the stress point move parallel to OB. Now, there are four different regimes within $ABCDEF$ and the hardening law for each regime can easily be formulated if the Masing hypothesis is extended by proper scaling of stress and strain components. In fact, for stress paths directed in the interior of $ABCDEF$, we have

region ABF: $\varepsilon_1^p - \varepsilon_{1B}^p = f(\sigma_1)$, $\varepsilon_2^p = \varepsilon_{2B}^p$,

region BEF: $\varepsilon_1^p - \varepsilon_{1B}^p = 2f\left(\dfrac{\sigma_1 - \sigma_2 + \sigma_{2B}}{2}\right)$, $\varepsilon_2^p - \varepsilon_{2B}^p = -(\varepsilon_1^p - \varepsilon_{1B}^p)$,

region BDE: $\varepsilon_{2B}^p - \varepsilon_2^p = 2f\left(\dfrac{\sigma_{2B} - \sigma_2}{2}\right)$, $\varepsilon_1^p = \varepsilon_{1B}^p$,

region BCD: $\varepsilon_{1B} - \varepsilon_1^p = f(\sigma_1)$, $\varepsilon_2^p = \varepsilon_{2B}^p$. (49)

[10–13], can accurately simulate uniaxial hardening curve, they cannot adequately predict plastic behaviour for more complex cyclic loading since the steady cyclic state would correspond to alternative plastic flow and no progressive flow could be predicted. Such models are used to study cyclic behaviour of structural elements but their general use seems questionable.

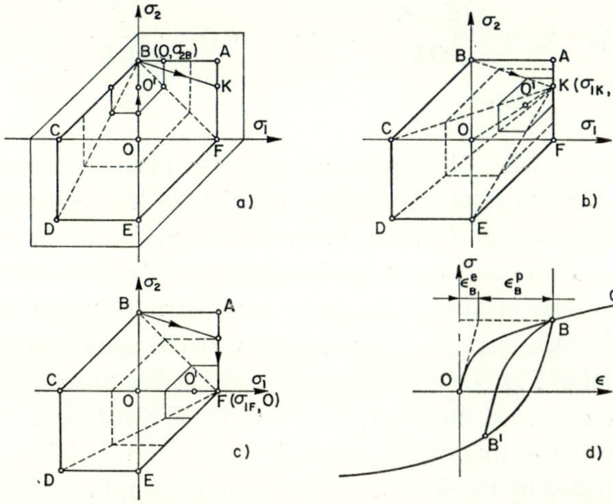

Fig. 4. Field of hardening moduli for the loading program *OBKF*, a) after reaching *B*, b) after reaching *K*, c) after reaching *F*, d) hardening curve and Masing reverse loading curve.

It is seen that the stress-strain relationships vary discontinuously when passing from one to the other region. For stress paths directed along corner régimes *BF*, *BE*, *BD*, there is no unique relation between rates of stress and plastic strain since the vector $\boldsymbol{\varepsilon}^p$ may rotate between normals to adjacent sides.[1])

Consider the loading path *OBKF*. Upon reaching *B*, the value of strain ε_{2B}^p can be obtained from the hardening curve of Fig. 4d, whereas $\varepsilon_{1B}^p = -\frac{1}{2}\varepsilon_{2B}^p$. Along the path *BK*, the stress is related to plastic strain by the first relationship (5). The field of hardening moduli upon reaching *K* is shown in Fig. 4b. The centers of all hexagons lie on the line *OO'K* and there are five different regimes within *ABDCEF* for which finite stress-strain relations can be derived, i.e.

$$KEF:\quad \varepsilon_{2K}^p - \varepsilon_2^p = \frac{KF}{OB} f\left[(\sigma_1 - \sigma_2 - \sigma_{2K})\frac{OB}{KF}\right],\ \varepsilon_1^p - \varepsilon_{1K}^p = $$
$$= -(\varepsilon_2^p - \varepsilon_{2K}^p),$$

[1]) This drawback is analogous to that discussed by Ziegler [3] for his hardening rule. To predict uniquely plastic strain for stress paths *BF*, *BE*, *BD*, we can for instance, postulate that $|\varepsilon_1/\varepsilon_2| = \frac{1}{2}$ for corners *B* and *E*, $|\varepsilon_2/\varepsilon_1| = \frac{1}{2}$ for corners *C* and *F*, and $|\varepsilon_1/\varepsilon_2| = 1$ for *D* and *A*.

$$KDE: \quad \varepsilon_{2K}^{p} - \varepsilon_{2}^{p} = \frac{OE+KF}{OB}\, f\left[(\sigma_{2K}-\sigma_{2})\, \frac{OB}{OE+KF}\right],$$

$$\varepsilon_{1}^{p} = \varepsilon_{1K}^{p} \tag{50}$$

and similar relations for other regimes. Thus the plastic strain at F can be obtained from the first relationship (50). The field of hardening moduli upon reaching F is shown in Fig. 4c. When the stress point moves beyond $ABCDEF$, the portion BC of the hardening curve of Fig. 4d should be used in evaluating plastic strain. These finite stress-strain relationships can be generalized to hardening rules other than those based upon Masing equation.

For the model of Fig. 4, some phenomena occurring during cyclic loading may be discussed at least qualitatively. Consider, for instance, the following problem: upon initial loading to point B, the material exhibits anisotropic properties described by the field of hardening moduli of Fig. 4a. What would be a loading programme that could restore the isotropy of the material? From Fig. 4a it follows that such programme should consist of alternating stress $\pm\sigma_2$ with a monotonically decreasing amplitude. It is interesting to note that tests of this kind reported by Krempl [27] confirm this general conclusion.

Concluding remarks

The mathematical structure of the plasticity theory with internal state parameters should account for local irreversibility conditions which require additional scalar equalities and discontinuously changing rate equations for state parameters. A similar mathematical structure can be used in constructing models for viscoplastic behaviour.

References

[1] Melan, E., Zur Plastizität des räumlichen Kontinuums, *Ing Arch.*, *9* (1938) 116–126.

[2] Prager, W., A new method of analysing stress and strain in work-hardening plastic solids, *Journal of Applied Mechanics*, *23* (1956) 493.

[3] Ziegler, H., A modification of Prager's hardening rule, Quarterly of Applied Mathematics, *17* (1959) 55–60.

[4] Eisenberg, M. and A. Philips, On non-linear kinematic hardening, Acta Mechanica, *6* (1968).

[5] Baltov, A. and A. Sawczuk, A rule of anisotropic hardening, Acta Mechanica, *1/2* (1965) 81–92.

[6] Taylor, G. I. Plastic strain in metals, *Journal of the Institute of Metals, 62* (1938).

[7] S. B. Batdorf and B. Budiansky, Polyaxial stress-strain relations of strain-hardening metal, *Journal of Applied Mechanics, 21* (1954) 323–326.

[8] T. H. Lin and M. Ito, Theoretical plastic distorsion of a polycrystalline aggregate under combined and reversed stress, *Journal of the Mechanics and Physics of Solids, 13* (1965) 103–115.

[9] Hill, R. The essential structure of consitutive laws for metal composites and polycrystals, *Journal of the Mechanics and Physics of Solids, 15* (1967) 79–95.

[10] Besseling, J. F., Theory of elastic, plastic and creep deformation of an initially isotropic material showing anisotropic strain hardening, creep recovery and secondary creep, *Journal of Applied Mechanics, 25* (1958).

[11] Iwan, W. D., On a class of models for the yielding behaviour of continuous and composite systems, *Journal of Applied Mechanics, 34* (1967).

[12] Wells, C. H. and P. R. Paslay, A small strain plasticity theory for planar slip materials, *Journal of Applied Mechanics, 36* (1969).

[13] Kafka, V., The general theory of isothermal elastic-plastic deformation of polycrystals based on an analysis of the microscopic state of stress, *Zeitschrift fur Angewandte Mathematik und Mechanik, 48* (1968) 265–282.

[14] Vakulenko, A. A., On relations between stress and strain in inelastic media, in 'Problems of Elasticity and Plasticity' ,(in Russian), Leningrad University, *1* (1961).

[15] Kröner, E., Dislocations: A new concept in the continuum theory of plasticity, *Journal of Mathematics and Physics, 42* (1963).

[16] Kratochvíl, J. and O. W. Dillon, Thermodynamics of elastic-plastic materials as a theory with internal state variables, *Journal of Applied Physics, 40* (1969) 3207–3218.

[17] Prager, W., Non-isothermal plastic deformation, *Proceedings Koninkl. Akademie van Wetenschappen, Series B, 61* (1958) 176–182.

[18] Fung, Y. C., Foundations of solid mechanics, Prentice Hall, Inc., (1965).

[19] Ziegler, H., Plastizität ohne Thermodynamik?, *Zeitschrift für Angewandte Mathematik und Physik, 21* (1970) 789–806.

[20] Mróz, Z., On the description of anisotropic work hardening, *Journal of Mechanics and Physics of Solids, 15* (1967) 163–175.

[21] Mróz, Z., An attempt to describe the behaviour of metals under cyclic loads using a more general work hardening model, *Acta Mechanica, 7* (1969) 199–212.

[22] Frederick, C. O. and P. J. Armstrong, Convergent internal stresses and steady cyclic states of stress, *The Journal of Strain Analysis, 1* (1966) 154–160.

[23] Mróz, Z., On the theory of steady plastic cycles in structures, Proc. First International Conference on Structural Mechanics in Reactor Technology, Berlin (1971).

[24] Ivlev, V. D. On theory of composite media, *Doklady Akademii Nauk SSSR, 8* (1963).

[25] Prager, W., Composite stress strain relations for elastoplastic solids, Proceedings of IUTAM Symposium 'Irreversible Aspects of Continuum Mechanics', Vienna 1966, Springer Verlag.

[26] Goss, Cz. and Z. Mróz, On composite models of plastic work hardening, (in Polish), Mechanika Teoretyczna i Stosowana, 1972. (in press).

[27] Krempl, E., Cyclic plasticity: Some properties of the hysteresis curve of structural metals at room temperature, *Journal of Basic Engineering*, (1971).

[28] Mróz, Z., On forms of constitutive laws for elastic-plastic solids, *Archiwum Mechaniki Stosowanej, 18* (1966) 3–35.

[29] Haythornthwaite, R. M., A more rational approach to strain hardening data, in 'Engineering Plasticity', Ed. J. Heyman, F. A. Leckie, Perg. Press, (1968).

On large elastic-plastic deformations

Th. Lehmann

Ruhr-University, Bochum, BRD

1. Introduction

In the analysis of large plastic deformations elastic strains may be neglected. However, this is not possible, if second-order effects need to be taken into account, since in this case the elastic strains change the kinematics of the process and thus affect the stresses.

In our considerations the material is assumed to be a classical continuum. This means that the deformations of the material will be determined completely either by the displacements of the material points in a reference space or by the changes of the metric of a body-fixed coordinate system. We shall prefer the second one, as more suitable to the problem under consideration.

The constitutive equations for the elastic and the plastic part of the deformations cannot be treated independently, even when elastic behaviour is assumed to be unaffected by the preceding plastic deformations as is done here. The constitutive equations are coupled because the stresses are acting in a body deformed simultaneously by the elastic and the plastic strains. Since the stress-strain relationship for the plastic part of the deformations has to be formulated as an incremental law, the stress-strain relations for the elastic part must be defined in a similar manner. On account of this requirement two problems arise:

1. to separate the strain rates into an elastic and a plastic part in an appropriate manner (kinematical problem),
2. to derive the stress-strain relationships for the elastic part as an incremental law[1]) (constitutive problem).

There are different ways to solve the problems.

[1]) The plastic stress-strain relations may be derived in the usual manner.

2. The kinematical problem

Consider a body-fixed coordinate system ξ^i be introduced deforming with the body. The metric of this coordinate system is a function of ξ^i and of the time t

$$g_{ik} = g_{ik}(\xi^r, t). \tag{1}$$

In the undeformed state, at time t_0, the metric

$$g_{ik} = g_{ik}(\xi^r, t_0) = \overset{\circ}{g}_{ik}(\xi^r). \tag{2}$$

Furthermore, a space-fixed coordinate system x^α will be used for some purposes. The body-fixed coordinate system ξ^i at time t_0 may coincide with the space-fixed coordinate system x^α [1]). In the following lower case Latin indices will refer to entities related to the body-fixed coordinate-system whereas Greek indices will denote those referred to the space-fixed system. Starting from the deformation gradient[2])

$$\frac{\partial x^\alpha}{\partial \xi^i} = x^\alpha{}_{,i}; \qquad x^\alpha = x^\alpha(\xi^i, t) \tag{3}$$

a first definition of the total strain rate may be expressed following Lee [1] and [2] as

$$\begin{aligned} d^\alpha_\beta &= \operatorname{sym}\left\{\frac{D}{dt}(x^\alpha{}_{,i})\xi^i{}_{,\beta}\right\} \\ &= \tfrac{1}{2}\{(\dot{x}^\alpha{}_{,i})\xi^i{}_{,\beta} + g_{\beta\rho}(\dot{x}^\rho{}_{,i})\xi^i{}_{,\mu}\, g^{\mu\alpha}\} \\ &= \tfrac{1}{2}\{v^\alpha|_\beta + v_\beta|^\alpha\} \end{aligned} \tag{4a}$$

In these formulas the following notation is used:

d^α_β = strain rate,

$\frac{D}{dt} = \dot{(\ )}$ = material derivative with respect to time at constant ξ^i,

$|_\beta$ = covariant differentiation with respect to x^β,

v^α = velocity of the material points ($v^\alpha = \dot{x}^\alpha$),

sym { } = the symmetrical part of { }.

[1]) This is not a necessary requirement but it simplifies the analysis.

[2]) $,\alpha$ denotes partial differentiation with respect to x^α, correspondingly $,i$ means partial differentiation with respect to ξ^i.

Transforming d^α_β to the body-fixed coordinate system

$$d^i_k = \xi^i{}_{,\alpha}\, d^\alpha_\beta x^\beta{}_{,k}. \tag{4b}$$

An equivalent definition of the total strain rate is[1])

$$\begin{aligned} d_{ik} &= \tfrac{1}{2}\dot{g}_{ik} = -\tfrac{1}{2}g_{ir}\dot{g}^{rs}g_{sk}, \\ d^i_k &= \tfrac{1}{2}g^{ir}\dot{g}_{rk} = -\tfrac{1}{2}\dot{g}^{ir}g_{rk}. \end{aligned} \tag{5}$$

A number of possibilities exist to resolve the total strain rate d^i_k into its elastic and plastic parts. The total strain will be considered using two steps, the first step leads from the undeformed state into an intermediate configuration by plastic deformation, and the second step transfers the intermediate configuration into the state of total deformation by elastic strains. The state of total deformation is compatible, provided that the body is treated as a classical continuum as here is the case. Intermediate configurations, in general, do not satisfy the compatibility requirement.
From the point of view of an observer located at the space-fixed coordinate system the total deformation can be decomposed by means of a multiplicative splitting of the deformation gradient (3)

$$x^\alpha_{,i} = \frac{\partial x^\alpha}{\partial x^M}\,\frac{\partial x^M}{\partial \xi^i} = x^\alpha_{,M}x^M_{,i}, \tag{6}$$

where $x^M_{,i}$ is the matrix prescribing the mapping from the undeformed state onto the intermediate configuration.

In view of (6) the following additive decomposition of the strain rate may be obtained (see Lee [1] and Freund [2])

$$\begin{aligned} d^\alpha_\beta &= \operatorname{sym}\{(\dot{x}^\alpha{}_{,i})\xi^i{}_{,\beta}\} = \operatorname{sym}\{(\dot{\overline{x^\alpha{}_{,M}x^M{}_{,i}}})\xi^i{}_{,\beta}\} \\ &= \operatorname{sym}\{(\dot{x}^\alpha{}_{,M})\underbrace{x^M{}_{,i}\,\xi^i{}_{,\beta}}_{x^M{}_{,\beta}}\} + \operatorname{sym}\{x^\alpha{}_{,M}(\dot{x}^M{}_{,i})\xi^i{}_{,\beta}\} \equiv \\ &\equiv \underset{E}{d^\alpha_\beta} + \underset{P}{d^\alpha_\beta} \end{aligned} \tag{7}$$

From the point of view of an observer attached to the body-fixed coordinate system it seems to be more natural to start with a splitting of the changes of the metric. This may be done in two different ways.

[1]) Operating within the body-fixed coordinate system the material derivation is the same as partial differentiation with respect to time: $\dot{(\,)} = \partial/\partial t$.

One way consists in an additive splitting of the changes of the metric by setting (see Green and Naghdi [3] or Sedov [5],

$$\tfrac{1}{2}(g_{ik}-\overset{\circ}{g}_{ik}) = \tfrac{1}{2}(g_{ik}-\overset{*}{g}_{ik})+\tfrac{1}{2}(\overset{*}{g}_{ik}-\overset{\circ}{g}_{ik}). \tag{8}$$

Where $\overset{*}{g}_{ik}$ belongs to an intermediate (incompatible) configuration and using a material derivation it follows from (8) that

$$d_{ik} = \tfrac{1}{2}\dot{g}_{ik} = \tfrac{1}{2}\frac{D}{dt}(g_{ik}-\overset{*}{g}_{ik})+\tfrac{1}{2}\frac{D}{dt}(\overset{*}{g}_{ik}-\overset{\circ}{g}_{ik}) \equiv \underset{E}{d_{ik}}+\underset{P}{d_{ik}} \tag{9a}$$

$$d^i_k = \tfrac{1}{2}g^{ir}\dot{g}_{rk} = \tfrac{1}{2}g^{ir}(\dot{g}_{rk}-\dot{\overset{*}{g}}_{rk})+\tfrac{1}{2}g^{ir}\dot{\overset{*}{g}}_{rk} \equiv \underset{E}{d^i_k}+\underset{P}{d^i_k}. \tag{9b}$$

For the second way of splitting one may start from the symmetric quantities

$$q^i_k = \overset{\circ}{g}{}^{ir}g_{rk} \qquad (q^{-1})^i_k = g^{ir}\overset{\circ}{g}_{rk}, \tag{10}$$

which transform the metric of the undeformed body into that of the deformed body and vice versa

$$\begin{aligned} g_{ik} &= q^r_i\overset{\circ}{g}_{rk} = \overset{\circ}{g}_{ir}q^r_k \\ \overset{\circ}{g}_{ik} &= (q^{-1})^r_i\, g_{rk} = g_{ir}(q^{-1})^r_k. \end{aligned} \tag{11}$$

These quantities can be divided into two multiplicative, non-commutative components as follows

$$q^i_k = \overset{\circ}{g}{}^{ir}\overset{*}{g}_{rm}\overset{*}{g}{}^{ms}g_{sk} \equiv \underset{P}{q^i{}_{.m}}\underset{E}{q^m_k} \tag{12a}$$

i.e. $\underset{P}{q^i{}_{.m}} \equiv \overset{\circ}{g}{}^{ir}\overset{*}{g}_{rm}$ whereas $\underset{E}{q^m_k} \equiv \overset{*}{g}{}^{ms}g_{sk}$
analogously

$$(q^{-1})^i_k = g^{ir}\overset{*}{g}_{rm}\overset{*}{g}{}^{ms}\overset{\circ}{g}_{sk} \equiv \underset{E}{(q^{-1})^i_m}\underset{P}{(q^{-1})^m{}_{.k}}. \tag{12b}$$

In this representation the tensors $\underset{E}{q^i_k}$ and $\underset{E}{(q^{-1})^i_k}$ are symmetric, but the tensors $\underset{P}{q^i{}_{.k}}$ and $\underset{P}{(q^{-1})^i{}_{.k}}$ are not.

The connection between the total strain rate and the material derivative of the quantities q^i_k and $(q^{-1})^i_k$ is given by[1])

$$d^i_k = \tfrac{1}{2}(q^{-1})^i_r(\dot{q})^r{}_{.k} = -\tfrac{1}{2}(\dot{q^{-1}})^i{}_{.r}q^r_k. \tag{13}$$

[1]) It is observed that, in general, the material derivation of a symmetric tensor related to the body-fixed coordinate system in mixed notation leads to an unsymmetric tensor.

Substituting the decomposition (12) into (13)

$$d^i_k = \tfrac{1}{2}(q^{-1})^i_r(\underset{P}{q}{}^r{}_{.m}\underset{E}{q}{}^m_k)^{\bullet} = \tfrac{1}{2}(q^{-1})^i_r\{(\underset{P}{\dot{q}})^r{}_{.m}\underset{E}{q}{}^m_k + \underset{P}{q}{}^r{}_{.m}(\underset{E}{\dot{q}})^m{}_{.k}\}$$

$$= -\tfrac{1}{2}((\underset{E}{q}{}^{-1})^i_m(\underset{P}{q}{}^{-1})^m{}_{.r})^{\bullet}q^r_k = -\tfrac{1}{2}\{(\underset{E}{\dot{q}}{}^{-1})^i{}_{.m}(\underset{P}{q}{}^{-1})^m_{.r} + (\underset{E}{q}{}^{-1})^i_m(\underset{P}{\dot{q}}{}^{-1})^m{}_{.r}\}q^r_k.$$

Applying the relationships

$$(\underset{P}{q}{}^{-1})^i_r q^r{}_{.m} = (\underset{E}{q}{}^{-1})^i_m, \qquad (\underset{P}{q}{}^{-1})^m{}_{.r}q^r_k = \underset{E}{q}{}^m_k,$$

and further observing that the right hand side of the above equations may be rewritten as a sum of two symmetric tensors

$$d^i_k = \tfrac{1}{4}\{(\underset{E}{q}{}^{-1})^i_r(\underset{E}{\dot{q}})^r{}_{.k} + g_{km}(\underset{E}{q}{}^{-1})^m_r(\underset{E}{\dot{q}})^r{}_{.s}g^{si}\}$$

$$+\tfrac{1}{4}\{(q^{-1})^i_r(\underset{P}{\dot{q}})^r{}_{.m}\underset{E}{q}{}^m_k + g_{kl}(q^{-1})^l_r(\underset{P}{\dot{q}})^r{}_{.m}\underset{E}{q}{}^m_s g^{si}\} \equiv \underset{E}{d}{}^i_k + \underset{P}{d}{}^i_k. \tag{14a}$$

The expressions in curly brackets represent $\underset{E}{d}{}^i_k$ and $\underset{P}{d}{}^i_k$ respectively.

The following can also be obtained

$$d^i_k = -\tfrac{1}{4}\{(\underset{E}{\dot{q}}{}^{-1})^i{}_{.r}\underset{E}{q}{}^r_k + g_{km}(\underset{E}{\dot{q}}{}^{-1})^m{}_{.r}\underset{E}{q}{}^r_s g^{si}\}$$

$$-\tfrac{1}{4}\{(\underset{E}{q}{}^{-1})^i_m(\underset{P}{\dot{q}}{}^{-1})^m{}_{.r}q^r_k + g_{kl}(\underset{E}{q}{}^{-1})^l_m(\underset{P}{\dot{q}}{}^{-1})^m{}_{.r}q^r_s g^{si}\} \equiv \underset{E}{d}{}^i_k + \underset{P}{d}{}^i_k. \tag{14b}$$

A comparison between the first terms on the right hand side in (14a) and (14b) and expressions (13) valid for the total strain rate justifies to define those terms as elastic strain rates. Expressing the quantities q^i_k etc. in terms of metric tensors the formulas (14a) and (14b) may be rewritten in different forms. For example, the plastic strain rate may be expressed as

$$\underset{P}{d}{}^i_k = +\tfrac{1}{4}\{\overset{*}{g}{}^{ir}\dot{\overset{*}{g}}_{rk} + g^{im}\dot{\overset{*}{g}}_{mr}\overset{*}{g}{}^{rs}g_{sk}\}. \tag{15}$$

The comparison of this expression with the corresponding term in equation (9b) shows that different ways of decomposition of the strain rate lead to different results. This is also true for the comparison between the decomposition according to equation (7) and other ways of decomposing the total strain rate. It is not possible to select the most appropriate manner of decomposition on kinematical grounds alone since it affects also the form of the constitutive equations.

3. The constitutive problem

When the constitutive equations are referred to the space-fixed coordinate system, it may be adequate to decompose the strain rate according to equations (6) and (7). Such an anlaysis was carried out by Lee [1] and Freund [2], leading to a rather complicated system of equations.

Green and Naghdi [3] as well as Sedov [5] employed a decomposition corresponding to the equations (8) and (9). However, their analysis appears to be incomplete since no incremental law was given for the elastic part of elastic-plastic deformations. On the other hand the plastic part of the deformations was treated very extensively. Therefore these studies may be regarded as detailed analyses of the constitutive equations for the plastic part.

In the present paper the deformations are decomposed according to equations (12) and (14). This procedure leads to a more convenient treatment of the constitutive problem in comparison with other methods of decomposition. Compared with the procedure suggested by Lee [1] and Freund [2] the advantage of the method employed is that the rotations of the reference frame with respect to the elements are eliminated right from the beginning by using the body-fixed coordinate system as the reference frame. Moreover, the additive decomposition of the covariant strain measure used by Green and Naghdi [3] as well as by Sedov [5] leads to some difficulties if the finite elastic stress-strain relation is transformed into an incremental law. These difficulties can be avoided by applying the decomposition of the deformations defined in equations (12) and (14).

The actual stress referred to the body-fixed coordinate system is σ_k^i. The Kirchhoff stress tensor s_k^i is expressed in terms of σ_k^i as

$$s_k^i = \frac{\mathring{\rho}}{\rho}\,\sigma_k^i \tag{16}$$

where ρ denotes the current mass density and $\mathring{\rho}$ the mass density at the reference time t_0. This tensor is known to be more suitable for defining the constitutive equations.

If the rate of work per unit mass is denoted by $\dot{w}$, such that

$$\mathring{\rho}\dot{w} = s_k^i d_i^k = s^{ik} d_{ik} \tag{17}$$

the rate of work can be split up into the elastic and plastic parts,

$$\mathring{\rho}\dot{w} = s_k^i \underset{E}{d_i^k} + s_k^i \underset{P}{d_i^k} = \mathring{\rho}\underset{E}{\dot{w}} + \mathring{\rho}\underset{P}{\dot{w}} \tag{18}$$

The rate of elastic work appearing above can be expressed as follows

$$\overset{\circ}{\rho}\dot{w}_E = \tfrac{1}{2} s^i_k (\underset{E}{q}^{-1})^k_r (\underset{E}{\dot{q}})^r_{\cdot i} \tag{19a}$$

$$= -\tfrac{1}{2} s^i_k \underset{E}{q}^r_i (\underset{E}{\dot{q}}^{-1})^k_{\cdot r}. \tag{19b}$$

Hence from thermodynamical considerations if follows (see [1] to [5]), that the stress can be derived from the free energy function φ per unit mass (Helmholtz-function). Introducing the Almansi-Hamel elastic strain tensor

$$\underset{(A)}{\varepsilon^i_k} = \tfrac{1}{2}\{\overset{*}{g}{}^{ir} g_{rk} - \delta^i_k\} = \tfrac{1}{2}\{\underset{E}{q}^i_k - \delta^i_k\} \tag{20a}$$

and the Cauchy-Green strain tensor

$$\underset{(C)}{\varepsilon^i_k} = \tfrac{1}{2}\{\delta^i_k - g^{ir} \overset{*}{g}_{rk}\} = \tfrac{1}{2}\{\delta^i_k - (\underset{E}{q}^{-1})^i_k\} \tag{20b}$$

it follows that

$$s^i_r (\underset{E}{q}^{-1})^r_k = \overset{\circ}{\rho} \frac{\partial \varphi}{\partial \underset{(A)}{\varepsilon^k_i}} = \overset{\circ}{\rho} \frac{\partial \varphi}{\partial \underset{E}{q}^k_i} \tag{21a}$$

or

$$s^r_k \underset{E}{q}^i_r = \overset{\circ}{\rho} \frac{\partial \varphi}{\partial \underset{(C)}{\varepsilon^k_i}} = \overset{\circ}{\rho} \frac{\partial \varphi}{\partial (\underset{E}{q}^{-1})^k_i}, \tag{21b}$$

respectively. Multiplying these formulas by $\underset{E}{q}^i_k$ or $(\underset{E}{q}^{-1})^i_k$, respectively, s^i_k may be calculated. If the elastic behaviour is assumed to be independent of the plastic deformation, s^i_k may be obtained as a function of $\underset{E}{q}^i_k$ (or $(\underset{E}{q}^{-1})^i_k$) and the temperature T,

$$s^i_k = s^i_k(\underset{E}{q}^r_s, T) \quad \text{or} \quad s^i_k = s^i_k((\underset{E}{q}^{-1})^r_s, T), \tag{22}$$

provided the specific free energy φ is known as a function of the corresponding strain tensor ε^i_k, and thus of $\underset{E}{q}^i_k$ or $(\underset{E}{q}^{-1})^i_k$ and of the temperature T [1]).

[1]) Instead of (21a) and (21b) other relations can be obtained, if different definitions of the elastic strain rate are employed; e.g. starting from $\underset{E}{d}_{ik} = \tfrac{1}{2}(\dot{g}_{ik} - \dot{\overset{*}{g}}_{ik})$ according to equation (9a) we get

$$s^{ik} = \overset{\circ}{\rho} \frac{\partial \varphi}{\partial \underset{(C)}{\varepsilon_{ik}}} \quad \text{with} \quad \underset{(C)}{\varepsilon_{ik}} = \tfrac{1}{2}(g_{ik} - \overset{*}{g}_{ik}).$$

It is only necessary to choose such pairs of stress and strain tensors, which are related one to another (see [6]). In every case we can rewrite eventually the respective relations in the form (22).

By inversion of equation (22)

$$\underset{E}{q^i_k} = q^i_k(s^r_s, T) \quad \text{or} \quad \underset{E}{(q^{-1})^i_k} = (q^{-1})^i_k(s^r_s, T). \tag{23}$$

This relationship may also be deduced directly, starting from the specific free enthalpy treated as a function of stress and temperature (see [1] to [5]).

If (23) and its material derivative is used for the definition of elastic strain rate corresponding to equation (14a) or (14b), the required incremental law governing elastic deformations is obtained

$$\begin{aligned}\underset{E}{d^i_k} &= \tfrac{1}{4}\{\underset{E}{(q^{-1})^i_r}\underset{E}{(\overset{\bullet}{q})^r_{.k}} + \underset{E}{(q^{-1})_{kr}}\underset{E}{(\overset{\bullet}{q})^r_{.s}}g^{si}\}\\ &= -\tfrac{1}{4}\{\underset{E}{(\overset{\bullet}{q^{-1}})^i_{.r}}\underset{E}{q^r_k} + g_{km}\underset{E}{(\overset{\bullet}{q^{-1}})^m_{.r}}\underset{E}{q^{ri}}\}\\ &= \underset{E}{d^i_k}(s^r_s, (\overset{\bullet}{s})^r_{.s}, T, \dot{T}, g_{lm}).\end{aligned} \tag{24}$$

In some special cases it may be more convenient to start with a finite relationship between any one of the elastic strain tensors on one side, and both stresses and temperature on the other. Differentiating this relationship with respect to time in order to get an expression for the elastic strain rate, it may be advantageous to use the covariant time derivation denoted by $|_0$. This differentiation with respect to the body-fixed coordinate system yields if applied to a second rank tensor

$$\begin{aligned}a^i_{.k}|_0 &= (\overset{\bullet}{a})^i_{.k} + d^i_r a^r_{.k} - d^r_k a^i_{.r}\\ &= g^{im}\{\dot{a}_{mk} - d^r_m a_{rk} - d^r_k a_{mr}\}\\ &= g_{km}\{\dot{a}^{im} + d^i_r a^{rm} + d^m_r a^{ir}\}.\end{aligned} \tag{25}$$ [1]

When the tensor is symmetric, the covariant differentiation is also symmetric and interchangeable by raising or lowering indices since

$$g_{ik}|_0 = 0 \tag{26}$$

[1]) This time derivative is sometimes called 'co-rotational differentiation'. In a space-fixed coordinate system following holds true

$$a^\alpha_{.\beta}|_0 = (\overset{\bullet}{a})^\alpha_{.\beta} + \omega^\alpha_{.\rho}a^\rho_{.\beta} - \omega^\rho_{.\beta}a^\alpha_{.\rho} \quad \text{with} \quad \omega^\alpha_{.\beta} = \tfrac{1}{2}\{v^\alpha|_\beta - v_\beta|^\alpha\}.$$

Its meaning as a covariant differentiation becomes obvious, as soon as we introduce the time as the fourth coordinate in an appropriate manner [7].

in contrast to the material derivative of the metric tensor. The covariant differentiation of a scalar is equivalent to the material differentiation.

Two special cases will be dealt with as examples. In the first one may start with the tensorially linear stress-strain relation

$$\underset{(A)}{\varepsilon^i_k} = \frac{1}{2G}\left\{s^i_k - \frac{\nu}{1+\nu} s^r_r \delta^i_k\right\} + \alpha(T-T_0)\delta^i_k \tag{27}$$

Since

$$\underset{(A)}{(\dot{\varepsilon})^{\,i}_{.k}} = \tfrac{1}{2}\underset{E}{(\dot{q})^{i}_{\ .k}} \tag{28}$$

$$\underset{E}{d^i_k} = \operatorname{sym}\tfrac{1}{2}\{\underset{E}{(q^{-1})^i_r}\underset{E}{(\dot{q})^r_{\ .k}}\}$$

$$= \frac{1}{2G}\left[\operatorname{sym}\{\underset{E}{(q^{-1})^i_r}(\dot{s})^r_{.k}\} - \frac{\nu}{1+\nu}\underset{E}{(q^{-1})^i_k}(\dot{s}^m_m)\right] + \alpha\underset{E}{(q^{-1})^i_k}\dot{T}. \tag{29}$$

The second example refers to a tensorially linear equation involving the strain tensor $\underset{(C)}{\varepsilon^i_k}$:

$$\underset{(C)}{\varepsilon^i_k} = \frac{1}{2G}\left\{s^i_k - \frac{\nu}{1+\nu} s^r_r \delta^i_k\right\} + \alpha(T-T_0)\delta^i_k \tag{30}$$

The procedure is analogous. Remembering that

$$\underset{(C)}{(\dot{\varepsilon})^{i}_{\ .k}} = -\tfrac{1}{2}\underset{E}{(\dot{q})^{i}_{\ .k}} \tag{31}$$

it follows that

$$\underset{E}{d^i_k} = -\operatorname{sym}\tfrac{1}{2}\{\underset{E}{(\dot{q}^{-1})^{i}_{\ .r}}\underset{E}{q^r_k}\}$$

$$= \frac{1}{2G}\left\{\operatorname{sym}\{\underset{E}{q^r_k}(\dot{s})^{i}_{\ .r}\} - \frac{\nu}{1+\nu}\underset{E}{q^i_k}(\dot{s}^m_m)\right\} + \alpha\underset{E}{q^i_k}\dot{T}. \tag{32}$$

The problem of defining the incremental stress-strain relationships for the plastic part of deformations will not be discussed here in detail. Since it is not yet possible to derive a complete and consistent set of state variables, one is not able to deduce a theory of plasticity through thermodynamic considerations only. Usually the theory of plastic potential is applied involving some additional assumptions (see [1] to [3]). This theory leads to the statement

$$\underset{P}{d^i_k} = \dot{\lambda}\frac{\partial F}{\partial s^k_i}, \tag{33}$$

where F stands for the yield condition and $\dot{\lambda}$ is a positive scalar multiplier. One of the consequences of this theory is the so-called normality rule, which states that the rate of plastic strain, is normal to the hyper-surface representing the yield condition in the stress space. But experimental evidence [8, 9] shows that for more complicated loading paths this normality rule does not hold. Having this in mind, it seems to be an obvious generalisation to establish an incremental stress-strain relation of the form

$$\underset{P}{d^i_k} = \dot{\lambda}\frac{\partial F}{\partial s^k_i} + \kappa^{ir}_{ks}\{s^s_r|_0 - \tfrac{1}{3}(\dot{s}^m_m)\delta^s_r\}, \tag{34}$$

which simplifies for the isotropic case to

$$\underset{P}{d^i_k} = \dot{\lambda}\frac{\partial F}{\partial s^k_i} + \kappa\{s^i_k|_0 - \tfrac{1}{3}(\dot{s}^m_m)\delta^i_k\}. \tag{35}$$

Stress-strain relationships of this kind are found to be in good agreement with experimental results regarding second-order effects [10].

4. Approximations for the elastic part of the deformation

In view of the fact, that for most materials, the ratio $\mathring{\rho}/\rho$ remains about equal to one even at large plastic deformation the Kirchhoff stress tensor s^i_k may be replaced by the spatial stress tensor σ^i_k with involved small error. For materials with weak hardening however the small differences due to the replacement of s^i_k by σ^i_k may become significant, since the yield condition is altered. In the special cases treated in section 3. (equations (29) and (32)) one may introduce further approximations. Since in most cases the quantities

$$\underset{E}{q^i_k} = \delta^i_k + 2\underset{(A)}{\varepsilon^i_k}$$

and

$$(\underset{E}{q}^{-1})^i_k = \delta^i_k - 2\underset{(C)}{\varepsilon^i_k}$$

deviate but little from unity they can be replaced by δ^i_k.

With these approximations the incremental laws (29) and (32) for the elastic part of deformations reduce to identical expressions, namely

$$\underset{E}{d^i_k} = \frac{1}{2G}\left\{ s^i_k|_0 - \frac{\nu}{1+\nu}(\dot{s}^r_r)\delta^i_k \right\} + \alpha \dot{T}\delta^i_k \tag{36a}$$

or

$$\underset{E}{d^i_k} = \frac{1}{2G}\left\{ \sigma^i_k|_0 - \frac{\nu}{1+\nu}(\dot{\sigma}^r_r)\delta^i_k \right\} + \alpha \dot{T}\delta^i_k, \tag{36b}$$

which in fact represent linear hypo-thermoelastic stress-strain relationships. The difference between these two formulas is caused by the use of the Kirchhoff stress tensor in (36a) and the usual spatial stress tensor in (36b).

Using the stress-strain relationships for the plastic part of the type (35) or

$$d^i_k = \dot{\lambda}\frac{\partial F(\sigma^i_k)}{\partial \sigma^k_i} + \kappa\{\sigma^i_k|_0 - \tfrac{1}{3}(\dot{\sigma}^r_r)\delta^i_k\}, \tag{37}$$

the total strain rate

$$d^i_k = \dot{\lambda}\frac{\partial F}{\partial s^k_i} + \left(\kappa + \frac{1}{2G}\right) s^i_k|_0 - \left(\frac{\kappa}{3} + \frac{1}{2G}\frac{\nu}{1+\nu}\right)(\dot{s}^r_r)\delta^i_k \tag{38a}$$

or (with σ^i_k instead of s^i_k)

$$d^i_k = \dot{\lambda}\frac{\partial F}{\partial \sigma^k_i} + \left(\kappa + \frac{1}{2G}\right) \sigma^i_k|_0 - \left(\frac{\kappa}{3} + \frac{1}{2G}\frac{\nu}{1+\nu}\right)(\dot{\sigma}^r_r)\delta^i_k. \tag{38b}$$

In these approximate formulas the second term of the plastic deformation and the term describing the elastic behaviour, have the same structure. Therefore it appears unnecessary to divide the total deformation into the elastic and the plastic parts during loading processes. During unloading, the elastic deformations have to be treated separately. This may be done exactly by a finite or an incremental law or approximately by assuming a hypo-elastic behaviour.

5. Numerical results

In order to compare exact and approximate expressions some simple iso-thermal loading and unloading processes, produced by normal and by shear stresses have been analysed. The acting stresses and the geometric quantities

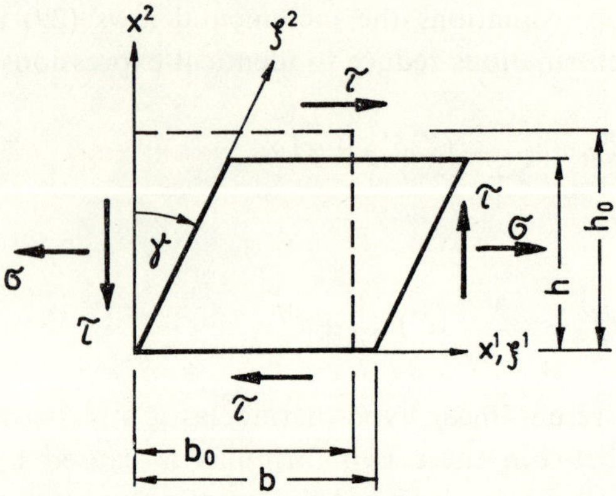

Fig. 1. Stresses and strains in tension and shear

$$\varepsilon_1 = \ln\frac{b}{b_0}, \quad \varepsilon_2 = \ln\frac{h}{h_0}, \quad \varepsilon_3 = \ln\frac{d}{d_0} \quad \vartheta = \text{tang}\,\gamma; \quad d\text{: thickness}$$

involved in this problem are indicated in Fig. 1. The yield condition is assumed in the form[1])

$$(s^i_k - \tfrac{1}{3}s^r_r\delta^i_k)(s^k_i - \tfrac{1}{3}s^m_m\delta^k_i) - (\mathring{k}^2 + 2B\underset{P}{w}) = 0. \tag{39}$$

The adopted material constants are

$$G = 2700\,\frac{\text{kp}}{\text{mm}^2}; \qquad B = 8\,\frac{\text{kp}}{\text{mm}^2}; \qquad \mathring{k} = 12\,\frac{\text{kp}}{\text{mm}^2}; \qquad \nu = 0.3.$$

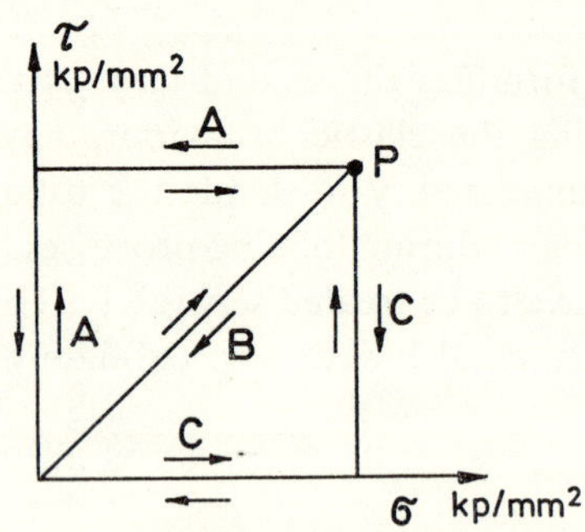

Fig. 2. Loading and unloading paths

[1]) This approach leads in pure tension to an approximately linear hardening, if the logarithmic strain is used as strain measure. Therefore in Fig. 1 the logarithmic strain is introduced.

In Fig. 2 different loading andun loading paths are shown to or from the same stress point.

The loading processes are analysed using the exact elastic strain rate equation (29) and the plastic strain rate corresponding to the formulas (39) and (33). The calculation is repeated with s_k^i replaced by σ_k^i in these formulas (approximation I). In a second approximation (II) the exact elastic stress-strain relations are replaced by hypo-elastic stress-strain relations according to equation (36b).

TABLE 1

Initial state: $\sigma = 0 \quad \tau = 0 \quad \varepsilon_1 = \varepsilon_2 = \varepsilon_3 = 0 \quad \vartheta = 0$
Final state: $\sigma = \tau = 7{,}3774$ kp/mm²

	Loading – path	$10^3 \cdot \vartheta$	$10^3 \cdot \varepsilon_1$	$10^3 \cdot \varepsilon_2$	$10^3 \cdot \varepsilon_3$
	A	10,7863	3,6998	−1,6431	−1,6412
Exact	B	10,7670	3,7173	−1,6518	−1,6499
	C	10,7594	3,7231	−1,6547	−1,6529
Approximation I	A	10,0025	3,4441	−1,5153	−1,5134
	B	9,9864	3,4583	−1,5224	−1,5206
	C	9,9800	3,4631	−1,5248	−1,5230
Approximation II	A	10,0142	3,4352	−1,5015	−1,5133
	B	10,0000	3.4495	−1,5087	−1,5205
	C	9,9953	3,4543	−1,5111	−1,5228

Table 1 shows that replacing s_k^i by σ_k^i (approximation I) leads to considerable deviations due to the changes of the yield condition in combination with the given weak hardening. On the other hand, approximating the elastic behaviour by a hypo-elastic stress-strain relationship causes only slight differences as the comparison between approximation I and approximation II shows for elastic-plastic loading processes.

The unloading is first treated exactly by the use of (27). In this case the unloading process is independent of the path. In the second case the Kirchhoff stresses are replaced by the usual stresses and, at the same time, the unloading process is assumed to be hypo-elastic (equation (36b)). Owing to this, the process becomes dependent on the unloading path. Therefore we

TABLE II

Initial state: $\sigma = \tau = 7{,}3774$ kp/mm² $\quad \varepsilon_1 = 3{,}4495\ 10^{-3}$
$\varepsilon_2 = -1{,}5087\ 10^{-3} \quad \varepsilon_3 = -1{,}5205\ 10^{-3} \quad \vartheta = 10^{-2}$
Final state: $\sigma = \tau = 0$

	Loading path	$10^3 \cdot \vartheta$	$10^3 \cdot \varepsilon_1$	$10^3 \cdot \varepsilon_2$	$10^3 \cdot \varepsilon_3$
Exact	A, B, C	7,25488	2,40304	−1,19322	−1,20496
Approximation	A	7,25401	2,40050	−1,19531	−1,20519
	B	7,25401	2,40051	−1,19532	−1,20519
	C	7,25401	2,40051	−1,19532	−1,20519

have to distinguish between different unloading paths. Table 2 shows, however, that the differences between exact and approximate solutions are very small. In the elastic unloading as well as in the elastic loading replacing s^i_k by σ^i_k is found to have little influence. From the results presented the conclusion can be drawn that in elastic-plastic loading it is reasonable to describe the elastic behaviour by the hypo-elastic stress-strain relations rather than to replace the Kirchhoff stresses s^i_k by the usual stresses σ^i_k. In unloading, however, as well as in elastic loading in general we can approximate the real elastic behaviour by using σ^i_k instead of s^i_k and by introducing at the same time the hypo-elastic stress-strain relations no appreciable error being thus made.

References

[1] Lee, E. H., Elastic-plastic deformation at finite strains, *Journ. Appl. Mech.*, *36* (1969) 1.

[2] Freund, L. B., Constitutive equations for elastic-plastic materials at finite strain, *Int. Journ. Sol. Struct.*, *6* (1970) 1193.

[3] Green, A. E. and P. M. Naghdi, A general theory of an elastic-plastic continuum, *Arch. Rat. Mech. Anal.*, *4* (1965) 251.

[4] Green, A. E. and J. E. Adkins, Large elastic deformation and nonlinear continuum mechanics, Clarendon Press, Oxford, 1960.

[5] Sedov, L. I., Foundations of the nonlinear mechanics of Continua, Pergamon Press, 1966.

[6] Macvean, D. B., Die Elementararbeit in einem Kontinuum und die Zuordnung von Spannungs- und Verzerrungstensoren, ZAMP, 19 (1968) 157.

[7] Lehmann, Th., Formänderungen eines klassischen Kontinuums in vierdimensionaler Darstellung, *Proc. XI Intern. Congr. Appl. Mech., München*, 1964, Springer-Verlag Berlin/Heidelberg/New York, (1966) 376.

[8] Naghdi, P. M., I. C. Rowly and C. W. Beadle, Experiments concerning the yield surface and the assumption of linearity in the plastic stress-strain relations, *Journ. Appl. Mech.*, *22* (1955) 416.

[9] Shiratori, E. and K. Ikegani, Studies of the anisotropic yield condition, *Journ. Mech. Phys. Sol.*, *17* (1969) 473.

[10] Lehmann, Th., Einige Bemerkungen zu einer allgemeinen Klasse von Stoffgesetzen für große elasto-plastische Formänderungen, (to be published in Ing. Arch., 1972).

Continuum theory of plastic deformation of granular media

V. N. Nikolaevskii

Institute of Physics of the Earth, USSR Academy of Sciences, Moscow

Continuum model of granular media is based on the following assumptions: the stress tensor and plastic strain rate tensor are coaxial, two independent scalar functions enter the stress-strain rate law for isotropic materials, the Mises-Schleicher yield condition is valid, plastic volume changes exist, the angle of internal friction and dilatancy rate depend on plastic volumetric strain as a state parameter, plastic and elastic strains are additive. As a result the plastic potential does not coincide with the yield function.

Plane plastic flow, spherically symmetric deformation and triaxial uniform compression are discussed. The conclusions of the theory are compared with experimental facts. The Drucker-Prager model is a particular and extreme case of the proposed theory.

Introduction

The constitutive relationships for plastic deformation of granular media are among most important problems of modern mechanics of continua. Substantial achievements in the application of continuum methods in soil mechanics are limited by the frame of elasticity theory and by the solutions of the so-called statically determinate problems [1], when the velocity analysis is not necessary.

The current situation in soil mechanics was discussed by Roscoe [2] from the point of view of engineering practice. The complex reological behavior of sands was compared with the existing theories. Roscoe showed that well-known methods of plasticity and physical interpretation of the Coulomb limit state have to be modified. At many instances such a comparison led to a conclusion that plasticity theories were not valid for soils at all. These judgements are caused by discrepancies between experimental data and models based on both the principle of normality of plastic strain-rate vector and the yield surface of Mohr-Coulomb's type. Such model was developed first by Drucker and Prager [3]. Although models of this type account for the dilatancy effect discovered by Reynolds [4], i.e. considerable volume changes of a granular medium under shear, the predicted rate of dilatancy is too high. Besides, there are some difficulties with predicted and observed slip planes [2, 5–8] and with dissipation processes in cohesionless soils [9, 10].

However, it has to be remembered, that the methods of plasticity are not restricted to the principle of normality. As it was mentioned in [11] the general theory includes also the cases when the plastic potential does not coincide with the yield function. Therefore our aim is to find such a plastic potential that properly accounts for the effect of dilatancy although the yield condition of the Coulomb-Mohr type (exactly, Mises-Schleicher) remains valid. The strain-hardening effect are also essential because the properties of granular media change due to the plastic deformation. One may hope that an appropriate model accounting for these effects may reassess the methods of plasticity theory for soil mechanics.

Of course, a phenomenological model is not the only approach. One can begin the analysis of plastic strains with a discussion of elementary interactions and sliding between individual grains of the considered media. In this way one may develop a continuum theory of media with microstructure, which will be to some extent analogous to the theory of dislocations developed for plastic metals.

It is well-known that in the theory of plasticity there are several approaches to the development of constitutive equations. It is possible to use some variational principles [12] or principles of thermodynamics of irreversible processes [13, 14], as an alternative one may construct directly the isotropic or anisotropic relations between dynamical and kinematical tensors [15], another way is to introduce some requirement for a dissipative function [16] or to formulate relations between basic invariants of tensors [17] and finally one may combine some methods [18]. But in any case it is impossible to avoid a choice of physical hypothesis and its interpretation in accordance with experimental data. The problem of constitutive equations is thus not a pure mathematical one.

In this paper viscous properties of soils which are due to the presence of clay components will not be discussed. The appropriate effects of relaxation may be included in the considered model in the same way as it was done by Olszak and Perzyna [19] for Drucker-Prager's model.

Basic concepts

According to principles of incremental plasticity [11] we adopt the relationship

$$de^p_{ij}/dt \equiv \varepsilon_{ij} = f(\sigma_{ij}; \alpha, Y, \Lambda, \chi, \ldots) \tag{1}$$

between the plastic strain-rate tensor ε_{ij} and the stress tensor σ_{ij}. Here f denotes a function; $\alpha, Y, \Lambda, \chi, \ldots$ are scalar parameters.

For linear isotropic case such a tensor relationship has the form:

$$\varepsilon_{ij} = -H\dot{\zeta}\delta_{ij} - p\dot{\zeta}\delta_{ij} + (\sigma_{ij} + p\delta_{ij})\dot{\lambda} \tag{2}$$

which contains two scalar factors ζ, $\dot{\lambda}$ and one additive scalar function H. All the above mentioned scalars can be functions of state parameters. Here $p = -\frac{1}{3}\sigma_{ij}\delta_{ij} = -\frac{1}{3}J_1$ is a mean pressure, δ_{ij} is the Kronecker symbol.

Experiments [2] show that the tensors ε_{ij} and σ_{ij} are coaxial, unlike e^p_{ij} and σ_{ij}. This fact supports the validity of the initial hypothesis (1) and its particular form (2).

The independend scalar factors $\dot{\lambda}$, $\dot{\zeta}$ will be determined by the introduction of two limit conditions. The first one is the usual Mises-Schleicher yield condition

$$\Phi_\sigma(\sigma_{ij}; \alpha, Y, \chi, \ldots) \equiv \sqrt{J^1_2} + \tfrac{1}{3}\alpha J_1 - Y = 0 \tag{3}$$

$$J^1_2 = \tfrac{1}{6}\{(\sigma_{11} - \sigma_{22})^2 + (\sigma_{11} - \sigma_{33})^2 + (\sigma_{22} - \sigma_{33})^2 + \\ + 6(\sigma^2_{12} + \sigma^2_{23} + \sigma^2_{13})\},$$

where $\alpha(\chi)$, $Y(\chi)$ are the coefficients of internal friction and cohesion, respectively and χ is a parameter of hardening or softening.

The second condition is the dilatancy rule representing a kinematical restriction on the strain-rate components

$$\Phi_\varepsilon(\varepsilon_{ij}; \Lambda, \ldots) \equiv I_1 - 2\Lambda\sqrt{I^1_2} = 0 \tag{4}$$

$$I_1 = \varepsilon_{11} + \varepsilon_{22} + \varepsilon_{33} = \varepsilon, \qquad 2\sqrt{I^1_2} = \dot{\gamma}$$

$$I^1_2 = \tfrac{1}{6}\{(\varepsilon_{11} - \varepsilon_{22})^2 + (\varepsilon_{11} - \varepsilon_{33})^2 + (\varepsilon_{22} - \varepsilon_{33})^2 + \\ + \varepsilon(\varepsilon^2_{12} + \varepsilon^2_{23} + \varepsilon^2_{13})\}$$

The rate of dilatancy Λ is also a function of χ. In the particular case $\Lambda \equiv 0$ the requirement (4) becomes the usual plastic incompressibility condition $\varepsilon \equiv 0$.

On substituting the relations (2) into the dilatancy condition (4) and requiring that the result coincide with the yield condition (3) the following relationships are obtained

$$\dot{\zeta} = -(\tfrac{2}{3})\Lambda\alpha\dot{\lambda}, \quad \alpha H = Y \tag{5}$$

The constitutive law (2) can be rewritten in the form

$$de_{ij}^{p} = (\sigma_{ij} + (\tfrac{2}{3})\Lambda\alpha H\delta_{ij} + (1 + \tfrac{2}{3}\Lambda\alpha)p\delta_{ij})d\lambda, \tag{6}$$

where $\varepsilon_{ij}dt = de_{ij}^{p}$ is an increment of plastic strain e_{ij}^{p} and $d\lambda > 0$ if the yield condition (3) is satisfied and the active loading, [11] takes place. The case of isotropic hardening or softening will be considered here and the plastic volumetric strain $e^{p} = e_{ij}^{p}\delta_{ij}$ will be chosen as a parameter χ. Then for mobilized internal friction and mobilized dilatancy rate $\alpha = \alpha(e^{p})$, $\Lambda = \Lambda(e^{p})$. Condition (3) corresponds to the straight line of the Mohr-Coulomb type on the plane p, $\sqrt{J_2^1}$.

The increments of total strains de_{ij} specified in terms of velocity u_j; i.e.

$$de_{ij} = \frac{1}{2}\left(\frac{\partial u_i}{\partial x_j} + \frac{\partial u_j}{\partial x_i}\right)dt \tag{7}$$

are equal to a sum of elastic and plastic strain increments:

$$de_{ij} = de_{ij}^{e} + de_{ij}^{p} \tag{8}$$

For elastic strains Hooke's law can be used, therefore

$$de_{ij}^{e} = Q_{ijkl}d\sigma_{kl} \tag{9}$$

where Q_{ijkl} is the tensor of elastic moduli

$$Q_{ijkl} = \frac{1+\nu}{E}\delta_{ik}\delta_{jl} - \frac{\nu}{E}\delta_{ij}\delta_{kl}$$

and ν stands for Poison's ratio, E for Young's modulus.

Thus, the constitutive law for total strains has the form:

$$\frac{de_{ij}}{dt} = Q_{ijkl}\frac{\tilde{d}\sigma_{kl}}{dt} + \frac{\partial\Psi}{\partial\sigma_{ij}}\dot{\lambda} \tag{10}$$

where $\tilde{d}/dt$ denotes the following derivative [20]

$$\frac{\tilde{d}\sigma_{ij}}{dt} = \frac{d\sigma_{ij}}{dt} - \sigma_{ik}\Omega_{jk} - \sigma_{jk}\Omega_{ik} \tag{11}$$

$$\Omega_{ij} = \partial u_i/\partial x_j - \partial u_j/\partial x_i$$

and Ψ is the plastic potential

$$de_{ij}^{p} = \frac{\partial \Psi}{\partial \sigma_{ij}}\, d\lambda \tag{12}$$

corresponding to the constitutive plastic relationships (6).

The law (10) has to be supplemented by the equations of mass and momentum balance:

$$\frac{\partial \rho}{\partial t} + u_j \frac{\partial \rho}{\partial x_j} + \rho \frac{\partial u_j}{\partial x_j} = 0 \tag{13}$$

$$\rho \frac{\partial u_i}{\partial t} + \rho u_j \frac{\partial u_i}{\partial x_j} = \frac{\partial \sigma_{ij}}{\partial x_j} + F_i \tag{14}$$

where ρ is the density of a medium, and F_i stands for body forces.

Plastic potential

A potential field Ψ = const will be found such that

$$\frac{de_{ij}^{p}}{d\lambda} = \frac{\partial \Psi}{\partial \sigma_{ij}} \equiv \sigma_{ij} + \tfrac{2}{3}\Lambda\alpha H \delta_{ij} - \tfrac{1}{3}(1 + \tfrac{2}{3}\Lambda\alpha)\sigma_{kl}\delta_{kl}\delta_{ij} \tag{15}$$

On integrating the right side of equation (15)

$$\Psi(\sigma_{ij}) = J_2^1 - 2\alpha\Lambda(Hp + \tfrac{1}{2}p^2) + \text{const} \tag{16}$$

On the plane p, $\sqrt{J_2^1}$ the equipotentials appear to be the second order curves

$$(\sqrt{J_2^1})^2 - \alpha\Lambda(p + H)^2 = \text{const} \tag{17}$$

The constant in the right-hand side of (17) can be determined in terms of coordinates p_0, $(\sqrt{J_2^1})_0$ of the point of intersection of the equipotential (17) with Mohr-Coulomb's line (3):

$$\sqrt{J_2^1} - \alpha(p + H) = 0 \tag{18}$$

Then equation (17) may be expressed in the form

$$(\sqrt{J_2^1})^2 - \alpha\Lambda(p + H)^2 = (\sqrt{J_2^1})_0(p + H)(\alpha - \Lambda) \tag{19}$$

We shall discuss experimental data [2] for sands as typical granular media. It is known [2, 21] that there exists such a critical state (or critical porosity, or critical density) of a sand that the rate of dilatancy Λ is equal to zero. The sands with the density smaller than the critical one are named loose and with the bigger density are called dense. The rate of dilatancy Λ of a loose sand is negative, i.e. volume of loose sand decreases during shearing. On the other hand volume of a dense sand increases due to shear, $\Lambda > 0$. Because the dilatancy rate varies in shearing, it is possible to think of a critical state as a certain limit state of sand. In this sense other plastic states are not limit states although the yield condition (3) is satisfied since plastic deformation takes place[1]. It is worthwile to note that internal friction in a loose sand is smaller than in a dense sand and the critical porosity is a function of pressure.

We shall try to interpret these effects by using the plastic volumetric strain e^p as a parameter of hardening or softening, by assuming that there is such a critical strain e^p_C where $\Lambda(e^p_C) = 0$. A straight line $\sqrt{J_2^1} = \text{const}$ parallel to the p-axis corresponds to this state as an equipotential.

For dense states when $\Lambda > 0$ the equipotentials are hyperbolas according to (19). For loose states $\Lambda < 0$ the equipotentials are ellipses. If one superimposes the plane p, $\sqrt{J_2^1}$ on the plane e^p, γ, it is possible to illustrate the normality of the projections de^p and $d\gamma$ of vector de^p_{ij} to the plastic equipotentials (Fig. 1). The dilatancy condition (4) shows the relative values of these projections i.e. (4) replaces the principle of normality of vector de^p_{ij} to the yield surface (18).

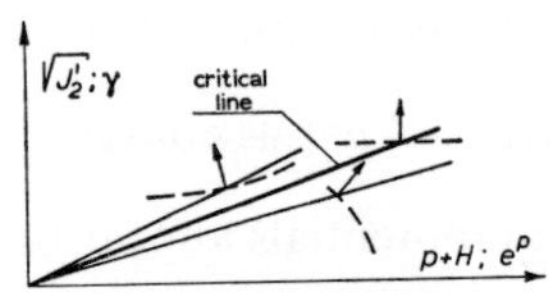

Fig. 1. Yield surfaces and plastic equipotentials.

Only in the particular case when hyperbola (19) degenerates into the straight line (18), i.e. when $\Lambda = \alpha$, the plastic potential and the yield surface coincide and vector de^p_{ij} is normal to the Mises-Schleicher yield surface which corresponds to the Drucker-Prager model. Experimental data [2] show that such a case is nonrealistic since always $\alpha > \Lambda$.

It is interesting to note the qualitative description [25] of 'yield surfaces' for 'consolidation, softening and critical states'. It was suggested in [38] and

[1]) According to the model points of stress plane under Mohr-Coulomb's line correspond to elastic state. In reality in this states elastic and plastic strains are of the same order or plastic strains are much less then elastic ones [22, 23]. The mechanism of dissipation of seismic waves is connected with these small plastic strains [24].

[39] that plastic potential surfaces coincide with the yield surfaces which have the density as a parameter of hardening. At the points of intersections of these equipotentials with the critical state line the components de^p are equal to zero and at the points of intersection with the p-axis $d\gamma = 0$. Hydrostatic compression therefore leads to an irreversible deformation of soils. From this point of view these potentials are equivalent to equipotentials used in [40].

According to the model proposed here the yield function (3) and the dilatancy function (4) do not need to coincide, in general. The model predicts plastic incompressibility under hydrostatic compression which corresponds to some experiments [26] where the crushing of grains was excluded. It seems to be reasonable to neglect such irreversible strains which are of the order of elastic strains (see previous footnote).

Plastic flows under spherical symmetry

It is assumed that the following conditions for the stress tensor hold true

$$\sigma_{11} = \sigma_r, \quad \sigma_{22} = \sigma_{33} = \sigma_\theta, \quad \sigma_{12} = \sigma_{23} = \sigma_{13} = 0 \tag{20}$$

where r is a radial coordinate. For the rigid-plastic model, the strain rates are as follows:

$$\varepsilon_{11} = \varepsilon_r = \frac{\partial u}{\partial r}, \quad \varepsilon_{22} = \varepsilon_{33} = \varepsilon_\theta = \frac{u}{r}, \quad \varepsilon_{12} = \varepsilon_{23} = \varepsilon_{13} = 0 \tag{21}$$

Hence the yield condition (3) takes the form

$$\Phi_\sigma = \frac{1}{\sqrt{3}}(\sigma_r - \sigma_\theta)x + \frac{\alpha}{3}(\sigma_r + 2\sigma_\theta) - Y = 0 \tag{22}$$

where $x = \operatorname{sgn}(\sigma_r - \sigma_\theta)$. The dilatancy condition becomes

$$\Phi_\varepsilon = (\varepsilon_r + 2\varepsilon_\theta) - \frac{2\Lambda}{\sqrt{3}}(\varepsilon_r - \varepsilon_\theta)x = 0 \tag{23}$$

Introducing the velocity u into (23) the following differential equation is obtained

$$\frac{\partial u}{\partial r} + 2\frac{u}{r} = \frac{2\Lambda}{\sqrt{3}}\left(\frac{\partial u}{\partial r} - \frac{u}{r}\right)x \tag{24}$$

If the rate of dilatancy is constant it may be shown [27] that the integral of this equation has the form

$$u = \frac{C(t)}{r^n}, \qquad n = \frac{2\sqrt{3} + 2\Lambda x}{\sqrt{3} - 2\Lambda x}, \tag{25}$$

where $C(t)$ is an arbitrary function.

In the particular case of incompressible media $\Lambda = 0$ and the integral (25) determines the velocity field

$$u = \frac{C(t)}{r^2} \qquad (n = 2) \tag{26}$$

However, experiments [28] on contained explosions have shown that the real velocity field is governed by the kinematical integral (25) with $n = 1.5 - 1.8$. A possible explanation of this effect is the following [27]. Under a compression wave a soil element is compacted and at advanced plastic flow this compacted soil begins to become looser according to dilatancy effect. In the region of flow where $u/r > \partial u/\partial r$ the equality $x = -1$ is satisfied and therefore $n = (2\sqrt{3} - 2\Lambda)/(\sqrt{3} + 2\Lambda) < 2$ if $\Lambda > 0$. The range $1.5 - 1.8$ for n corresponds to values $\Lambda = 0.18 - 0.09$ which are smaller than value $\alpha \sim 0.5$ commonly found for sands.

The requirement of constant dilatancy rate in the whole region of plastic flow is clearly a simplification of the reality. In actual conditions the rate of dilatancy Λ diminishes and even approaches zero. As a result the density of soil decreases only until the density of a critical state is reached as a finite value. Therefore the kinematic integral (25) is valid for separate spherical zones and corresponds to the results of tests [29], which have shown the variations of a degree n in expression (25).

Plane plastic deformations

Consider the important particular case[1]) when

$$de^p_{33} = de^p_{13} = de^p_{23} = 0 \tag{27}$$

Then

$$\sigma_{33} = -\frac{6\Lambda\alpha H}{6 - 2\alpha\Lambda} + \frac{3 + 2\alpha\Lambda}{6 - 2\alpha\Lambda}(\sigma_{11} + \sigma_{22}), \qquad \sigma_{13} = \sigma_{23} = 0 \tag{28}$$

[1]) More correct approach requires $u_3(x_1; x_2; x_3 = \text{const}) \equiv 0$ but for a perfect plastic model the condition (27) is sufficient.

Accordingly, the invariants J_1, J_2^1 can be expressed by

$$J_1 = -\frac{6\alpha\Lambda H}{6 - 2\alpha\Lambda} + \frac{9}{6 - 2\alpha\Lambda}(\sigma_{11} + \sigma_{22})$$

$$J_2^1 = \tfrac{1}{4}(\sigma_{11} - \sigma_{22})^2 + \sigma_{12}^2 + \frac{3\alpha^2\Lambda^2}{(6 - 2\alpha\Lambda)^2}(\sigma_{11} + \sigma_{22} - 2H)^2$$

Therefore the yield condition (3) can be transformed to the formc orresponding to the plane problem, i.e. to the Mohr-Coulomb condition:

$$\sqrt{(J_2^1)_*} + \tfrac{1}{2}\alpha_*(J_1)_* - \alpha_* H_* = 0 \tag{29}$$

where new variables

$$\alpha_* = \alpha\frac{\sqrt{3(3 - \Lambda^2)}}{3 - \alpha\Lambda}, \qquad H_* = H \tag{30}$$

$$(J_1)_* = \sigma_{11} + \sigma_{22} = -2p_*, \qquad (J_2^1)_* = \tfrac{1}{4}(\sigma_{11} - \sigma_{22})^2 + \sigma_{12}^2 \tag{31}$$

are introduced.

In a similar manner the dilatancy condition (4) can be transformed to become

$$(I_1)_* = 2\Lambda_*\sqrt{(I_2^1)_*} \tag{32}$$

where the variables corresponding to the plane problem are introduced

$$\Lambda_* = \Lambda\sqrt{3/(3 - \Lambda^2)} \tag{33}$$

$$(I_1)_* = \varepsilon_{11} + \varepsilon_{22}, \quad (I_2^1)_* = \tfrac{1}{4}(\varepsilon_{11} - \varepsilon_{22})^2 + \varepsilon_{12}^2$$

The constitutive relations (6) now take the form

$$de_{ij}^p = (\sigma_{ij} + \alpha_*\Lambda_* H\delta_{ij} + (1 + \alpha_*\Lambda_*)p_*\delta_{ij})d\lambda \tag{34}$$

However, the values of Λ and α are of the order of 10^{-1} and the following estimations are valid:

$$\alpha_* \approx \alpha, \qquad \Lambda_* \approx \Lambda \tag{35}$$

Next consider the plane ideal-plastic flow of a granular medium of the type:

$$u_1 = \frac{U}{h}x_2, \quad u_2 = \frac{V}{h}x_2, \quad u_3 \equiv 0 \tag{36}$$

in the layer $0 \leqq x_2 \leqq h$. Here U, V are constants interconnected due to the dilatancy condition (32). Indeed, one has

$$\varepsilon_{12} = U(2h)^{-1}, \quad \varepsilon_{22} = Vh^{-1}, \quad \varepsilon_{11} = 0$$

$$V^2 = \frac{\Lambda^2_*}{1 - \Lambda^2_*} U^2 \tag{37}$$

The uniform stress distribution in the layer is

$$\sigma_{12} = \frac{U}{2h} \frac{1}{\lambda}, \quad \sigma_{22} = -\frac{1 - \alpha_* \Lambda_*}{\alpha_* \Lambda_*} \frac{V}{h} \frac{1}{2\lambda} - \alpha_* \Lambda_* H$$

$$\sigma_{11} = -\frac{V}{h} \frac{1}{2\lambda} \frac{1 + \alpha_* \Lambda_*}{\alpha_* \Lambda_*} - \alpha_* \Lambda_* H \tag{38}$$

If the stress components $P = -\sigma_{22}$, $Q = \sigma_{12}$ are measured at the boundaries $x_2 = 0$, $x_2 = h$, then the relation

$$Q = \frac{\alpha_* \sqrt{1 - \Lambda_*^2}}{1 - \alpha_* \Lambda_*} (P + \alpha_* \Lambda_* H) \tag{39}$$

follows. This relation can be interpreted in terms of Coulomb's law

$$K = \frac{\alpha_* \sqrt{1 - \Lambda_*^2}}{1 - \alpha_* \Lambda_*}, \qquad C = \frac{\alpha_*^2 \Lambda_* \sqrt{1 - \Lambda_*^2}}{1 - \alpha_* \Lambda_*} H \tag{40}$$

K, C are a fictitious coefficient of internal friction and a fictitious internal cohesion, respectively. Only in the case of critical state $\Lambda_* = 0$, and therefore $K = \alpha_* = \sin \varphi$ where φ is an angle of internal solid friction, $C = 0$. In Figs. 2, and 3 the curves for K and C are shown as functions of Λ_* when

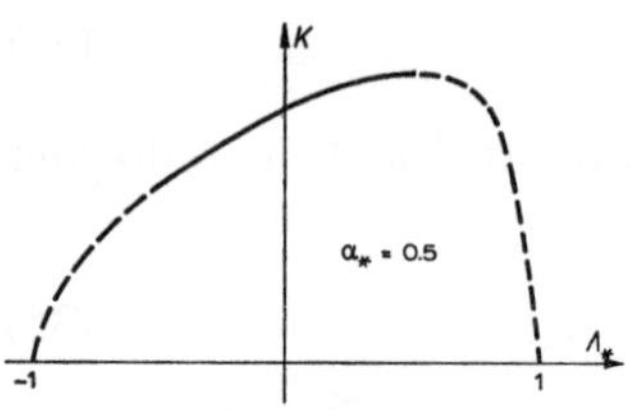

Fig. 2.

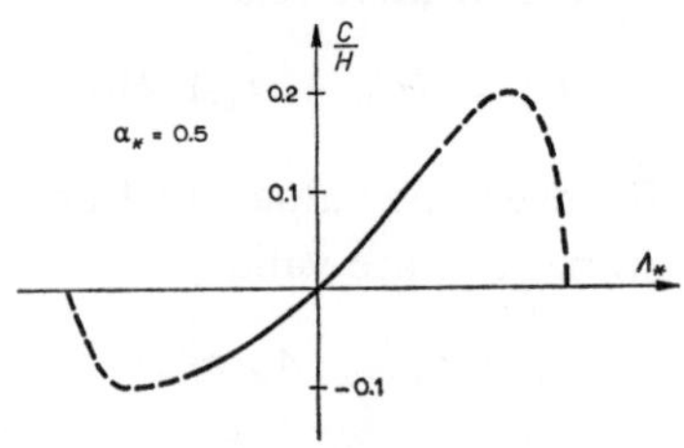

Fig. 3.

Fig. 2. Fictitious coefficient of internal friction versus dilatancy rate.

Fig. 3. Fictitious cohesion versus dilatancy rate.

$\alpha_* = \text{const}$, $H = \text{const}$. Even in this case K is seen to increase with increasing Λ_* and the characteristic peak value of resistance is observed to exist. The coefficient C may be even negative for loose sands due to the negative dilatancy effect.

It may be seen that a need for interpretation of experimental results from direct shear tests exists.

Triaxial compression[1])

If one intends to interpret the results of triaxial compression test assuming a uniform stress distribution, the following conditions will be fulfilled

$$\sigma_{11} = \sigma_1(t), \quad \sigma_{22} \equiv \sigma_{33} = \sigma_2, \quad \sigma_{12} = \sigma_{13} = \sigma_{23} = 0 \tag{41}$$

Then the yield condition takes the form

$$x(\sigma_1 - \sigma_2) + (\alpha/\sqrt{3})(\sigma_1 + 2\sigma_2) = Y, \quad x = \text{sgn}(\sigma_1 - \sigma_2) \tag{42}$$

and for plastic strain increments

$$de_1^p = \frac{2\sqrt{3} + 2x\Lambda}{3\sqrt{3}}(\sigma_1 - \sigma_2)d\lambda, \quad de_2^p = \frac{-\sqrt{3} + 2x\Lambda}{3\sqrt{3}}(\sigma_1 - \sigma_2)d\lambda \tag{43}$$

The sum of expressions (43) gives

$$d\lambda = \frac{\sqrt{3}}{2x\Lambda(\sigma_1 - \sigma_2)}\, de^p \tag{44}$$

From the yield condition (42) the relation

$$de^p = \frac{(x\sqrt{3} + \alpha)d\sigma_1 - (x\sqrt{3} - 2\alpha)d\sigma_2}{3(dY/de^p) - (\sigma_1 + 2\sigma_2)(d\alpha/de^p)} \tag{45}$$

is obtained such that the hardening effects $Y(e^p)$, $\alpha(e^p)$ are accounted for.

According to relations (43), (44), the plastic strain increments de_i^p can be expressed in the terms of $d\sigma_1$, $d\sigma_2$. Elastic parts of strains are connected with $d\sigma_1$, $d\sigma_2$ by Hooke's law (9). Now we restrict the discussion to the 'standard' triaxial test where $d\sigma_2 = 0$ and $Y \equiv 0$. Thus

[1]) The analysis was performed by N. M. Syrnikov.

$$de_1 = \left(\frac{1}{E} + \frac{2\sqrt{3} + 2x\Lambda}{2} \frac{-x\sqrt{3} - \alpha}{2x\Lambda(\sigma_1 + 2\sigma_2)(d\alpha/de^p)}\right) d\sigma_1$$

$$de_2 = \left(-\frac{\nu}{E} - \frac{\sqrt{3} - 2x\Lambda}{3} \frac{-x\sqrt{3} - \alpha}{2x\Lambda(\sigma_1 + 2\sigma_2)(d\alpha/de^p)}\right) d\sigma_2 \tag{46}$$

$$de^p = \frac{-x\sqrt{3} - \alpha}{(\sigma_1 + 2\sigma_2)(d\alpha/de^p)} d\sigma_1$$

During compression along the axis 1 the conditions $\sigma_1 < \sigma_2 < 0$, $x = -1$ are fulfilled on assuming the simpliest hardening relations

$$\alpha = \alpha_c - \Pi(e^p - e_c^p)^m, \quad \Lambda = -\Gamma(e^p - e_c^p)^m \tag{47}$$

and the following numerical values

$$\alpha_c = 0.762, \quad \Pi = 18.5 \cdot 10^4, \quad \Gamma = 1.25 \cdot 10^4$$

$$m = 3, \quad \nu = 0.3, \quad E = 3000 \text{ kg/cm}^2$$

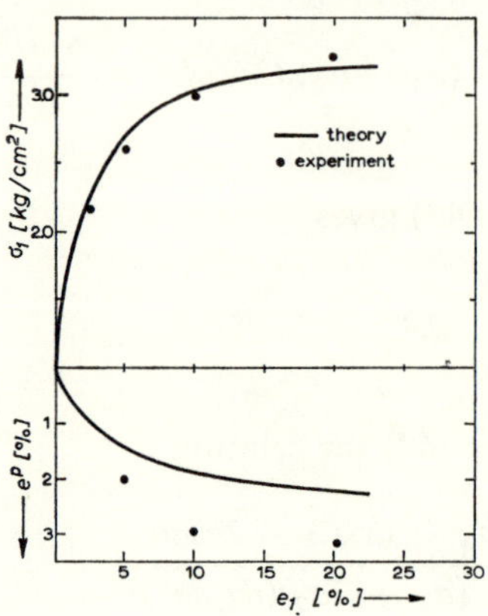

Fig. 4. Comparison of theory and experiment for triaxial standard test conditions.

it becomes possible to fit in Fig. 4 the experimental results from [20] with the initial data:

$$(\sigma_1)_c = -3.32 \text{ kg/cm}^2, \; e_c = -3.3\%, \; e_c^p = -3.26\%, \; \sigma_2 \equiv -1 \text{ kg/cm}^2$$

Some hypothetical relations $\alpha_*(\Lambda_*)$, $d_*(e^p)$ can be deduced from [31–37].

References

[1] Sokolovskii, V. V., Statics of soil media. GTTI Moscow, 1954; Butterworth London, 1960.

[2] Roscoe, K. H., The influence of strains in soil mechanics. Geotechnique, *20* (1970) 129–170.

[3] Drucker, D. C. and W. Prager, Soil mechanics and plastic analysis or limit design. *Quart. Appl. Math.*, *10* (1952) 157–165.

[4] Reynolds, O., On the dilatancy of media composed of rigid particles in contact. *Phil. Mag.*, *Ser.* 5, *20* (1885) 469–481.

[5] Shield, R. T., Mixed boundary value problems in soil mechanics. *Quart. Appl. Math.*, *11* (1953) 61–75.

[6] Drescher, A., K. Kwaszczyńska and Z. Mróz, Statics and kinematics of the granular medium in the case of wedge indentation. *Arch. Mech. Stos.*, *19* (1967) 99–113.

[7] Jeske, T., On the kinematics of a granular medium in the case of limit equilibrium state under plane strain conditions. *Arch. Mech. Stos.*, *20* (1968) 211–224.

[8] Nikolaevskii, V. N. and N. M. Syrnikov, On plane limit flow of a granular dilating medium (in Russian). Izvestia A.N. SSSR, Mekh. Tverd. Tela (MTT) *2* (1970) 159–166.

[9] Chadwick, P., A. D. Cox and H. G. Hopkins, Mechanics of deep underground explosions. *Phil. Trans. Roy. Soc. of London, A256*, 1964.

[10] Mandl, G. and R. Fernández Luque, Fully developed plastic shear flow of granular materials. *Geotechnique*, *20* (1970) 277–307.

[11] Olszak, W., Z. Mróz and P. Perzyna, Recent trends in the development of the theory of plasticity. Pergamon-Press – PWN, Oxford-Warszawa, 1963.

[12] Sedov, L. I., Mathematical models of developing of new continuum models (in Russian). *Advances Math. Sci.* (YMH), *20*, No. 5 (125) (1965) 121–180.

[13] Vakulenko, A. A., On stress-strain laws for nonelastic media. *Doklady* AN SSSR (DAN), *118* (1958) 665–668.

[14] Ziegler, H., An attempt to generalize Onsager's principle and its significance for reological problems. ZAMP, *96* (1958) 748–763.

[15] Kachanov, L. M., Fundamentals of the the theory of plasticity (in Russian). 2nd ed., Nauka, Moscow, 1969.

[16] Ivlev, D. D., On dissipative function in a theory of plastic media (in Russian). *Doklady* AN SSSR (DAN), *176* (1967) 1037–1040.

[17] Thomas, T. Y., Interdependance of the yield condition and the stress-strain relations for plastic flows. *Proc. Nat. Acad. Sci. US*, *40* (1954) 593–597.

[18] Sawczuk, A. and P. Stutz, On formulation of stress-strain relations for soils at failure. ZAMP, *19* (1968) 770–778.

[19] Olszak, W. and P. Perzyna, Les equations constitutives et phenomenes de relaxation dans les milieux elasto/visco-plastique. Rend. Semin. mat. a fis. Milano, *38* (1968) 12–18.

[20] Sedov, L. I., Concept of different rates of tensor variations, (in Russian). *Prikl. Math. Mekh. 24* (1960).

[21] Casagrande, A., Characteristics of cohesionless soils affecting the stability of slopes and earth fills. J. Boston Soc. Civ. Engrs, Jan., 1936.

[22] Drescher, A. and W. Bojanowski, On the influence of stress path upon the mechanical properties of granular materials, (in Polish). *Arch. Inzyn. Lad.*, *14* (1968) 351–365.

[23] Porooshasb, H. B., I. Holubec and A. N. Sherbourne, Yielding and flow of sand in triaxial compression. Canad. Geotechnical J., part 1: *3*, No. 4, 1966; parts 2–3: *4*, No. 4, 1967.

[24] Nikolaevskii, V. N., Monochromatic waves in an elastic medium with local effects of solid friction, (in Russian). Mekh. Tverd. Tela (MTT), No. 4 (1968) 85–92.

[25] Mróz, Z. and A. Drescher, Limit plasticity approach to some cases of flow of bulk of bulk solids. Trans. ASME, J. Enging for Industry, 68-MH-23 (1968) 1–8.

[26] Ko, N. Y. and R. F. Scott, Deformation of sand in hydrostatic compression, *J. Soil Mech. Found. Div.* SM3, PNC ASCE *93* (1967).

[27] Nikolaevskii, V. N., On relationships of volume and shear plastic strains and on shock waves in soft soils, (in Russian). Doklady AN SSSR (DAN), *177* (1967) 542–545.

[28] Dokuchaev, M. M., V. N. Rodionov and A. N. Romashov, Excavating explosion, (in Russian). *Acad. Sci. Printing House*, Moscow 1963.

[29] Zelmanov, I. L., O. S. Kolkov, A. M. Tikhomirov and A. F. Shazukevich, Motion of sandy soil during contained explosion, (in Russian). Physics of Combustion and Explosion (F.G.V.), No. 1 (1968) 116–121.

[30] Vesié, A. S. and G. W. Clough, Behaviour of granular materials under high stresses. *J. Soil Mech. Found. Div.*, SM3, Proc ASCE, (1968).

[31] Nikolaevskii, V. N. and E. F. Afanasiev, On some examples of media with microstructure of continuous particles. *Int. J. Solids and Struct.*, *5* (1969) 671–678.

[32] Ericksen, G. L. and C. Truesdell, Exact theory of stress and strain in rods and shells. *Arch. Ration Mach. Analysis*, *1* (1958) 295–323.

[33] Orowan, E., Dilatancy and the seismic focal mechanism. Rev. of Geophysics, *4*, No. 3, Aug. 1966.

[34] Nikolaevskii, V. N., On some generalization of Coulomb's limit condition for perfect granular continua, (in Russian). *Prikl. Mekhanika*, *5*, No. 3 (1969).

[35] Nikolaevskii, V. N., On formulation of constitutive relationships for plane flow of Coulomb's continuum, (in Russian). *Prikl. Math. Mekh.* (PMM), *32* (1968).

[36] de Josselin de Jong, G., Statics and kinematics in the failable zone of a granular material. Delft, Waltman, 1959.

[37] Rowe, P. W., The stress-dilatancy relation for static equilibrium of an assembly of particles in contacts. *Proc. Roy. Soc.*, *A269* (1962) 500–527.

[38] Mróz, Z. and K. Kwaszczyńska, Some boundary-value problems for granular density-hardening media, (in Polish). Rozprawy Inzynierskie, *19* (1971) 15–42.

[39] Drescher, A., Experimental verification of density hardening media, (in Polish). IBTP Report 9, 1971, Warszawa.

[40] Diduh, B. I. and V. A. Ioselevich, On development of plastic-hardening theory for soils, (in Russian). Mekh. Tverd. Tela (MTT), No. 2 (1970) 155–158.

On constructing models of cohesionless media by specifying the dissipation function

I. A. Berezhnoi, D. D. Ivlev and V. B. Tchadov (U.S.S.R.)

Constructing of models of plastic behaviour by prescribing the dissipation function was considered by several writers [1–4].

In [5] the dissipation function was considered for solids whose yield condition depends on the first invariant of the stress tensor, i.e. for plastic media whose properties depend on hydrostatic pressure. Similar relations were used in [6] to describe yielding of granular media and soils.

In this paper models of mechanical behaviour of cohesionless media will be developed starting with an appropriate form of the dissipation function.

In order to clearly set forth the basic ideas, let us consider first the plane strain case. The yielding condition is of the form

$$[(\sigma_x - \sigma_y)^2 + 4\tau_{xy}^2]^{\frac{1}{2}} + a(\sigma_x + \sigma_y) = 2k, \qquad a, k = \text{const.} \tag{1}$$

where σ_x, σ_y, τ_{xy} are the components of the stress tensor σ_{ij}.

The associated flow rule yields

$$\begin{aligned} &\varepsilon_x = \lambda[(\Sigma)^{-1}(\sigma_x - \sigma_y) + a], \qquad \varepsilon_y = \lambda[(\Sigma)^{-1}(\sigma_y - \sigma_x) + a], \\ &\varepsilon_{xy} = 2\lambda(\Sigma)^{-1}\tau_{xy}, \qquad \Sigma = [(\sigma_x - \sigma_y)^2 + 4\tau_{xy}]^{\frac{1}{2}} \end{aligned} \tag{2}$$

where ε_x, ε_y, ε_{xy} are the components of the strain rate tensor ε_{ij}.

The relationship (2) results in

$$\varepsilon_x + \varepsilon_y = 2\lambda a, \qquad \frac{\varepsilon_x - \varepsilon_y}{\sigma_x - \sigma_y} = \frac{\varepsilon_{xy}}{\tau_{xy}} = 2\lambda(\Sigma)^{-1}, \tag{3}$$

hence we obtain

$$2\lambda = \frac{1}{a}(\varepsilon_x + \varepsilon_y), \quad 2\lambda = [(\varepsilon_x - \varepsilon_y)^2 + 4\varepsilon_{xy}^2]^{\frac{1}{2}}. \tag{4}$$

Eqs. (4) provide the coupling between the shear strain rate and the rat of volume change

$$[(\varepsilon_x - \varepsilon_y)^2 + 4\varepsilon_{xy}^2]^{\frac{1}{2}} - \frac{1}{a}(\varepsilon_x + \varepsilon_y) = 0, \tag{5}$$

which is a characteristic property of deformation of granular media. For incompressible media, $a = 0$, we have $\varepsilon_x + \varepsilon_y = 0$. The relations given by the flow rule can eventually be expressed in the form

$$[(\varepsilon_x - \varepsilon_y)^2 + 4\varepsilon_{xy}^2]^{\frac{1}{2}} - \frac{1}{a}(\varepsilon_x + \varepsilon_y) = 0, \quad \frac{\varepsilon_x - \varepsilon_y}{\sigma_x - \sigma_y} = \frac{\varepsilon_{xy}}{\tau_{xy}}. \tag{6}$$

Eqs. (1), (6) constitute a set of governing equations. The second equation (6) represents the isotropy condition for the medium.

Let us now consider the dissipation function

$$D = D(\varepsilon_{ij}) = \sigma_x\varepsilon_x + \sigma_y\varepsilon_y + 2\tau_{xy}\varepsilon_{xy}. \tag{7}$$

Using (3) and (1) we arrive at the following relation

$$D(\varepsilon_{ij}) = 2k\lambda. \tag{8}$$

On account of (4) the dissipation function (8) has a twin analytical expression. It was shown in [5] that for cohesionless soils, $k = 0$, the dissipation $D = 0$.

We shall now consider the problem of constructing a model for cohesionless medium by specifying the dissipation function. Eq. (5) will be used for this purpose. The dissipation function assumes two forms

$$D = \frac{k}{a}(\varepsilon_x + \varepsilon_y) \quad \text{or} \quad D = k[(\varepsilon_x + \varepsilon_y)^2 + 4\varepsilon_{xy}^2]^{\frac{1}{2}}. \tag{9}$$

In order to obtain (1) and (6) we shall use the associated loading rule $\sigma_{ij} = (\partial D/\partial \varepsilon_{ij})$ and take Eq. (5) into consideration. Let us assume the dissipation function as follows

$$D = \frac{k}{a}(\varepsilon_x + \varepsilon_y) + \nu\left\{[(\varepsilon_x - \varepsilon_y)^2 + 4\varepsilon_{xy}^2]^{\frac{1}{2}} - \frac{1}{a}(\varepsilon_x + \varepsilon_y)\right\} \tag{10}$$

where ν is the Lagrangian multiplier.

According to the associated loading rule, we obtain from (10)

$$\begin{aligned} \sigma_x &= \frac{k}{a} + \nu\left[(J)^{-1}(\varepsilon_x - \varepsilon_y) - \frac{1}{a}\right], \\ \sigma_y &= \frac{k}{a} + \nu\left[(J)^{-1}(\varepsilon_y - \varepsilon_x) - \frac{1}{a}\right], \\ \tau_{xy} &= 2\nu(J)^{-1}\varepsilon_{xy}, \quad J = [(\varepsilon_x - \varepsilon_y)^2 + 4\varepsilon_{xy}^2]^{\frac{1}{2}}, \end{aligned} \tag{11}$$

hence, eventually,

$$\sigma_x + \sigma_y = \frac{2}{a}(k - v), \qquad \frac{\sigma_x - \sigma_y}{\varepsilon_x - \varepsilon_y} = \frac{\tau_{xy}}{\varepsilon_{xy}} = 2v(J)^{-1} \tag{12}$$

and therefore the Lagrangian multiplier assumes double form

$$2v = 2k - a(\sigma_x + \sigma_y), \quad 2v = [(\sigma_x - \sigma_y)^2 + 4\tau_{xy}^2]^{\frac{1}{2}}. \tag{13}$$

Eqs. (13) immediately result in the corresponding yield condition.

The shear dilatation relationship is prescribed a priori as in (5), whereas the isotropy relations are determined by (12). It is seen that by prescribing a form of the dissipation function and accounting for the dilatation relationship we may eventually obtain the relations (1), (6) defining the behaviour of cohesionless media.

The expression

$$D = k[(\varepsilon_x - \varepsilon_y)^2 + 4\varepsilon_{xy}^2]^{\frac{1}{2}} + v\left\{[(\varepsilon_x - \varepsilon_y)^2 + 4\varepsilon_{xy}^2]^{\frac{1}{2}} - \frac{1}{a}(\varepsilon_x - \varepsilon_y)\right\}, \tag{14}$$

can easily be seen to lead to the relations (1), (6) as well.

Let us now generalize the obtained results. Assume that the dissipation function is expressed by

$$\sigma_{ij}\varepsilon_{ij} = D_1(\varepsilon) = D_2(\varepsilon'_{ij}) \tag{15}$$

where $\varepsilon = \frac{1}{3}\varepsilon_{ii}$, $\varepsilon'_{ij} = \varepsilon_{ij} - \delta_{ij}\varepsilon$.

The functions D_1, D_2 are homogeneous of the order one in their arguments. It is easily seen that in this case $D_1(\varepsilon) = A\varepsilon$, $A = \text{const}$. Let us assume the following form of the dissipation function considered as the primitive concept of the theory

$$\sigma_{ij}\varepsilon_{ij} = D_1(\varepsilon) + v[D_1(\varepsilon) - D_2(\varepsilon'_{ij})], \tag{16}$$

where v is the Lagrangian multiplier.

The associated loading rule yields the components of the stress tensor

$$\sigma_{ij} = \frac{\partial D_1}{\partial \varepsilon_{ij}} + v\left(\frac{\partial D_1}{\partial \varepsilon_{ij}} - \frac{\partial D_2}{\partial \varepsilon_{ij}}\right) \tag{17}$$

or, in an explicit form,

$$\sigma_x = \tfrac{1}{3}A\delta_{ij} + v\left[\tfrac{1}{3}A\delta_{ij} - \frac{1}{3}\left(2\frac{\partial D_2}{\partial \varepsilon'_x} - \frac{\partial D_2}{\partial \varepsilon'_y} - \frac{\partial D_2}{\varrho \varepsilon_2}\right)\right], \ldots, \quad \tau_{xy} = v\frac{\partial D_2}{\partial \varepsilon_{xy}}, \ldots, \tag{18}$$

The expressions not shown in (18) can be derived by the circular permutation of the indices.

Eq. (18) yields

$$\sigma = \tfrac{1}{3}A(1+\nu),\ \sigma'_x = -\frac{\nu}{3}\left(2\frac{\partial D_2}{\partial \varepsilon'_x} - \frac{\partial D_2}{\partial \varepsilon'_y} - \frac{\partial D_2}{\partial \varepsilon'_z}\right),\ldots,\quad \tau_{xy} = \nu\frac{\partial D_2}{\partial \varepsilon_{xy}},\ldots, \tag{19}$$

where $\sigma = \frac{1}{3}\sigma_{ii}$, $\sigma'_x = \sigma_x - \sigma, \ldots$.

From (19) we find

$$\sigma'_x = \left(\frac{1}{3} - \frac{\sigma}{A}\right)\left(2\frac{\partial D_2}{\partial \varepsilon'_x} - \frac{\partial D_2}{\partial \varepsilon'_y} - \frac{\partial D_2}{\partial \varepsilon'_z}\right),\ldots\quad \tau_{xy} = \left(\frac{1}{3} - \frac{\sigma}{A}\right)\frac{\partial D_2}{\partial \varepsilon_{xy}},\ldots \tag{20}$$

The derivatives $(\partial D/\partial \varepsilon'_{ij})$ are functions, homogeneous of the order zero in the components ε'_{ij}. Therefore (19) gives the yield condition in terms of the stress tensor

$$f(\sigma, \sigma'_{ij}) = 0. \tag{21}$$

Let us now derive the flow rule associated with (21). Clearly

$$\frac{\partial f}{\partial \sigma_{ij}}\,\mathrm{d}\sigma_{ij} = 0.$$

Making use of (16) we conclude that

$$\varepsilon_{ij}\,\mathrm{d}\sigma_{ij} + \sigma_{ij}\,\mathrm{d}\varepsilon_{ij} = \left[\frac{\partial D_1}{\partial \varepsilon_{ij}} + \nu\left(\frac{\partial D_1}{\partial \varepsilon_{ij}} - \frac{\partial D_2}{\partial \varepsilon_{ij}}\right)\right]\mathrm{d}\varepsilon_{ij}. \tag{22}$$

From (22) and (17) it follows that

$$\varepsilon_{ij}\,\mathrm{d}\sigma_{ij} = 0 \tag{23}$$

and therefore (23) and (21) result in

$$\left(\mu\frac{\partial f}{\partial \sigma_{ij}} - \varepsilon_{ij}\right)\mathrm{d}\sigma_{ij} = 0. \tag{24}$$

Since this must hold for any $\mathrm{d}\sigma_{ij}$ the associated flow rule follows as

$$\varepsilon_{ij} = \mu\frac{\partial f}{\partial \sigma_{ij}}. \tag{25}$$

As an example let the dissipation function $D = \sqrt{\varepsilon'_{ij}\varepsilon'_{ij}}$ be considered. It can readily be seen that $\partial D_2/\partial\varepsilon_{ij} = \varepsilon'_{ij}/\sqrt{\varepsilon'_{ij}\varepsilon'_{ij}}$.

Then, according to (20)

$$\sigma'_{ij} = \left(\frac{1}{3} - \frac{\sigma}{A}\right)\frac{\varepsilon'_{ij}}{\sqrt{\varepsilon'_{ij}\varepsilon'_{ij}}}. \qquad (26)$$

On squaring Eq. (26) and making some rearrangements, we arrive at the following condition of yielding

$$\sigma + A(\sigma'_{ij}\sigma'_{ij})^{\frac{1}{2}} = A/3. \qquad (27)$$

This yield condition represents a cone in the space of principal stresses.

References

[1] Ziegler, H., Some extremum principles in irreversible thermodynamics, Progress in Solid Mech. *4*, 91–193, North-Holland, Amsterdam, 1963.
[2] Ivlev, D. D., On the dissipation function in the theory of plastic solids (in Russian), Doklady AN SSSR, *176* (1967) No. 5.
[3] Sedov, L. I., Mechanics of continuous media (in Russian), Nauka, Moscow, 1970.
[4] Berezhnoi, I. A., D. D. Ivlev and E. V. Makarov, in Problems of mechanics of deformable solids (in Russian), Sudpromgiz, 1970.
[5] Sawczuk, A. and P. Stutz, On formulation of stress-strain relations for solids at failure, ZAMP, *11* (1968) 770–778.
[6] Sokolovskii, V. V., Statics of granular media (in Russian), Moscow, 1954.